Lecture Notes in Computer Science 14349

Founding Editors

Gerhard Goos
Juris Hartmanis

The series Lecture Notes in Computer Science (LNCS), including its subseries Lecture Notes in Artificial Intelligence (LNAI) and Lecture Notes in Bioinformatics (LNBI), has established itself as a medium for the publication of new developments in computer science and information technology research, teaching, and education.

LNCS enjoys close cooperation with the computer science R & D community, the series counts many renowned academics among its volume editors and paper authors, and collaborates with prestigious societies. Its mission is to serve this international community by providing an invaluable service, mainly focused on the publication of conference and workshop proceedings and postproceedings. LNCS commenced publication in 1973.

Xiaohuan Cao · Xuanang Xu · Islem Rekik ·
Zhiming Cui · Xi Ouyang
Editors

Machine Learning in Medical Imaging

14th International Workshop, MLMI 2023
Held in Conjunction with MICCAI 2023
Vancouver, BC, Canada, October 8, 2023
Proceedings, Part II

Springer

Editors
Xiaohuan Cao ⓘ
Shanghai United Imaging Intelligence Co.,
Ltd.
Shanghai, China

Islem Rekik ⓘ
Imperial College London
London, UK

Xi Ouyang ⓘ
Shanghai United Imaging Intelligence Co.,
Ltd.
Shanghai, China

Xuanang Xu ⓘ
Rensselaer Polytechnic Institute
Troy, NY, USA

Zhiming Cui ⓘ
ShanghaiTech University
Shanghai, China

ISSN 0302-9743 ISSN 1611-3349 (electronic)
Lecture Notes in Computer Science
ISBN 978-3-031-45675-6 ISBN 978-3-031-45676-3 (eBook)
https://doi.org/10.1007/978-3-031-45676-3

This Springer imprint is published by the registered company Springer Nature Switzerland AG
The registered company address is: Gewerbestrasse 11, 6330 Cham, Switzerland

Paper in this product is recyclable.

Preface

The 14th International Workshop on Machine Learning in Medical Imaging (MLMI 2023) was held in Vancouver, Canada, on October 8, 2023, in conjunction with the 26th International Conference on Medical Image Computing and Computer Assisted Intervention (MICCAI 2023).

As artificial intelligence (AI) and machine learning (ML) continue to significantly influence both academia and industry, MLMI 2023 aims to facilitate new cutting-edge techniques and their applications in the medical imaging field, including, but not limited to medical image reconstruction, medical image registration, medical image segmentation, computer-aided detection and diagnosis, image fusion, image-guided intervention, image retrieval, etc. MLMI 2023 focused on major trends and challenges in this area and facilitated translating medical imaging research into clinical practice. Topics of interests included deep learning, generative adversarial learning, ensemble learning, transfer learning, multi-task learning, manifold learning, and reinforcement learning, along with their applications to medical image analysis, computer-aided diagnosis, multi-modality fusion, image reconstruction, image retrieval, cellular image analysis, molecular imaging, digital pathology, etc.

The MLMI workshop has attracted original, high-quality submissions on innovative research work in medical imaging using AI and ML. MLMI 2023 received a large number of submissions (139 in total). All the submissions underwent a rigorous double-blind peer-review process, with each paper being reviewed by at least two members of the Program Committee, composed of 89 experts in the field. Based on the reviewing scores and critiques, 93 papers were accepted for presentation at the workshop and chosen to be included in two Springer LNCS volumes, which resulted in an acceptance rate of 66.9%. It was a tough decision and many high-quality papers had to be rejected due to the page limitation.

We are grateful to all Program Committee members for reviewing the submissions and giving constructive comments. We also thank all the authors for making the workshop very fruitful and successful.

October 2023

Xiaohuan Cao
Xuanang Xu
Islem Rekik
Zhiming Cui
Xi Ouyang

Organization

Workshop Organizers

Xiaohuan Cao   Shanghai United Imaging Intelligence Co., Ltd.,
China
Xuanang Xu   Rensselaer Polytechnic Institute, USA
Islem Rekik   Imperial College London, UK
Zhiming Cui   ShanghaiTech University, China
Xi Ouyang   Shanghai United Imaging Intelligence Co., Ltd.,
China

Steering Committee

Dinggang Shen   ShanghaiTech University, China/Shanghai United
Imaging Intelligence Co., Ltd., China
Pingkun Yan   Rensselaer Polytechnic Institute, USA
Kenji Suzuki   Tokyo Institute of Technology, Japan
Fei Wang   Visa Research, USA

Program Committee

Reza Azad   RWTH Aachen University, Germany
Ulas Bagci   Northwestern University, USA
Xiaohuan Cao   Shanghai United Imaging Intelligence, China
Heang-Ping Chan   University of Michigan Medical Center, USA
Jiale Cheng   South China University of Technology, China
Cong Cong   University of New South Wales, Australia
Zhiming Cui   ShanghaiTech University, China
Haixing Dai   University of Georgia, USA
Yulong Dou   ShanghaiTech University, China
Yuqi Fang   University of North Carolina at Chapel Hill, USA
Yuyan Ge   Xi'an Jiaotong University, China
Hao Guan   University of North Carolina at Chapel Hill, USA
Hengtao Guo   Rensselaer Polytechnic Institute, USA
Yu Guo   Tianjin University, China
Minghao Han   Fudan University, China

Shijie Huang	ShanghaiTech University, China
Yongsong Huang	Tohoku University, Japan
Jiayu Huo	King's College London, UK
Xi Jia	University of Birmingham, UK
Caiwen Jiang	ShanghaiTech University, China
Xi Jiang	University of Electronic Science and Technology of China, China
Yanyun Jiang	Shandong Normal University, China
Ze Jin	Tokyo Institute of Technology, Japan
Nathan Lampen	Rensselaer Polytechnic Institute, USA
Junghwan Lee	Georgia Institute of Technology, USA
Gang Li	University of North Carolina at Chapel Hill, USA
Yunxiang Li	UT Southwestern Medical Center, USA
Yuxuan Liang	Rensselaer Polytechnic Institute, USA
Mingquan Lin	Weill Cornell Medicine, USA
Jiameng Liu	ShanghaiTech University, China
Mingxia Liu	University of North Carolina at Chapel Hill, USA
Muran Liu	ShanghaiTech University, China
Siyuan Liu	Dalian Maritime University, China
Tao Liu	Fudan University, China
Xiaoming Liu	United Imaging Research Institute of Intelligent Imaging, China
Yang Liu	King's College London, UK
Yuxiao Liu	ShanghaiTech University, China
Zhentao Liu	ShanghaiTech University, China
Lei Ma	ShanghaiTech University, China
Diego Machado Reyes	Rensselaer Polytechnic Institute, USA
Runqi Meng	ShanghaiTech University, China
Janne Nappi	Massachusetts General Hospital, USA
Mohammadreza Negahdar	Genentech, USA
Chuang Niu	Rensselaer Polytechnic Institute, USA
Xi Ouyang	Shanghai United Imaging Intelligence, China
Caner Ozer	Istanbul Technical University, Turkey
Huazheng Pan	East China Normal University, China
Yongsheng Pan	ShanghaiTech University, China
Linkai Peng	Southern University of Science and Technology, China
Saed Rezayi	University of Georgia, USA
Hongming Shan	Fudan University, China
Siyu Cheng	ShanghaiTech University, China
Xinrui Song	Rensselaer Polytechnic Institute, USA
Yue Sun	University of North Carolina at Chapel Hill, USA

Minhui Tan	Southern Medical University, China
Wenzheng Tao	University of Utah, USA
Maryam Toloubidokhti	Rochester Institute of Technology, USA
Bin Wang	Northwestern University, USA
Haoshen Wang	ShanghaiTech University, China
Linwei Wang	Rochester Institute of Technology, USA
Qianqian Wang	Liaocheng University, China
Sheng Wang	Shanghai Jiao Tong University, China
Xingyue Wang	ShanghaiTech University, China
Jie Wei	Northwestern Polytechnical University, China
Han Wu	ShanghaiTech University, China
Mengqi Wu	University of North Carolina at Chapel Hill, USA
Qingxia Wu	United Imaging Research Institute of Intelligent Imaging, China
Chenfan Xu	Shanghai University, China
Xuanang Xu	Rensselaer Polytechnic Institute, USA
Kai Xuan	Nanjing University of Information Science and Technology, China
Junwei Yang	University of Cambridge, UK
Xin Yang	Chinese University of Hong Kong, China
Yuqiao Yang	Tokyo Institution of Technology, Japan
Linlin Yao	Shanghai Jiao Tong University, China
Xin You	Shanghai Jiao Tong University, China
Qinji Yu	Shanghai Jiao Tong University, China
Renping Yu	Zhengzhou University, China
Jiadong Zhang	ShanghaiTech University, China
Lintao Zhang	University of North Carolina at Chapel Hill, USA
Shaoteng Zhang	Northwestern Polytechnical University, China
Xiao Zhang	Northwest University, China
Xukun Zhang	Fudan University, China
Yi Zhang	Sichuan University, China
Yuanwang Zhang	ShanghaiTech University, China
Zheyuan Zhang	Northwestern University, USA
Chongyue Zhao	University of Pittsburgh, USA
Yue Zhao	Chongqing University of Posts and Telecommunications, China
Yushan Zheng	Beihang University, China
Zixu Zhuang	Shanghai Jiao Tong University, China

Contents – Part II

Contents – Part I

GEMTrans: A General, Echocardiography-Based, Multi-level Transformer Framework for Cardiovascular Diagnosis

Masoud Mokhtari[1], Neda Ahmadi[1], Teresa S. M. Tsang[2],
Purang Abolmaesumi[1(✉)], and Renjie Liao[1]

[1] Electrical and Computer Engineering, University of British Columbia, Vancouver, BC, Canada
{masoud,nedaahmadi,purang,purang}@ece.ubc.ca
[2] Vancouver General Hospital, Vancouver, BC, Canada
t.tsang@ubc.ca

Abstract. Echocardiography (echo) is an ultrasound imaging modality that is widely used for various cardiovascular diagnosis tasks. Due to inter-observer variability in echo-based diagnosis, which arises from the variability in echo image acquisition and the interpretation of echo images based on clinical experience, vision-based machine learning (ML) methods have gained popularity to act as secondary layers of verification. For such safety-critical applications, it is essential for any proposed ML method to present a level of explainability along with good accuracy. In addition, such methods must be able to process several echo videos obtained from various heart views and the interactions among them to properly produce predictions for a variety of cardiovascular measurements or interpretation tasks. Prior work lacks explainability or is limited in scope by focusing on a single cardiovascular task. To remedy this, we propose a **G**eneral, **E**cho-based, **M**ulti-Level **Trans**former (GEMTrans) framework that provides explainability, while simultaneously enabling multi-video training where the inter-play among echo image patches in the same frame, all frames in the same video, and inter-video relationships are captured based on a downstream task. We show the flexibility of our framework by considering two critical tasks including ejection fraction (EF) and aortic stenosis (AS) severity detection. Our model achieves mean absolute errors of 4.15 and 4.84 for single and dual-video EF estimation and an accuracy of 96.5% for AS detection, while providing informative task-specific attention maps and prototypical explainability.

Keywords: Echocardiogram · Transformers · Explainable Models

1 Introduction and Related Works

Echocardiography (echo) is an ultrasound imaging modality that is widely used to effectively depict the dynamic cardiac anatomy from different standard views

---

Supplementary Information The online version contains supplementary material available at https://doi.org/10.1007/978-3-031-45676-3_1.

[23]. Based on the orientation and position of the obtained views with respect to the heart anatomy, different measurements and diagnostic observations can be made by combining information across several views. For instance, Apical Four Chamber (A4C) and Apical Two Chamber (A2C) views can be used to estimate ejection fraction (EF) as they depict the left ventricle (LV), while Parasternal Long Axis (PLAX) and Parasternal Short Axis (PSAX) echo can be used to detect aortic stenosis (AS) due to the visibility of the aortic valve in these views.

Challenges in accurately making echo-based diagnosis have given rise to vision-based machine learning (ML) models for automatic predictions. A number of these works perform segmentation of cardiac chambers; Liu et al. [14] perform LV segmentation using feature pyramids and a segmentation coherency network, while Cheng et al. [3] and Thomas et al. [24] use contrastive learning and GNNs for the same purpose, respectively. Some others introduce ML frameworks for echo view classification [8,10] or detecting important phases (e.g. end-systole (ES) and end-diastole (ED)) in a cardiac cycle [7]. Other bodies of work focus on making disease prediction from input echo. For instance, Duffy et al. [6] perform landmark detection to predict LV hypertrophy, while Roshanitabrizi et al. [20] predict rheumatic heart disease from Doppler echo using an ensemble of transformers and convolutional neural networks. In this paper, however, we focus on ejection fraction (EF) estimation and aortic stenosis severity (AS) detection as two example applications to showcase the generality of our framework and enable its comparison to prior works in a tractable manner. Therefore, in the following two paragraphs, we give a brief introduction to EF, AS and prior automatic detection works specific to these tasks.

EF is a ratio that indicates the volume of blood pumped by the heart and is an important indicator of heart function. The clinical procedure to estimating this ratio involves finding the ES and ED frames in echo cine series (videos) and tracing the LV on these frames. A high level of inter-observer variability of 7.6% to 13.9% has been reported in clinical EF estimates [18]. Due to this, various ML models have been proposed to automatically estimate EF and act as secondary layers of verification. More specifically, Esfeh et al. [13] propose a Bayesian network that produces uncertainty along with EF predictions, while Reynaud et al. [19] use BERT [4] to capture frame-to-frame relationships. Recently, Mokhtari et al. [16] provide explainability in their framework by learning a graph structure among frames of an echo.

The other cardiovascular task we consider is the detection of AS, which is a condition in which the aortic valve becomes calcified and narrowed, and is typically detected using spectral Doppler measurements [17,21]. High inter-observer variability, limited access to expert cardiac physicians, and the unavailability of spectral Doppler in many point-of-care ultrasound devices are challenges that can be addressed through the use of automatic AS detection models. For example, Huang et al. [11,12] predict the severity of AS from single echo images, while Ginsberg et al. [9] adopt a multitask training scheme.

Our framework is distinguished from prior echo-based works on multiple fronts. First, to the best of our knowledge, our model is the first to produce attention maps on the patch, frame and video levels for echo data, while allowing multiple videos to be processed simultaneously as shown in Fig. 1. Second, unlike prior works that are proposed for a single cardiovascular task, our framework is general and can be modified for a variety of echo-based metrics. More concretely, the versatility of our framework comes from its multi-level attention mechanism. For example, for EF, temporal attention between the echo frames is critical to capture the change in the volume of the LV, while for AS, spatial attention to the valve area is essential, which is evident by how the model is trained to adapt its learned attention to outperform all prior works. Lastly, we provide patch and frame-level, attention-guided prototypical explainability.

Our contributions are summarized below:

- We propose GEMTrans, a general, transformer-based, multi-level ML framework for vison-based medical predictions on echo cine series (videos).
- We show that the task-specific attention learned by the model is effective in highlighting the important patches and frames of an input video, which allows the model to achieve a mean absolute error of 4.49 on two EF datasets, and a detection accuracy score of 96.5% on an AS dataset.
- We demonstrate how prototypical learning can be easily incorporated into the framework for added multi-layer explainability.

2   Method

2.1   Problem Statement

The input data for both EF and AS tasks compose of one or multiple B-mode echo videos denoted by $X \in \mathbb{R}^{K \times T \times H \times W}$, where K is the number of videos per sample, T is the number of frames per video, and H and W are the height and width of each grey-scale image frame. For **EF Estimation**, we consider both the single video (A4C) and the dual video (A2C and A4C) settings corresponding to $K = 1$ and $K = 2$, respectively. Our datasets consist of triplets $\{x_{ef}^i, y_{ef}^i, y_{seg}^i\}$, where $i \in [1, ..., n]$ is the sample number. $y_{ef}^i \in [0, 1]$ is the ground truth EF value, $y_{seg}^i \in \{0, 1\}^{H \times W}$ is the binary LV segmentation mask, and $x^i \in \mathbb{R}^{K \times T \times H \times W}$ are the input videos defined previously. The goal is to learn an EF estimation function $f_{ef} : \mathbb{R}^{K \times T \times H \times W} \mapsto \mathbb{R}$. For **AS Classification**, we consider the dual-video setting (PLAX and PSAX). Here, our dataset $D_{as} = \{X_{as}, Y_{as}\}$ consists of pairs $\{x_{as}^i, y_{as}^i\}$, where x_{as}^i are the input videos and $y_{as}^i \in \{0, 1\}^4$ is a one-hot label indicating healthy, mild, moderate and severe AS cases. Our goal is to learn $f_{as} : \mathbb{R}^{2 \times T \times H \times W} \mapsto \mathbb{R}^4$ that produces a probability over AS severity classes.

2.2   Multi-Level Transformer Network

As shown in Fig. 1, we employ a three-level transformer network, where the levels are tasked with patch-wise, frame-wise and video-wise attention, respectively.

   Spatial Transformer Encoder (STE) captures the attention among patches within a certain frame and follows ViT's [5] architecture. As shown

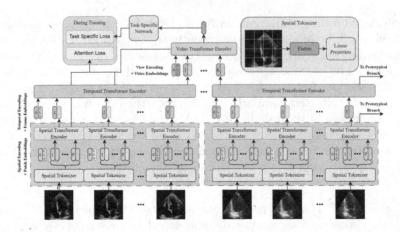

Fig. 1. GEMTrans Overview - The multi-level transformer network processes one or multiple echo videos and is composed of three main components. *Spatial Transformer Encoder (STE)* produces attention among patches in the same image frame, while *Temporal Transformer Encoder (TTE)* captures the temporal dependencies among the frames of each video. Lastly, *Video Transformer Encoder (VTE)* produces an embedding summarizing all available data for a patient by processing the learned embedding of each video. Different downstream tasks can then be performed using this final learned embedding. During training, both the final prediction and the attention learned by different layers of the framework can be supervised (not all connections are shown for cleaner visualization).

in Eqs. (1) and (2), the Spatial Tokenizer (ST) first divides the image into non-overlapping $p \times p$ sized patches before flattening and linearly projecting each patch:

$$\hat{x}_{k,t} = [\hat{x}_{k,t,1}, \hat{x}_{k,t,2}, ..., \hat{x}_{k,t,HW/p^2}] = f_{\text{patch}}(x_{k,t}, p); \tag{1}$$

$$x'_{k,t} = [x'_{k,t,1}, x'_{k,t,2}, ..., x'_{k,t,HW/p^2}] = f_{\text{lin}}(\text{flatten}(\hat{x}_{k,t})), \tag{2}$$

where p is the patch size, $k \in [1, ..., K]$ is the video number, $t \in [1, ..., T]$ is the frame number, $f_{\text{patches}} : \mathbb{R}^{H \times W} \mapsto \mathbb{R}^{HW/p^2 \times p \times p}$ splits the image into equally-sized patches, and $f_{\text{lin}} : \mathbb{R}^{p^2} \mapsto \mathbb{R}^d$ is a linear projection function that maps the flattened patches into d-dimensional embeddings. The obtained tokens from the ST are then fed into a transformer network [25] as illustrated in Eqs. (3) to (6):

$$h_{k,t}^0 = [\text{cls}_{\text{spatial}}; x'_{k,t}] + E_{\text{pos}}; \tag{3}$$

$$h_{k,t}^{\prime l} = \text{MHA}(\text{LN}(h_{k,t}^{l-1})) + h_{k,t}^{l-1}, \quad l \in [1, ..., L]; \tag{4}$$

$$h_{k,t}^l = MLP(LN(h_{k,t}^{\prime l})) + h_{k,t}^{\prime l}, \quad l \in [1, ..., L]; \tag{5}$$

$$z_{k,t} = LN(h_{k,t,0}^L), \tag{6}$$

where $\text{cls}_{\text{spatial}} \in \mathbb{R}^d$ is a token similar to the [*class*] token introduced by Devlin *et al.* [4], $E_{\text{pos}} \in \mathbb{R}^d$ is a learnable positional embedding, MHA is a multi-head attention network [25], LN is the LayerNorm operation, and MLP is a multi-layer perceptron. The obtained result $z_{k,t} \in \mathbb{R}^d$ can be regarded as an embedding summarizing the t_{th} frame in the k_{th} video for a sample.

Temporal Transformer Encoder (TTE) accepts as input the learned embeddings of the STE for each video $z_{k,1...T}$ and performs similar operations outlined in Eqs. (3) to (6) to generate a single embedding $v_k \in \mathbb{R}^d$ representing the whole video from the k_{th} view. **Video Transformer Encoder (VTE)** is the same as TTE with the difference that each input token $v_k \in \mathbb{R}^d$ is a representation for a complete video from a certain view. The output of VTE is an embedding $u^i \in \mathbb{R}^d$ summarizing the data available for patient i. This learned embedding can be used for various downstream tasks as described in Sec. 2.5.

2.3 Attention Supervision

For EF, the ED/ES frame locations and their LV segmentation are available. This intermediary information can be used to supervise the learned attention of the transformer. Therefore, inspired by Stacey *et al.* [22], we supervise the last-layer attention that the cls token allocates to other tokens. More specifically, for spatial attention, we penalize the model for giving attention to the region outside the LV, while for temporal dimension, we encourage the model to give more attention to the ED/ES frames. More formally, we define $\text{ATTN}_{\text{cls}}^{\text{spatial}} \in [0,1]^{HW/p^2}$ and $\text{ATTN}_{\text{cls}}^{\text{temporal}} \in [0,1]^T$ to be the L_{th}-layer, softmax-normalized spatial and temporal attention learned by the MHA module (see Eq. (4)) of STE and TTE, respectively. The attention loss is defined as

$$y'_{\text{seg}} = \text{OR}(y'^{\text{ed}}_{\text{seg}}, y'^{\text{es}}_{\text{seg}}); \tag{7}$$

$$L_{\text{attn, s}}^{\text{spatial}} = \begin{cases} (\text{ATTN}_{\text{cls},s}^{\text{spatial}} - 0)^2, & \text{if } y'_{\text{seg},s} = 0 \text{ (outside LV)} \\ 0, & \text{otherwise}; \end{cases} \tag{8}$$

$$L_{\text{attn, t}}^{\text{temporal}} = \begin{cases} (\text{ATTN}_{\text{cls},s}^{\text{temporal}} - 1)^2, & \text{if } t \in [\text{ED}, \text{ES}] \\ 0, & \text{otherwise}; \end{cases} \tag{9}$$

$$L_{\text{attn}} = \lambda_{\text{temporal}} \Sigma_{t=1}^T L_{\text{attn, t}}^{\text{temporal}} + \lambda_{\text{spatial}} \Sigma_{s=1}^{HW/p^2} L_{\text{attn, s}}^{\text{spatial}}, \tag{10}$$

where $y'^{\text{ed}}_{\text{seg}}, y'^{\text{es}}_{\text{seg}} \in \{0,1\}^{HW/p^2}$ are the coarsened versions (to match patch size) of y_{seg} at the ED and ES locations, OR is the bit-wise logical *or* function, ed/es indicate the ED/ES temporal frame indices, and $\lambda_{\text{temporal}}, \lambda_{\text{spatial}} \in [0,1]$ control the effect of spatial and temporal losses on the overall attention loss. *A figure is provided in the supp. material for further clarification.*

2.4 Prototypical Learning

Prototypical learning provides explainability by presenting training examples (prototypes) as the reasoning for choosing a certain prediction. As an example, in the context of AS, patch-level prototypes can indicate the most important patches in the training frames that correspond to different AS cases. Due to the multi-layer nature of our framework and inspired by Xue *et al.* [26], we can expand this idea and use our learned attention to filter out uninformative details prior to learning patch and frame-level prototypes. As shown in Fig. 1, the STE and TTE's attention information are used in prototypical branches to learn these prototypes. It must be noted that we are not using prototypical learning to improve performance, but rather to provide added explainability. For this reason, prototypes are obtained in a post-processing step using the pretrained transformer framework. *Prototypical networks are provided in the supp. material.*

2.5   Downstream Tasks and Optimization

The output embedding of VTE denoted by $u \in \mathbb{R}^{n \times d}$ can be used for various downstream tasks. We use Eq. (11) to generate predictions for EF and use an L2 loss between $\hat{y}_{ef}$ and y_{ef}^{i} for optimization denoted by L_{ef}. For AS severity classification, Eq. (12) is used to generate predictions and a cross-entropy loss is used for optimization shown as L_{as}:

$$\hat{y}_{ef} = \sigma(\text{MLP}(u)) \quad \hat{y}_{ef} \in \mathbb{R}^{n \times 1}; \tag{11}$$

$$\hat{y}_{as} = \text{softmax}(\text{MLP}(u)) \quad \hat{y}_{as} \in \mathbb{R}^{n \times 4}; \tag{12}$$

$$L_{overall} = L_{ef \text{ or } as} + L_{attn}, \tag{13}$$

where σ is the Sigmoid function. Our overall loss function is shown in Eq. (13).

3   Experiments

3.1   Implementation

Our code-base, pre-trained model weights and the corresponding configuration files are provided at https://github.com/DSL-Lab/gemtrans. All models were trained on four 32 GB NVIDIA Tesla V100 GPUs, where the hyper-parameters are found using Weights & Biases random sweep [2]. Lastly, we use the ViT network from [15] pre-trained on ImageNet-21K for the STE module.

3.2   Datasets

We compare our model's performance to prior works on three datasets. In summary, for the single-video case, we use the EchoNet Dynamic dataset that consists of $10,030$ AP4 echo videos obtained at Stanford University Hospital [18] with a training/validation/test (TVT) split of $7,465$, $1,288$ and $1,277$. For the dual-video setting, we use a private dataset of $5,143$ pairs of AP2/AP4 videos with a TVT split of $3,649$, 731, 763. For the AS severity detection task, we use a private dataset of PLAX/PSAX pairs with a balanced number of healthy, mild, moderate and severe AS cases and a TVT of $1,875$, 257, and 258. For all datasets, the frames are resized to 224×224. Our use of this private dataset is approved by the University of British Columbia Research Ethics Board, with an approval number H20-00365.

3.3 Results

Quantitative Results: In Tables 1 and 2, we show that our model outperforms all previous work in EF estimation and AS detection. For EF, we use the mean absolute error (MAE) between the ground truth and the predicted EF values, and R^2 correlation score. For AS, we use an accuracy metric for four-class AS severity prediction and binary detection of AS. The results show the flexibility of our model to be adapted for various echo-based tasks while achieving high levels of performance. **Qualitative Results:** In addition to having the superior quantitative performance, we show that our model provides explainability through its task-specific learned attention (Fig. 2a) and the learned prototypes (Fig. 2b). *More results are shown in the supp. material.* **Ablation Study:** We show the effectiveness of our design choices in Table 3, where we use EF estimation for the EchoNet Dynamic dataset as our test-bed. It is evident that our attention loss described in Sec. 2.3 is effective for EF when intermediary labels are available, while it is necessary to use a pre-trained ViT as the size of medical datasets in our experiments are not sufficiently large to build good inductive bias.

Table 1. Quantitative results for EF on the test set - LV Biplane dataset results for models not supporting multi-video training are indicated by "-". MAE is the Mean Absolute Error and R^2 indicates variance captured by the model.

Model	EchoNet Dynamic		LV. Biplane	
	MAE [mm] ↓	R^2 Score ↑	MAE [mm] ↓	R^2 Score ↑
Ouyang et al. [18]	7.35	0.40	–	–
Reynaud et al. [19]	5.95	0.52	–	–
Esfeh et al. [13]	4.46	0.75	–	–
Thomas et al. [24]	4.23	**0.79**	–	–
Mokhtari et al. [16]	4.45	0.76	5.12	0.68
Ours	**4.15**	**0.79**	**4.84**	**0.72**

Table 2. Quantitative results for AS on the test set - *Severity* is a four-class classification task, while *Detection* involves the binary detection of AS.

Model	Accuracy [%] ↑	
	Severity	Detection
Huang *et al.* [12]	73.7	94.1
Bertasius *et al.* [1]	75.3	94.8
Ginsberg *et al.* [9]	74.4	94.2
Ours	**76.2**	**96.5**

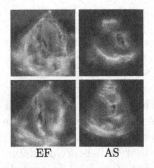

EF   AS

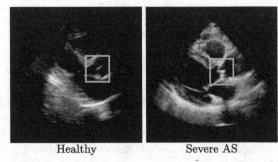

Healthy   Severe AS

(a) **Learned Patch-Level Attention** - We visualize the learned attention of STE, where for EF, the model is focusing on the walls of the LV, while for AS, the model learns to attend to the valve area, which is clinically correct.

(b) **Learned Patch-Level Prototypes** - Learned prototypes use STE's attention to properly focus on the valve area for a healthy and severe AS case. We can see that in the healthy case, the aortic valve is thin and not calcified. However, in the severe case, the calcification of aortic valve is apparent (i.e. the valve appears bright in the image). *Frame-level and EF prototypes presented in supp. material.*

Fig. 2. Explainability through learned attention and prototypes.

Table 3. Ablation study on the validation set of EchoNet Dynamic - We see that both spatial and temporal attention supervision are effective for EF estimation, while the model does not converge without pretraining the ViT. MAE is the Mean Absolute Error and R^2 indicates variance captured by the model.

Model	MAE [mm] ↓	R^2 Score ↑
No Spatial Attn. Sup.	4.42	0.77
No Temporal Attn. Sup.	4.54	0.76
No ViT Pretraining	5.61	0.45
Ours	**4.11**	**0.80**

4   Conclusions and Future Work

In this paper, we introduced a multi-layer, transformer-based framework suitable for processing echo videos and showed superior performance to prior works on two challenging tasks while providing explainability through the use of prototypes and learned attention of the model. Future work will include training a large multi-task model with a comprehensive echo dataset that can be disseminated to the community for a variety of clinical applications.

References

1. Bertasius, G., Wang, H., Torresani, L.: Is space-time attention all you need for video understanding? In: Meila, M., Zhang, T. (eds.) Proceedings of the 38th International Conference on Machine Learning. Proceedings of Machine Learning Research, vol. 139, pp. 813–824. PMLR (2021)
2. Biewald, L.: Experiment tracking with weights and biases (2020)
3. Cheng, L.H., Sun, X., van der Geest, R.J.: Contrastive learning for echocardiographic view integration. In: Wang, L., Dou, Q., Fletcher, P.T., Speidel, S., Li, S. (eds.) MICCAI 2022. LNCS, vol. 13434, pp. 340–349. Springer, Cham (2022)
4. Devlin, J., Chang, M.W., Lee, K., Toutanova, K.: Bert: pre-training of deep bidirectional transformers for language understanding (2019)
5. Dosovitskiy, A., et al.: An image is worth 16x16 words: transformers for image recognition at scale. In: International Conference on Learning Representations (2021)
6. Duffy, G., et al.: High-throughput precision phenotyping of left ventricular hypertrophy with cardiovascular deep learning. JAMA Cardiol. 7(4), 386–395 (2022)
7. Fiorito, A.M., Østvik, A., Smistad, E., Leclerc, S., Bernard, O., Lovstakken, L.: Detection of cardiac events in echocardiography using 3D convolutional recurrent neural networks. In: IEEE International Ultrasonics Symposium, pp. 1–4 (2018)
8. Gao, X., Li, W., Loomes, M., Wang, L.: A fused deep learning architecture for viewpoint classification of echocardiography. Inf. Fusion 36, 103–113 (2017)
9. Ginsberg, T., et al.: Deep video networks for automatic assessment of aortic stenosis in echocardiography. In: Noble, J.A., Aylward, S., Grimwood, A., Min, Z., Lee, S.-L., Hu, Y. (eds.) ASMUS 2021. LNCS, vol. 12967, pp. 202–210. Springer, Cham (2021). https://doi.org/10.1007/978-3-030-87583-1_20
10. Gu, A.N., et al.: Efficient echocardiogram view classification with sampling-free uncertainty estimation. In: Noble, J.A., et al. (eds.) ASMUS 2021. LNCS, vol. 12967, pp. 139–148. Springer, Cham (2021). https://doi.org/10.1007/978-3-030-87583-1_14
11. Huang, Z., Long, G., Wessler, B., Hughes, M.C.: A new semi-supervised learning benchmark for classifying view and diagnosing aortic stenosis from echocardiograms. In: Proceedings of the 6th Machine Learning for Healthcare Conference (2021)
12. Huang, Z., Long, G., Wessler, B., Hughes, M.C.: Tmed 2: a dataset for semi-supervised classification of echocardiograms (2022)
13. Kazemi Esfeh, M.M., Luong, C., Behnami, D., Tsang, T., Abolmaesumi, P.: A deep Bayesian video analysis framework: towards a more robust estimation of ejection fraction. In: Martel, A.L., et al. (eds.) MICCAI 2020. LNCS, vol. 12262, pp. 582–590. Springer, Cham (2020). https://doi.org/10.1007/978-3-030-59713-9_56

14. Liu, F., Wang, K., Liu, D., Yang, X., Tian, J.: Deep pyramid local attention neural network for cardiac structure segmentation in two-dimensional echocardiography. Med. Image Anal. **67**, 101873 (2021)
15. Melas-Kyriazi, L.: Vit pytorch (2020). https://github.com/lukemelas/PyTorch-Pretrained-ViT
16. Mokhtari, M., Tsang, T., Abolmaesumi, P., Liao, R.: EchoGNN: explainable ejection fraction estimation with graph neural networks. In: Wang, L., Dou, Q., Fletcher, P.T., Speidel, S., Li, S. (eds.) Medical Image Computing and Computer Assisted Intervention. MICCAI 2022, vol. 13434, pp. 360–369. Springer Nature Switzerland, Cham (2022). https://doi.org/10.1007/978-3-031-16440-8_35
17. Otto, C.M., et al.: 2020 ACC/AHA guideline for the management of patients with valvular heart disease: executive summary. J. Am. Coll. Cardiol. **77**(4), 450–500 (2021)
18. Ouyang, D., et al.: Video-based AI for beat-to-beat assessment of cardiac function. Nature **580**, 252–256 (2020)
19. Reynaud, H., Vlontzos, A., Hou, B., Beqiri, A., Leeson, P., Kainz, B.: Ultrasound video transformers for cardiac ejection fraction estimation. In: de Bruijne, M., et al. (eds.) MICCAI 2021. LNCS, vol. 12906, pp. 495–505. Springer, Cham (2021). https://doi.org/10.1007/978-3-030-87231-1_48
20. Roshanitabrizi, P., et al.: Ensembled prediction of rheumatic heart disease from ungated doppler echocardiography acquired in low-resource settings. In: Wang, L., Dou, Q., Fletcher, P.T., Speidel, S., Li, S. (eds.) MICCAI 2022. LNCS, vol. 13431, pp. 602–612. Springer, Cham (2022). https://doi.org/10.1007/978-3-031-16431-6_57
21. Spitzer, E., et al.: Aortic stenosis and heart failure: disease ascertainment and statistical considerations for clinical trials. Card. Fail. Rev. **5**, 99–105 (2019)
22. Stacey, J., Belinkov, Y., Rei, M.: Supervising model attention with human explanations for robust natural language inference. In: Proceedings of the AAAI Conference on Artificial Intelligence, vol. 36, no. 10, pp. 11349–11357 (2022)
23. Suetens, P.: Fundamentals of Medical Imaging, 2nd edn. Cambridge University Press, Cambridge (2009)
24. Thomas, S., Gilbert, A., Ben-Yosef, G.: Light-weight spatio-temporal graphs for segmentation and ejection fraction prediction in cardiac ultrasound. In: Wang, L., Dou, Q., Fletcher, P.T., Speidel, S., Li, S. (eds.) MICCAI 2022. LNCS, vol. 13434, pp. 380–390. Springer, Cham (2022). https://doi.org/10.1007/978-3-031-16440-8_37
25. Vaswani, A., et al.: Attention is all you need. In: Guyon, I., et al. (eds.) Advances in Neural Information Processing Systems, vol. 30. Curran Associates (2017)
26. Xue, M., et al.: Protopformer: concentrating on prototypical parts in vision transformers for interpretable image recognition. ArXiv (2022)

Unsupervised Anomaly Detection in Medical Images with a Memory-Augmented Multi-level Cross-Attentional Masked Autoencoder

Yu Tian[1]([✉]), Guansong Pang[5], Yuyuan Liu[2], Chong Wang[2], Yuanhong Chen[2], Fengbei Liu[2], Rajvinder Singh[3], Johan W. Verjans[2,3,4], Mengyu Wang[1], and Gustavo Carneiro[6]

[1] Harvard Ophthalmology AI Lab, Harvard University, Boston, USA
`ytian11@meei.harvard.edu`
[2] Australian Institute for Machine Learning, University of Adelaide, Adelaide, Australia
[3] Faculty of Health and Medical Sciences, University of Adelaide, Adelaide, Australia
[4] South Australian Health and Medical Research Institute, Adelaide, Australia
[5] Singapore Management University, Singapore, Singapore
[6] Centre for Vision, Speech and Signal Processing, University of Surrey, Guildford, UK

Abstract. Unsupervised anomaly detection (UAD) aims to find anomalous images by optimising a detector using a training set that contains only normal images. UAD approaches can be based on reconstruction methods, self-supervised approaches, and Imagenet pre-trained models. Reconstruction methods, which detect anomalies from image reconstruction errors, are advantageous because they do not rely on the design of problem-specific pretext tasks needed by self-supervised approaches, and on the unreliable translation of models pre-trained from non-medical datasets. However, reconstruction methods may fail because they can have low reconstruction errors even for anomalous images. In this paper, we introduce a new reconstruction-based UAD approach that addresses this low-reconstruction error issue for anomalous images. Our UAD approach, the memory-augmented multi-level cross-attentional masked autoencoder (MemMC-MAE), is a transformer-based approach, consisting of a novel memory-augmented self-attention operator for the encoder and a new multi-level cross-attention operator for the decoder. MemMC-MAE masks large parts of the input image during its reconstruction, reducing the risk that it will produce low reconstruction errors because anomalies are likely to be masked and cannot be reconstructed. However, when the anomaly is not masked, then the normal patterns stored in the encoder's memory combined with the decoder's multi-level cross-attention will constrain the accurate reconstruction of the anomaly. We show that our method achieves SOTA anomaly detection and localisation on colonoscopy, pneumonia, and covid-19 chest x-ray datasets.

Keywords: Pneumonia · Covid-19 · Colonoscopy · Unsupervised Learning · Anomaly Detection · Anomaly Segmentation · Vision Transformer

ⓒ The Author(s), under exclusive license to Springer Nature Switzerland AG 2024
X. Cao et al. (Eds.): MLMI 2023, LNCS 14349, pp. 11–21, 2024.
https://doi.org/10.1007/978-3-031-45676-3_2

1   Introduction

Detecting and localising anomalous findings in medical images (e.g., polyps, malignant tissues, etc.) are of vital importance [1,4,7,12–15,17–19,27,29,30,32, 34]. Systems that can tackle these tasks are often formulated with a classifier trained with large-scale datasets annotated by experts. Obtaining such annotation is often challenging in real-world clinical datasets because the amount of normal images from healthy patients tend to overwhelm the amount of anomalous images. Hence, to alleviate the challenges of collecting anomalous images and learning from class-imbalanced training sets, the field has developed unsupervised anomaly detection (UAD) models [3,31] that are trained exclusively with normal images. Such UAD strategy benefits from the straightforward acquisition of training sets containing only normal images and the potential generalisability to unseen anomalies without collecting all possible anomalous sub-classes.

Current UAD methods learn a one-class classifier (OCC) using only normal/healthy training data, and detect anomalous/disease samples using the learned OCC [3,8,11,16,22,25,26,33,36]. UAD methods can be divided into: 1) reconstruction methods, 2) self-supervised approaches, and 3) Imagenet pre-trained models. Reconstruction methods [3,8,16,25,36] are trained to accurately reconstruct normal images, exploring the assumption that the lack of anomalous images in the training set will prevent a low error reconstruction of an test image that contains an anomaly. However, this assumption is not met in general because reconstruction methods are indeed able to successfully reconstruct anomalous images, particularly when the anomaly is subtle. Self-supervised approaches [28,31,34] train models using contrastive learning, where pretext tasks must be designed to emulate normal and anomalous image changes for each new anomaly detection problem. Imagenet pre-trained models [5,24] produce features to be used by OCC, but the translation of these models into medical image problems is not straightforward. Reconstruction methods are able to circumvent the aforementioned challenges posed by self-supervised and Imagenet pre-trained UAD methods, and they can be trained with a relatively small amount of normal samples. However, their viability depends on an acceptable mitigation of the potentially low reconstruction error of anomalous test images.

In this paper, we introduce a new UAD reconstruction method, the Memory-augmented Multi-level Cross-attention Masked Autoencoder (MemMC-MAE), designed to address the low reconstruction error of anomalous test images. MemMC-MAE is a transformer-based approach based on masked autoencoder (MAE) [9] with of a novel memory-augmented self-attention encoder and a new multi-level cross-attention decoder. MemMC-MAE masks large parts of the input image during its reconstruction, and given that the likelihood of masking out an anomalous region is large, then it is unlikely that it will accurately reconstruct that anomalous region. However, there is still the risk that the anomaly is not masked out, so in this case, the normal patterns stored in the encoder's memory combined with the correlation of multiple normal patterns in the image, utilised by the decoder's multi-level cross-attention can explicitly constrain the accurate anomaly reconstruction to produce high reconstruction error (high anomaly

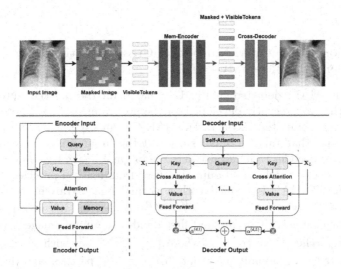

Fig. 1. Top: overall MemMC-MAE framework. Yellow tokens indicate the unmasked visible patches, and blue tokens indicate the masked patches. Our memory-augmented transformer encoder only accepts the visible patches/tokens as input, and its output tokens are combined with dummy masked patches/tokens for the missing pixel reconstruction using our proposed multi-level cross-attentional transformer decoder. **Bottom-left:** proposed memory-augmented self-attention operator for the transformer encoder, and **bottom-right:** proposed multi-level cross-attention operator for the transformer decoder.

score). The encoder's memory is also designed to address the MAE's long-range 'forgetting' issue [20], which can be harmful for UAD due to the poor reconstruction based on forgotten normality patterns and 'unwanted' generalisability to subtle anomalies during testing. Our contributions are summarised as:

- To the best of our knowledge, this is the first memory-based UAD method that relies on MAE [9];
- A new memory-augmented self-attention operator for our MAE transformer encoder to explicitly encode and memorise the normality patterns; and
- A novel decoder architecture that uses the learned multi-level memory-augmented encoder information as prior features to a cross-attention operator.

Our method achieves better anomaly detection and localisation accuracy than most competing approaches on the UAD benchmarks using the public Hyper-Kvasir colonoscopy dataset [2], pneumonia [10] and Covid-X [37] Chest X-ray (CXR) dataset.

2   Method

2.1   Memory-Augmented Multi-level Cross-Attentional Masked Autoencoder (MemMC-MAE)

Our MemMC-MAE, depicted in Fig. 1, is based on the masked autoencoder (MAE) [9] that was recently developed for the pre-training of models to be used in downstream computer vision tasks. MAE has an asymmetric architecture, with a encoder that takes a small subset of the input image patches and a smaller/lighter decoder that reconstructs the original image based on the input tokens from visible patches and dummy tokens from masked patches.

Our MemMC-MAE is trained with a normal image training set, denoted by $\mathcal{D} = \{\mathbf{x}_i\}_{i=1}^{|\mathcal{D}|}$, where $\mathbf{x} \in \mathcal{X} \subset \mathbb{R}^{H \times W \times R}$ (H: height, W: width, R: number of colour channels). Our method first divides the input image $\mathbf{x}$ into non-overlapping patches $\mathcal{P} = \{\mathbf{p}_i\}_{i=1}^{|\mathcal{P}|}$, where $\mathbf{p} \in \mathbb{R}^{\hat{H} \times \hat{W} \times R}$, with $\hat{H} << H$ and $\hat{W} << W$. We then randomly mask out 75% of the $|\mathcal{P}|$ patches, and the remaining visible patches $\mathcal{P}^{(v)} = \{\mathbf{p}_v\}_{v=1}^{|\mathcal{P}^{(v)}|}$ (with $|\mathcal{P}^{(v)}| = 0.25 \times |\mathcal{P}|$) are used by the MemMC-MAE to encode the normality patterns of those patches, and all $|\mathcal{P}^{(v)}|$ encoded visible patches and $|\mathcal{P}| - |\mathcal{P}^{(v)}|$ dummy masked patches are used as the input of a new multi-level cross-attention decoder to reconstruct the image.

The training of MemMC-MAE is based on the minimisation of the mean squared error (MSE) loss between the input and reconstructed images at the pixels of the masked patches of the training images. The approach is evaluated on a testing set $\mathcal{T} = \{(\mathbf{x}, y, \mathbf{m})_i\}_{i=1}^{|\mathcal{T}|}$, where $y \in \mathcal{Y} = \{\text{normal}, \text{anomalous}\}$, and $\mathbf{m} \in \mathcal{M} \subset \{0,1\}^{H \times W \times 1}$ denotes the segmentation mask of the lesion in the image $\mathbf{x}$. When testing, we also mask 75% of the image and the patch-wise reconstruction error indicates anomaly localisation, and the mean reconstruction error of all patches is used to detect image-wise anomaly. Below we provide details on the major contributions of MemMC-MAE, which are the memory-augmented transformer encoder that stores the long-term normality patterns of the training samples, and the new multi-level cross-attentional transformer decoder to leverage the correlation of features from the encoder to reconstruct the missing normal pixels.

Memory-augmented Transformer Encoder (Fig. 1 - Bottom Left). We modify the encoder from the transformer with our a novel memory-augmented self-attention, by extending the keys and values of the self-attention operation with learnable memory matrices that store normality patterns, which are updated via back-propagation. To this end, the proposed self-attention (SA) module for layer $l \in \{0, ..., L-1\}$ is defined as:

$$\mathbf{X}^{(l+1)} = f_{SA}\big(\mathbf{W}_Q^{(l)}\mathbf{X}^{(l)}, [\mathbf{W}_K^{(l)}\mathbf{X}^{(l)}, \mathbf{M}_K^{(l)}], [\mathbf{W}_V^{(l)}\mathbf{X}^{(l)}, \mathbf{M}_V^{(l)}]\big), \qquad (1)$$

where $\mathbf{X}^{(0)}$ is the encoder input matrix containing $|\mathcal{P}^{(v)}|$ patch tokens formed from the visible image patches transformed through the linear projection $\mathbf{W}^{(0)}$, with $|\mathcal{P}^{(v)}|$ being the number of visible tokens/patches, $\mathbf{X}^{(l)}, \mathbf{X}^{(l+1)}$ are the input and output of layer l, $\mathbf{W}_Q^{(l)}, \mathbf{W}_K^{(l)}, \mathbf{W}_V^{(l)}$ are the linear projections of the encoder's layer l for query, key and value of the self-attention operator, respectively, and $\mathbf{M}_K^{(l)}, \mathbf{M}_V^{(l)}$ are the layer l learnable memory matrices that are concatenated with $\mathbf{W}_K \mathbf{X}^{(l)}$ and $\mathbf{W}_V \mathbf{X}^{(l)}$ using the operator $[.,.]$. The self-attention operator $f_{SA}(.)$ follows the standard ViT [6] and transformer [35], which computes a weighted sum of value vectors according to the cosine similarity distribution between query and key. Such memory-augmented self-attention aims to store normal patterns that are not encoded in the feature $\mathbf{X}^{(l)}$, forcing the decoder to reconstruct anomalous input patches into normal output patches during testing.

Multi-level Cross-Attention Transformer Decoder (Fig. 1 - Bottom Right). Our transformer decoder computes the cross-attention operation using the outputs from all encoder layers and the decoder layer output from the self-attention operator (see Fig. 1 - Bottom right). More formally, the layer $d \in \{0, ..., D - 1\}$ of our decoder outputs

$$\mathbf{Y}^{(d+1)} = \sum_{l=1}^{L} \alpha^{(d,l)} \times f_{SA}\big(f_{SA}(\mathbf{Y}^{(d)}, \mathbf{Y}^{(d)}, \mathbf{Y}^{(d)}), \mathbf{W}_K^{(d)} \mathbf{X}^{(l)}, \mathbf{W}_V^{(d)} \mathbf{X}^{(l)}\big), \quad (2)$$

where $\mathbf{Y}^{(d)}$ and $\mathbf{Y}^{(d+1)}$ represent the input and output of the decoder layer d containing $|\mathcal{P}|$ tokens (i.e., $|\mathcal{P}^{(v)}|$ tokens from the visible patches of the encoder and $|\mathcal{P}| - |\mathcal{P}^{(v)}|$ dummy tokens from the masked patches), $\mathbf{X}^{(l)}$ denotes the output from encoder layer $l-1$, and $\mathbf{W}_K^{(d)}, \mathbf{W}_V^{(d)}$ are the linear projections of the layer d of the decoder for the key and value of the self-attention operator, respectively. Note that all $|\mathcal{P}|$ input tokens for the decoder are attached with positional embeddings. The multi-level cross-attention results in (2) are fused together with a weighted sum operation using the weight $\alpha^{(l,d)}$, which is computed based on a linear projection layer and sigmoid function to control the weight of different layers' cross-attention results, as in

$$\alpha^{(d,l)} = \sigma\left(\mathbf{W}_\alpha^{(d,l)}\left(\left[f_{SA}(\mathbf{Y}^{(d)}, \mathbf{Y}^{(d)}, \mathbf{Y}^{(d)}), \mathbf{Y}^{(d+1)}\right]\right)\right), \quad (3)$$

where $\sigma(.)$ is the sigmoid function, and $\mathbf{W}_\alpha^{(d,l)}$ denotes a learnable weight matrix. Such fusion mechanism enforces the correlation of multiple normal patterns in the image present at different levels of encoding information to contribute at different decoding layers by adjusting their relative importance using the self-attention output from $f_{SA}(.)$ and cross-attention output $\mathbf{Y}^{(d+1)}$.

2.2 Anomaly Detection and Segmentation

We compute the anomaly score [3] with multi-scale structural similarity (MS-SSIM) [38]. The anomaly scores are pooled from 10 different random seeds for masking image patches with a fixed 75% masking ratio, which enables a more robust anomaly detection and localisation. The anomaly localisation mask is obtained by computing the mean MS-SSIM scores for all patches, and the anomaly detection relies on the mean MS-SSIM scores from the patches [3].

3 Experiments and Results

Datasets and Evaluation Measures. Three disease screening datasets are
used in our experiments. We test anomaly detection on the CXR images of the
pneumonia chest X-ray dataset [10] and Covid-X dataset [37], and both anomaly
detection and localisation on the colonoscopy images of the Hyper-Kvasir
dataset [2]. The publicly available **pneumonia chest X-ray dataset** [10], consisting of normal and pneumonia-affected images, was obtained from a total
of 6,480 patients. In accordance with [39], we structured the anomaly detection dataset such that the training set encompasses 1,349 normal images, and
the testing set comprises 234 normal and 390 pneumonia images. Each chest
X-ray image has been resized to the standardized dimensions of 256×256 pixels. **Covid-X** [37] has a training set with 1,670 Covid-19 positive and 13,794
Covid-19 negative CXR images, but we only use the 13,794 Covid-19 negative
CXR images for training. The test set contains 400 CXR images, consisting
of 200 positive and 200 negative images, each image with size 299×299 pixels. **Hyper-Kvasir** is a large-scale public gastrointestinal dataset. The images
were collected from the gastroscopy and colonoscopy procedures from Baerum
Hospital in Norway, and were annotated by experienced medical practitioners.
The dataset contains 110,079 images from unhealthy and healthy patients, out
of which, 10,662 are labelled. Following [31], 2,100 normal images are selected,
from which we use 1,600 for training and 500 for testing. The testing set also
contains 1,000 anomalous images with their segmentation masks. Detection is
assessed with area under the ROC curve (AUC), and localisation is evaluated
with intersection over union (IoU).

Implementation Details. For the transformer, we follow ViT-B [6,9] for
designing the encoder and decoder, consisting of stacks of transformer blocks.

Table 1. Anomaly detection AUC test results on Pneumonia and Covid-X Chest
X-ray datasets and Hyper-Kvasir colonoscopy dataset. CCD+IGD* [31] requires at
least 2×longer training time than other approaches in the table because of a two-stage
self-supervised pre-training and fine-tuning.

Methods	Publication	Pneumonia	Covid-X	Hyper-Kvasir
DAE [21]	ICANN'11	0.599	0.557	0.705
OCGAN [23]	CVPR'18	0.703	0.612	0.813
F-anoGAN [25]	IPMI'17	0.755	0.669	0.907
ADGAN [16]	ISBI'19	0.627	0.659	0.913
MS-SSIM [3]	AAAI'22	0.695	0.634	0.917
PANDA [24]	CVPR'21	0.657	0.629	0.937
PaDiM [5]	ICPR'21	0.663	0.614	0.923
IGD [3]	AAAI'22	0.734	0.699	0.939
CCD+IGD* [31]	MICCAI'21	0.775	0.746	**0.972**
Ours		**0.879**	**0.917**	**0.972**

Inspired by U-Net [40] for medical segmentation, we add residual connections to transfer information from earlier to later blocks for both the encoder and decoder. Each encoder block contains a memory-augmented self-attention block and an MLP block with LayerNorm (LN). Each decoder block contains a multi-level cross-attention block and an MLP block with LayerNorm (LN). We also adopt a linear projection layer after the encoder to match the different width between encoder and decoder [9]. We add positional embeddings (with the sine-cosine version) to both the encoder and decoder input tokens. RandomResizedCrop is used for data augmentation during training. Our method is trained for 2000 epochs in an end-to-end manner using the Adam optimiser with a weight decay of 0.05 and a batch size of 256. The learning rate is set to 1.5e-3. In the beginning, we warm up the training process for 5 epochs. The method is implemented in PyTorch and runs on an NVIDIA 3090 GPU. The overall training time is around 22 h, and the mean inference time takes 0.21 s per image.

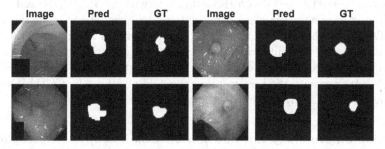

Fig. 2. Segmentation results of our proposed method on Hyper-Kvasir [2], with our predictions (Pred) and ground truth annotations (GT).

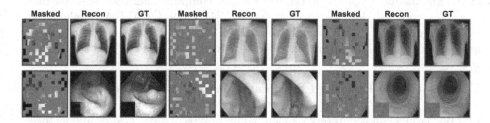

Fig. 3. Reconstruction of testing images from Covid-X (Top) and Hyper-Kvasir (Bottom). For each triplet, we show the masked image (left), our MemMC-MAE reconstruction (middle), and the ground-truth (right). Normal testing images are marked with green boxes, and anomalous ones are marked with red boxes. (Color figure online)

Evaluation on Anomaly Detection on Chest X-Ray and Colonoscopy. We compare our method with nine competing UAD approaches: DAE [21], OCGAN [23], f-anogan [25], ADGAN [16], MS-SSIM autoencoder [3], PANDA [24], PaDiM [5], CCD [31] and IGD [3]. We apply the same experimental setup (i.e., image pre-processing, training strategy, evaluation methods) to these methods above as the one for our approach for fair comparison.

Table 2. Ablation study on Covid-X of the encoder's memory-augmented operator (Mem-Enc) and the decoder's multi-level cross-attention (MC-Dec).

MAE	Mem-Enc	MC-Dec	AUC - Covid	AUC - Hyper
✓			0.799	0.915
✓	✓		0.862	0.956
✓	✓	✓	**0.917**	**0.972**

Table 3. Anomaly localisation: Mean IoU test results on Hyper-Kvasir on 5 groups of 100 images.

Methods	Localisation - IoU
IGD [3]	0.276
PaDiM [5]	0.341
CAVGA-R_u [36]	0.349
CCD + IGD [31]	0.372
Ours	**0.419**

The quantitative comparison results for anomaly detection are shown in Table 1 for Pneumonia, Covid-X, and Hyper-Kvasir benchmarks. Our MemMC-MAE achieves the best AUC results on three datasets with 87.9%, 91.7% and 97.2%, respectively. On pneumonia chest x-ray dataset, our model surpasses the previous SOTA approaches by a minimum 10.4% AUC and a maximum 28% AUC. On Covid-X, our result outperforms all competing methods by a large margin with an improvement of 17.1% over the second best approach. For Hyper-Kvasir, our result is on par with the best result in the field produced by CCD+IGD [31], which has a training time 2× longer than our approach.

Evaluation on Anomaly Localisation on Colonoscopy. We compare our anomaly localisation results on Table 3 with four recently proposed UAD baselines: IGD [3], PaDiM [5], CCD [31] and CAVGA-R_u [36]. The results of these methods on Table 3 are from [31]. Following [31], we randomly sample five groups of 100 anomalous images from the test set and compute the mean segmentation IoU. The proposed MemMC-MAE surpasses IGD, PaDiM, CAVGA-R_u and CCD by a minimum of 4.7% and a maximum of 14.3% IoU, illustrating the effectiveness of our model in localising anomalous tissues.

Visualisation of Predicted Segmentation. The visualisation of polyp segmentation results of MemMC-MAE on Hyper-Kvasir [2] is shown in Fig. 2. Notice that our model can accurately segment colon polyps of various sizes and shapes.

Visualisation of Reconstructed Images. Figure 3 shows the reconstructions produced by MemMC-MAE on Covid-X (Top) and Hyper-Kvasir (Bottom) testing images. Notice that our method can effectively reconstruct the anomalous images with polyps/covid as normal images by automatically removing the polyps or blurring the anomalous regions, leading to larger reconstruction errors for those anomalies. The normal images are accurately reconstructed with smaller reconstruction errors than the anomalous images.

Ablation Study. Table 2 shows the contribution of each component of our proposed method on Covid-X and Hyper-Kvasir testing set. The baseline MAE [9] achieves 79.9% and 91.5% AUC on the two datasets, respectively. Our method obtains a significant performance gain by adding the memory-augmented self-attention operator to the transformer encoder (Mem-Enc). Adding the proposed multi-level cross-attention operator into the decoder (MC-Dec) further boosts the performance on both datasets.

4   Conclusion

We proposed a new UAD reconstruction method, called MemMC-MAE, for anomaly detection and localisation in medical images, which to the best of our knowledge, is the first memory-based UAD method using MAE. MemMC-MAE introduced a novel memory-augmented self-attention operator for the MAE encoder and a new multi-level cross-attention for the MAE decoder to address the large reconstruction error of anomalous images that plague UAD reconstruction methods. The resulting anomaly detector showed SOTA anomaly detection and localisation accuracy on three public medical datasets. Despite the remarkable performance, the results can potentially improve if we use MemMC-MAE as a pre-training approach for other UAD methods, which we plan to explore in the future.

References

1. Baur, C., Wiestler, B., Albarqouni, S., Navab, N.: Scale-space autoencoders for unsupervised anomaly segmentation in brain MRI. In: Martel, A.L., et al. (eds.) MICCAI 2020. LNCS, vol. 12264, pp. 552–561. Springer, Cham (2020). https://doi.org/10.1007/978-3-030-59719-1_54
2. Borgli, H., et al.: Hyperkvasir, a comprehensive multi-class image and video dataset for gastrointestinal endoscopy. Sci. Data 7(1), 1–14 (2020)
3. Chen, Y., Tian, Y., Pang, G., Carneiro, G.: Deep one-class classification via interpolated gaussian descriptor. arXiv preprint arXiv:2101.10043 (2021)
4. Chen, Y., et al.: Bomd: bag of multi-label descriptors for noisy chest x-ray classification. arXiv preprint arXiv:2203.01937 (2022)
5. Defard, T., Setkov, A., Loesch, A., Audigier, R.: Padim: a patch distribution modeling framework for anomaly detection and localization. arXiv preprint arXiv:2011.08785 (2020)
6. Dosovitskiy, A., et al.: An image is worth 16x16 words: transformers for image recognition at scale. arXiv preprint arXiv:2010.11929 (2020)
7. Fan, D.-P., et al.: PraNet: parallel reverse attention network for polyp segmentation. In: Martel, A.L., et al. (eds.) MICCAI 2020. LNCS, vol. 12266, pp. 263–273. Springer, Cham (2020). https://doi.org/10.1007/978-3-030-59725-2_26
8. Gong, D., et al.: Memorizing normality to detect anomaly: memory-augmented deep autoencoder for unsupervised anomaly detection. In: ICCV, pp. 1705–1714 (2019)
9. He, K., Chen, X., Xie, S., Li, Y., Dollár, P., Girshick, R.: Masked autoencoders are scalable vision learners. arXiv preprint arXiv:2111.06377 (2021)

10. Kermany, D.S., et al.: Identifying medical diagnoses and treatable diseases by image-based deep learning. Cell **172**(5), 1122–1131 (2018)
11. Li, C.L., et al.: Cutpaste: self-supervised learning for anomaly detection and localization. In: CVPR, pp. 9664–9674 (2021)
12. Litjens, G., et al.: A survey on deep learning in medical image analysis. Med. Image Anal. **42**, 60–88 (2017)
13. Liu, F., Tian, Y., Cordeiro, F.R., Belagiannis, V., Reid, I., Carneiro, G.: Noisy label learning for large-scale medical image classification. arXiv preprint arXiv:2103.04053 (2021)
14. Liu, F., et al.: Self-supervised mean teacher for semi-supervised chest x-ray classification. arXiv preprint arXiv:2103.03629 (2021)
15. Liu, F., et al.: ACPL: anti-curriculum pseudo-labelling for semi-supervised medical image classification. In: CVPR (2022)
16. Liu, Y., et al.: Photoshopping colonoscopy video frames. In: ISBI, pp. 1–5 (2020)
17. Liu, Y., et al.: Translation consistent semi-supervised segmentation for 3d medical images. arXiv preprint arXiv:2203.14523 (2022)
18. Luo, Y., et al.: Harvard glaucoma fairness: a retinal nerve disease dataset for fairness learning and fair identity normalization. arXiv preprint arXiv:2306.09264 (2023)
19. LZ, C.T.P., et al.: Computer-aided diagnosis for characterisation of colorectal lesions: a comprehensive software including serrated lesions. Gastrointest. Endosc. (2020)
20. Martins, P.H., Marinho, Z., Martins, A.F.: Infinity-former: infinite memory transformer. arXiv preprint arXiv:2109.00301 (2021)
21. Masci, J., Meier, U., Cireşan, D., Schmidhuber, J.: Stacked convolutional auto-encoders for hierarchical feature extraction. In: Honkela, T., Duch, W., Girolami, M., Kaski, S. (eds.) ICANN 2011. LNCS, vol. 6791, pp. 52–59. Springer, Heidelberg (2011). https://doi.org/10.1007/978-3-642-21735-7_7
22. Pang, G., Shen, C., van den Hengel, A.: Deep anomaly detection with deviation networks. In: Proceedings of the 25th ACM SIGKDD International Conference on Knowledge Discovery and Data Mining, pp. 353–362 (2019)
23. Perera, P., Nallapati, R., Xiang, B.: Ocgan: one-class novelty detection using gans with constrained latent representations. In: CVPR, pp. 2898–2906 (2019)
24. Reiss, T., Cohen, N., Bergman, L., Hoshen, Y.: Panda: adapting pretrained features for anomaly detection and segmentation. In: Proceedings of the IEEE/CVF Conference on Computer Vision and Pattern Recognition, pp. 2806–2814 (2021)
25. Schlegl, T., et al.: f-anogan: fast unsupervised anomaly detection with generative adversarial networks. Med. Image Anal. **54**, 30–44 (2019)
26. Seeböck, P., et al.: Exploiting epistemic uncertainty of anatomy segmentation for anomaly detection in retinal oct. IEEE Trans. Med. Imaging **39**(1), 87–98 (2019)
27. Shi, M., et al.: Artifact-tolerant clustering-guided contrastive embedding learning for ophthalmic images in glaucoma. IEEE J. Biomed. Health Inf. (2023)
28. Sohn, K., Li, C.L., Yoon, J., Jin, M., Pfister, T.: Learning and evaluating representations for deep one-class classification. arXiv preprint arXiv:2011.02578 (2020)
29. Tian, Y., et al.: Few-shot anomaly detection for polyp frames from colonoscopy. In: Martel, A.L., et al. (eds.) MICCAI 2020. LNCS, vol. 12266, pp. 274–284. Springer, Cham (2020). https://doi.org/10.1007/978-3-030-59725-2_27
30. Tian, Y., et al.: Contrastive transformer-based multiple instance learning for weakly supervised polyp frame detection. In: Wang, L., Dou, Q., Fletcher, P.T., Speidel, S., Li, S. (eds.) MICCAI 2022. LNCS, pp. 88–98. Springer, Cham (2022). https://doi.org/10.1007/978-3-031-16437-8_9

31. Tian, Y., et al.: Constrained contrastive distribution learning for unsupervised anomaly detection and localisation in medical images. In: de Bruijne, M., et al. (eds.) MICCAI 2021. LNCS, vol. 12905, pp. 128–140. Springer, Cham (2021). https://doi.org/10.1007/978-3-030-87240-3_13
32. Tian, Y., et al.: One-stage five-class polyp detection and classification. In: 2019 IEEE 16th International Symposium on Biomedical Imaging (ISBI 2019), pp. 70–73. IEEE (2019)
33. Tian, Y., et al.: Pixel-wise energy-biased abstention learning for anomaly segmentation on complex urban driving scenes. arXiv preprint arXiv:2111.12264 (2021)
34. Tian, Y., et al.: Self-supervised multi-class pre-training for unsupervised anomaly detection and segmentation in medical images. arXiv preprint arXiv:2109.01303 (2021)
35. Vaswani, A., et al.: Attention is all you need. Adv. Neural Inf. Process. Syst. **30** (2017)
36. Venkataramanan, S., Peng, K.-C., Singh, R.V., Mahalanobis, A.: Attention guided anomaly localization in images. In: Vedaldi, A., Bischof, H., Brox, T., Frahm, J.-M. (eds.) ECCV 2020. LNCS, vol. 12362, pp. 485–503. Springer, Cham (2020). https://doi.org/10.1007/978-3-030-58520-4_29
37. Wang, L., Lin, Z.Q., Wong, A.: Covid-net: a tailored deep convolutional neural network design for detection of covid-19 cases from chest x-ray images. Sci. Rep. **10**(1), 1–12 (2020)
38. Wang, Z., et al.: Multiscale structural similarity for image quality assessment. In: The Thrity-Seventh Asilomar Conference on Signals, Systems and Computers, 2003. vol. 2, pp. 1398–1402. IEEE (2003)
39. Zhao, H., et al.: Anomaly detection for medical images using self-supervised and translation-consistent features. IEEE Trans. Med. Imaging **40**(12), 3641–3651 (2021)
40. Zhou, Z., Rahman S., Md Mahfuzur, Tajbakhsh, N., Liang, J.: UNet++: a nested U-net architecture for medical image segmentation. In: Stoyanov, D., et al. (eds.) DLMIA/ML-CDS -2018. LNCS, vol. 11045, pp. 3–11. Springer, Cham (2018). https://doi.org/10.1007/978-3-030-00889-5_1

LMT: Longitudinal Mixing Training, a Framework to Predict Disease Progression from a Single Image

Rachid Zeghlache[1,2]($\boxtimes$), Pierre-Henri Conze[1,3], Mostafa El Habib Daho[1,2], Yihao Li[1,2], Hugo Le Boité[5], Ramin Tadayoni[5], Pascal Massin[5], Béatrice Cochener[1,2,4], Ikram Brahim[1,6], Gwenolé Quellec[1], and Mathieu Lamard[1,2]

[1] LaTIM UMR 1101, Inserm, Brest, France
rachid.zeghlache@univ-brest.fr
[2] University of Western Brittany, Brest, France
[3] IMT Atlantique, Brest, France
[4] Ophtalmology Department, CHRU Brest, Brest, France
[5] Lariboisière Hospital, AP-HP, Paris, France
[6] LBAI UMR 1227, Inserm, Brest, France

Abstract. Longitudinal imaging is able to capture both static anatomical structures and dynamic changes in disease progression toward earlier and better patient-specific pathology management. However, conventional approaches rarely take advantage of longitudinal information for detection and prediction purposes, especially for Diabetic Retinopathy (DR). In the past years, Mix-up training and pretext tasks with longitudinal context have effectively enhanced DR classification results and captured disease progression. In the meantime, a novel type of neural network named Neural Ordinary Differential Equation (NODE) has been proposed for solving ordinary differential equations, with a neural network treated as a black box. By definition, NODE is well suited for solving time-related problems. In this paper, we propose to combine these three aspects to detect and predict DR progression. Our framework, Longitudinal Mixing Training (LMT), can be considered both as a regularizer and as a pretext task that encodes the disease progression in the latent space. Additionally, we evaluate the trained model weights on a downstream task with a longitudinal context using standard and longitudinal pretext tasks. We introduce a new way to train time-aware models using t_{mix}, a weighted average time between two consecutive examinations. We compare our approach to standard mixing training on DR classification using OPHDIAT a longitudinal retinal Color Fundus Photographs (CFP) dataset. We were able to predict whether an eye would develop a severe DR in the following visit using a single image, with an AUC of 0.798 compared to baseline results of 0.641. Our results indicate that our longitudinal pretext task can learn the progression of DR disease and that introducing t_{mix} augmentation is beneficial for time-aware models.

Keywords: Disease progression · mix-up training · diabetic retinopathy · time-aware model · predictive medicine

© The Author(s), under exclusive license to Springer Nature Switzerland AG 2024
X. Cao et al. (Eds.): MLMI 2023, LNCS 14349, pp. 22–32, 2024.
https://doi.org/10.1007/978-3-031-45676-3_3

1   Introduction

According to the International Diabetes Federation, by 2045, diabetes will impact 700 million individuals globally, with over one-third suffering from Diabetic Retinopathy (DR) [20]. DR is the leading cause of vision loss worldwide and is caused by high blood sugar damaging retinal vessels, leading to swelling and leakage [15]. Color fundus photographs (CFP) are used clinically to detect DR. Severity is graded in five classes (0, 1, 2, 3 and 4) using the International Clinical DR (ICDR) scale: 0 is no apparent DR, 1 is mild non-proliferative DR (NPDR), 2 is moderate NPDR, 3 is severe NPDR, and 4 is proliferative DR (PDR). Early detection and treatment, particularly in mild to moderate NPDR, may slow DR progression and reduce blindness incidence. Very few papers try to predict the progression of DR using a single CFP [2,18]. Despite its difficulty, this task is crucial for better patient follow-up management.

Recently, longitudinal pretext tasks (LPT) have emerged to encode disease progression, such as Longitudinal Self-Supervised Learning (LSSL) introduced by Rivail et al. [17] using a Siamese network to predict time lapses between consecutive retinal Optical Coherence Tomography (OCT) scans. Zhao et al. [27] proposed a theoretical framework for LSSL using an auto-encoder and an alignment term that forces the topology of the latent space to change in the direction of longitudinal changes. An extension was proposed in [16] to create a smooth trajectory field and a dynamic graph was computed to connect nearby subjects and enforce maximally aligned progression directions. Authors in [26] successfully used LSSL in the context of DR to predict the change from grades {0,1} to {2,3,4}, referred to as Moderate+, between two consecutive CFPs. However, since these approaches are self-supervised, it is unclear whether they learn the disease progression or the longitudinal changes.

Neural Ordinary Differential Equation (NODE) is a new type of neural networks that parameterizes the continuous dynamics of ordinary differential equations. It is ideally suited for solving time-related problems. Time-aware models, including NODEs, have achieved state-of-the-art performance in various tasks related to irregularly-sampled time series data or disease progression [1,9,19,25]. However, NODEs remain challenging to train. Authors in [8] proposed a simple yet efficient technique to regularize NODEs by randomly solving ODE for longer time points.

In recent years, mixing augmentation training has been successful in computer vision [12,22,24]. Mix-up [22] performs on the training set linearly, mixing a random pair of examples and their corresponding labels. Manifold Mix-up [24] extends such principle to linear interpolation to the hidden representation. Based on two one-hot labels, the mix-up training generates soft labels that model the relationship between two classes. One-hot labels describe the intra-class relationship, while Mix-up labels describe the inter-class relationship.

These soft labels modulate the learned decision boundaries, providing the model with more information. In this sense, Mix-up training can be seen as a pretext task. Going further, longitudinal-based Mix-up can be considered as a pretext task that captures the disease's (presumed) linear progression. Motivated by that, we propose a LPT using Manifold Mix-up (MM) [24], which we consider to be more suitable form of Mix-up training for our objective because manifold Mix-up has proven to provide more informative hidden representations. In order to learn feature representations embedded with disease progression that can be reapplied in tackling longitudinal-based problems, especially for the challenging task of prediction the disease progression based on a single image. We introduce t_{mix} an intermediate time in the course of the disease progression, t_{mix} is the weighted average time between two consecutive examinations. t_{mix} is used to obtain soft labels based on a severity profile. Additionally, it acts as a data augmentation for training time-aware models. To the best of our knowledge, this work is the first to automatically assess DR and forecast its progression using Mix-up training and time-aware models, using only one CFP.

2   Methods

2.1   Preliminary

Let $\mathcal{V}$ be the set of consecutive patient-specific image pairs for the collection of all CFP images. $\mathcal{V}$ contains all $(x_{t_i}, x_{t_{i+1}})$ that are from the same patient where x_{t_i} is scanned before $x_{t_{i+1}}$ with $i \in [0, m-2]$, m being the number of visits for a given eye. Figure 1 displays a simplified architecture for clarity. We define backbone $g_{1:n}$ with n layers, where $g_{1:k}$ denotes the part of the neural network mapping the input data to the hidden representation at layer k. h_l represents a classification or regression head l, (y, y') one-hot labels, $\text{Beta}(\alpha, \alpha)$ the Beta distribution and $\ell(.)$ the Binary Cross-Entropy (BCE) loss. We define the mixing operator by $\text{Mix}_\lambda(a, b) = \lambda \cdot a + (1 - \lambda) \cdot b$ with $\lambda \sim \text{Beta}(\alpha, \alpha)$ where $\lambda \in [0, 1]$. **Mix-up** was introduced in [22] as a simple regularization method to minimize overfitting in deep neural networks. It linearly interpolates a mini-batch of random examples and their labels to transform the training set.
Manifold Mix-up is an extension of Mix-up to hidden representations [24]. During training, a random layer k from a set of eligible layers S in a neural network is selected. It processes two random data mini-batches (x, y) and (x', y'), until reaching layer k. The Mix-up is then performed on these two intermediate mini-batches $(g_k(x), y)$ and $(g_k(x'), y')$, continuing the forward pass with the mixed representation until the ending layer n. These mixed representations are then fed to the classification head h_l and projected to the number of classes.
Neural Ordinary Differential Equations (NODEs) approximate unknown ordinary differential equations by a neural network [5] that parameterizes the continuous dynamics of hidden units $\mathbf{z} \in \mathbb{R}^n$ over time with $\mathbf{t} \in \mathbb{R}$. NODEs are able to model the instantaneous rate of change of $\mathbf{z}$ with respect to $\mathbf{t}$ using a neural network u with parameters θ.

$$\lim_{h \to 0} \frac{\mathbf{z}_{t+h} - \mathbf{z}_t}{h} = \frac{d\mathbf{z}}{dh} = u(t, \mathbf{z}, \boldsymbol{\theta}) \tag{1}$$

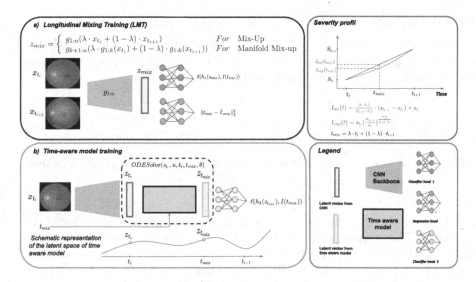

Fig. 1. Illustration of Longitudinal Mixing Training in a) and time-aware model training using t_{mix} in b). Fig. 1a) and Fig. 1.b) can be trained simultaneously or independently.

The analytical solution of Eq. 1 is given by:

$$\mathbf{z}_{t_1} = \mathbf{z}_{t_0} + \int_{t_0}^{t_1} u(t, \mathbf{z}, \boldsymbol{\theta}) dt = \text{ODESolve}(\mathbf{z}(t_0), u, t_0, t_1, \theta) \qquad (2)$$

where $[t_0, t_1]$ represents the time horizon for solving the ODE, u being a neural network, and θ is the trainable parameters of u. By using a black-box ODE solver introduced in [5], we can solve the Initial Value Problem (IVP) and calculate the hidden state at any desired time using Eq. 2. We can differentiate the solutions of the ODE solver with respect to the parameters θ, the initial state $\mathbf{z}_{t_0}$ at initial time t_0, and the solution at time t. This can be achieved by using the adjoint sensitivity method [5]. Through the latent representation of a given image, we define an IVP that aims to solve the ODE from t_i to a terminal time t_{i+1}:

$$\dot{z}(t) = u(z(t), t, \theta), \text{with the initial value } z(t_i) = z_{t_i} \qquad (3)$$

2.2   Longitudinal Mixing Training (LMT)

We denote s_{t_i} the severity grade of image x_{t_i}. $I(t)$ is the severity interpolation function between two consecutive longitudinal pairs. In the linear case, between x_{t_i} and $x_{t_{i+1}}$ we have $I_{lin}(t) = \frac{(t-t_i)}{(t_{i+1}-t_i)} \cdot (s_{t_{i+1}} - s_{t_i}) + s_{t_i}$. In conventional Mix-up training, the labels are mixed. Instead, we propose to mix the time between consecutive pairs $t_{mix} = \text{Mix}_\lambda(t_i, t_{i+1})$, then used this t_{mix} to evaluate $I(t)$ and uses this signal as supervision. Motivated by the assumption that the progression of DR is a slow process, we tested another monotonic disease progression profile expressed as $I_{exp}(t) = s_{t_i}(\frac{s_{t_{i+1}}}{s_{t_i}})^{\frac{t-t_i}{t_{i+1}-t_i}}$. During training, we one-hot encode the value of interpolation at t_{mix} to get our soft label. Depending on the mix-up method, we have as latent representation of the mixed pair:

$$z_{mix} = \begin{cases} g_{1:n}(\text{Mix}_\lambda(x_{t_i}, x_{t_{i+1}})) & for \text{ Mix-Up} \\ g_{k+1:n}(\text{Mix}_\lambda(g_{1:k}(x_{t_i}), g_{1:k}(x_{t_{i+1}}))) & for \text{ Manifold Mix-up} \end{cases} \quad (4)$$

During the training of our LMM, we add a **time consistency** loss, as follows:

$$L_{t_{mix}} = \| t_{mix} - \tilde{t}_{mix} \|_2^2 \quad (5)$$

with $\tilde{t}_{mix} = h_2(z_{mix})$, where h_2 is regression head that predicts the value of the current t_{mix} for a given pair. This term is inspired by [17] and motivated by authors in [13], who used Manifold Mix-up coupled with SSL loss to enhance the quality of the feature extraction. For the training of the LMM, the total loss is:

$$L = \underset{(x_{t_i}, x_{t_{i+1}}) \sim \mathcal{V}}{\mathbb{E}} \underset{\lambda \sim \text{Beta}(\alpha, \alpha)}{\mathbb{E}} \underset{k \sim \mathcal{S}}{\mathbb{E}} \ell(h_1(z_{mix}), I(t_{mix})) + L_{t_{mix}} \quad (6)$$

Concerning the use of t_{mix} for the NODE; instead of solving the ODE from t_i to a terminal time t_{i+1} using Eq. 3, we solve to the intermediate time t_{mix} (see Fig. 1.b) then use this t_{mix} to evaluate $I(t)$ and take this signal as supervision for training. Note that this approach could be applied to any time-aware model.

3   Experiments and Results

Dataset. The proposed models were trained and evaluated on OPHDIAT [14], a large CFP database collected from the Ophthalmology Diabetes Telemedicine network consisting of examinations acquired from 101,383 patients between 2004 and 2017. Out of 763,848 interpreted CFP images, 673,017 were assigned a DR severity grade, and the others were non-gradable. Patients range in age from 9 to 91, and image sizes vary from 1440 × 960 to 3504 × 2336 pixels. Each examination includes at least two eye images. To limit consecutive pairs without progression, 10412 patients were selected with at least one severity change. Each patient had 2-5 scans, averaging 3.43, spanning an average interval of 4.86 years. This dataset was further divided into training (60%), validation (20%), and test (20%) based on patients. We randomly selected one image per eye for each examination, resulting in 49578 pairs. Our longitudinal downstream task is to predict whether an eye without DR at the initial visit was later graded as having Mild+ DR within two years. 8,111 patients and 13,936 eyes fit this criterion for the training. For the DR assignment, a specific test set consisted of patients assigned the same grade from two ophthalmologists, resulting in 9,734 eyes of 4,996 patients. Except for the registration, we followed the same image processing performed in [26]. All the timestamp were normalized by 2×365.

Implementation Details. In our basic architecture, we employed a stack of 2 pre-activated residual blocks (ReLu+BN). In each residual block, the residual feature map was calculated using a series of three 3×3 convolutions, the first of which always halves the number of the feature maps employed at the present scale. Our encoder comprised seven levels; the first six levels are composed of two residual blocks and the latter deals with only one residual block. This provides a final latent representation of size $64 \times 4 \times 4$. The last three layers of our backbone were used for the eligible layers S. The different networks were trained for 200 epochs by the AdamW optimizer, OneCycleLR as a scheduler, weight decay of 10^{-4}, and a batch size of 128, using an NVIDIA A6000 GPU with the PyTorch framework. A grid search was performed for several key hyper-parameters of all Mix-up algorithms and SSL, including $\alpha \in \{0.2, 0.5, 1.0, 2.0, 3.0, 5.0, 10.0\}$, learning rate $\in \{10^{-2}, 10^{-3}, 10^{-4}\}$ and three different seed values. Concerning the NODE, we used the Pytorch package Torchdiffeq [4]. This library provides ODE solvers, and backpropagation through ODE solutions and a support of the adjoint method for constant memory cost. Our NODE is a combination of dense layers followed by the tanh activation function and the adjoint method with "dopri5" as a solver. The loss in each task was used to monitor the model's performance on the validation set, the best one was kept.

Experiments for DR Severity Assessment Using Mixing Training. Usually, a new permutation is applied in mixing training at each batch. Since we only use fixed consecutive pairs, the network looks at fewer examples. To perform fair comparisons, we tried multiple permutations that matched the distribution of longitudinal pairs. We report the Quadratic-weighted Kappa for different scenarios of mixing training in Table 1.

Experiments to Evaluate the Quality of Feature Extraction. For the downstream task, we provided the results from both linear evaluation and fine-tuning. The performance was evaluated with the Area Under the receiver operating characteristic Curve (AUC) and reported Table 2. The linear evaluation was conducted by training a linear layer on top of the pre-trained and frozen encoder and trained using the same set-up that was tried with LMM. For the classical feature extractor, we used AE, VAE [11], and SimCLR [6]. For longitudinal SSL, we used longitudinal Siamese [17], LSSL [27], and LNE [16].

Experiments Using t_{mix} for Time-Aware Models. Another time-aware model T-LSTM, introduced in [1], was used for this experiment in order to demonstrate the effectiveness of t_{mix}. Both time-aware model take as input the latent representation z_{t_i} and the time difference between x_{t_i} and $x_{t_{i+1}}$ (Δ_t) in order to predict the latent representation of $z_{t_{i+1}}$. For the NODE, it was performed by the mean of IVP defined in Eq. 3 while for T-LSTM [1], the LSTM gates were modulated by Δ_t to produce the future $z_{t_{i+1}}$. We tested three training set-ups:

Table 1. Comparison of the best Kappa for Mix-up training for DR severity assessment.

	Kappa	α
Mix-up ⋆	0.7646	0.2
Manifold Mix-up ⋆	0.7747	2.0
Mix-up	0.7314	0.2
Manifold Mix-up	0.7342	2.0
Longitudinal Mix-up (LM) + $I_{lin}(t)$	0.7339	0.2
Longitudinal Manifold Mix-up (LMM) + $I_{lin}(t)$	**0.7511**	2.0
Longitudinal Mix-up (LM) + $I_{exp}(t)$	0.5595	0.5
Longitudinal Manifold Mix-up (LMM) + $I_{exp}(t)$	**0.7350**	2.0

Table 2. Results on linear evaluation and fine-tuning of pre-trained model using AUC

	AUC (Mild+ DR within 2 years)	
Weights	Frozen	Fine-tuned
Random	-	0.584
MM [24] ($\alpha = 2.0$)	0.564	0.595
AE	0.531	0.569
VAE [11]	0.510	0.575
SimCLR [6]	0.544	0.558
L-Siamese [17]	0.562	0.593
LSSL [27]	0.579	0.602
LNE [16]	0.570	0.595
Ours (LMM $\alpha = 2.0$)	**0.613**	**0.627**

1. Using one image and the time of the next examination, we trained the model presented in Fig. 1.b) with the loss $\ell(h_3(z_{t_{i+1}}, s_{t+1}))$.
2. Similarly to (1), we used image x_t but instead of giving the time of the next examination, t_{mix} was used and trained with the loss $\ell(h_3(z_{t_{mix}}, I(t_{mix}))$.
3. We used (2) with our LMM, i.e., we use (a) and (b) simultaneously (Fig. 1).

Table 3. Comparison of AUCs for the next visit for time-aware model training with and without t_{mix} with linear progression assumption.

	AUC Mild+DR	AUC moderate+DR	AUC severe+DR	Best α
(1) NODE [5]	0.584	0.617	0.641	-
(1) T-LSTM [1]	0.608	0.646	0.677	-
(2) NODE + t_{mix} (ours)	0.632	0.695	0.725	2.0
(2) T-LSTM + t_{mix} (ours)	0.610	0.661	0.725	0.5
(3) NODE+LMM (ours)	**0.657**	**0.721**	**0.798**	2.0

The best results for the longitudinal task are obtained with LMM (Table 2), indicating that it effectively captures disease progression. Moreover, the longitudinal task performed better than classical feature extraction methods, which is aligned with [7,16,17,25–27]. Concerning the use of t_{mix}, results in Table 3 indicate that it is beneficial for both time-aware models. We believe t_{mix} plays the role of data augmentation in the context of disease progression for Time-Aware models and is regarded as a method to regularize the training of NODE like [8]. The fact that the set-up (3) performs better than other configurations supports the idea that LMM and t_{mix} are beneficial to solve time-related problems.

Only LMM adapts $I_{exp}(t)$ (Fig. 2), indicating that with LMM, other types of severity profiles could be used. We suspect that LM suffers from manifold intrusion [10]. Mixing longitudinal pairs create an existing severity grade in the

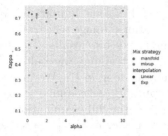

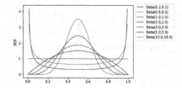

Fig. 2. Best value of Kappa when alpha varies for LM and LMM for the two profiles.

Fig. 3. Beta distribution for alpha values used.

dataset, making training more challenging when using Mix-up. LMM performs better than regular MM, according to Table 1, when the profile of severity progression was supposed to be linear between two consecutive exams, suggesting that longitudinal pairs are more informative than random pairs with the same label distribution. The more α increases, the more the severity interpolation is taken into consideration because when $\lambda \leq 0.1 \Rightarrow t_{mix} \simeq t_1$ and $\lambda = 0.5 \Rightarrow t_{mix} = \frac{t_0 + t_1}{2}$ and finally $\lambda \geq 0.9 \Rightarrow t_{mix} \simeq t_0$, as illustrated by the Beta distribution in Fig. 3. In Fig. 2 for MML, we found that as alpha increases, Kappa differences between severity profiles increase. This could indicate that $I_{exp}(t)$ is sub-optimal for DR progression. In addition, for $\alpha = 10$, the LMM is almost trained with the center of the severity interpolation $I(t)$ as a label and yet, according to Fig. 2, is able to assign DR with a Kappa of 0.75 (second best value of Kappa in all experiments). This could suggest, like in the original MM [24], that the LMM is able to disentangle factors of variations, such as the one responsible for encoding the disease progression. When alpha is low, there is a higher chance of sampling values of λ closer to either 0 or 1 from the Beta distribution. Time-aware models are then practically trained for two tasks: 1- predict the severity grade of the current image ($t_{mix} \simeq t_i \Rightarrow I_{lin}(t_{mix}) \simeq s_{t_i}$) and 2- predict the next visit grade based on the last exam ($t_{mix} \simeq t_{i+1} \Rightarrow I_{lin}(t_{mix}) \simeq s_{t_{i+1}}$). As a result, the model receives more information, which could explain the increase in performance. However, according to our experiments, T-LSTM does not perform well when α is high. Only the NODE gains from having diverse time point $t_{mix} \in [t_i, t_{i+1}]$ during training, showing that it can successfully change its hidden dynamic when time varies, in line with the conclusions of [5,19].

4 Discussion and Conclusion

In this paper, we proposed straightforward modifications to Manifold Mix-up training. This adaptation aims to enhance training of time-aware models for disease progression by introducing t_{mix}. The results are encouraging and may help clinicians to choose the best DR screening intervals. Our framework is general and could be easily extended to other Mix-up training [12,23] and time-aware models [3, 19] or to other diseases. However our work has some limitations. We did not register images between consecutive examinations. Image registration is a critical step, as mentioned in [21], and could enhances our results. To overcome the lack of grade diversity for a given pair during training, we could train with all potential pairs of a patient in a follow-up, as done in [7]. We made a strong assumption on the disease progression by supposing one common severity profile, yet observing good results. Since we can access the full examination, we could use a more accurate interpolation function to better fit the DR progression. We hope this work will benefit the fields of longitudinal analysis and disease progression.

Acknowledgments. The work takes place in the framework of the ANR RHU project Evired. This work benefits from State aid managed by the French National Research Agency under the "Investissement d'Avenir" program bearing the reference ANR-18-RHUS-0008.

References

1. Baytas, I., Xiao, C., Zhang, X., Wang, F., Jain, A., Zhou, J.: Patient subtyping via time-aware LSTM networks. In: Proceedings of the 23rd ACM SIGKDD International Conference on Knowledge Discovery and Data Mining, pp. 65–74 (2017). https://doi.org/10.1145/3097983.3097997
2. Bora, A., et al.: Predicting the risk of developing diabetic retinopathy using deep learning. Lancet Digit. Health **3**(1), e10–e19 (2021)
3. Brouwer, E.D., Simm, J., Arany, A., Moreau, Y.: GRU-ODE-Bayes: continuous modeling of sporadically-observed time series. In: Advances in Neural Information Processing Systems (2019)
4. Chen, R.T.Q.: Torchdiffeq (2018). https://github.com/rtqichen/torchdiffeq
5. Chen, R.T.Q., Rubanova, Y., Bettencourt, J., Duvenaud, D.: Neural ordinary differential equations (2018). https://doi.org/10.48550/ARXIV.1806.07366. https://arxiv.org/abs/1806.07366
6. Chen, T., Kornblith, S., Norouzi, M., Hinton, G.: A simple framework for contrastive learning of visual representations (2020). https://doi.org/10.48550/ARXIV.2002.05709. https://arxiv.org/abs/2002.05709
7. Emre, T., Chakravarty, A., Rivail, A., Riedl, S., Schmidt-Erfurth, U., Bogunović, H.: TINC: temporally informed non-contrastive learning for disease progression modeling in retinal OCT volumes. In: Wang, L., Dou, Q., Fletcher, P.T., Speidel, S., Li, S. (eds.) Medical Image Computing and Computer Assisted Intervention – MICCAI 2022. MICCAI 2022. LNCS, vol. 13432. Springer, Cham (2022). https://doi.org/10.1007/978-3-031-16434-7_60
8. Ghosh, A., Behl, H.S., Dupont, E., Torr, P.H.S., Namboodiri, V.: STEER: simple temporal regularization for neural odes (2020). https://doi.org/10.48550/ARXIV.2006.10711. https://arxiv.org/abs/2006.10711
9. Gruffaz, S., Poulet, P.E., Maheux, E., Jedynak, B., Durrleman, S.: Learning Riemannian metric for disease progression modeling. In: Ranzato, M., Beygelzimer, A., Dauphin, Y., Liang, P., Vaughan, J.W. (eds.) In: Advances in Neural Information Processing Systems, vol. 34, pp. 23780–23792. Curran Associates, Inc. (2021). https://proceedings.neurips.cc/paper/2021/file/c7b90b0fc23725f299b47c5224e6ec0d-Paper.pdf

10. Guo, H., Mao, Y., Zhang, R.: MixUp as locally linear out-of-manifold regularization (2018). https://doi.org/10.48550/ARXIV.1809.02499. https://arxiv.org/abs/1809.02499

11. Kingma, D.P., Welling, M.: Auto-encoding variational Bayes (2013). https://doi.org/10.48550/ARXIV.1312.6114. https://arxiv.org/abs/1312.6114

12. Liu, Z., et al.: AutoMix: unveiling the power of mixup for stronger classifiers (2021). https://doi.org/10.48550/ARXIV.2103.13027. https://arxiv.org/abs/2103.13027

13. Mangla, P., Singh, M., Sinha, A., Kumari, N., Balasubramanian, V.N., Krishnamurthy, B.: Charting the right manifold: Manifold mixup for few-shot learning. In: Proceedings of the IEEE/CVF Winter Conference on Applications of Computer Vision, pp. 2218-2227 (2020)

14. Massin, P., et al.: OPHDIAT: a telemedical network screening system for diabetic retinopathy in the Île-de-France. Diabetes Metab. **34**(3), 227–234 (2008)

15. Ogurtsova, K., et al.: IDF diabetes atlas: global estimates for the prevalence of diabetes for 2015 and 2040. Diabetes Res. Clin. Pract. **128**, 40–50 (2017)

16. Ouyang, J., et al.: Self-supervised longitudinal neighbourhood embedding. In: de Bruijne, M., et al. (eds.) MICCAI 2021. LNCS, vol. 12902, pp. 80–89. Springer, Cham (2021). https://doi.org/10.1007/978-3-030-87196-3_8

17. Rivail, A., et al.: Correction to: modeling disease progression in retinal OCTs with longitudinal self-supervised learning. In: Rekik, I., Adeli, E., Park, S.H. (eds.) PRIME 2019. LNCS, vol. 11843, pp. C1–C1. Springer, Cham (2019). https://doi.org/10.1007/978-3-030-32281-6_19

18. Rom, Y., Aviv, R., Ianchulev, S., Dvey-Aharon, Z.: Predicting the future development of diabetic retinopathy using a deep learning algorithm for the analysis of non-invasive retinal imaging. medRxiv (2022). https://doi.org/10.1101/2022.03.31.22272079. https://www.medrxiv.org/content/early/2022/07/05/2022.03.31.22272079

19. Rubanova, Y., Chen, R.T.Q., Duvenaud, D.: Latent odes for irregularly-sampled time series (2019). https://doi.org/10.48550/ARXIV.1907.03907. https://arxiv.org/abs/1907.03907

20. Saeedi, P., et al.: Global and regional diabetes prevalence estimates for 2019 and projections for 2030 and 2045: results from the international diabetes federation diabetes atlas, 9th edition. Diabetes Res. Clin. Pract. **157**, 107843 (2019)

21. Saha, S.K., Xiao, D., Bhuiyan, A., Wong, T.Y., Kanagasingam, Y.: Color fundus image registration techniques and applications for automated analysis of diabetic retinopathy progression: a review. Biomed. Signal Process. Control **47**, 288–302 (2019)

22. Thulasidasan, S., Chennupati, G., Bilmes, J.A., Bhattacharya, T., Michalak, S.E.: On mixup Training: improved calibration and predictive uncertainty for deep neural networks. In: Neural Information Processing Systems (2019)

23. Venkataramanan, S., Kijak, E., Amsaleg, L., Avrithis, Y.: AlignMixup: improving representations by interpolating aligned features. In: Proceedings of the IEEE/CVF Conference on Computer Vision and Pattern Recognition, pp. 19174–19183 (2022)

24. Verma, V., et al.: Manifold Mixup: better representations by interpolating hidden states. In: International Conference on Machine Learning, pp. 6438–6447 (2018)

25. Zeghlache, R., et al.: Longitudinal self-supervised learning using neural ordinary differential equation. In: International Workshop on Predictive Intelligence in Medicine. Vancouver (Canada), Canada (2023). https://imt-atlantique.hal.science/hal-04171357

26. Zeghlache, R., et al.: Detection of diabetic retinopathy using longitudinal self-supervised learning. In: Antony, B., Fu, H., Lee, C.S., MacGillivray, T., Xu, Y., Zheng, Y. (eds.) Ophthalmic Medical Image Analysis. OMIA 2022. LNCS, vol. 13576. Springer, Cham (2022). https://doi.org/10.1007/978-3-031-16525-2_5
27. Zhao, Q., Liu, Z., Adeli, E., Pohl, K.M.: Longitudinal self-supervised learning. Med. Image Anal. **71**, 102051 (2021). https://doi.org/10.1016/j.media.2021.102051

Identifying Alzheimer's Disease-Induced Topology Alterations in Structural Networks Using Convolutional Neural Networks

Feihong Liu[1,2], Yongsheng Pan[2], Junwei Yang[2,3], Fang Xie[4], Xiaowei He[1],
Han Zhang[2], Feng Shi[5], Jun Feng[1(✉)], Qihao Guo[6(✉)],
and Dinggang Shen[2,5,7(✉)]

[1] School of Information Science and Technology, Northwest University, Xi'an, China
fengjun@nwu.edu.cn
[2] School of Biomedical Engineering, ShanghaiTech University, Shanghai, China
[3] Department of Computer Science and Technology, University of Cambridge,
Cambridge, UK
[4] PET Center, Huashan Hospital, Fudan University, Shanghai, China
[5] Shanghai United Imaging Intelligence Co. Ltd., Shanghai, China
[6] Department of Gerontology, Shanghai Jiao Tong University Affiliated Sixth
People's Hospital, Shanghai, China
qhguo@sjtu.edu.cn,
[7] Shanghai Clinical Research and Trial Center, Shanghai, China
dgshen@shanghaitech.edu.cn

Abstract. Identifying topology alterations in white matter connectivity has emerged as a promising avenue for exploring potential markers of Alzheimer's disease (AD). However, conventional graph learning methods struggle to accurately represent the subtle and heterogeneous topology alterations caused by AD, leading to marginal classification accuracy. In this study, we address this issue through a two-fold approach. Firstly, to more reliably capture AD-induced alterations, we collect multi-shell high-angular resolution diffusion MRI data and construct a topology tensor to incorporate multiple edge-based attributes. Secondly, we propose a novel CNN framework called REST-Net, utilizing lightweight convolutional kernels to integrate the multiple attributes, enhancing its capacity for topology representation. With extensive experiments, REST-Net outperforms seven state-of-the-art graph learning methods for binary and tertiary classification tasks. Of utmost importance, the white matter connections identified by REST-Net guide the selection of target bundles for further analysis, which can potentially provide valuable insights for clinical and pharmacological investigations.

Keywords: Alzheimer's disease · Structural networks · Graph learning

1 Introduction

Diffusion magnetic resonance imaging (dMRI) is widely employed to observe the wiring diagram of white-matter (WM) connectivity non-invasively, allowing the

X. Cao et al. (Eds.): MLMI 2023, LNCS 14349, pp. 33–42, 2024.
https://doi.org/10.1007/978-3-031-45676-3_4

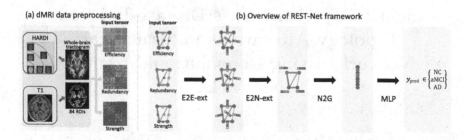

Fig. 1. Proposed framework overview. (a) Topological attributes are extracted to construct the input tensor of REST-Net. (b) REST-Net is designed with two types of lightweight kernels to customize information flow. In REST-Net, E2E-ext and E2N-ext layers generate edge-based and node-based features, where the lightweight kernels integrate features across multiple attributes. Additionally, an N2G layer reads out graph-level features to predict the graph label.

characterization of the disconnection symptom in Alzheimer's disease (AD) [4]. Recent studies have highlighted two key advantages: *i*) the topological attributes of structural networks may serve as markers in the pre-clinical stage of AD [9,28], offering promising potential for the early diagnosis; *ii*) network topology provides a holistic view that enhances the understanding of AD progression, complementing other imaging modalities that reveal local changes, such as of β-Amyloid (Aβ)/Tau deposition, grey matter (GM)/WM atrophy, etc. [16,21]. However, the AD-induced topology alterations are subtle and heterogeneous [6, 27] (*i.e.*, only a small number of attributes and brain regions are associated with AD, and they exhibit large subject-specific variability) necessitating further methodological investigations in representing network topology.

The representation of network topology remains a challenging area in the graph learning field. Although graph convolutional neural networks (GCNNs) have found widespread use in analyzing functional networks [3], they are less reliable in extracting topology features [26]. To address this limitation, several supervised algorithms (*e.g.*, graph isomorphism network (GIN) [26] and Brain-NetCNN (BNC) [10]) as well as unsupervised algorithms (*e.g.*, Graph2Vector (G2Vec) [13] and Line-Graph2Vector (GL2Vec) [5]) have been proposed for accurately extracting topology feature. However, their capacity to characterize topology alterations caused by AD has yet to be evaluated, and these algorithms could merely achieve marginal classification accuracy.

In addition to methodological challenges, the angular resolution of dMRI data also determines the topology representation of brain networks. Due to the limitations in tractography, low angular resolution data demonstrate inferior representation capacity than multi-shell high angular resolution diffusion MRI (HARDI) data in terms of capturing reliable WM connectivity patterns. Despite the potential benefits of HARDI data in investigating AD, there has been a relative scarcity of studies that have exploited its advantages.

In this study, we address the above mentioned limitations through a two-fold approach. Firstly, we collect multi-shell HARDI data to obtain more reliable structural networks and construct a topology tensor, redundancy-efficiency-strength tensor (REST), constructed from three edge-based topological attributes. Secondly, we propose a novel CNN framework called REST-Net, which utilizes two types of lightweight convolutional kernels to integrate complementary information across the three attributes and enhance the capacity for topology representation. Extensive experimental results demonstrate that REST-Net outperforms seven SOTA graph learning algorithms in terms of accuracy (F1), specificity, and sensitivity on binary and tertiary classification tasks. Furthermore, the WM connections identified by REST-Net with high predictive power guide the selection of target fiber bundles for further analysis, which should potentially provide valuable insights for clinical and pharmacological investigations.

2 Method

2.1 Construction of Structural Networks

We utilize a Human Connectome Project (HCP) standard pipeline for processing HARDI data, which are detailed in Sect. 3.2. In brief, a cortical parcellation incorporating both subcortical nuclei and cortical cortexes (Desikan-Killiany-Tourville atlas, DKT) is conducted [25], as shown in Fig. 1(a); networks are constructed with vertices defined by the 84 GM regions of interest (ROIs) and with edges defined by WM connections. In this study, structural networks are modeled by the undirected graph, $\mathcal{G} = (\mathcal{V}, \mathcal{E}, \mathcal{W})$. $\mathcal{V} = \{v_1, \cdots, v_{84}\}$ denotes vertices, $\mathcal{E} \subseteq \mathcal{V} \times \mathcal{V}$ denotes edges, and $\mathcal{W} = (w_{(i,j)})$ denotes the obtained fiber counts with SIFT2 [20]. The adjacency matrix $\mathcal{A} = (a_{(i,j)})$ is defined as:

$$a_{(i,j)} = \begin{cases} 1, & \text{if } w_{(i,j)} > 200, \\ 0, & \text{otherwise}, \end{cases} \qquad (1)$$

where 200 is chosen as the optimal threshold by maximizing the performance in a binary classification task, i.e., normal control (NC) - AD.

2.2 REST Construction

To enhance the capacity of tolology representation, we manually extract topological attributes across three scales, i.e., node efficiency (*efficiency* for short), path redundancy (*redundancy* for short), and *strength*. From $\mathcal{A}$, efficiency is derived by the inverse shortest path length of a pair of vertices [17]; redundancy is defined as the amount of the shortest paths between two nodes [2]; strength is defined by obtained fiber counts $w_{(i,j)}$ with SIFT2. From coarse to fine, efficiency treats multiple routes of the shortest path as single route; redundancy goes further by discerning and quantifying each alternative route; and at the finest scale, strength measures the fiber amount along the path. Note that, both efficiency and redundancy are obtained with NetworkX [18].

As demonstrated in Fig. 1(a), the three attributes are denoted by three matrices (*i.e.*, $\mathbb{E}$, $\mathbb{R}$, and $\mathbb{S}$), with a size of 84×84. By concatenating them, we construct REST (denoted by $\mathbb{T}$) with a size of $84 \times 84 \times 3$.

2.3 REST-Net Framework

Inspired by the merit of lightweight kernels in ShuffleNet which customize network information flow and improve the capacity of feature representation [29], we design REST-Net with two types of lightweight convolutional kernels. One is for propagating features along edges, and the other is for integrating features. As shown in Fig. 1(b), REST-Net composes of four different layers, *i.e.*, extended edge-to-edge (E2E-ext), extended edge-to-node (E2N-ext), node-to-graph (N2G), and multi-layer perceptron (MLP). The input tensor of REST-Net is denoted by $\mathcal{T}^0$ ($84 \times 84 \times 3 \times 1$).

2.4 REST-Net Layers

E2E-ext layer composes of three kinds of convolutional kernels, *i.e.*, f_r, f_c, and f_f. The f_r and f_c convolve the row and column of $\mathcal{T}$, respectively; and f_f convolves across $\mathbb{E}$, $\mathbb{R}$, and $\mathbb{S}$ three attributes for integration. The three kernels result in three 1×1 vectors, and eventually, they are added as the updated feature of an edge v_i-v_j,

$$\mathcal{T}^l_{(i,j,k)} = f_r\left(\mathcal{T}^{l-1}_{(i,:,k)}\right) + f_c\left(\mathcal{T}^{l-1}_{(:,j,k)}\right) + f_f\left(\mathcal{T}^{l-1}_{i,j,:}\right). \tag{2}$$

Specifically, f_r and f_c in E2E-ext layer extract the edge features of v_i-v_j by aggregating neighbor edges connecting to nodes v_i and v_j, respectively. Meanwhile, f_f in E2E-ext layer weighted-averages the three attributes of v_i-v_j for integration. By this, the information flows of feature propagation and integration are segragated. Finally, E2E-ext layer outputs $\mathcal{T}^l$ with a size of ($84 \times 84 \times 3 \times 32$), the integrated edge features. In this work, two E2E-ext layers are employed to propagate features across two-hop neighbor edges.

E2N-ext layer composes of f_r and f_f. In E2N-ext, f_r and f_f execute in two separate steps: Firstly, f_r extracts node features of v_i, attribute by attribute,

$$\mathcal{T}^{l-1}_{(i,k)} = f_r\left(\mathcal{T}^{l-1}_{(i,:,k)}\right), \tag{3}$$

and secondly, f_f integrates the concatenated features of node v_i, resulting in $\mathcal{T}^l_i$ with a size of (84×64),

$$\mathcal{T}^l_i = f_f\left(\mathcal{T}^{l-1}_{(i,1)} \frown \mathcal{T}^{l-1}_{(i,2)} \frown \mathcal{T}^{l-1}_{(i,3)}\right). \tag{4}$$

N2G layer composes of 2-D kernels $f_{2\text{-}D}$ for graph-level feature readout, which is the same as [10]. Eventually, MLP predicts the labels with three (1×256, 1×128, and 1×30) **FC** layers. In REST-Net, f_r, f_c, f_f, and $f_{2\text{-}D}$ are ($1 \times 84 \times 1$), ($84 \times 1 \times 1$), ($1 \times 1 \times 3$), and (84×64) sized kernels.

3   Experiments

3.1   Dataset

A cohort of 217 subjects was recruited in our experiments, including 81 NC, 69 amnestic mild cognitive impairment (aMCI), and 67 AD subjects. The diagnosis was made by two experienced neurologists and checked by an expert neurologist. This study was approved by the institutional ethics review board of Huashan Hospital Fudan University and Shanghai Jiao Tong University Affiliated Sixth People's Hospital and written informed consents were obtained from all subjects prior to any experiment. All MRI data were acquired on a 3.0 T Siemens Prisma scanner, including T1 data with a voxel size of $0.8 \times 0.8 \times 0.8 \, \text{mm}^3$ and HARDI data with a voxel size of $1.5 \times 1.5 \times 1.5 \, \text{mm}^3$. The HARDI data compose of 7 $b0$ volumes and 92 diffusion-weighted volumes that are obtained in 2 shells ($1500 \, \text{s/mm}^{-2}$ and $3000 \, \text{s/mm}^{-2}$).

In our experiments, we chose the leave-one-out (LOO) cross-validation strategy. In each fold, 216 subjects were for training while 1 subject was for testing. During training, the subjects of minority class were randomly oversampled. After having tested all subjects, the mean accuracy (ACC) or F1 (for the tertiary classification task), specificity (SPE), and sensitivity (SEN) scores were obtained.

3.2   Preprocessing Pipeline

The T1 and HARDI data were processed using FSL [7], ANTs [24], and MRtrix3 [23], largely following the HCP pipeline. For T1 data, 84 cortical and subcortical ROIs were used [12,25]. For dMRI data, we implemented bias- and distortion-correction, denoising, co-registration, and rotation of B-matrix; next for tracking fibers, multi-shell multi-tissue constrained spherical deconvolution was used for deriving fiber orientation distribution functions [8], and a probabilistic tracking method was used [22], with seeds located on the interface of gray matter and white matter. After weighting streamlines using SIFT2 [20], we constructed the structural networks with weighted fiber counts. Eventually, the extracted $\mathbb{E}$, $\mathbb{R}$, and $\mathbb{S}$ were organized as a 3D topological attribute tensor $\mathcal{T}^0$ with a size of $84 \times 84 \times 3 \times 1$, which is the input of REST-Net.

3.3   Comparison Methods

To evaluate topology representation capacity, seven SOTA graph learning (both non-deep learning and deep learning) algorithms were employed in three classification tasks, *i.e.*, NC-aMCI, NC-AD, and NC-aMCI-AD. The unsupervised algorithms are Graph2Vector (G2Vec) [13] and Line-Graph2Vector (GL2Vec) [5]; and supervised algorithms include GCNN [11], BNC, graph isomorphism network (GIN) [26], GCNN-fusion, and BNC-fusion. These supervised methods were implemented on TensorFlow [1].

G2Vec and **GL2Vec** are kernel-based methods that extract topological features using rooted subgraphs. In this paper, each graph was represented by a

(1×64)-sized feature vector. Next, SVM classifiers were also trained using LOO strategy [15], and hyperparameters were optimized via GridSearchCV [14].

GCNN propagates features across nodes. In our experiments, the input of GCNN was $\mathcal{A}$ without node features. GCNN was implemented with two graph convolutional layers (with the size of 128 and 64, respectively) and two FC layers (with the size of 1024 and 128 respectively). We employed the cross-entropy loss, L2 regularization, and AdamOptimizer with a learning rate of 0.1. Meanwhile, **GCNN-fusion** has the same settings but with the input node features, *i.e.*, the concatenated rows of $\mathbb{E}$, $\mathbb{R}$, and $\mathbb{S}$, with the size of 84×252 ($252 = 84 \times 3$).

BNC employs the same network architecture as [10] and employs a cross-entropy loss, L2 regularization, and AdamOptimizer with a learning rate of 0.0001. **BNC-fusion** has the input of $\mathcal{T}^0$ instead of $\mathcal{A}$, and its information flow is not customized.

GIN is one variant of GCNNs, designed for improving discriminative power toward network topology with the input of adjacency matrix $\mathcal{A}$. In our experiments, we employed cross-entropy loss, L2 regularization, and AdamOptimizer with a learning rate of 0.1.

3.4   Experimental Settings

The settings of REST-Net were the same with BNC but with the input of $\mathcal{T}^0$. It is worth noting here, all supervised methods were with a dropout rate of 0.5, and other parameters were optimized on the classification tasks of NC-AD. All methods with fixed parameters executed NC-aMCI and NC-aMCI-AD tasks.

3.5   Results

Table 1 demonstrates ACC (F1), SPE, and SEN scores achieved by 8 algorithms, and they were obtained from the three classification tasks. It is worth noting here, ACCs were obtained from the binary classification tasks, while F1 scores were obtained from the tertiary classification task. From Table 1, we can find two distinctive observations: *i*) REST-Net outperforms other 7 SOTA graph learning algorithms; *ii*) unsupervised methods, typically GL2Vec, can achieve better performance in the NC-aMCI classification task than GCNN, BNC, and GIN. In addition, the result comparisons among BNC, BNC-fusion, and our proposed REST-Net can also be regarded as an ablation study, which demonstrates the advantage of our proposed $\mathbb{T}$ as well as the information flow customization strategy.

More supportive observations can be noticed if considering the commonalities among the eight algorithms. Their commonalities could lie in three aspects: *Case i*, $\mathcal{A}$-based or $\mathcal{T}$-based input; *Case ii*, node-based or edge-based methods; *Case iii*, lightweight kernels or conventional kernels. By this, we can have three different views for observing the results, which are listed as follows.

Case i: Comparing those $\mathcal{A}$-based methods (G2Vec, GL2Vec, GCNN, BNC, and GIN), the three $\mathcal{T}$-based methods (GCNN-fusion, BNC-fusion, and REST-Net) are with the input of topological attributes. We can find that GCNN-fusion,

Table 1. Classification performance of the three classification tasks.

G2Vec	NC-aMCI	NC-AD	NC-aMCI-AD
ACC (F1)	51.33	52.94	38.23 (46.59, 35.04, 33.06)
SPE	47.83	52.24	69.30 (60.29, 70.27, 77.33)
SEN	54.32	53.62	38.42 (50.62, 34.78, 29.85)
GL2Vec	NC-aMCI	NC-AD	NC-aMCI-AD
ACC (F1)	58.00	58.11	38.64 (47.40, 36.87, 31.67)
SPE	56.52	53.42	69.58 (62.50, 68.91, 77.33)
SEN	59.26	62.67	38.89 (50.62, 37.68, 28.35)
GCNN	NC-aMCI	NC-AD	NC-aMCI-AD
ACC (F1)	48.67	66.22	40.83 (44.69, 29.85, 47.93)
SPE	51.85	76.54	70.09 (57.35, 69.59, 83.33)
SEN	44.93	53.73	40.55 (49.38, 28.99, 43.28)
BNC	NC-aMCI	NC-AD	NC-aMCI-AD
ACC (F1)	46.00	64.86	35.63 (38.64, 32.31, 35.94)
SPE	45.68	67.90	67.60 (55.15, 72.97, 74.67)
SEN	46.38	61.19	35.58 (41.97, 30.43, 34.32)
GIN	NC-aMCI	NC-AD	NC-aMCI-AD
ACC (F1)	47.20	54.66	37.69 (38.55, 35.11, 39.41)
SPE	39.13	46.27	68.67 (61.03, 73.65, 71.33)
SEN	46.91	54.32	37.71 (39.51, 33.33, 40.30)
GCNN-fusion	NC-aMCI	NC-AD	NC-aMCI-AD
ACC (F1)	54.00	74.32	45.49 (45.24, 34.53, 56.69)
SPE	61.73	**81.48**	72.30 (63.97, 68.92, 84.00)
SEN	44.93	65.67	45.14 (46.91, 34.78, 53.73)
BNC-fusion	NC-aMCI	NC-AD	NC-aMCI-AD
ACC (F1)	61.33	77.03	53.91 (57.14, 43.97, 60.61)
SPE	64.20	76.54	76.87 (75.00, 72.30, 83.33)
SEN	57.97	77.61	53.81 (56.79, 44.92, 59.70)
Proposed	NC-aMCI	NC-AD	NC-aMCI-AD
ACC (F1)	**64.00**	81.08	**54.77** (59.39, 43.97, 60.94)
SPE	**65.43**	80.25	**77.30** (74.26, 72.30, 85.33)
SEN	**62.32**	82.09	**54.54** (60.49, 44.93, 58.21)

BNC-fusion, and REST-Net collectively demonstrate better ACC (F1), SPE, and SEN scores than other methods. The results affirm the efficacy of the attributes directly associated with AD which alleviate difficulties in learning representative alteration features attributed to AD. We also observe that GL2Vec even outperforms GCNN-fusion, potentially due to its superior capacity in representing disconnection syndromes through edge-based analysis.

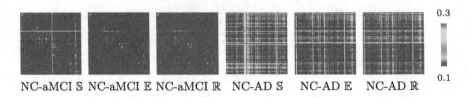

NC-aMCI $\mathbb{S}$ NC-aMCI $\mathbb{E}$ NC-aMCI $\mathbb{R}$ NC-AD $\mathbb{S}$ NC-AD $\mathbb{E}$ NC-AD $\mathbb{R}$

Fig. 2. CAM values indicate the predicative edges, *i.e.*, WM connections.

Case ii: To further assess the superiority of edge-based and node-based methods, we compared four node-based graph learning algorithms (G2Vec, GCNN, GCNN-fusion, and GIN) with other four edge-based algorithms that are designed to leverage edge features: GL2Vec, BNC, and BNC-fusion, and REST-Net. The eight algorithms can be grouped into four pairs, and it is evident that GL2Vec, BNC-fusion, and REST-Net outperform their counterparts, G2Vec, GCNN-fusion, and GIN. Despite a lower accuracy of BNC compared to GCNN, it demonstrates a higher sensitivity, also suggesting its proficiency in identifying aMCI and AD subjects. Overall, these findings underscore the superiority of methods that utilize edge-based features, potentially due to they are more effective in delineating the disconnection symptoms associated with AD.

Case iii: To validate the efficacy of customizing information flow, we compared BNC-fusion with REST-Net which utilizes lightweight kernels to integrate features across multiple attributes. It is evident that REST-Net achieves superior performance in Table 1, possibly because the multiple attributes can cross-examine each other and ensure the integration of their complementary information. This work suggests the necessity of customizing the information flow.

To demonstrate the effectiveness of our proposed REST-Net in distinguishing aMCI and AD from NC, we employed the gradient-guided class activation map (Grad-CAM) algorithm [19] to generate visual maps that highlight the predictive features at the intermediate hidden layers. We obtained the CAM maps subject by subject, and they are averaged as shown in Fig. 2, which illustrates the predictive edges of the first E2E-ext layer. The predictive edges reveal that the REST-Net architecture relies more on $\mathbb{S}$ than $\mathbb{E}$ and $\mathbb{R}$. Notably, edges in $\mathbb{S}$, $\mathbb{E}$, and $\mathbb{R}$ connected to specific ROIs such as the *precuneus, superior parietal lobule, postcentral gyrus*, and *supramarginal gyrus* collectively exhibit the highest predictability in differentiating aMCI from NC. Results in Fig. 2 also demonstrate more predictive edges for differentiating AD from NC compared to those in NC-aMCI classification task. These predictive edges are largely consistent with previous investigations [16, 28], while also reveal the absence of connections associated with the entorhinal cortex and hippocampus. These discoveries prompt further exploration through bundle-based analysis to validate their implications.

4   Conclusion and Discussions

In this paper, we first collected multi-shell HARDI data and then presented REST and REST-Net for identifying AD-induced topology alterations. REST-Net exploits the REST as input to reduce difficulties in representing topology from the adjacency matrix and employs those light-weight kernels to extract the integrative features in representing the subtle and heterogeneous topology alterations. Extensive experiments demonstrate the superiority of REST-Net compared with seven SOTA graph learning methods, which support the effectiveness of our proposed framework. Specifically, our study highlights the importance of network topology representation and reveals the potential of AI-aided topological marker discovery. This study's highest predictive edges involve the precuneus, superior parietal lobule, postcentral gyrus, and supramarginal gyrus. This could be labor-dense work, even only given a graph with a coarse cortical parcellation with 84 nodes for humans if without AI assistance. However, we must note that REST-Net is only suitable for edge-based topological attributes and may not be suitable for analyzing node-based attributes. Further studies are needed to investigate the applicability of REST and REST-Net in specific contexts. Overall, our work provides a promising tool to identify predictive white matter connections to the field of topology-based analysis on AD.

Acknowledgements. This work was supported in part by National Natural Science Foundation of China (No. 62203355, 62131015, 62073260, 12271434), Science and Technology Commission of Shanghai Municipality (STCSM) (No. 21010502600), and Shanghai Pujiang Program (No. 21PJ1421400).

References

1. Abadi, M., et al.: TensorFlow: large-scale machine learning on heterogeneous distributed systems. arXiv preprint arXiv:1603.04467 (2016)
2. Arenaza-Urquijo, E.M., Vemuri, P.: Resistance vs resilience to Alzheimer disease: clarifying terminology for preclinical studies. Neurology **90**(15), 695–703 (2018)
3. Bessadok, A., Mahjoub, M.A., Rekik, I.: Graph neural networks in network neuroscience. TPAMI (2022)
4. Catani, M., Ffytche, D.H.: The rises and falls of disconnection syndromes. Brain **128**(10), 2224–2239 (2005)
5. Chen, H., Koga, H.: GL2vec: graph embedding enriched by line graphs with edge features. In: Gedeon, T., Wong, K.W., Lee, M. (eds.) ICONIP 2019. LNCS, vol. 11955, pp. 3–14. Springer, Cham (2019). https://doi.org/10.1007/978-3-030-36718-3_1
6. De Strooper, B., Karran, E.: The cellular phase of Alzheimer's disease. Cell **164**(4), 603–615 (2016)
7. Jenkinson, M., et al.: FSL. NeuroImage **62**(2), 782–790 (2012)
8. Jeurissen, B., et al.: Multi-tissue constrained spherical deconvolution for improved analysis of multi-shell diffusion MRI data. Neuroimage **103**, 411–426 (2014)
9. Jonkman, L.E., et al.: Relationship between β-amyloid and structural network topology in decedents without dementia. Neurology **95**(5), e532–e544 (2020)

10. Kawahara, J., et al.: BrainNetCNN: convolutional neural networks for brain networks; towards predicting neurodevelopment. Neuroimage **146**, 1038–1049 (2017)
11. Kipf, T.N., Welling, M.: Semi-supervised classification with graph convolutional networks. arXiv preprint arXiv:1609.02907 (2016)
12. Klein, A., Tourville, J.: 101 labeled brain images and a consistent human cortical labeling protocol. Front. Neurosci. **6**(171), 1–12 (2012)
13. Narayanan, A., et al.: Graph2vec: learning distributed representations of graphs. arXiv preprint arXiv:1707.05005 (2017)
14. Pedregosa, F., et al.: Scikit-learn: machine learning in Python. J. Mach. Learn. Res. **12**, 2825–2830 (2011)
15. Platt, J., et al.: Probabilistic outputs for support vector machines and comparisons to regularized likelihood methods. Adv. Large Margin Classif. **10**(3), 61–74 (1999)
16. Reid, A.T., Evans, A.C.: Structural networks in Alzheimer's disease. Eur. Neuropsychopharmacol. **23**(1), 63–77 (2013)
17. Rubinov, M., Sporns, O.: Complex network measures of brain connectivity: uses and interpretations. Neuroimage **52**(3), 1059–1069 (2010)
18. Schult, D.A.: Exploring network structure, dynamics, and function using NetworkX. In: Proceedings of SciPy (2008)
19. Selvaraju, R.R., Cogswell, M., Das, A., Vedantam, R., Parikh, D., Batra, D.: Grad-CAM: visual explanations from deep networks via gradient-based localization. In: Proceedings of ICCV, pp. 618–626 (2017)
20. Smith, R.E., et al.: Sift2: enabling dense quantitative assessment of brain white matter connectivity using streamlines tractography. Neuroimage **119**, 338–351 (2015)
21. Taylor, N.L., Shine, J.M.: A whole new world: embracing the systems-level to understand the indirect impact of pathology in neurodegenerative disorders. J. Neurol. **270**(4), 1969–1975 (2023)
22. Tournier, J.D., et al.: Improved probabilistic streamlines tractography by 2nd order integration over fibre orientation distributions. In: Proceedings of the International Society for Magnetic Resonance in Medicine, vol. 1670 (2010)
23. Tournier, J.D., et al.: MRtrix3: a fast, flexible and open software framework for medical image processing and visualisation. Neuroimage **202**, 116137 (2019)
24. Tustison, N.J., et al.: N4ITK: Improved N3 bias correction. TMI **29**(6), 1310–1320 (2010)
25. Xiao, B., et al.: Weakly supervised confidence learning for brain MR image dense parcellation. In: Suk, H.-I., Liu, M., Yan, P., Lian, C. (eds.) MLMI 2019. LNCS, vol. 11861, pp. 409–416. Springer, Cham (2019). https://doi.org/10.1007/978-3-030-32692-0_47
26. Xu, K., et al.: How powerful are graph neural networks? In: Proceedings of ICLR (2018)
27. Yang, Z., et al.: A deep learning framework identifies dimensional representations of Alzheimer's Disease from brain structure. Nat. Commun. **12**, 7065 (2021)
28. Yu, M., et al.: The human connectome in Alzheimer disease-relationship to biomarkers and genetics. Nat. Rev. Neurol. **17**(9), 545–563 (2021)
29. Zhang, X., et al.: ShuffleNet: an extremely efficient convolutional neural network for mobile devices. In: Proceedings of CVPR, pp. 6848–6856 (2018)

Specificity-Aware Federated Graph Learning for Brain Disorder Analysis with Functional MRI

Junhao Zhang[1], Xiaochuan Wang[1], Qianqian Wang[2], Lishan Qiao[1,3(✉)], and Mingxia Liu[2(✉)]

[1] School of Mathematics Science, Liaocheng University, Liaocheng 252000, Shandong, China
qiaolishan@lcu.edu.cn
[2] Department of Radiology and BRIC, University of North Carolina at Chapel Hill, Chapel Hill, NC 27599, USA
mingxia_liu@med.unc.edu
[3] School of Computer Science and Technology, Shandong Jianzhu University, Jinan 250101, Shandong, China

Abstract. Resting-state functional magnetic resonance imaging (rs-fMRI) provides a non-invasive solution to explore abnormal brain connectivity patterns caused by brain disorders. Graph neural network (GNN) has been widely used for fMRI representation learning and brain disorder analysis, thanks to its potent graph representation abilities. Training a generalizable GNN model often requires large-scale subjects from different medical centers/sites, but the traditional centralized utilization of multi-site data unavoidably encounters challenges related to data privacy and storage. Federated learning (FL) can coordinate multiple sites to train a shared model without centrally integrating multi-site fMRI data. However, previous FL-based methods for fMRI analysis usually ignore specificity of each site, including factors such as age, gender, and population. To this end, we propose a specificity-aware federated graph learning (SFGL) framework for fMRI-based brain disorder diagnosis. The proposed SFGL consists of a shared branch and a personalized branch, where the parameters of the shared branch are sent to a server and the parameters of the personalized branch remain in each local site. In the shared branch, we employ a graph isomorphism network and a Transformer to learn dynamic representations from fMRI data. In the personalized branch, vectorized representations of demographic information (*i.e.*, gender, age, and education) and functional connectivity network are integrated to capture specificity of each site. We aggregate representations learned by shared branches and personalized branches for classification. Experimental results on two fMRI datasets with a total of $1,218$ subjects demonstrate that SFGL outperforms several state-of-the-art methods.

Keywords: Functional MRI · Federated learning · Specificity

X. Cao et al. (Eds.): MLMI 2023, LNCS 14349, pp. 43–52, 2024.
https://doi.org/10.1007/978-3-031-45676-3_5

1   Introduction

Resting-state functional magnetic resonance imaging (rs-fMRI) provides a non-invasive tool that aids in identifying abnormal or impaired brain functional connectivities by tracking related alterations in blood flow [1]. Functional connectivity networks (FCNs) derived from rs-fMRI data can naturally be modeled as graphs, where each node represents a brain region-of-interest (ROI) and each edge denotes the connection between two ROIs [2–4]. Due to the graph-structured nature of FCNs, graph neural network (GNN), which generalizes classical convolution operation from grid data to graph data, has shown great potential in fMRI representation learning and brain disorder analysis [5–8]. Training a deep learning model with good generalization ability usually requires a large number of subjects collected by different medical centers/sites [9,10].

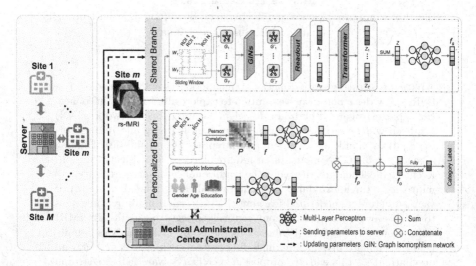

Fig. 1. Overview of specificity-aware federated graph learning (SFGL) framework, including multiple local clients/sites and a server for model aggregation (left panel). At each local client (right panel), our model contains a *shared branch* and a *personalized branch*, where parameters of the shared branch are sent to a server and parameters of the personalized branch remain at each local client to preserve site specificity.

Traditional learning-based methods need to send multi-site data to a central location/server, which inevitably encounters challenges related to data privacy and storage [11–13]. As a decentralized algorithm, federated learning (FL) has been employed as an effective solution to preserve data privacy and reduce data storage pressure [14]. It can allow multiple medical sites to collaboratively train a shared model without centrally sharing multi-site data [15,16]. However, existing FL-based approaches for fMRI analysis usually ignore the specificity of each site, such as the age, gender, and population characteristics of subjects [17].

To this end, we propose a specificity-aware federated graph learning (SFGL) framework for automated brain disease diagnosis with rs-fMRI data. As shown in Fig. 1, the proposed SFGL consists of a shared branch and a personalized branch at each local client/site. In the shared branch, we first divide the fMRI time series using a sliding window to construct dynamic FCNs, and then use a graph isomorphism network (GIN) and a Transformer [18] as the backbone to learn dynamic graph representations. In the personalized branch, we integrate vectorized representations of demographic information (*i.e.*, gender, age, and education) and FCN, obtaining personalized representations for each subject. Finally, representations learned by shared and personalized branches are fused, followed by a fully connected layer and a softmax layer for brain disease prediction. During the SFGL training, the parameters of the shared branch are sent to the central server, while the parameters of the personalized branch locally remain at each site. Therefore, the SFGL not only allows each site to share the knowledge learned from other sites without centrally integrating local data but also enables the model to be adapted to each site's data distribution for personalized local training. Experimental results on 1, 218 subjects from Autism Brain Imaging Data Exchange (ABIDE) [19] and REST-meta-MDD Consortium [20] validate the effectiveness of SFGL in fMRI-based brain disorder diagnosis.

2   Materials and Methodology

Materials and Data Preprocessing. Three sites from the public ABIDE [19] with rs-fMRI are used, including NYU, UM and UCLA. They respectively include 74 autism spectrum disorder patients (ASDs) and 98 normal controls (NCs), 47 ASDs and 73 NCs, 48 ASDs and 37 NCs. These three sites respectively use 3T Allegra scanner, 3T GE Signa scanner, and 3T Trio scanner. For NYU, the scanning parameters are set as follows: repetition time (TR) = 2, 000 ms, echo time (TE) = 15 ms, a total of 33 slices with a thickness of 4 mm. For UCLA, TR = 3, 000 ms, TE = 28 ms, a total of 34 slices with a thickness of 4 mm. For UM, TR = 2, 000 ms, TE = 30 ms, a total of 40 slices with a thickness of 3 mm.

Three sites from the public REST-meta-MDD Consortium [20] with rs-fMRI data are also used, including Site 20, Site 21, and Site 25. They respectively have 282 major depressive disorder patients (MDDs) and 251 NCs, 86 MDDs and 70 NCs, 89 MDDs and 63 NCs. The Site 20 and the Site 21 use a 3T Trio scanner, while Site 25 uses a 3T Verio scanner. For Site 20, the scanning parameters are set as follows: TR = 2, 000 *ms*, TE = 30 ms, a total of 32 slices with a thickness of 3 *mm*. For Site 21, TR = 2, 000 ms, TE = 30 ms, a total of 33 slices with a thickness of 3.5 mm. For Site 25, TR = 2, 000 ms, TE = 25 ms, a total of 36 slices with a thickness of 4 mm.

The fMRI data processing for these two datasets is performed using the Data Processing Assistant for Resting-State fMRI (DPARSF) pipeline [21]. Specifically, we first discard the first five and ten volumes for ABIDE and REST-meta-MDD Consortium data, respectively. Then, we perform head motion correction, spatial smoothing and normalization, bandpass filtering ($0.01 - 0.10\,Hz$)

of BOLD time series, and nuisance signals regression for these two datasets. Finally, based on the AAL atlas, each brain is divided into 116 ROIs.

Proposed Method. As shown in Fig. 1, the proposed SFGL is designed to collaboratively learn an fMRI representation learning and prediction model while keeping the training data in each local site based on multi-site fMRI data. With each site as a specific client, the SFGL consists of a shared branch, a personalized branch, and a federated aggregation strategy. More details are given as follows.

(1) Shared Branch at Client Side. To model temporal dynamics within fMRI series, we first divide the BOLD signal into T segments using sliding windows with a width of Γ and a stride of s. For each time window W_t ($t = 1, \cdots, T$), we calculate Pearson correlation (PC) coefficients between brain ROIs, obtaining an FCN $P(t)$. We measure node features using each row of $P(t)$, with the node feature matrix denoted as $X(t) = P(t)$. To remove noisy/redundant information, we empirically retain the top 30% strongest edges for each FCN $P(t)$ [18]. Thus, we can get a sparse adjacency matrix $A(t) \in \{0, 1\}$ for the time window t, where 1 represents there exists an edge between brain ROIs, and 0 denotes the edge is removed. In this way, we can obtain a dynamic network/graph sequence $G_t = (X(t), A(t))$ for each subject.

As shown in Fig. 1, we then take the constructed dynamic graphs as input of a spatio-temporal attention graph isomorphism network (STAGIN) [18] for feature extraction, including a graph isomorphism network (GIN) and a Transformer. We set the number of GIN layers as 2. For a specific time segment, the updated node feature matrix generated from two GIN layers can be expressed as:

$$X' = \sigma \left(\epsilon^1 I + A \right) \left(\sigma \left(\epsilon^0 I + A \right) X W^0 \right) W^1, \tag{1}$$

where σ is activation function, W^0 and W^1 are learnable weight matrices, ϵ^0 and ϵ^1 are learnable parameters. After that, the Sero readout module [18], a squeeze-excitation network based on multi-layer perceptron (MLP), is used to obtain graph-level features, formulated as:

$$Q = \text{Sigmoid}(P^1 \sigma(P^0 \cdot \text{MEAN}(X'))), \tag{2}$$

where Q is the temporal attention matrix, P^0 and P^1 are learnable matrices, and $\text{MEAN}(\cdot)$ represents the average operation across the node dimension. We obtain a series of spatial graph representations for each subject, represented as $h_t = X'(t)Q(t)$ ($t = 1, \cdots, T$). Finally, we put h_t into a single-head Transformer to get spatio-temporally dynamic graph representation sequence $\{Z_t\}_{t=1}^{T}$, and the final dynamic graph representation for each subject is defined as $Z = \sum_{t=1}^{T} Z_t$.

(2) Personalized Branch at Client Side. To capture the specificity information of each site, we design a personalized branch that integrates fMRI features and demographic information from imaging and non-imaging views. For fMRI feature extraction, we first compute the PC coefficients between ROIs throughout the entire time period, resulting in the PC matrix P for each subject. We

then flatten the upper triangular matrix of P into a vector and regard it as the hidden imaging feature f. The imaging feature for each subject is represented as $f' = fW_f$, where W_f is a learnable matrix. For the demographic information, it is first discretized and digitized (e.g., representing males as 0 and females as 1) to represent the information of each subject as a vector p. The non-imaging feature is defined as $p' = pW_p$. Finally, we concatenate the imaging feature f' and non-imaging feature p' to obtain a subject-level representation $f_p = f' \oplus p'$, where $\oplus$ denotes concatenation operation.

Denote dynamic representations generated by the shared branch as $f_s = ZW_s$, where W_s is a learnable weight matrix. At each client side, we fuse representations generated from shared and personalized branches to obtain a feature vector f_o for each subject, which is formulated as:

$$f_o = \gamma f_p + (1 - \gamma)f_s, \qquad (3)$$

where $\gamma \in [0, 1]$ is used to balance contributions of f_p and f_s. This feature vector is then fed into a fully connected layer and a softmax layer for prediction.

(3) Federated Aggregation at Server Side. For the local model at the m-th ($m = 1, \cdots, M$) site, the mapping function of shared and personalized branches is denoted as $\Phi_{\phi_m}(\cdot)$, and $\Theta_{\theta_m}(\cdot)$, respectively. ϕ_m and θ_m represent the learnable parameters of $\Phi_{\phi_m}(\cdot)$ and $\Theta_{\theta_m}(\cdot)$, respectively. During the r-th round of server-client communication, the following parameter update procedure is executed: $\phi_m^{r+1} \leftarrow \phi_m^r - \eta \nabla L_m$, $\theta_m^{r+1} \leftarrow \theta_m^r - \eta \nabla L_m$, where η is learning rate and L_m is the loss function of m-th site. For each site, parameters ϕ_m^{r+1} of the shared branch are sent to the server, while parameters θ_m^{r+1} of the personalized branch are only updated locally. Denote n_m as the number of samples at Site m. On the server side, we perform weighted aggregation of $\{\phi_m^{r+1}\}_{m=1}^M$ from all M sites as:

$$\phi^{r+1} = \sum_{m=1}^{M} \frac{n_m}{N} \phi_m^{r+1}, \qquad (4)$$

where $N = \sum_{m=1}^{M} n_m$. Then, each site receives the aggregated ϕ^{r+1} and also updates its own parameter ϕ_m^{r+1}, initiating the next round of communication.

Implementation Details. For the proposed SFGL method, the parameters are listed as follows. We set the communication rounds $R = 10$, local training epochs $E = 5$, slide window length $\Gamma = 30$, slide window stride $s = 2$, and $\gamma = 0.8$ in Eq. (3). In addition, we set the batch size to 4 and the dropout rate to 0.5. This model is optimized using the Adam with a learning rate of $\eta = 0.001$.

3   Experiment

Experimental Settings. We use a 5-fold cross-validation (CV) strategy in the experiments. Five evaluation metrics are used here, including accuracy (ACC), precision (PRE), recall (REC), F1-Score (F1), and the area under curve (AUC).

Competing Methods. We compare the proposed SFGL method with three non-FL methods and five FL methods. The three non-FL strategies are as follows. (1) **Cross**: In this method, we use data from one site as the training set and the data from all the remaining sites as the test set. We use "tr<site>" to represent the site used for training. (2) **Single**: Each site trains and tests using its own dataset separately with a 5-fold CV. (3) **Mix**: Data from all sites are mixed for training and testing through the 5-fold CV. For a fair comparison, these three non-FL strategies merge the shared branch and the personalized branch of our SFGL into a unified workflow during execution. The five FL methods are introduced as follows. (1) **FedAvg** [14]: In this method, we send both shared branch and personalized branch parameters to the server for aggregation. (2) **FedProx** [22]: Regularization terms are added during parameter aggregation to alleviate parameter drift. (3) **MOON** [23]: In this approach, we calculate the cosine similarity between the representations generated by local models and the global model, maximizing the distance of local model representations in each communication round and minimizing the distance between representations of local model and global model. (4) **pFedMe** [24]: Moreau Envelopes are used as a regularized loss function to decouple the optimization process of local models from the learning process of the global model. (5) **LGFed** [25]: Only the parameters of the final fully connected layer are sent to the central server for aggregation, and the remaining layers of each site are not involved in aggregation during local training. We generally employed the default setting of all competing methods and meticulously worked to confirm that the dimensions of the input, the latent variable, and the training parameters are comparable to our SFGL.

Table 1. Results of eleven different methods in ASD vs. NC classification on the ABIDE dataset. Best results are shown in bold.

Method	ACC	PRE	REC	AUC	F1
trNYU	0.606(0.041)	0.596(0.065)	**0.748(0.222)**	0.633(0.064)	0.660(0.093)
trUCLA	0.570(0.038)	0.677(0.055)	0.516(0.207)	0.600(0.031)	0.583(0.108)
trUM	0.574(0.040)	0.562(0.029)	0.662(0.100)	0.587(0.043)	0.628(0.028)
Single	0.607(0.127)	0.628(0.177)	0.627(0.172)	0.596(0.130)	0.627(0.156)
Mix	0.607(0.024)	0.617(0.032)	0.630(0.126)	0.605(0.028)	0.681(0.042)
FedAvg	0.648(0.051)	0.701(0.048)	0.653(0.144)	0.690(0.072)	0.677(0.102)
FedProx	0.654(0.087)	0.664(0.163)	0.531(0.172)	0.667(0.089)	0.598(0.126)
MOON	0.641(0.071)	0.617(0.078)	0.673(0.190)	0.622(0.068)	0.644(0.113)
pFedMe	0.645(0.075)	0.681(0.094)	0.687(0.161)	0.668(0.099)	0.684(0.097)
LGFed	0.639(0.100)	0.659(0.128)	0.670(0.168)	0.665(0.120)	0.695(0.107)
SFGL (Ours)	**0.681(0.071)**	**0.714(0.117)**	0.721(0.174)	**0.719(0.085)**	**0.718(0.103)**

Experimental Results. In Table 1 and Table 2, we respectively report the classification results of SFGL and competing methods in two tasks on the ABIDE and REST-meta-MDD Consortium datasets. From Tables 1-2, we have the following interesting findings. *First*, results obtained by FL methods generally

Table 2. Results of eleven different methods in MDD vs. NC classification on the REST-meta-MDD Consortium. Best results are shown in bold.

Method	ACC	PRE	REC	AUC	F1
trSite20	0.578(0.042)	0.567(0.058)	0.628(0.108)	0.562(0.041)	0.579(0.072)
trSite21	0.569(0.016)	0.570(0.046)	0.674(0.225)	0.573(0.021)	0.620(0.107)
trSite25	0.566(0.023)	0.604(0.022)	0.571(0.067)	0.556(0.018)	0.586(0.028)
Single	0.575(0.065)	0.605(0.052)	0.656(0.128)	0.578(0.076)	0.629(0.078)
Mix	0.580(0.039)	0.615(0.008)	0.606(0.175)	0.606(0.030)	0.611(0.115)
FedAvg	0.600(0.050)	0.624(0.043)	0.654(0.113)	0.628(0.065)	0.639(0.064)
FedProx	0.610(0.051)	0.648(0.066)	0.663(0.145)	0.657(0.056)	0.656(0.084)
MOON	0.608(0.074)	**0.655(0.053)**	0.611(0.167)	0.662(0.081)	0.632(0.122)
pFedMe	0.607(0.062)	0.615(0.084)	0.582(0.173)	0.602(0.073)	0.598(0.115)
LGFed	0.595(0.073)	0.631(0.057)	0.628(0.112)	0.639(0.064)	0.633(0.060)
SFGL (Ours)	**0.629(0.047)**	0.643(0.034)	**0.726(0.097)**	**0.678(0.063)**	**0.682(0.046)**

exhibit show significant improvements compared to non-FL methods. For example, for ASD vs. NC classification on ABIDE, our method achieves an improvement of more than 7.5% in terms of ACC compared to three methods trained on a single site (*i.e.*, trNYU, trUCLA, and trUM). This clearly demonstrates that FL enables multiple sites to "share" the knowledge they have learned, allowing each site to receive external information to assist its own learning process and enhance classification results. *Second*, our proposed SFGL method outperforms the other five competing FL methods in most cases. For instance, the AUC value of SFGL is 67.8%, surpassing the second-best result (AUC = 65.7%) achieved by FedProx by 2.1%. This could be due to our model's ability to capture dynamic graph information through the shared branch and extract site-specific information through the personalized branch.

Table 3. Results of the SFGL under different hyperparameters (*i.e.*, γ and E) on two datasets, with best results shown in bold.

Hyperparameter		ASD vs. NC on ABIDE		MDD vs. NC on REST-meta-MDD	
		ACC	AUC	ACC	AUC
γ	0.4	0.647(0.083)	0.689(0.107)	0.596(0.060)	0.650(0.047)
	0.5	0.653(0.086)	0.668(0.118)	0.615(0.055)	0.656(0.039)
	0.6	0.642(0.100)	0.702(0.085)	0.606(0.095)	0.651(0.063)
	0.7	0.646(0.066)	0.700(0.081)	0.608(0.068)	0.667(0.057)
	0.8	**0.681(0.071)**	**0.719(0.085)**	**0.629(0.047)**	**0.678(0.063)**
	0.9	0.638(0.111)	0.688(0.086)	0.586(0.069)	0.666(0.052)
E	1	0.589(0.085)	0.650(0.111)	0.597(0.075)	0.647(0.090)
	5	**0.681(0.071)**	**0.719(0.085)**	**0.629(0.047)**	**0.678(0.063)**
	10	0.636(0.085)	0.703(0.091)	0.616(0.060)	0.670(0.054)
	15	0.634(0.073)	0.685(0.096)	0.617(0.051)	0.655(0.054)
	20	0.606(0.100)	0.684(0.081)	0.544(0.056)	0.563(0.047)

4   Discussion

Influence of Key Hyperparameters. To investigate the influence of hyper-parameters, we record the results of our SFGL with different parameter values, including balancing coefficient γ and local epochs E. The results of SFGL with different parameter values in two classification tasks are reported in Table 3.

From Table 3, we can see that SFGL produces the best results in two tasks with $\gamma = 0.8$, and cannot yield good results when the value of γ is too large or too small. When γ is too large, the SFGL becomes overly dependent on the personalized branch, thus failing to take full advantage of the support from other sites in identifying dynamic information embedded within the BOLD signal. When a very small γ (*e.g.*, 0.4), the SFGL becomes excessively dependent on the shared branch, which could lead to a failure to preserve the unique characteristics (specificity) of each site. Besides, our SFGL produces the best performance with local epochs $E = 5$. Employing an excessively large number of local training epochs (*e.g.*, $E = 20$), the proposed SFGL fails to achieve satisfactory results in the two classification tasks. This may occur as an excessive number of local updates could result in the model gravitating towards local optima, instead of striving to reach the global optima.

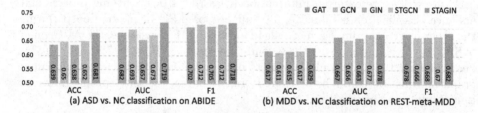

Fig. 2. Results of SFGL using different backbones in the shared branch on two datasets.

Influence of Different Backbones in Shared Branch. In the main experiments, we use STAGIN (*i.e.*, GIN+Transformer) as the backbone in the shared branch to extract dynamic graph features from input fMRI data. To investigate the impact of different backbones on the performance of our SFGL, we employ several classic graph neural networks as backbones in SFGL, including GAT [26], GCN [27], GIN [28], and STGCN [5]. The experimental results of SFGL with five different backbones in the shared branch are reported in Fig. 2.

It can be observed from Fig. 2 that GNNs with spatio-temporal feature learning techniques (*i.e.*, STGCN and STAGIN) attain a superior classification accuracy compared to static GNNs (*e.g.*, GAT). This implies that extracting dynamic information from fMRI can aid in the diagnosis of brain disorders. Moreover, it is evident that STAGIN generally surpasses STGCN in performance. This could likely be due to the adaptive aggregation of node features via GIN and the employment of the Transformer module in STAGIN.

5   Conclusion and Future Work

A specificity-aware federated graph learning (SFGL) framework is designed for multi-site fMRI representation learning and brain disorder diagnosis. The SFGL consists of a shared branch, a personalized branch, and a federated aggregation strategy, where the parameters of the personalized branch are not shared with the server to preserve the specificity of each site. Features learned by shared and personalized branches are aggregated for classification. Results on two public multi-site fMRI datasets demonstrate the effectiveness of SFGL in fMRI-based brain disorder identification when compared with state-of-the-art methods.

In the current work, we focus only on the correlations between ROIs within each subject, ignoring the correlations between subjects [29]. Considering both types of correlations may provide richer information to boost the performance of fMRI-based brain disease diagnosis. Besides, it is interesting to explore the use of adaptive weights instead of fixed weights to merge features from the shared and personalized branch, which will also be our future work.

Acknowledgment. L. Qiao was supported in part by National Natural Science Foundation of China (Nos. 61976110, 62176112, 11931008) and Natural Science Foundation of Shandong Province (No. ZR202102270451).

References

1. Khosla, M., Jamison, K., Ngo, G.H., Kuceyeski, A., Sabuncu, M.R.: Machine learning in resting-state fMRI analysis. Magn. Reson. Imaging **64**, 101–121 (2019)
2. Saeidi, M., et al.: Decoding task-based fMRI data with graph neural networks, considering individual differences. Brain Sci. **12**(8), 1094 (2022)
3. ElGazzar, A., Thomas, R., Van Wingen, G.: Benchmarking graph neural networks for fMRI analysis. arXiv preprint arXiv:2211.08927 (2022)
4. Jiang, H., Cao, P., Xu, M., Yang, J., Zaiane, O.: Hi-GCN: a hierarchical graph convolution network for graph embedding learning of brain network and brain disorders prediction. Comput. Biol. Med. **127**, 104096 (2020)
5. Gadgil, S., Zhao, Q., Pfefferbaum, A., Sullivan, E.V., Adeli, E., Pohl, K.M.: Spatiotemporal graph convolution for resting-state fMRI analysis. In: Martel, A.L., et al. (eds.) MICCAI 2020. LNCS, vol. 12267, pp. 528–538. Springer, Cham (2020). https://doi.org/10.1007/978-3-030-59728-3_52
6. Fang, Y., Wang, M., Potter, G.G., Liu, M.: Unsupervised cross-domain functional MRI adaptation for automated major depressive disorder identification. Med. Image Anal. **84**, 102707 (2023)
7. Yao, D., Sui, J., Wang, M., Yang, E., Jiaerken, Y., Luo, N., Yap, P.T., Liu, M., Shen, D.: A mutual multi-scale triplet graph convolutional network for classification of brain disorders using functional or structural connectivity. IEEE Trans. Med. Imaging **40**(4), 1279–1289 (2021)
8. Wang, M., Huang, J., Liu, M., Zhang, D.: Modeling dynamic characteristics of brain functional connectivity networks using resting-state functional mri. Med. Image Anal. **71**, 102063 (2021)
9. Neyshabur, B., Bhojanapalli, S., McAllester, D., Srebro, N.: Exploring generalization in deep learning. In: Advances in Neural Information Processing Systems 30 (2017)

10. Lian, C., Liu, M., Pan, Y., Shen, D.: Attention-guided hybrid network for dementia diagnosis with structural MR images. IEEE Trans. Cybern. **52**(4), 1992–2003 (2020)
11. Goddard, M.: The EU general data protection regulation (GDPR): European regulation that has a global impact. Int. J. Mark. Res. **59**(6), 703–705 (2017)
12. Act, A.: Health insurance portability and accountability act of 1996. Public Law **104**, 191 (1996)
13. Guan, H., Liu, M.: Domain adaptation for medical image analysis: a survey. IEEE Trans. Biomed. Eng. **69**(3), 1173–1185 (2021)
14. McMahan, B., Moore, E., Ramage, D., Hampson, S., y Arcas, B.A.: Communication-efficient learning of deep networks from decentralized data. In: Artificial Intelligence and Statistics, PMLR, pp. 1273–1282 (2017)
15. Pillutla, K., Malik, K., Mohamed, A.R., Rabbat, M., Sanjabi, M., Xiao, L.: Federated learning with partial model personalization. In: International Conference on Machine Learning, PMLR, pp. 17716–17758 (2022)
16. Li, X.C., et al.: Federated learning with position-aware neurons. In: Proceedings of the IEEE/CVF Conference on Computer Vision and Pattern Recognition, pp. 10082–10091 (2022)
17. Li, X., Gu, Y., Dvornek, N., Staib, L.H., Ventola, P., Duncan, J.S.: Multi-site fMRI analysis using privacy-preserving federated learning and domain adaptation: Abide results. Med. Image Anal. **65**, 101765 (2020)
18. Kim, B.H., Ye, J.C., Kim, J.J.: Learning dynamic graph representation of brain connectome with spatio-temporal attention. Adv. Neural. Inf. Process. Syst. **34**, 4314–4327 (2021)
19. Di Martino, A., et al.: The autism brain imaging data exchange: towards a large-scale evaluation of the intrinsic brain architecture in autism. Mol. Psychiatry **19**(6), 659–667 (2014)
20. Yan, C.G., et al.: Reduced default mode network functional connectivity in patients with recurrent major depressive disorder. Proc. Natl. Acad. Sci. **116**(18), 9078–9083 (2019)
21. Yan, C., Zang, Y.: DPARSF: a MATLAB toolbox for "pipeline" data analysis of resting-state fMRI. Front. Syst. Neurosci. **13** (2010)
22. Li, T., Sahu, A.K., Zaheer, M., Sanjabi, M., Talwalkar, A., Smith, V.: Federated optimization in heterogeneous networks. Proc. Mach. Learn. Syst. **2**, 429–450 (2020)
23. Li, Q., He, B., Song, D.: Model-contrastive federated learning. In: Proceedings of the IEEE/CVF Conference on Computer Vision and Pattern Recognition, pp. 10713–10722 (2021)
24. T Dinh, C., Tran, N., Nguyen, J.: Personalized federated learning with moreau envelopes. In: Advances in Neural Information Processing Systems 33, pp. 21394–21405 (2020)
25. Liang, P.P., et al.: Think locally, act globally: federated learning with local and global representations. arXiv preprint arXiv:2001.01523 (2020)
26. Velickovic, P., et al.: Graph attention networks. Stat. **1050**(20), 10–48550 (2017)
27. Kipf, T.N., Welling, M.: Semi-supervised classification with graph convolutional networks. arXiv preprint arXiv:1609.02907 (2016)
28. Xu, K., Hu, W., Leskovec, J., Jegelka, S.: How powerful are graph neural networks? arXiv preprint arXiv:1810.00826 (2018)
29. Liu, M., Zhang, D., Shen, D.: Relationship induced multi-template learning for diagnosis of Alzheimer's disease and mild cognitive impairment. IEEE Trans. Med. Imaging **35**(6), 1463–1474 (2016)

3D Transformer Based on Deformable Patch Location for Differential Diagnosis Between Alzheimer's Disease and Frontotemporal Dementia

Huy-Dung Nguyen[(✉)], Michaël Clément, Boris Mansencal, and Pierrick Coupé

University of Bordeaux, CNRS, Bordeaux INP, LaBRI, UMR 5800, 33400 Talence,
France
huy-dung.nguyen@u-bordeaux.com

Abstract. Alzheimer's disease and Frontotemporal dementia are common types of neurodegenerative disorders that present overlapping clinical symptoms, making their differential diagnosis very challenging. Numerous efforts have been done for the diagnosis of each disease but the problem of multi-class differential diagnosis has not been actively explored. In recent years, transformer-based models have demonstrated remarkable success in various computer vision tasks. However, their use in disease diagnostic is uncommon due to the limited amount of 3D medical data given the large size of such models. In this paper, we present a novel 3D transformer-based architecture using a deformable patch location module to improve the differential diagnosis of Alzheimer's disease and Frontotemporal dementia. Moreover, to overcome the problem of data scarcity, we propose an efficient combination of various data augmentation techniques, adapted for training transformer-based models on 3D structural magnetic resonance imaging data. Finally, we propose to combine our transformer-based model with a traditional machine learning model using brain structure volumes to better exploit the available data. Our experiments demonstrate the effectiveness of the proposed approach, showing competitive results compared to state-of-the-art methods. Moreover, the deformable patch locations can be visualized, revealing the most relevant brain regions used to establish the diagnosis of each disease.

Keywords: Deformable Patch Location · 3D Transformer · Differential diagnosis · Alzheimer's Disease · Frontotemporal Dementia

1 Introduction

Alzheimer's disease (AD) and Frontotemporal dementia (FTD) are the two most prevalent types of neurodegenerative disorders. They are the main cause of cognitive impairment and dementia [2]. Therefore, their differential diagnosis is crucial for determining appropriate interventions and treatment plans. However, these

X. Cao et al. (Eds.): MLMI 2023, LNCS 14349, pp. 53–63, 2024.
https://doi.org/10.1007/978-3-031-45676-3_6

diseases share several overlapping symptoms such as memory loss and behavior changes, making their differential diagnosis challenging even when they have different clinical diagnostic criteria [28]. Indeed, several studies have demonstrated the limitations of cognitive tests in distinguishing patients with FTD from those with AD [13,38]. Furthermore, cognitively normal (CN) people may also exhibit some changes in behavior and memory as a result of the natural aging process. Consequently, an automatic tool for multi-class diagnosis (*i.e.*, AD *vs.* FTD *vs.* CN) is highly valuable in a real clinical context.

Several works have reported that AD and FTD are associated with brain structure atrophy [27,29], which can be visualized using structural magnetic resonance imaging (sMRI) [9,24]. This modality has been used to extract structure volumes [9] or used as input of convolutional neural networks (CNN) [11,26] for differential diagnosis. In recent years, transformer-based models appear to be a promising alternative to CNN-based models in computer vision tasks. However, their application in disease diagnostic (*e.g.*, differential diagnosis) is still limited due to their computational demands and data requirements.

To alleviate computation problems, classification can be considered as a 2D problem. Lyu *et al.* and Jang *et al.* used 2D features extracted from MRI, both using a vision transformer (ViT) [8] for AD classification [15,19]. However, the lack of spatial information in such 2D approaches may not be optimal. Regarding 3D methods, for AD diagnosis, Li *et al.* downsampled the input image before feeding it to their transformer [16], Zhang *et al.* reduced the feature map dimension by setting a big patch size for embedding [40]. However, these strategies may reduce the details of local regions. For natural image classification, other techniques to reduce computation are local attention [17] and deformable attention [37]. The idea of both methods is to reduce the size of the attention matrix by decreasing the number of query, key, and value points. In the case of deformable attention mechanism, key points can be visualized for better interpretation.

Transformer-based models are known to require a large amount of data to achieve high performance [8]. In medical imaging, the limited number of labeled sMRI makes it difficult to train these models effectively. In this situation, data augmentation plays an important role in the model generalization. While data augmentation has been shown to be effective for transformer in natural image classification [33], its effectiveness in medical imaging has not been investigated.

In this paper, we first propose a 3D transformer-based architecture using a deformable patch location (DPL) module for the problem of multi-class differential diagnosis (*i.e.*, AD *vs.* FTD *vs.* CN). In the backbone, we employ local attention [17] instead of global one to reduce the computation. Our DPL module is inspired from the deformable attention [37], however, unlike the original model, deformable points in DPL are determined for each sub-volume of the image rather than being shared across the entire image. Second, to alleviate data scarcity, we propose an efficient combination of various data augmentation techniques. The exploration of data augmentation for 3D transformer-based classification using sMRI has remained relatively unexplored until now, and our strategy aims to fill this gap. Moreover, our data augmentation allows a multi-

Table 1. Number of participants.

	Dataset	CN	AD	FTD
In-domain	ADNI2	180	149	
	NIFD	136		150
Out-of-domain	NACC	2182	485	37

scale prediction, improving our model performance. Finally, we propose to combine our transformer-based method with a support vector machine (SVM) using structure volumes to even better exploit the limited training data. As a result, our framework shows competitive results compared to state-of-the-art methods for multi-class differential diagnosis.

2 Materials and Method

2.1 Datasets and Preprocessing

Table 1 describes the number of participants used in this study. The data consisted of 3319 subjects from multiple studies: the Alzheimer's Disease Neuroimaging Initiative (ADNI) [14], the Frontotemporal lobar Degeneration Neuroimaging Initiative (NIFD)[1] and the National Alzheimer's Coordinating Center (NACC) [3]. We only used T1-weighted MRIs at the baseline acquired with 3 T machines. For the NIFD dataset, we only selected the behavior variant, progressive non-fluent aphasia, and semantic variant sub-types. The ADNI2 and NIFD datasets constituted our in-domain dataset while the NACC constituted our out-of-domain dataset. The in-domain dataset was used to perform a 10-fold cross-validation. The out-of-domain was used as an external dataset for evaluating the generalization capacity of the trained models.

The T1w MRI was preprocessed in 5 steps, which included (1) denoising [22], (2) inhomogeneity correction [35], (3) affine registration into MNI152 space ($181 \times 217 \times 181$ voxels at $1\,\text{mm} \times 1\,\text{mm} \times 1\,\text{mm}$) [1], (4) intensity standardization [21] and (5) intracranial cavity (ICC) extraction [23]. After that, we cropped at the image center a volume of size $144 \times 168 \times 144$ voxels to remove empty spaces. The brain structure volumes (*i.e.*, normalized volume in % of ICC) were measured using a brain segmentation predicted by AssemblyNet [7]. These volume features were used as input for our SVM.

2.2 Method

Overview. Figure 1 shows an overview of our proposed model. Our model is composed of four parts: a volume embedding (VE), N blocks of a patch multi-head self-attention (P-MSA) followed by a shift patch multi-head self-attention

---

[1] Available at https://ida.loni.usc.edu/.

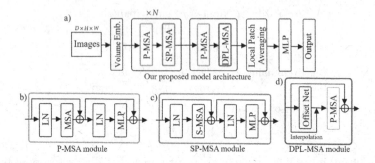

Fig. 1. The architecture of our proposed model

(SP-MSA - the main building block of Swin [17]), a deformable patch location multi-head self-attention module (DPL-MSA) and a local patch averaging layer followed by a multi-layer perceptron (MLP). Intuitively, the VE module encodes an MRI to a 3D volume of tokens. The N blocks of P-MSA and SP-MSA process these tokens as a attention-based feature extractor. Then, the DPL-MSA block predicts deformable patch locations and performs attention on them. While standard transformer-based approaches perform a global average of all the patches together [17], in our method we perform local average of patches in the same area (*i.e.*, sub-volume). To this end, we divide the brain feature map into 27 sub-volumes ($3 \times 3 \times 3$ areas evenly distributed along 3 axis). This is because different brain locations may be affected by a disease differently, thus should be weighted differently in the model decision. Finally, we use an MLP for classification.

Volume Embedding. We start with a preprocessed image of size $144 \times 168 \times 144$ (at $1\,\mathrm{mm}^3$) (see Fig. 1a). The VE module uses a CNN (similar to [34]) to embed the input into token vectors (with an embedding dimension of 96). This results in a 96-channel 3D feature map of size $36 \times 42 \times 36$.

Feature Extractor. The obtained 3D feature map is fed into three (P-MSA + SP-MSA) blocks. The details of each block are presented in Fig. 1b,c. Our implementation of these blocks is based on [17]. The local attention size is set to $6 \times 7 \times 6$. By using attention mechanism, the size of feature maps remain unchanged.

DPL Block. Taking the output of the feature extractor, we first update the feature map with a P-MSA module (see Fig. 1a). We then split it into $6 \times 6 \times 6$ reference patches of size $(p_x, p_y, p_z) = (6 \times 7 \times 6)$. Their centers are denoted as: $(x_{ct}^i, y_{ct}^i, z_{ct}^i)$. The coordinates of these points are normalized in $[0, 1]$. Each reference patch is used as input of an offset network (see Fig. 1d) to predict the offset logits $(\delta_x^i, \delta_y^i, \delta_z^i)$. The deformable patch center $(x_{Dct}^i, y_{Dct}^i, z_{Dct}^i)$ is then calculated by: $x_{Dct}^i = x_{ct}^i + \tanh \delta_x^i/(2 \times p_x)$ (idem for y_{Dct}^i and z_{Dct}^i). Based on the deformable patch centers, we interpolate our feature map to obtain the corresponding deformable patches of size $p_x \times p_y \times p_z$. After that, we apply a

P-MSA module to these deformable patches. Finally, a shortcut from reference patches is added to the output of the P-MSA module (see Fig. 1d).

Local Patch Averaging. We consider the obtained 96-channel 3D brain feature map (of size $36 \times 42 \times 36$) as a $3 \times 3 \times 3$ areas of size $12 \times 14 \times 12$ voxels, which are evenly distributed along 3 dimensions. We first average each deformable patch to a 96-channel mean token (of size $1 \times 1 \times 1$). Then, all the mean tokens located in a same area are averaged. Finally, we concatenate the obtained tokens and feed it into a MLP for classification.

2.3   Data Augmentation

In this part, we describe our combination of data augmentation techniques. We start with mixup, which has been known to reduce overfitting in various applications [39]. Following this, we apply a series of affine transformations, including rotation and scaling, commonly used in medical imaging applications [12,25]. To further enhance our augmentation process, we randomly crop images at an arbitrary position (with a probability p) and resize them to match the input resolution. This technique, similar to "Random resized crop" in 2D imaging [32], mitigates overfitting and allows evaluation at both global and local views of an image. During inference, we ensemble predictions from multiple views to improve the model performance. In Sect. 3.1, we demonstrate the importance of each of these techniques on our framework accuracy.

2.4   Validation Framework and Ensembling

When evaluating our models, we made two predictions for each image: one for the whole image and one for a crop of that image. The cropping position was selected from nine cropping positions: a center crop and eight crops at corners. For each trained model, the crop position that produced the lowest loss on the validation set was selected. Finally, we averaged the two obtained results.

To further exploit the limited amount of training data, we combined (*i.e.*, average) the transformer prediction with SVM prediction based on brain structures volumes (see Sect. 2.5).

2.5   Implementation Details

The offset network consisted of 3 layers: 3D convolution with 24 channels, kernel = (6, 7, 6), GELU activation [10] and another 3D convolution with 3 channels, kernel = 1. For data augmentation, rotation range was $\pm 0.05 rad$ and scale range was [0.9, 1.1], the crop size was (132, 154, 132), the probability $p = 0.7$. The model was trained for 300 epochs using AdamW optimizer [18], cosine learning rate scheduler (start at 3e-4 and end at 5e-5). To train the SVM models, we used a grid search of three kernels (linear, polynomial, and gaussian) and 50 values of the hyper-parameter C in $[10^{-2}, 10^2]$ on the validation for tuning hyper-parameters. The SVM models used the same train/validation/test (70%/20%/10%) splits of in-domain data during cross-validation than our deep learning models.

3   Experimental Results

In this study, we first performed a 10-fold cross-validation on in-domain dataset. This resulted in 20 models (10 Transformers and 10 SVM models). We concatenated the prediction of 10 test folds to compute the global in-domain performance. For out-of-domain evaluation, we averaged all 10 predictions to estimate the model performance. We used 3 metrics to assess the model performance: accuracy (ACC), balanced accuracy (BACC) and area under curve (AUC).

3.1   Ablation Study

Performance Study. In this part, we studied the impact of each contribution on our model performance. These factors could be organized into 4 groups: Input type (2D/3D), architecture (local patch averaging, non linear volume embedding), validation framework (multi-scale prediction) and ensemble (combination with SVM). The used data augmentation schema was described in Sect. 2.3. Table 2 showed the results of the comparison.

First, we implemented a basic 2D transformer-based architecture (exp. 1) and its 3D version (exp. 2) to see if the spatial information from 3D input is valuable. We observed that the 3D version was better than the 2D version in all metrics. Second, using local patch averaging (exp. 3) improved our model performance, confirming the effectiveness of assigning different weights to different brain areas. Third, the nonlinear volume embedding (exp. 4) could also improve the performance of transformer, which was inline with [34]. Then, the DPL module demonstrated an improvement in performance across almost all metrics (exp. 5). Finally, the multi-scale prediction (exp. 6) and ensembling (exp. 7) increased even more our model performance in both in-domain and out-of-domain data.

Data Augmentation Study. Table 3 shows the contribution of each data augmentation technique to our model performance. The ensembling with SVM was removed for analysis and the multi-scale evaluation was applied only when multi-crop was used. First, without any data augmentation, the obtained result (exp. 1) was lower than in other experiments. Second, combining different augmentations (exp. 2, 3, 4) progressively improved the model's generalization. This showed the effectiveness of our data augmentation for medical imaging applications.

3.2   Comparison with State-of-the-Art Methods

In this section, we compare our results with current state-of-the-art methods for the multi-class diagnosis AD *vs.* FTD *vs.* CN. Hu *et al.* proposed an CNN-based architecture inspired by Resnet which processes the whole 3D MRI for classification [11]. Ma *et al.* used a MLP with cortical thickness (Cth) and brain structure volumes extracted from a 3D MRI [20]. They also used a generative adversarial network to generate new data to prevent over-fitting. More recently, Nguyen *et al.* used a large number of CNN to grade brain regions. The grading values were then averaged for each brain structure and used as input of a MLP

Table 2. Ablation study of the model performance. Results obtained using the data augmentation described in Sect. 2.3. Gray text, symbols: that option is the same as in the previous experiment. Red, Blue: best, second result.

No.	2D/3D	Local patch averaging	Nonlinear VE	DPL module	Multi-scale prediction	Combination with SVM	In-domain ACC	BACC	AUC	Out-of-domain ACC	BACC	AUC
1	2D	✗	✗	✗	✗	✗	68.8	64.1	81.1	77.4	63.3	78.4
2	3D	✗	✗	✗	✗	✗	78.4	74.7	90.1	81.5	75.2	87.8
3	3D	✓	✗	✗	✗	✗	82.9	79.5	92.7	85.4	78.2	89.3
4	3D	✓	✓	✗	✗	✗	83.6	80.3	92.5	86.6	79.7	89.9
5	3D	✓	✓	✓	✗	✗	83.4	80.7	93.4	87.1	80.1	90.5
6	3D	✓	✓	✓	✓	✗	85.2	82.5	94.1	87.7	80.7	91.0
7	3D	✓	✓	✓	✓	✓	86.2	83.4	94.5	89.3	82.8	91.6

Table 3. Ablation study of the data augmentation. Gray symbols: that option is the same as in the previous experiment. Red, Blue: best, second result.

No.	Mixup	Rand. affine	Multi crops	In-domain ACC	BACC	AUC	Out-of-domain ACC	BACC	AUC
1	✗	✗	✗	74.6	69.0	87.8	84.3	73.3	87.3
2	✓	✗	✗	77.6	72.0	88.4	84.8	76.0	87.4
3	✓	✓	✗	82.1	78.9	91.5	86.2	78.6	90.0
4	✓	✓	✓	85.2	82.5	94.1	87.7	80.7	91.0

for classification [26]. For a fair comparison, we reimplemented these methods and trained them under the same training setting as our method and on the same data. Table 4 shows the results of the comparison.

Overall, our method presented most of the time the best performance in all metrics (*i.e.*, ACC, BACC and AUC) and for both in-domain and out-of-domain data. Moreover, our method was the only method based on the transformer mechanism. This suggested that transformer-based methods can obtain competitive results compared to CNN-based networks even with a limited amount of data.

3.3 Visualization of Deformable Patch Location

Figure 2 shows the centers of deformable patch locations for patients with AD and FTD. For each patient group, the patch center positions are calculated as the averaged center locations from our ten models. To enrich visual comprehension,

Table 4. Comparison with state-of-the-art methods. Red, Blue: best, second result.

Method	In-domain			Out-of-domain		
	ACC	BACC	AUC	ACC	BACC	AUC
CNN on intensities [11]	76.3	72.5	90.0	85.2	68.8	86.5
MLP on Cth and volumes [20]	77.1	75.9	86.4	69.1	74.6	87.5
3D Grading [26]	86.0	84.7	93.8	87.1	81.6	91.6
Our method	86.2	83.4	94.5	89.3	82.8	91.6

we utilized GradCAM to attribute an importance score within the range of $[0, 1]$ to each patch. Patches obtaining an importance score above 0.3 are displayed. Furthermore, a higher importance score is visually represented by a larger circle, and the warmth of the circle's color increases with the score.

The obtained results were coherent with the current knowledge about these diseases. Indeed, for AD patients, the structures that obtained higher score were the left hippocampus [30], bilateral entorhinal cortex, bilateral ventricle [6] and parietal lobe [31]. In FTD patients, the frontal pole [4], superior frontal gyrus [5] and left temporal cortex [36] were highlighted.

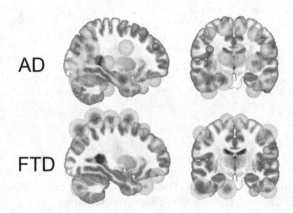

Fig. 2. Visualization of deformable patch locations. The importance of each patch was estimated with GradCAM. Warmer color, larger radius mean higher importance score.

4   Conclusion

Our study presents a novel 3D transformer model, which incorporates a deformable patch location module for the differential diagnosis between cognitively normal subjects, patients with Alzheimer's disease and patients with

Frontotemporal dementia. The proposed module enhances the model's accuracy and provides useful visualizations that reveal insights into each disease. To address the problem of limited training data, we designed a combination common data augmentations for training transformer models using 3D MRI. Furthermore, we proposed to combine both our deep learning model and an SVM using brain structure volumes to even better exploit the limited data. As a result, our framework showed competitive performance compared to state-of-the-art methods.

References

1. Avants, B.B., et al.: A reproducible evaluation of ANTs similarity metric performance in brain image registration. Neuroimage **54**, 2033–2044 (2011)
2. Bang, J., et al.: Frontotemporal dementia. The Lancet **386**, 1672–1682 (2015)
3. Beekly, D.L., et al.: The National Alzheimer's Coordinating Center (NACC) database: the uniform data set. Alzheimer Disease Associat. Disord. **21**, 249–258 (2007)
4. Boeve, B.F., et al.: Advances and controversies in frontotemporal dementia: diagnosis, biomarkers, and therapeutic considerations. Lancet Neurol. **21**, 258–272 (2022)
5. Brambati, S.M., et al.: A tensor based morphometry study of longitudinal gray matter contraction in FTD. Neuroimage **35**(3), 998–1003 (2007)
6. Coupé, P., et al.: Lifespan changes of the human brain in Alzheimer's disease. Sci. Rep. **9**, 3998 (2019)
7. Coupé, P., et al.: AssemblyNet: a large ensemble of CNNs for 3D whole brain MRI segmentation. Neuroimage **219**, 117026 (2020)
8. Dosovitskiy, A., et al.: An image is worth 16x16 words: transformers for image recognition at scale. arXiv:2010.11929 (2020)
9. Du, A.T., et al.: Different regional patterns of cortical thinning in Alzheimer's disease and frontotemporal dementia. Brain **130**, 1159–1166 (2006)
10. Hendrycks, D., Gimpel, K.: Gaussian error linear units (GELUS). arXiv preprint arXiv:1606.08415 (2016)
11. Hu, J., et al.: Deep learning-based classification and voxel-based visualization of frontotemporal dementia and Alzheimer's disease. Front. Neurosci. **14**, 626154 (2021)
12. Hussain, Z., Gimenez, F., Yi, D., Rubin, D.: Differential data augmentation techniques for medical imaging classification tasks. In: AMIA Annual Symposium Proceedings, vol. 2017, p. 979 (2017)
13. Hutchinson, A.D., et al.: Neuropsychological deficits in frontotemporal dementia and Alzheimer's disease: a meta-analytic review. J. Neurol. Neurosurg. Psychiatry **78**, 917–928 (2007)
14. Jack, C.R., et al.: The Alzheimer's disease neuroimaging initiative (ADNI): MRI methods. J. Magn. Reson. Imaging **27**, 685–691 (2008)
15. Jang, J., Hwang, D.: M3t: three-dimensional Medical image classifier using multiplane and multi-slice transformer. In: Proceedings of the IEEE/CVF Conference on Computer Vision and Pattern Recognition, pp. 20718–20729 (2022)
16. Li, C., et al.: Trans-ResNet: integrating transformers and CNNs for Alzheimer's disease classification. In: 2022 IEEE 19th International Symposium on Biomedical Imaging (ISBI), pp. 1–5 (2022)

17. Liu, Z., et al.: Swin transformer: hierarchical vision transformer using shifted windows. In: Proceedings of the IEEE/CVF International Conference on Computer Vision, pp. 10012–10022 (2021)
18. Loshchilov, I., Hutter, F.: Decoupled weight decay regularization. arXiv preprint arXiv:1711.05101 (2017)
19. Lyu, Y., et al.: Classification of Alzheimer's disease via vision transformer. In: Proceedings of the 15th International Conference on Pervasive Technologies Related to Assistive Environments, pp. 463–468 (2022)
20. Ma, D., et al.: Differential diagnosis of frontotemporal dementia, Alzheimer's disease, and normal aging using a multi-scale multi-type feature generative adversarial deep neural network on structural magnetic resonance images. Front. Neurosci. **14**, 853 (2020)
21. Manjón, J.V., et al.: Robust MRI brain tissue parameter estimation by multistage outlier rejection. Magn. Reson. Med. **59**, 866–873 (2008)
22. Manjón, J.V., et al.: Adaptive non-local means denoising of MR images with spatially varying noise levels: spatially adaptive nonlocal denoising. J. Magn. Reson. Imaging **31**, 192–203 (2010)
23. Manjón, J.V., et al.: Nonlocal intracranial cavity extraction. Int. J. Biomed. Imaging **2014**, 1–11 (2014)
24. Möller, C., et al.: Alzheimer disease and behavioral variant frontotemporal dementia: automatic cassification based on cortical atrophy for single-subject diagnosis. Radiology **279**, 838–848 (2016)
25. Nalepa, J., Marcinkiewicz, M., Kawulok, M.: Data augmentation for brain-tumor segmentation: a review. Front. Comput. Neurosci. **13**, 83 (2019)
26. Nguyen, H., et al.: Interpretable differential diagnosis for Alzheimer's disease and frontotemporal dementia. In: Medical Image Computing and Computer Assisted Intervention, pp. 61–69 (2022)
27. Pini, L., et al.: Brain atrophy in Alzheimer's disease and aging. Ageing Res. Rev. **30**, 25–48 (2016)
28. Rascovsky, K., et al.: Sensitivity of revised diagnostic criteria for the behavioural variant of frontotemporal dementia. Brain **134**, 2456–2477 (2011)
29. Rosen, H.J., et al.: Patterns of brain atrophy in frontotemporal dementia and semantic dementia. Neurology **58**(2), 198–208 (2002)
30. Schuff, N., et al.: MRI of hippocampal volume loss in early Alzheimer's disease in relation to ApoE genotype and biomarkers. Brain **132**, 1067–1077 (2009)
31. Silhan, D., et al.: The parietal atrophy score on brain magnetic resonance imaging is a reliable visual scale. Curr. Alzheimer Res. **17**(6), 534–539 (2020)
32. Touvron, H., Vedaldi, A., Douze, M., Jégou, H.: Fixing the train-test resolution discrepancy. Adv. Neural Inf. Processing Syst. **32** (2019)
33. Touvron, H., et al.: Training data-efficient image transformers and distillation through attention. In: International Conference on Machine Learning, pp. 10347–10357 (2021)
34. Touvron, H., Cord, M., El-Nouby, A., Verbeek, J., Jégou, H.: Three things everyone should know about vision transformers. In: Avidan, S., et al. (eds.) Computer Vision – ECCV 2022: 17th European Conference, Tel Aviv, Israel, 23–27 October 2022, Proceedings, Part XXIV, pp. 497–515. Springer, Cham (2022). https://doi.org/10.1007/978-3-031-20053-3_29
35. Tustison, N.J., et al.: N4ITK: improved N3 bias correction. IEEE Trans. Med. Imaging **29**, 1310–1320 (2010)
36. Whitwell, J.L., et al.: Distinct anatomical subtypes of the behavioural variant of frontotemporal dementia: a cluster analysis study. Brain **132**, 2932–2946 (2009)

37. Xia, Z., et al.: Vision transformer with deformable attention. In: Conference on Computer Vision and Pattern Recognition, pp. 4794–4803 (2022)
38. Yew, B., et al.: Lost and forgotten? Orientation versus memory in Alzheimer's disease and frontotemporal dementia. J. Alzheimer's Dis. JAD **33**, 473–481 (2013)
39. Zhang, H., et al.: mixup: beyond empirical risk minimization. arXiv:1710.09412 (2018)
40. Zhang, S., et al.: 3D Global Fourier Network for Alzheimer's disease diagnosis using structural MRI. In: Medical Image Computing and Computer Assisted Intervention, pp. 34–43 (2022)

Consisaug: A Consistency-Based Augmentation for Polyp Detection in Endoscopy Image Analysis

Ziyu Zhou[1], Wenyuan Shen[2], and Chang Liu[3(✉)]

[1] Shanghai Jiao Tong University, Shanghai, China
zhouziyu@sjtu.edu.cn
[2] Carnegie Mellon University, Pittsburgh, PA, USA
wenyuan2@andrew.cmu.edu
[3] SenseTime Research, Shanghai, China
liuchang@sensetime.com

Abstract. Colorectal cancer (CRC), which frequently originates from initially benign polyps, remains a significant contributor to global cancer-related mortality. Early and accurate detection of these polyps via colono-scopy is crucial for CRC prevention. However, traditional colonoscopy methods depend heavily on the operator's experience, leading to suboptimal polyp detection rates. Besides, the public database are limited in polyp size and shape diversity. To enhance the available data for polyp detection, we introduce Consisaug, an innovative and effective methodology to augment data that leverages deep learning. We utilize the constraint that when the image is flipped the class label should be equal and the bonding boxes should be consistent. We implement our Consisaug on five public polyp datasets and at three backbones, and the results show the effectiveness of our method. All the codes are available at (https://github.com/Zhouziyuya/Consisaug).

Keywords: Colonoscopy · Polyp detection · Image augmentation

1 Introduction

Colonoscopy, while essential for colorectal cancer (CRC) screening, is expensive, resource-demanding, and often met with patient reluctance. Unfortunately, up to 26% of colonoscopies may miss lesions and adenomas [1], as they heavily rely on the expertise of the endoscopist. In routine examinations, distinguishing between neoplastic and non-neoplastic polyps poses challenges, especially for less experienced endoscopists using current equipment [2,3]. Besides, object detection labeling involves the expertise of the endoscopist assigning both a category and a bounding box location to each object in an image. This process is time-consuming, with an average of 10 s per object [4]. Consequently, object detection labeling incurs significant costs, demands extensive time commitments, and requires substantial effort.

Recently, there has been a great interest in deep learning in CRC screening. Various studies have developed models for automatic polyp segmentation [5,6], polyp detection [7,8] aiming to reduce the access barrier to pathological services. However, the deficiency of training data seriously impedes the development of polyp detection techniques. The existing fully-annotated databases, including CVC-ClinicDB [9], ETIS-Larib [10], CVC-ColonDB [11], Kvasir-Seg [12] and LDPolypVideo [13], are very limited in polyp size and shape diversity, which are far from the significant complexity in the actual clinical situation. Therefore, in this paper we want to find out an augmentation to fully use the dataset itself. By this motivation, we put forward an consistency-based augmentation to improve the performance of polyp detection which use Student-Teacher model to distill knowledge. Following we coarsely describe the consistency regularization and Student-Teacher model used in our architecture.

Consistency regularization is a method that has seen wide applications in semi-supervised learning, unsupervised learning, and self-supervised learning. The core idea behind consistency regularization is to encourage the model to produce similar outputs for similar inputs, thereby leveraging unlabeled data to improve generalization performance [14–16]. Student-Teacher models have been a focal point of research in the field of machine learning and specifically in the domain of knowledge distillation, where information is transferred from one machine learning model (the teacher) to another (the student). The objective is to leverage the capabilities of a large, complex model (the teacher) and distill this knowledge into a smaller, simpler model (the student), thereby optimizing computational efficiency without compromising the performance significantly [17,18].

Through this work, we have made the following contributions:

- We propose a straightforward yet effective augmentation scheme that take advantage of the polyp image's intrinsic flipping consistency property;
- We novelly combine flipping consistency with Student-Teacher architecture which show great effectiveness in polyp detection;
- The proposed consistency constraint augmentation for polyp detection works well on multiple datasets and backbones and effective for not only in-domain samples but cross-domain samples.

2   Method

The Consisaug to be presented works similarly depending on whether it is for a CNN-based or a transformer-based object detector. The overall structure is depicted in Fig. 1. The proposed structure is the combination of the Student-Teacher model and an object detection algorithm. To allow one-to-one correspondence of target objects, an original image I is added to initial augmentation to get image x. And x is added to our flipping augmentations to get the flip one x'. As shown in Fig. 1, a paired bounding box should represent the same class and their localization information should be consistent.

In the Student branch, the labeled samples are trained using supervised loss in typical object detection approaches. The consistency loss is additionally used to combine the two outputs of the teacher and student model. In this section, the Student-Teacher model, consistency loss for localization and for classification will be introduced respectively.

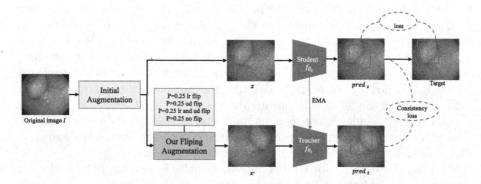

Fig. 1. Overall structure of our proposed method.

2.1   The Student-Teacher Model

The framework of Consisaug used for this work shares the same overall structure as recent knowledge distillation approaches. There are two parts of the model *Student* and *Teacher* shown in Fig. 1. The *Student* model outputs predictions $f(I)$ of input images: $f(I) \triangleq f_{\theta_s}(I)$ and the same as *Teacher* model: $f(\hat{I}) \triangleq f_{\theta_t}(\hat{I})$, $f(I)$ and $f(\hat{I})$ are a paired of bounding boxes which should represent the same class and consistent localization, I and $\hat{I} \in R^{C \times H \times W}$ are the two input images of the models after respective augmentations. Differently, in the student branch, the original image I_0 will be randomly added initial augmentation including addnoise, multi-scale, flip and etc. to get image I, while in the teacher branch, I will be randomly added our flipping augmentations. The two models have the same architecture and the same initial weight. Besides, the weights of *Student* θ_s are updated by back-propagation and the weights of *Teacher* θ_t are updated by exponential moving average as Eq. 1. More precisely, the temperature $\tau \in [0, 1]$ is given and updated after each iteration.

$$\theta_t \leftarrow \tau\theta_s + (1 - \tau)\theta_t \tag{1}$$

2.2   Consistency Loss for Localization

We denote $f_{loc}^k(I)$ as the output localization of the rediction box. The localization result for the k-th candidate box $f_{loc}^k(I)$ consists of $[\Delta cx, \Delta cy, \Delta w, \Delta h]$, which represent the displacement of the center and scale coefficient of a candidate box, respectively. The output $f_{loc}^k(I)$ and its flipping version $f_{loc}^k(\hat{I})$ require a simple modification to be equivalent to each other. Since our flipping transformations make the coordinate offset move into the opposite direction, a negation should be applied to correct them. And we take the left and right flipping as an example in Eq. 2

$$\Delta cx^k \iff -\Delta c\hat{x}^{k'}$$
$$\Delta cy^k, \Delta w^k, \Delta h^k \iff \Delta cyy^{k'}, \Delta \hat{w}^{k'}, \Delta \hat{h}^{k'}$$

(2)

The localization consistency loss used for a single pair of bounding boxes in our method is given as below:

$$l_{con\_loc}(f_{loc}^k(I), f_{loc}^{k'}(\hat{I})) = \frac{1}{4}\left(\left\|\Delta cx^k - \left(-\Delta c\hat{x}\hat{x}^{k'}\right)\right\|^2 + \left\|\Delta cy^k - \Delta c\hat{y}\hat{k}^{k'}\right\|^2 \right.$$
$$\left. + \left\|\Delta w^k - \Delta \hat{w}^{k'}\right\|^2 + \left\|\Delta h^k - \Delta \hat{h}^{k'}\right\|^2\right)$$

(3)

The overall consistency loss for localization is then obtained from the average of loss values from all bounding box pairs:

$$\mathcal{L}_{con-l} = \mathbb{E}_k\left[l_{con\_loc}\left(f_{loc}^k(I), f_{loc}^{k'}(\hat{I})\right)\right]$$

(4)

2.3   Consistency Loss for Classification

As for consistency loss for classification, we use Jensen-Shannon divergence (JSD) instead of the L_2 distance as the consistency regularization loss. L_2 distance treats all the classes equal, while in our flipping consistency circumstance irrelevant classes with low probability should not effect the classification performance much. JSD is a weaker constraint loss which is suitable in the consistency setting. The classification consistency loss is defined as below:

$$l_{con\_cls}\left(f_{cls}^k(I), f_{cls}^{k'}(\hat{I})\right) = JS\left(f_{cls}^k(I), f_{cls}^{k'}(\hat{I})\right)$$

(5)

where JS denotes Jensen-Shannon divergence and $f_{cls}^k(I)$ is the model prediction class of the k-th box in image I. The overall consistency loss for classification of a pair of flipping images can be clarified as:

$$\mathcal{L}_{con-c} = \mathbb{E}_k\left[l_{con\_cls}\left(f_{cls}^k(I), f_{cls}^{k'}(\hat{I})\right)\right]$$

(6)

The total consistency loss is the sum of location and classification consistency loss:

$$\mathcal{L}_{con} = \mathcal{L}_{con-l} + \mathcal{L}_{con-c}$$

(7)

Consequently, the final loss $\mathcal{L}$ is composed of the original object detector's fully supervised loss $\mathcal{L}_s$ and our consistency loss $\mathcal{L}_{con}$:

$$\mathcal{L} = \mathcal{L}_s + \mathcal{L}_{con}$$

(8)

3   Experiments and Results

3.1   Implementation Details

Datasets and Baselines. Experiments are conducted on 5 public polyp datasets LDPolypVideo [13], CVC-ColonDB [11], CVC-ClinicDB [9], Kvasir-Seg [12] and ETIS-Larib [10]. The first one is the largest-scale challenging colonoscopy polyp detection dataset and the others are standard benchmarks for polyp segmentation. We summarize the 5 datasets by listing their parameters in Table 1. We use the official train test split on LDPolypVideo dataset and split 10% data from train set for validation. As for datasets with no officially released partition, we split 80% for training and 10% for validating and testing respectively.

We train our Consisaug on three baseline models to evaluate the effectiveness of our method: yolov5 [19], SSD [20] and detr [21]. The first two are CNN-based models and the third one is Transformer-based model. All experiments have been done under the similar setting of the official codes and both codes are implemented based on Pytorch. The evaluation metrics used in our experiments are recall, precision, mAP50, F1-score, F2-score and the last three is vital in image detection. In detail, we use AdamW [22] optimizer with a cosine learning rate schedule, linear warm up of 10 epochs while the overall epoch is 100, and 0.0001 for the maximum learning rate value. The batch size is 32 and the image size is 640. We train with single Nvidia RTX3090 24G GPU to proceed each experiment.

Table 1. Summary of public annotated colonoscopy datasets.

Dataset	Label	Resolution	N_{images}	N_{videos}	N_{polyps}
LDPolypVideo	Bounding box	560×480	33884	160	200
CVC-ClinicDB	Mask	384×288	612	29	29
CVC-ColonDB	Mask	574×500	380	15	15
ETIS-Larib	Mask	1225×966	196	34	44
Kvasir-Seg	Mask	Various	1000	N/A	N/A

3.2   Results

Consisaug Outperforms the Vanilla Version on Different Backbones. To demonstrate the effectiveness of our method, we train the polyp detection on three backbones yolov5 [19], SSD [20], and DETR [21], and all are trained on the LDPolypVideo dataset [13], which has the largest size and diversity among the publicly released polyp datasets. The vanilla version model is trained using the official code and hyper-parameters, while the Consisaug version is trained using our method, which is reconstructed with the student-teacher model and our consistency-based augmentation. The results are shown in Table 2. All methods are trained three times, and the best results for each baseline are bolded. From the results, we can conclude that our Consisaug can enhance the polyp detection not only on CNN-based backbones (yolov5, SSD) but transformer-based backbone (DETR) from the three evaluation indexes mAP50, F1-score, and F2-score. Moreover, our method can also improve the recall, which is vital for lesion detection in medical image analysis.

Table 2. The polyp detection results on LDPolypVideo dataset. The best results for each baseline are bolded.

Baseline	Method	Recall	Precision	mAP50	F1-score	F2-score
yolov5	Vanilla	0.378 ± 0.008	$\mathbf{0.578 \pm 0.012}$	0.510 ± 0.017	0.457 ± 0.010	0.406 ± 0.007
	Consisaug	$\mathbf{0.453 \pm 0.004}$	0.575 ± 0.015	$\mathbf{0.540 \pm 0.024}$	$\mathbf{0.507 \pm 0.018}$	$\mathbf{0.473 \pm 0.011}$
SSD	Vanilla	0.658 ± 0.028	0.152 ± 0.006	0.515 ± 0.013	0.248 ± 0.009	0.396 ± 0.011
	Consisaug	$\mathbf{0.667 \pm 0.024}$	$\mathbf{0.155 \pm 0.003}$	$\mathbf{0.527 \pm 0.014}$	$\mathbf{0.251 \pm 0.010}$	$\mathbf{0.401 \pm 0.006}$
DETR	Vanilla	0.584 ± 0.026	0.446 ± 0.013	0.468 ± 0.011	0.506 ± 0.017	0.550 ± 0.016
	Consisaug	$\mathbf{0.629 \pm 0.030}$	$\mathbf{0.480 \pm 0.013}$	$\mathbf{0.504 \pm 0.015}$	$\mathbf{0.544 \pm 0.016}$	$\mathbf{0.592 \pm 0.022}$

Consisaug Shows Effectiveness on Different Colonoscopy Datasets. To further verify the validity of our Consisaug method, we train the vanilla version and our Consisaug on other datasets. All the experiments are implemented on yolov5, and the results are shown in Table 3. Consisaug outperforms the vanilla version in at least four mAP50, F1-score, and F2-score metrics on five datasets.

Table 3. The polyp detection results based on yolov5 baseline for different datasets. The best results for each dataset are bolded.

Dataset	Method	Precision	Recall	mAP50	F1-score	F2-score
LDPolypVideo	Vanilla	0.378 ± 0.008	$\mathbf{0.578 \pm 0.012}$	0.510 ± 0.017	0.457 ± 0.010	0.406 ± 0.007
	Consisaug	$\mathbf{0.453 \pm 0.004}$	0.575 ± 0.015	$\mathbf{0.540 \pm 0.024}$	$\mathbf{0.507 \pm 0.018}$	$\mathbf{0.473 \pm 0.011}$
CVC-ClinicDB	Vanilla	0.933 ± 0.002	0.781 ± 0.006	0.865 ± 0.009	0.850 ± 0.003	0.807 ± 0.007
	Consisaug	$\mathbf{0.967 \pm 0.002}$	$\mathbf{0.932 \pm 0.013}$	$\mathbf{0.963 \pm 0.004}$	$\mathbf{0.949 \pm 0.011}$	$\mathbf{0.939 \pm 0.015}$
CVC-ClolonDB	Vanilla	$\mathbf{0.997 \pm 0.003}$	0.789 ± 0.013	0.891 ± 0.005	0.881 ± 0.007	0.823 ± 0.001
	Consisaug	0.970 ± 0.001	$\mathbf{0.842 \pm 0.001}$	$\mathbf{0.916 \pm 0.002}$	$\mathbf{0.901 \pm 0.004}$	$\mathbf{0.865 \pm 0.001}$
ETIS-Larib	Vanilla	$\mathbf{0.982 \pm 0.003}$	0.350 ± 0.005	$\mathbf{0.634 \pm 0.009}$	0.516 ± 0.007	0.402 ± 0.003
	Consisaug	0.800 ± 0.0005	$\mathbf{0.450 \pm 0.012}$	0.629 ± 0.003	$\mathbf{0.576 \pm 0.002}$	$\mathbf{0.493 \pm 0.004}$
Kvasir-Seg	Vanilla	$\mathbf{0.937 \pm 0.006}$	0.730 ± 0.008	0.848 ± 0.001	$\mathbf{0.821 \pm 0.007}$	0.764 ± 0.002
	Consisaug	0.879 ± 0.003	$\mathbf{0.762 \pm 0.002}$	$\mathbf{0.857 \pm 0.005}$	0.816 ± 0.002	$\mathbf{0.783 \pm 0.006}$

Consisaug Transcends the Vanilla Version on Cross-domain Datasets. We also validate our method on cross-domain colonoscopy datasets detection. The vanilla and Consisaug versions are all trained on LDPolypVideo dataset yolov5 backbone. We test the two checkpoints on the other four whole datasets and the results are shown in Table 4. The results show the transferability of our model which is trained on one domain and tested on the other domains. Our method's performance exceeds the vanilla version on all datasets from the metric of mAP50 and surpasses the vanilla version on three datasets from the F1-score and F2-score.

Table 4. Cross-domain polyp detection results. The four different datasets are test using vanilla yolov5 model and our Consisaug yolov5 model. The two models are all trained on LDPolypVideo dataset. The best results for each dataset are bolded.

Dataset	Method	Precision	Recall	mAP50	F1-score	F2-score
CVC-ClinicDB	Vanilla	**0.783**	0.598	0.716	0.678	0.628
	Consisaug	0.782	**0.652**	**0.746**	**0.711**	**0.674**
CVC-ClolonDB	Vanilla	0.780	0.578	0.701	0.664	0.610
	Consisaug	**0.769**	**0.639**	**0.721**	**0.698**	**0.661**
ETIS-Larib	Vanilla	0.725	**0.625**	0.713	**0.671**	**0.643**
	Consisaug	**0.852**	0.548	**0.719**	0.667	0.590
Kvasir-Seg	Vanilla	0.706	0.638	0.707	0.670	0.651
	Consisaug	**0.755**	**0.632**	**0.736**	**0.688**	**0.653**

Qualitative Results. In Fig. 2, we provide the polyp detection results of our Consisaug on LDPolypVideo test set. Our method can locate the polyp tissues in many challenging cases, such as small targets, motion blur and reflection images, polyps between colon folds, etc. But there are also some failure cases detection for the low image quality or the targets hidden in the dark.

4   Ablation Study

Table 5. The ablation study results on LDPolypVideo dataset yolov5 baseline. The best results for each baseline are bolded.

Sup loss	Consisaug	Flip aug	Recall	Precision	mAP50	F1-score	F2-score
✓	✗	✗	0.364 ± 0.004	0.589 ± 0.002	0.501 ± 0.001	0.450 ± 0.003	0.394 ± 0.007
✓	✗	✓	0.378 ± 0.008	0.578 ± 0.012	0.510 ± 0.017	0.457 ± 0.010	0.406 ± 0.007
✓	✓	✗	0.386 ± 0.008	0.554 ± 0.003	0.515 ± 0.003	0.455 ± 0.009	0.411 ± 0.011
✓	✓	✓	**0.453 ± 0.004**	**0.575 ± 0.015**	**0.540 ± 0.014**	**0.507 ± 0.018**	**0.473 ± 0.011**

In this section, we test the component of our Consisaug to provide deeper insight into our model. The supervised loss with flipping augmentations shown in Table 5 is the vanilla version used in Sect. 3.2. The ablation studies can be split into four combinations: (a) the model only uses supervised loss; (b) the model uses supervised loss and flipping augmentations; (c) the model uses supervised loss

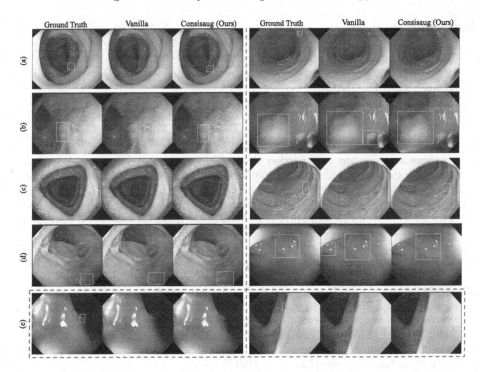

Fig. 2. There are three columns for each image set. The first column is the image with ground truth, the second column shows the detection results of vanilla model and the third column is the results of our Consisaug method. The qualitative results prove that our Consisaug can (a) detect small targets, (b) detect targets in motion blur and reflections images, (c) detect targets between colon folds, (d) reduce false positive samples. And in (e) there will also be some failure cases for the hard detecting polyps.

combining with our Consisaug and (d) the model uses supervised loss, flipping augmentations and Consisaug. Comparing (b) and (c) we can infer that our flipping consistency augmentation Consisaug is more effective than the pure flipping augmentations. And from the results in (d), combining our Consisaug with flipping augmentations can get the best performance which further proves the validity of our method.

5 Conclusion

We propose Consisaug, a novel Student-Teacher based augmentation for lesion detection task. Our approach takes advantage of the characteristics of colonoscopic surgery, in which the lens can be rotated at any angle in the body so the flip of the colonoscopy picture at any angle is the image state that can be obtained. Therefore, we leverage the peculiarity of the colonscopies and the flip detecting consistency to prove our method. Extensive experiments demonstrate that Consisaug is a valid augmentation across five datasets and three backbones.

References

1. Zhao, S.: Magnitude, risk factors, and factors associated with adenoma miss rate of tandem colonoscopy: a systematic review and meta-analysis. Gastroenterology **156**(6), 1661–1674 (2019)
2. Wadhwa, V., et al.: Physician sentiment toward artificial intelligence (AI) in colonoscopic practice: a survey of us gastroenterologists. Endosc. Int. Open **8**(10, E1379–E1384 (2020)
3. Dayyeh, B.K.A., et al.: ASGE technology committee systematic review and meta-analysis assessing the ASGE PIVI thresholds for adopting real-time endoscopic assessment of the histology of diminutive colorectal polyps. Gastrointest. Endosc. **81**(3), 502–e1 (2015)
4. Russakovsky, O., Li, L.-J., Fei-Fei, L.: Best of both worlds: human-machine collaboration for object annotation. In: Proceedings of the IEEE Conference on Computer Vision and Pattern Recognition, pp. 2121–2131 (2015)
5. Tomar, N.K., Shergill, A., Rieders, B., Bagci, U., Jha, D.: Transresu-net: transformer based resu-net for real-time colonoscopy polyp segmentation. arXiv preprint arXiv:2206.08985 (2022)
6. Fan, D.-P., et al.: PraNet: parallel reverse attention network for polyp segmentation. In: Martel, A.L., et al. (eds.) MICCAI 2020. LNCS, vol. 12266, pp. 263–273. Springer, Cham (2020). https://doi.org/10.1007/978-3-030-59725-2_26
7. Sun, X., et al.: MAF-net: multi-branch anchor-free detector for polyp localization and classification in colonoscopy. In: International Conference on Medical Imaging with Deep Learning, pp. 1162–1172. PMLR (2022)
8. Jiang, Y., Zhang, Z., Zhang, R., Li, G., Cui, S., Li, Z.: Yona: you only need one adjacent reference-frame for accurate and fast video polyp detection. arXiv preprint arXiv:2306.03686 (2023)
9. Bernal, J., Sánchez, F.J., Fernández-Esparrach, G., Gil, D., Rodríguez, C., Vilariño, F.: WM-DOVA maps for accurate polyp highlighting in colonoscopy: validation vs. saliency maps from physicians. Comput. Med. Imaging Graph. **43**, 99–111 (2015)
10. Silva, J., Histace, A., Romain, O., Dray, X., Granado, B.: Toward embedded detection of polyps in WCE images for early diagnosis of colorectal cancer. Int. J. Comput. Assist. Radiol. Surg. **9**, 283–293 (2014)
11. Bernal, J., Sánchez, J., Vilarino, F.: Towards automatic polyp detection with a polyp appearance model. Pattern Recogn. **45**(9), 3166–3182 (2012)
12. Borgli, H., et al.: Hyperkvasir, a comprehensive multi-class image and video dataset for gastrointestinal endoscopy. Sci. Data **7**(1), 283 (2020)
13. Ma, Y., Chen, X., Cheng, K., Li, Y., Sun, B.: LDPolypVideo benchmark: a large-scale colonoscopy video dataset of diverse polyps. In: de Bruijne, M., et al. (eds.) MICCAI 2021. LNCS, vol. 12905, pp. 387–396. Springer, Cham (2021). https://doi.org/10.1007/978-3-030-87240-3_37
14. Laine, S., Aila, T.: Temporal ensembling for semi-supervised learning. arXiv preprint arXiv:1610.02242 (2016)
15. Tarvainen, A., Valpola, H.: Mean teachers are better role models: weight-averaged consistency targets improve semi-supervised deep learning results. In: Advances in Neural Information Processing Systems, vol. 30 (2017)
16. Miyato, T., Maeda, S., Koyama, M., Ishii, S.: Virtual adversarial training: a regularization method for supervised and semi-supervised learning. IEEE Trans. Pattern Anal. Mach. Intell. **41**(8), 1979–1993 (2018)

17. Hinton, G., Vinyals, O., Dean, J.: Distilling the knowledge in a neural network. arXiv preprint arXiv:1503.02531 (2015)
18. Xie, Q., Luong, M.-T., Hovy, E., Le, Q.V.: Self-training with noisy student improves imagenet classification. In: Proceedings of the IEEE/CVF Conference on Computer Vision and Pattern Recognition, pp. 10687–10698 (2020)
19. Jocher, G., et al.: ultralytics/yolov5: v7. 0-yolov5 sota realtime instance segmentation. Zenodo (2022)
20. Liu, W., et al.: SSD: single shot multibox detector. In: Leibe, B., Matas, J., Sebe, N., Welling, M. (eds.) ECCV 2016. LNCS, vol. 9905, pp. 21–37. Springer, Cham (2016). https://doi.org/10.1007/978-3-319-46448-0_2
21. Carion, N., Massa, F., Synnaeve, G., Usunier, N., Kirillov, A., Zagoruyko, S.: End-to-end object detection with transformers. In: Vedaldi, A., Bischof, H., Brox, T., Frahm, J.-M. (eds.) ECCV 2020. LNCS, vol. 12346, pp. 213–229. Springer, Cham (2020). https://doi.org/10.1007/978-3-030-58452-8_13
22. Loshchilov, I., Hutter, F.: Decoupled weight decay regularization. arXiv preprint arXiv:1711.05101 (2017)

Cross-view Contrastive Mutual Learning Across Masked Autoencoders for Mammography Diagnosis

Qingxia Wu[1,2], Hongna Tan[3,4], Zhi Qiao[1,2], Pei Dong[1,2], Dinggang Shen[5],
Meiyun Wang[3,4](✉), and Zhong Xue[1,5](✉)

[1] United Imaging Research Institute of Intelligent Imaging, Beijing, China
zhong.xue@ieee.org
[2] United Imaging Intelligence (Beijing) Co. Ltd., Beijing, China
[3] Henan Provincial People's Hospital, Henan, China
[4] People's Hospital of Zhengzhou University, Henan, China
mywang@zzu.edu.cn
[5] Shanghai United Imaging Intelligence Co. Ltd., Shanghai, China

Abstract. Mammography is a widely used screening tool for breast cancer, and accurate diagnosis is critical for the effective management of breast cancer. In this study, we propose a novel cross-view mutual learning method that leverages a Cross-view Masked Autoencoder (CMAE) and a Dual-View Affinity Matrix (DAM) to extract cross-view features and facilitate malignancy classification in mammography. CMAE aims to extract the underlying features from multi-view mammography data without relying on lesion labeling information or multi-view registration. DAM helps overcome the limitations of single-view models and identifies unique patterns and features in each view, thereby improving the accuracy and robustness of breast tissue representations. We evaluate our approach on a large-scale in-house mammography dataset and demonstrate promising results compared to existing methods. Additionally, we perform an ablation analysis to investigate the influence of different loss functions on the performance of our method. The results show that all the proposed components contribute positively to the final performance. In summary, the proposed cross-view mutual learning method shows great potential for assisting malignant classification.

Keywords: Mammography diagnosis · Cross-view masked autoencoder · Contrastive learning · Classification

1 Introduction

Breast cancer is the most commonly diagnosed cancer and the leading cause of cancer death in women worldwide [6]. Thanks to the abundance of mammography screening data and recent advances in deep neural networks, the diagnostic performance of breast cancer on mammography has improved significantly, with

X. Cao et al. (Eds.): MLMI 2023, LNCS 14349, pp. 74–83, 2024.
https://doi.org/10.1007/978-3-031-45676-3_8

some studies demonstrating performance comparable to that of expert radiologists [13–15]. However, despite these improvements, there remains an ongoing area of investigation regarding the optimal approach to fully utilizing the information from multi-view mammography, which includes craniocaudal (CC) and mediolateral oblique (MLO) views. As depicted in Fig. 1, the MLO and CC views manifest discrete lesion features. Notably, both views present conspicuous yellow-marked calcifications. Nevertheless, the CC view conspicuously displays a red-marked mass lesion, which indicates a high likelihood of malignancy, given its concurrent calcifications. Previous research has shown that using multiple views in mammography outperforms models that rely on a single view [2,8,11,21], emphasizing the importance of developing new methods to fully leverage the information contained in multiple views. Nevertheless, some methods rely on lesion labeling information to build multi-view models, which can be inefficient in practice [9,19]. Therefore, further investigation is necessary to extract the relationship between multi-view data in high-dimensional space without relying on annotation. This paper aims to address this gap by fully utilizing the information contained in multi-view mammography for breast cancer diagnosis.

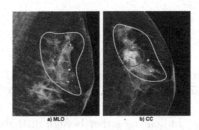

Fig. 1. The CC and MLO mammograms. The yellow ring shows calcifications, and the red ring shows mass. (Color figure online)

Recently, self-supervised learning approaches of Masked Autoencoder (MAE) has gained popularity due to its ability to enhance the performance of deep learning models in various applications [7,18]. This method involves reconstructing the original input data from various masked versions, allowing the model to learn meaningful representations of the data without explicit supervision. In medical imaging analysis, MAE has been applied to different tasks such as thorax disease classification, multi-organ segmentation, and medical vision-and-language pre-training and showed improved robustness and performance [5,20,23].

On the other hand, contrastive learning is a training technique that directly learns useful features from data by leveraging similarities and dissimilarities between samples [3,4,17]. By using a contrastive loss function, the technique aims to bring similar examples closer together while pushing dissimilar examples farther apart in the feature space. In classification, it can bring positive pairs closer and negative pairs apart, acting implicitly like feature clustering during training so as to focusing more on extracting features directly related to output responses. In mammography, this approach has been applied to improve breast cancer diagnosis and lesion detection in a pre-training and fine-tuning fashion [1, 10, 22].

In this study, we propose a cross-view contrastive mutual learning approach for mammography diagnosis. Our method is based on a Cross-view Masked Autoencoder (CMAE), which randomly masks the raw input images and trains the model to predict the missing parts. By leveraging CMAE, we aim to extract the underlying features from multi-view mammography data without relying on lesion labeling information. To extract cross-view information, we introduce a Dual-View Affinity Matrix (DAM) to learn the dual-view high-level representations by contrastive learning. DAM helps overcome the limitations of single-view models, identifies and integrates unique patterns and features from each view, thereby improving the accuracy and robustness of breast tissue representations.

The main contributions of this paper are summarized as follows. 1) We introduce CMAE and DAM that enable the model to learn cross-view unique information without explicit annotations or multi-view registration; 2) We show that our approach improves the overall malignant diagnostic performance on a large-scale in-house dataset, demonstrating the general applicability and efficacy of our method.

2   Method

We aim at exploiting the inherent cross-view (CC and MLO) characteristics of mammograms to diagnose malignancy. Denoting paired CC and MLO view mammograms as $\langle I_i^{cc}, I_i^{mlo} \rangle = \{\langle I_1^{cc}, I_1^{mlo} \rangle, \ldots, \langle I_N^{cc}, I_N^{mlo} \rangle\}$, alongside corresponding binary labels $\{y_1, \ldots, y_N\}$ indicating either benign or malignancy, our goal is to train a convolutional neural network to classify malignancy $f : \langle I_i^{mlo}, I_i^{mlo} \rangle \mapsto y_i$, by jointly using CC and MLO views. The overall framework is illustrated in Fig. 2, and detailed exposition of its fundamental components is as follows.

2.1   Cross-view Masked Autoencoder

Cross-view Masked Autoencoder (CMAE) endeavors to boost inherent dual-view characteristics of mammograms to acquire resilient representations capable of effectively handling incomplete information. It aims to reconstruct randomly masked data using encoder-decoder structure so that important local appearance information can be effectively extracted.

Masking Strategy. Given a paired CC and MLO mammograms $I_i = \langle I_i^{cc}, I_i^{mlo} \rangle$ we can mask them with random block masks, represented as $x_i^{cc} = f_m^{cc}(I_i^{cc}; \theta_m)$ and $x_i^{mlo} = f_m^{mlo}(I_i^{mlo}; \theta_m)$, where θ_m is the ratio of masked blocks within the image. We set different random seeds for masking I_i^{cc} and I_i^{mlo}, resulting in different masked regions for the two view images. During training, we adopt a high masking ratio (*i.e.*, 60%) and freeze the weight of the classifier to initialize the encoders and eliminate redundancy in the images thereby allowing the model

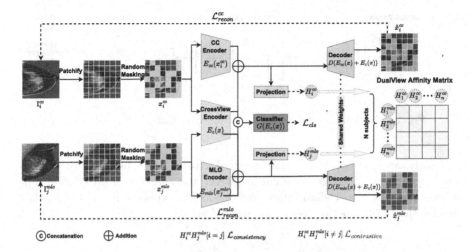

Fig. 2. Overview of the proposed cross-view mutual learning method.

to learn valuable features. Subsequently, we decrease the masking ratio (*i.e.*, 10%) and train all network components to acquire task-specific information.

Encoder Design. CMAE comprises three encoders: CC encoder, MLO encoder, and Cross-view encoder as shown in Fig. 3. The CC encoder and MLO encoder are designed to learn view position information, whereas the Cross-view encoder is intended to learn task-specific information. We employ the ConvNeXt V2 model [12,18] as the encoder, which treats masked images as 2D sparse arrays of pixels. To achieve this, the standard convolution layers in the encoder are converted to submanifold sparse convolution, allowing the model to operate only on visible patches and reducing the amount of computation required during training.

The CC encoder $E_{cc}(x^{cc}; \theta_{cc})$ and MLO encoder $E_{mlo}(x^{mlo}; \theta_{mlo})$ consist of two blocks of ConvNeXt V2-nano, and are parameterized by θ_{cc} and θ_{mlo}, respectively. The CC and MLO encoders aim to obtain the corresponding single-view hidden representations h_{cc} or h_{mlo}, respectively. The Cross-view encoder $E_c(x; \theta_c)$ consists of four blocks of ConvNeXt V2-nano, and is parameterized by θ_c. Its objective is to obtain the cross-view hidden representation h_c^{cc} and h_c^{mlo}.

Decoder Design. We use a lightweight, plain ConvNeXt block [12] as the decoder, say $D(h; \theta_d)$, and with a dimension set to 512. To ensure that the outputs of $E_{cc}(x)$ and $E_c(x)$ have the same size, we downsample the output of the CC encoder and add it to the output of Cross-view encoder. The resulting output is then passed through the decoder to obtain the reconstructed CC image, namely $\hat{x}_{cc} = D(E_{cc}(x_{cc}) + E_c(x_{cc})) = D(h_{cc} + h_c^{cc})$. We perform the same operation to obtain the reconstructed MLO image, namely $\hat{x}_{mlo} = D(E_{mlo}(x_{mlo}) + E_c(x_{mlo})) = D(h_{mlo} + h_c^{mlo})$.

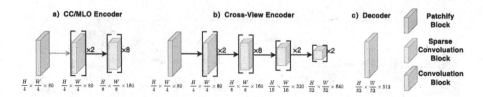

Fig. 3. The detailed architecture of Cross-view Masked Autoencoder.

Finally, we compute the mean squared error (MSE) between the reconstructed and target images. Similar to MAE, the target is a patch-wise normalized image of the original input, and the loss is applied only on the masked patches, the reconstruction loss is formulated as:

$$\mathcal{L}_{\text{recon}} = \frac{1}{2}(\mathcal{L}_{\text{recon}}^{cc} + \mathcal{L}_{\text{recon}}^{mlo}), \tag{1}$$

where $\mathcal{L}_{\text{recon}}^{cc}$ is the reconstruction loss for CC view, and $\mathcal{L}_{\text{recon}}^{mlo}$ is the reconstruction loss for MLO view.

2.2 Dual-View Contrastive Learning

In mammography diagnosis, it is common to capture two standard views of the breast - CC and MLO view. Although these views are complementary and offer distinct features, they both exhibit the same lesion. To improve lesion classification performance and account for the differences in appearance and context between the CC and MLO views, we propose a Dual-View Affinity Matrix (DAM) for dual-view contrastive learning.

To implement DAM, we consider the representations of CC and MLO views from the same breast as positive pairs, while the remaining combinations are treated as negative pairs. We use a nonlinear projection header, denoted as $\mathcal{H}$, to extract high-level CC and MLO view representations, namely H_i^{cc} and H_j^{mlo} respectively. By sampling K pairs from a mini-batch from two views, we generate $2K$ data points as inputs. Among these data points, the two image views corresponding to the same breast are taken as positive sample pairs (paired views), which yield K positive sample pairs, denoted as $\{H_i^{cc}, H_i^{mlo}\}_{i=1}^{K}$. For each positive pair, the remaining $2(K-1)$ data points in the mini-batch are regarded as negative pairs, denoted as $\{\{H_i^{cc}\}_{i\neq j}^{K-1} \cup \{H_j^{mlo}\}_{i\neq j}^{K-1}\}$. For the positive pairs, we used consistency loss to measure their distance:

$$\mathcal{L}_{\text{consistency}} = \|H_i^{cc} - H_i^{mlo}\|^2. \tag{2}$$

For the negative pairs, we used contrastive loss to measure their difference:

$$\mathcal{L}_{\text{contrastive}} = -\log \frac{\exp(\text{sim}(H_i^{cc}, H_i^{mlo})/\tau)}{\sum_{j=1}^{2N} \mathbb{1}_{[i\neq j]} \exp(\text{sim}(H_i^{cc}, H_j^{mlo})/\tau)}, \tag{3}$$

where $\text{sim}(\cdot, \cdot)$ is the cosine similarity between two representations. $\mathbb{1}_{[j \neq i]} \in \{0, 1\}$ is an indicator function equaling 1 when $j \neq i$ and $\tau = 0.1$ is the temperature parameter.

2.3 Cross-View Mutual Learning for Mammography Classification

To better deal with the cross-view information and capture view-invariant features, we propose a Cross-view Mutual Classifier (CVMC) for mammography classification. The CVMC consists of a binary classification function, denoted by $G(h; \theta)$, that maps hidden representations h to malignant predictions $\hat{y}$, and this function is parameterized by θ_g. The input to the CVMC is the concatenated output of $E_c(x_{cc})$ and $E_c(x_{mlo})$, The resulting output is a prediction $\hat{y}$ for the malignant class, which is given by $\hat{y} = G(E_c(x))$. We use weighted cross-entropy as the classification loss, which is formulated as:

$$\mathcal{L}_{cls} = -\sum (Wy\log(\hat{y}) + (1-y)\log(\hat{y})), \tag{4}$$

where y and $\hat{y}$ denote the ground truth and predicted probability. W is the weight for each category.

In summary, the total loss function of the proposed cross-view mutual learning method consists of four items: classification loss, reconstruction loss, contrastive loss, and consistency loss. It can be formulated as:

$$\mathcal{L}_{total} = \mathcal{L}_{cls} + \alpha\mathcal{L}_{recon} + \beta\mathcal{L}_{contrastive} + \gamma\mathcal{L}_{consistency}, \tag{5}$$

where β is a ramp-up function to adjust the weight value according to the epoch number, just like [16], and α, γ are scalars.

3 Results

3.1 Dataset

28234 mammograms from 14117 breasts (9758 benign and 4359 malignant confirmed by biopsy) are collected from our collaborative hospitals with IRB approval. The mammography scanners included Hologic, UIH, and GE. The acquisition parameters were as following: spacing $= 0.05 \sim 0.07\,\text{mm}$, resolution $= 4604 \times 5859 \sim 3328 \times 4096$. Each breast has both CC and MLO views. We randomly split mammograms into the training (24140), validation (1270), and testing (2824) set. Due to GPU limitation, we re-sampled images to $0.1\,\text{mm}$, normalized intensities using the z-score, and then resized them to 224×224 or 512×512 (see comparative results).

3.2 Experimental Setup

Our model is implemented in PyTorch with 2 NVIDIA Tesla A40 GPUs. We empirically set α and γ to 0.1. The network was trained for a total of 900 epochs.

During the first 800 epochs, the classifier's weight was frozen and a mask ratio of 0.6 was used. We employed the AdamW optimizer with a cosine learning rate scheduler, an initial learning rate of 0.00015, weight decay of 0.05, and warmup for 40 epochs. In this phase, minimal data augmentation is used, which includes only random resized cropping. For the subsequent 100 epochs, the masking ratio was reduced to 0.1, and the model was optimized using the AdamW optimizer with a learning rate of 0.0001 and weight decay of 0.001. In this phase, we employed a combination of data augmentation strategies, including random color jitter, random grayscaling, random cropping, and random horizontal flipping. The batch size is 64 per GPU. Finally, we evaluated the proposed method using the area under the curve (AUC), accuracy (ACC), and F1-score (F1) as metrics.

3.3   Result and Analysis

The performance metrics of our method are presented in Table 1. For an image size of 224, our approach attains AUC of 0.927, ACC of 0.885, and F1 of 0.802. Notably, our method achieves even higher performance with an image size of 512, as evidenced by AUC of 0.945, ACC of 0.909, and F1 of 0.840.

Comparison with Other Methods. The comparative results between the proposed method and the alternative methods are reported in Table 1. The methods compared include (1) A model employing ResNet50 as the backbone for classification, which averages the outputs of CC and MLO views to generate the final result. (2) A model that uses both the CC and MLO views as two-channel inputs to train a ResNet50. (3) A model similar to (1) to process each view but concatenates the last feature map to generate the final classification result. (4) A method proposed by Nan Wu et al. [19]. The original method outputs two classification tasks, namely benign vs non-benign and malignant vs non-malignant. Here we only compare our results with the metric for the malignant vs non-malignant task. As the original method involves two steps, with the first step requiring the contours of the lesions, we utilize the open-access weights of step one and train and fine-tune step two for comparison.

Ablation Study. We also performed an ablation study to evaluate the impact of different losses on the performance of the proposed method. The experiments performed are as follows: (1) Use only the classification loss ($\mathcal{L}_{cls}$). (2) Use classification loss and the reconstruction loss without ImageNet self-supervised masked autoencoder pre-training weights ($\mathcal{L}_{cls} + \mathcal{L}_{recon}$(w/o tf)). (3) Use the classification loss and reconstruction loss with ImageNet self-supervised masked autoencoder pre-training weights from [18] ($\mathcal{L}_{cls} + \mathcal{L}_{recon}$(w/ tf)). (4) Use classification loss, contrastive loss, and consistency loss ($\mathcal{L}_{cls} + \mathcal{L}_{contrastive} + \mathcal{L}_{consistency}$). (5) Use classification loss, contrastive loss, consistency loss, and reconstruction loss with pre-training weight but without patch-wise normalized original input and reconstructed images ($\mathcal{L}_{cls} + \mathcal{L}_{recon}$(w/o patch norm) + $\mathcal{L}_{contrastive}$ + $\mathcal{L}_{consistency}$). (6) Use classification loss, contrastive loss, consistency loss, and

Table 1. Comparison with other methods and results of ablation analysis.

Methods	AUC	ACC	F1
ResNet50 (two view average)	0.817	0.793	0.548
ResNet50 (two-channel input)	0.826	0.806	0.603
ResNet50 (feature map concatenation)	0.842	0.802	0.600
Nan Wu *et al.* [19]	0.894	0.854	0.735
Ablation study			
$\mathcal{L}_{cls}$	0.866	0.834	0.680
$\mathcal{L}_{cls} + \mathcal{L}_{recon}$(w/o tf)	0.874	0.807	0.566
$\mathcal{L}_{cls} + \mathcal{L}_{recon}$(w/ tf)	0.885	0.839	0.733
$\mathcal{L}_{cls} + \mathcal{L}_{contrastive} + \mathcal{L}_{consistency}$	0.910	0.8670	0.756
$\mathcal{L}_{cls} + \mathcal{L}_{recon}$(w/o patch norm) $+ \mathcal{L}_{contrastive} + \mathcal{L}_{consistency}$	0.920	0.870	0.756
$\mathcal{L}_{cls} + \mathcal{L}_{recon}$(w/ patch norm) $+ \mathcal{L}_{contrastive} + \mathcal{L}_{consistency}$ (ours-224)	0.927	0.885	0.802
$\mathcal{L}_{cls} + \mathcal{L}_{recon}$(w/ patch norm) $+ \mathcal{L}_{contrastive} + \mathcal{L}_{consistency}$ (ours-512)	0.945	0.909	0.840

reconstruction loss with pre-train weight and patch-wise normalization in the reconstructed image (ours-224). (7) The same setting as (6) except the input size is 512×512 (ours-512). All the aforementioned experiments except the last one were performed with an input image size of 224×224.

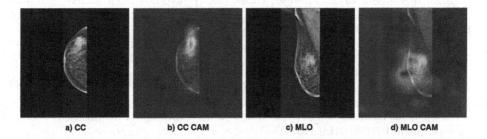

a) CC   b) CC CAM   c) MLO   d) MLO CAM

Fig. 4. The Gram-CAM of the output of Cross-view Encoder.

Representative Samples. We also present the representative samples from the testing set in Fig. 4. Specifically, a) and c) are the original CC and MLO mammograms, while b) and d) are the Grad-CAM plot of the last convolution layer from the Cross-view encoder. The highlighted region in red precisely denotes the lesion area, thereby affirming the fact that the cross-view encoder is capable of extracting the representations specific to malignancy.

4   Conclusion

We proposed a novel cross-view mutual learning approach for mammography classification. The method leverages CMAE and DAM to extract correlated features across different views, obviating the need for lesion labeling or dual-view registration, which are then used to improve the classification performance of mammography. Experimental results demonstrate the effectiveness of our approach, showing that it can help to enhance mammography classification tasks.

References

1. Cao, Z., et al.: Supervised contrastive pre-training for mammographic triage screening models. In: de Bruijne, M., et al. (eds.) MICCAI 2021. LNCS, vol. 12907, pp. 129–139. Springer, Cham (2021). https://doi.org/10.1007/978-3-030-87234-2_13
2. Carneiro, G., Nascimento, J., Bradley, A.P.: Automated analysis of unregistered multi-view mammograms with deep learning. IEEE Trans. Med. Imaging **36**(11), 2355–2365 (2017)
3. Chen, T., Kornblith, S., Norouzi, M., Hinton, G.: A simple framework for contrastive learning of visual representations. In: International Conference on Machine Learning, pp. 1597–1607. PMLR (2020)
4. Chen, T., Kornblith, S., Swersky, K., Norouzi, M., Hinton, G.E.: Big self-supervised models are strong semi-supervised learners. Adv. Neural. Inf. Process. Syst. **33**, 22243–22255 (2020)
5. Chen, Z., et al.: Multi-modal masked autoencoders for medical vision-and-language pre-training. In: Wang, L., Dou, Q., Fletcher, P.T., Speidel, S., Li, S. (eds.) Medical Image Computing and Computer Assisted Intervention – MICCAI 2022. MICCAI 2022. LNCS, vol. 13435. Springer, Cham (2022). https://doi.org/10.1007/978-3-031-16443-9_65
6. Giaquinto, A.N., et al.: Breast cancer statistics. CA: Cancer J. Clin. **72**(6), 524–541 (2022)
7. He, K., Chen, X., Xie, S., Li, Y., Dollár, P., Girshick, R.: Masked autoencoders are scalable vision learners. In: Proceedings of the IEEE/CVF Conference on Computer Vision and Pattern Recognition, pp. 16000–16009 (2022)
8. Kyono, T., Gilbert, F.J., Schaar, M.: Multi-view multi-task learning for improving autonomous mammogram diagnosis. In: Machine Learning for Healthcare Conference, pp. 571–591. PMLR (2019)
9. Li, H., Chen, D., Nailon, W.H., Davies, M.E., Laurenson, D.I.: Dual convolutional neural networks for breast mass segmentation and diagnosis in mammography. IEEE Trans. Med. Imaging **41**(1), 3–13 (2021)
10. Li, Z., et al.: Domain generalization for mammography detection via multi-style and multi-view contrastive learning. In: de Bruijne, M., et al. (eds.) MICCAI 2021. LNCS, vol. 12907, pp. 98–108. Springer, Cham (2021). https://doi.org/10.1007/978-3-030-87234-2_10
11. Liu, Y., Zhang, F., Chen, C., Wang, S., Wang, Y., Yu, Y.: Act like a radiologist: towards reliable multi-view correspondence reasoning for mammogram mass detection. IEEE Trans. Pattern Anal. Mach. Intell. **44**(10), 5947–5961 (2021)
12. Liu, Z., Mao, H., Wu, C.Y., Feichtenhofer, C., Darrell, T., Xie, S.: A convNet for the 2020s. In: Proceedings of the IEEE/CVF Conference on Computer Vision and Pattern Recognition, pp. 11976–11986 (2022)
13. Lotter, W., et al.: Robust breast cancer detection in mammography and digital breast tomosynthesis using an annotation-efficient deep learning approach. Nat. Med. **27**(2), 244–249 (2021)
14. McKinney, S.M., et al.: International evaluation of an AI system for breast cancer screening. Nature **577**(7788), 89–94 (2020)

15. Salim, M., et al.: External evaluation of 3 commercial artificial intelligence algorithms for independent assessment of screening mammograms. JAMA Oncol. **6**(10), 1581–1588 (2020)

16. Tarvainen, A., Valpola, H.: Mean teachers are better role models: weight-averaged consistency targets improve semi-supervised deep learning results. Advances in Neural Information Processing Systems 30 (2017)

17. Tian, Y., Sun, C., Poole, B., Krishnan, D., Schmid, C., Isola, P.: What makes for good views for contrastive learning? Adv. Neural. Inf. Process. Syst. **33**, 6827–6839 (2020)

18. Woo, S., et al.: ConvNeXt V2: co-designing and scaling convnets with masked autoencoders. arXiv preprint arXiv:2301.00808 (2023)

19. Wu, N., et al.: Deep neural networks improve radiologists' performance in breast cancer screening. IEEE Trans. Med. Imaging **39**(4), 1184–1194 (2019)

20. Xiao, J., Bai, Y., Yuille, A., Zhou, Z.: Delving into masked autoencoders for multi-label thorax disease classification. In: Proceedings of the IEEE/CVF Winter Conference on Applications of Computer Vision, pp. 3588–3600 (2023)

21. Yang, Z., et al.: MommiNet-v2: mammographic multi-view mass identification networks. Med. Image Anal. **73**, 102204 (2021)

22. You, K., Lee, S., Jo, K., Park, E., Kooi, T., Nam, H.: Intra-class contrastive learning improves computer aided diagnosis of breast cancer in mammography. In: Wang, L., Dou, Q., Fletcher, P.T., Speidel, S., Li, S. (eds.) Medical Image Computing and Computer Assisted Intervention – MICCAI 2022. MICCAI 2022. LNCS, vol. 13433. Springer, Cham (2022). https://doi.org/10.1007/978-3-031-16437-8_6

23. Zhou, L., Liu, H., Bae, J., He, J., Samaras, D., Prasanna, P.: Self pre-training with masked autoencoders for medical image analysis. arXiv preprint arXiv:2203.05573 (2022)

Modeling Life-Span Brain Age from Large-Scale Dataset Based on Multi-level Information Fusion

Nan Zhao[1,2,3], Yongsheng Pan[1], Kaicong Sun[1], Yuning Gu[1], Mianxin Liu[1],
Zhong Xue[3], Han Zhang[1], Qing Yang[1], Fei Gao[2], Feng Shi[3(✉)],
and Dinggang Shen[1,3,4(✉)]

[1] School of Biomedical Engineering, ShanghaiTech University, Shanghai, China
dgshen@shanghaitech.edu.cn
[2] School of Information Science and Technology, ShanghaiTech University,
Shanghai, China
[3] Shanghai United Imaging Intelligence Co., Ltd., Shanghai, China
feng.shi@uii-ai.com
[4] Shanghai Clinical Research and Trial Center, Shanghai, China

Abstract. Predicted brain age could be used to measure individual brain status over development and degeneration, which could also indicate the potential risk of age-related brain disorders. Although various techniques for the estimation of brain age have been developed, most approaches only cover a small age range, either young or elderly period, leading to limited applications. In this work, we propose a novel approach to build a brain age prediction model on a lifespan dataset with T1-weighted magnetic resonance imaging (MRI) scans. First, we utilize different neural networks to extract features from 1) an original 3D MRI scan associated with the brain maturing and aging process, 2) three (axial, coronal, and sagittal) 2D slices selected based on prior knowledge to provide possible white matter hypointensity information, and 3) volume ratios of different brain regions related to maturing and aging. Then, these extracted features of multiple levels are fused by the transformer-based cross-attention mechanism to predict the brain age. Our experiments are conducted on a total of 5376 subjects aged from 6 to 96 years from 8 cohorts. In particular, our model is built on 3372 healthy subjects and applied to 2004 subjects with brain disorders. Experimental results show that our method achieves a mean absolute error (MAE) of 2.72 years between estimated brain age and chronological age. Furthermore, when applying our model to age-related brain disorders, it turns out that both cerebral small vessel disease (SVD) and Alzheimer's disease (AD) groups demonstrate accelerated brain aging.

Keywords: Brain age prediction · MRI · CNN · Brain disorders

1 Introduction

The burden of age-related brain diseases, such as attention-deficit/hyperactivity disorder (ADHD), autism spectrum disorder (ASD), and Alzheimer's disease

X. Cao et al. (Eds.): MLMI 2023, LNCS 14349, pp. 84–93, 2024.
https://doi.org/10.1007/978-3-031-45676-3_9

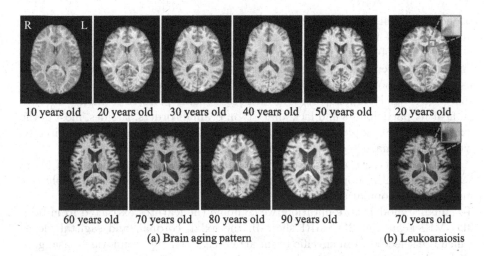

Fig. 1. The brain aging pattern (a) and the presence of leukoaraiosis (b). (a) 9 axial slices from brain scans of different subjects with chronological ages from 10 to 90 years old. (b) A subject (aged 70) demonstrates the existence of leukoaraiosis attributed to vascular aging, which is presented as white matter hypointensities surrounding the ventricular boundaries on T1-weighted MRI.

(AD), has been rising globally in the past decade according to the latest Global Burden of Disease Study. Existing works [1,2] show that individuals with certain brain disorders present abnormality of brain structures and deviate from normal brain maturing or aging trajectories. Thereby, recent studies [3–5] aim to develop a brain age estimation model on healthy individuals, and further investigate the relationship between brain age and diseases in individuals with brain disorders. This could be beneficial to quantify some heterogeneity within the disease, and identify individuals at risk of poor health outcomes [6].

The human brain will undergo a series of complex changes as age increases, including progressive (e.g., cell growth and myelination) and regressive (e.g., synaptic pruning) processes, and widespread atrophy [6,7]. Based on structural T1-weighted MRI data as shown in Fig. 1 (a), researchers [8] have observed that three anatomical features are highly related to brain aging course: 1) brain atrophy is a natural consequence of aging, involving the decrease of thickness of grey matter caused primarily by the neuronal loss; 2) leukoaraiosis is characterized by white matter hypointensities, arising due to vascular aging [9]; 3) ventricular dilation occurs in the aging brain. Additionally, it is proved that volumes of grey matter (GM), white matter (WM), and cerebrospinal fluid (CSF) undergo significant alterations during brain development and aging process [10]. Therefore, it is promising to estimate brain age utilizing sMRI data, and the estimated brain age can indicate the potential age-related health risks [11]. By calculating an individual's brain-predicted age gap [4] from the brain age and chronological age, we can identify deviations from normative brain development and aging trajectories. This discrepancy can be an initial step in facilitating early-stage diagnosis of age-related brain disorders.

There have been studies attempting to estimate brain age from neuroimaging data via deep learning algorithms [3,12–15]. However, the existing methods usually have poor generalizability due to limited number of subjects [16,17] and short age range [16,18–20]. Some studies [13–15] only perform predictive analysis on healthy participants and lack analysis on brain disorders. Besides, most of the existing methods exploit either 3D [13,14] or 2D [21–25] imaging data directly, which cannot make use of the information effectively.

In this paper, we aim to devise a model to estimate brain age from sMRI scans on a healthy population covering the entire lifespan, and then apply this model to the population with age-related brain diseases for predictive analysis. To enhance model representation, we combine the features extracted from both 3D sMRI scan and 2D sMRI slices in the axial, coronal, and sagittal views, which are selected from specific brain areas (e.g., centrum semiovale, basal ganglia, and mesencephalon) that may contain white matter hypointensity information. To enhance predictive performance, we integrate volume ratios of different brain regions derived from a pre-trained segmentation model as additional model input. This is based on the prior knowledge that volumes of brain regions are related to brain maturing and aging. The experimental results on sMRI scans across the entire lifespan demonstrate the effectiveness of our proposed method. Moreover, our analysis reveals that different brain diseases exhibit distinct trends in brain development, which can be indicated by the brain age gap [3].

2   Methods

Model Overview. The overall architecture of our proposed method is illustrated in Fig. 2 (a). It contains five pathways to extract comprehensive and effective information from an initial 3D sMRI scan, three selected 2D slices of different views, and also volumes and volume ratios of parcellated brain regions (GM, WM, CSF, and Desikan-Killiany (DK) 106 regions), dubbed as the general anatomical feature (GAF) in this work. The first pathway uses a 3D convolutional neural network (CNN) to extract features from the original 3D sMRI scan, while the next three pathways employ a shared 2D CNN to separately extract features from three (sagittal, coronal, and axial) distinct brain views, with their intersection point selected from specific brain regions (e.g., centrum semiovale, basal ganglia, and mesencephalon). These regions have been clinically confirmed to be potential locations for the occurrence of white matter hypointensity. Additionally, considering that volumes of different brain regions change during the brain maturing and aging process, the last pathway extracts the GM, WM, CSF, and also DK 106 brain regions from segmentation maps derived from a pre-trained segmentation network (VB NET) [26], and pass their volumes and volume ratios to the multilayer perceptron (MLP) for learning the representations. The learned representations from different pathways are ultimately fused through multi-head attention (MHA) mechanism and a concatenation operation.

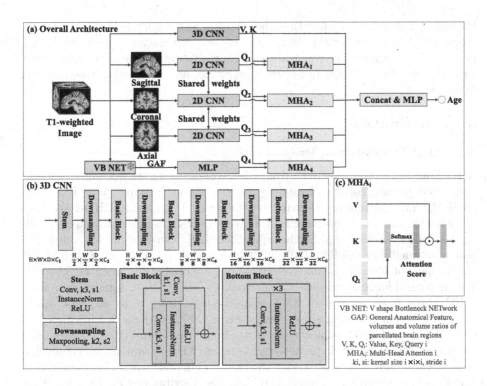

Fig. 2. Network Overview. 3D CNN architecture is shown in (b), and 2D CNNs have similar architecture as 3D CNN, but with different operation dimensions. The weights of VB-NET in (a) are frozen.

CNNs for Feature Extraction. Different from previous studies for brain age estimation [14, 21], we leverage CNNs to extract features from both the original 3D MRI scan and its three 2D slices for each subject. All these branches have the same architecture as illustrated in Fig. 2 (b), which contains four different modules, including shallow stem, downsampling, basic, and bottom blocks. Beginning with the shallow stem layer and progressing through downsampling layers, our model effectively captures multi-scale feature maps with diverse receptive fields. The shallow stem is a sequence of $3 \times 3 \times 3$ convolutional layer, instance normalization layer, and Rectified Linear Unit (ReLU) layer. Each downsampling layer is a $3 \times 3 \times 3$ max-pooling layer with a stride of 2 to reduce spatial dimensions of the preceding feature map. The channel numbers of features maps (denoted by C_i, $i \in \{1, 2, \ldots, 6\}$) are $\{1, 16, 32, 64, 128, 128\}$, respectively.

Feature Fusion of Multiple Pathways. These features from multiple pathways provide a multi-level description of brain maturation from different perspectives. Fusing these features can more effectively exploit information of brain developmental patterns. Considering that the 2D MRI slices and GAF are derived from the original MRI data, we use the features of the 3D CNN as the values V and keys K while the other features as the queries Q. The attention scores are determined by the similarity between two features measured by dot product. Furthermore, the raw features from the first pathway are concatenated with all weighted features from other pathways, and are further fused in two feed-forward layers.

3   Experiments

3.1   MRI Datasets and Preprocessing

MRI Datasets. We collected a comprehensive dataset from five public cohorts and three private cohorts, including Autism Brain Imaging Data Exchange (ABIDE) [27], Attention Deficit Hyperactivity Disorder (ADHD-200) [28], Alzheimer's Disease Neuroimaging Initiative (ADNI; adni.loni.usc.edu), Open Access Series of Imaging Studies (OASIS) [29], Consortium for reliability and reproducibility (CoRR) [30], *RENJI Hospital, HUASHAN Hospital,* and Consortium of Chinese Brain Molecular and Functional Mapping (*CBMFM*). The demographic information is outlined in Table 1. The dataset comprises sMRI scans from a total of 5376 subjects aged from 6 to 96 years, spanning 36 different scanning sites. Among them, 3372 subjects are healthy controls (HCs) and 2004 subjects are with brain disorders (BDs). These BDs include attention-deficit/hyperactivity disorder (ADHD), autism spectrum disorder (ASD), cerebral small vessel disease (SVD), mild cognitive impairment (MCI), and Alzheimer's disease (AD). The age distributions for HCs and BDs groups are shown in Fig. 3.

Preprocessing. The sMRI scans used for brain age estimation have isotropic or near isotropic resolution of $1 \times 1 \times 1$ mm^3. To minimize error accumulation and enhance model generalizability, we conducted a minimum preprocessing pipeline for each sMRI scan, including N4 bias correction [31], skull-stripping [26], and affine registration to the standard Montreal Neurological Institute (MNI) space [32]. After preprocessing, each 3D brain volume has a size of $182 \times 218 \times 182$ with an isotropic spatial resolution of 1 mm^3.

Data Normalization and Augmentation. To ensure consistency, linear Min-Max normalization was applied to each MRI data to scale the intensity value appropriately. In order to enhance the generalization capability, each 3D image underwent a series of randomized transformations, including horizontally flipping along coronal view, rotation within a range of -5 to 5 degrees, and scaling with a zoom factor ranging from 0.95 to 1.05. Subsequently, each image was precisely cropped to a size of $160 \times 192 \times 160$ mm^3 to remove the irrelevant background. For the extraction of 2D slices, we selected three slices from specified brain regions within the central 70 slices for the sagittal view, the central 80 slices for the coronal view, and the central 70 slices for the axial view. For each subject, the volumes of different brain regions in GAF are normalized to 0–1 divided by the maximum brain volume among all subjects.

Table 1. Demographic information of 8 cohorts. SD: Standard Deviation.

Cohort	Category	Total	HCs	BDs	Range	Mean ± SD	Male/Female
ABIDE	HC, ASD	1010	511	499	6–64	17.4 ± 8.1	870/140
ADHD-200	HC, ADHD	767	487	280	7–22	12.0 ± 3.2	478/289
ADNI	HC, MCI, AD	1348	565	783	55–96	73.9 ± 7.4	652/696
OASIS	HC, AD	716	634	82	42–89	65.2 ± 8.8	321/395
CoRR	HC	474	474	0	6–60	24.1 ± 10.5	258/216
RENJI	HC, SVD	297	37	260	41–84	65.3 ± 7.2	228/69
HUASHAN	HC, MCI	266	166	100	43–80	64.4 ± 7.3	103/163
CBMFM	HC	498	498	0	20–60	38.0 ± 11.9	228/270
Total	HC, ADHD, ASD, SVD, MCI, AD	5376	3372	2004	6–96	42.7 ± 25.3	3138/2238

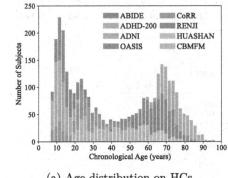

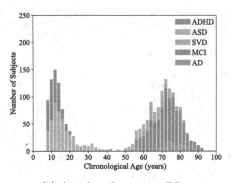

(a) Age distribution on HCs   (b) Age distribution on BDs

Fig. 3. Chronological age distributions for HCs and BDs groups from 8 cohorts with the age range from 6 to 96 years.

3.2 Experimental Setup and Platform

Our method was implemented with Pytorch and trained using Adam optimizer for 200 epochs with a mini-batch size of 8. The learning rate was initially set as 0.001 with a cosine annealing decay policy. A weight decay of 0.0001 was used to mitigate the overfitting problem. The training loss function was defined as the mean square error (MSE), as indicated in Eq. 1, where $\hat{\mathbf{y}} \in \mathbb{R}^n$ and $\mathbf{y} \in \mathbb{R}^n$ represent brain-predicted age and chronological age, respectively.

$$L = \frac{1}{n}\|\hat{\mathbf{y}} - \mathbf{y}\|_2^2 = \frac{1}{n}\sum_{i=1}^{n}|\hat{y}_i - y_i|^2 \tag{1}$$

Our experiments were conducted on an NVIDIA GPU of GA100, A100 PCIe 80GB, and a CPU with Inter(R) Xeon(R) Gold 5218R CPU@2.10 Hz and 400GB RAM. Our code was released at https://github.com/zhaonann/brain-age-estimation-from-sMRI.

Table 2. Comparisons of different algorithms for brain age estimation. The best results are shown in bold. (#Params: the number of parameters; MAE (years): mean absolute error; RMSE (years): root mean square error; r: Pearson correlation coefficient between brain-predicted age and chronological age)

Input Data	Method	#Params	MAE ↓	RMSE ↓	r ↑
3D MRI	3D ResNet-18 [33]	33.161M	3.14	4.51	0.984
	3D ResNet-34 [33]	63.470M	3.17	4.53	0.984
	SFCN [14]	2.956M	3.01	4.45	0.985
GAF	MLP	0.089M	5.07	6.95	0.964
2D slices	2D CNN	0.700M	3.84	5.19	0.979
3D MRI	3D CNN	1.647M	2.80	3.89	0.988
3D MRI+GAF	3D CNN+MLP	1.835M	2.74	3.87	0.988
3D MRI+2D slices	3D CNN+2D CNN	2.644M	2.73	3.88	**0.989**
3D MRI+2D slices+GAF (Ours)	3D CNN+2D CNN +MLP	2.849M	**2.72**	**3.84**	**0.989**

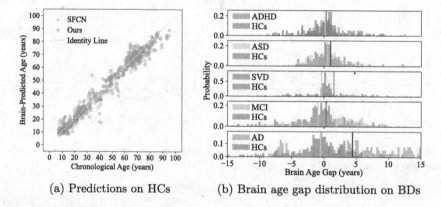

(a) Predictions on HCs   (b) Brain age gap distribution on BDs

Fig. 4. Predictive analysis of brain age. The gray histograms in (b) manifest the distribution of healthy controls (HCs) matched for age, sex, and scan protocols selected from validation and test subsets (with HCs for SVD also selected from the training set due to the limited number of subjects). The mean value of the brain age gap (i.e., chronological age subtracted from predicted brain age) for each group is shown with a solid line in the corresponding color.

3.3   Results

Performance Evaluation on Healthy Controls. We first evaluated our proposed model on the dataset with 3372 HCs. This dataset was randomly split into training, validation, and test sets in a ratio of 8:1:1. The test results for comparison methods were attained by picking the best one on the validation set.

We summarize the performance in Table 2. We can see that our model achieves the highest age prediction accuracy with a mean absolute error (MAE) of 2.72 years by resorting to the combination of 3D MRI data, 2D slices, and

GAF. Figure 4 (a) presents the estimated and chronological ages of healthy controls for our devised model and SFCN [14], which turns out to have a strong linear correlation, implying promising predictive accuracy. Meanwhile, the superiority of our method is evident in the estimates with fewer outliers. Notably, the predicted brain age for the younger group shows higher accuracy than that for the elder group. This is attributed to relatively small age values in the younger group and complied with clinical expectations of a lower estimated error in the age group. For the ablation study, we evaluate the effectiveness of different components including GAF and 2D slices, and demonstrate the results in Table 2. We can see that the use of 2D slices brings improvement and the combination of all proposed components obtains the best performance.

Predictive Analysis on Brain Disorders. So far we have validated our model on HCs. In this section, we apply it to BDs to disclose their brain developmental patterns. In Fig. 4 (b), we demonstrate the relationship between brain age and diseases measured by our designed model. The diseases include ADHD, ASD, SVD, MCI, and AD. For each of them and the case-control HCs, we depict the distribution histogram of the brain age gap and its mean value that is indicated by a solid line. Both SVD and AD groups exhibit advanced brain aging patterns, reflected in the positive brain age gap, in line with previous findings in [10,34]. The remarkable acceleration of brain aging in AD is supported by a mean brain age gap of 4.43 years. These observations suggested that a positive brain age gap is linked with cognitive decline and neurodegeneration. Moreover, no significant deviations in brain age gaps are found for ADHD, ASD, and MCI.

4   Conclusion

In this work, we perform brain age estimation on a large population across the lifespan by our proposed deep-learning based model. We have investigated brain status during individual growth and aging based on sMRI scans. In order to effectively extract brain features and incorporate prior knowledge for brain age estimation, we propose to integrate information from 3D images, selected 2D slices of different views, and volume ratios of different brain regions. Moreover, we have conducted predictive analysis on brain disorders using our proposed model. Accelerated aging in the brain is noted in individuals with SVD or AD.

Acknowledgments. This work was supported in part by National Natural Science Foundation of China (62131015), Science and Technology Commission of Shanghai Municipality (STCSM) (21010502600), Key R&D Program of Guangdong Province, China (2021B0101420006), STI2030-Major Projects (2022ZD0213100), The China Postdoctoral Science Foundation (Nos. BX2021333, 2021M703340), and National Key Research and Development Program of China (2022YFE0205700). Data collection and sharing for this project was funded by Shanghai Zhangjiang National Innovation Demonstration Zone Special Funds for Major Projects (ZJ2018-ZD-012), Shanghai Pilot Program for Basic Research (JCYJ-SHFY-2022-014), and Shanghai Pujiang Program (21PJ1421400).

References

1. Scheltens, P., Blennow, K., Breteler, M.M.B., de Strooper, B., Frisoni, G.B., Salloway, S., Van der Flier, W.M.: Alzheimer's disease. The Lancet **388**(10043), 505–517 (2016)
2. Rundek, T., Tolea, M., Ariko, T., Fagerli, E.A., Camargo, C.J.: Vascular cognitive impairment (VCI). Neurotherapeutics **19**(1), 68–88 (2022)
3. Franke, K., Ziegler, G., Klöppel, S., Gaser, C.: Estimating the age of healthy subjects from T1-weighted MRI scans using kernel methods: exploring the influence of various parameters. Neuroimage **50**(3), 883–892 (2010)
4. Kaufmann, T., et al.: Common brain disorders are associated with heritable patterns of apparent aging of the brain. Nat. Neurosci. **22**(10), 1617–1623 (2019)
5. Popescu, S.G., Glocker, B., Sharp, D.J., Cole, J.H.: Local brain-age: a U-net model. Front. Aging Neurosci. **13**, 761954 (2021)
6. Cole, J.H., Franke, K.: Predicting age using neuroimaging: innovative brain ageing biomarkers. Trends Neurosci. **40**(12), 681–690 (2017)
7. Franke, K., Gaser, C.: Longitudinal changes in individual BrainAGE in healthy aging, mild cognitive impairment, and Alzheimer's Disease. GeroPsych. **25**(4), 235–245 (2012)
8. Jégou, S.: How to Estimate the Age of Your Brain with MRI Data. https://medium.com/thelaunchpad/how-to-estimate-the-age-of-your-brain-with-mri-data-c60df60da95d (2019), (Accessed 9 July 2023)
9. Wei, K., et al.: White matter hypointensities and hyperintensities have equivalent correlations with age and CSF β-amyloid in the nondemented elderly. Brain Behav. **9**(12), e01457 (2019)
10. Bethlehem, R.A., et al.: Brain charts for the human lifespan. Nature **604**(7906), 525–533 (2022)
11. Cole, J.H., et al.: Brain age predicts mortality. Mol. Psychiatry **23**(5), 1385–1392 (2018)
12. Cole, J.H., et al.: Predicting brain age with deep learning from raw imaging data results in a reliable and heritable biomarker. Neuroimage **163**, 115–124 (2017)
13. He, S., Grant, P.E., Ou, Y.: Global-local transformer for brain age estimation. IEEE Trans. Med. Imaging **41**(1), 213–224 (2022)
14. Peng, H., Gong, W., Beckmann, C.F., Vedaldi, A., Smith, S.M.: Accurate brain age prediction with lightweight deep neural networks. Med. Image Anal. **68**, 101871 (2021)
15. Baecker, L., et al.: Brain age prediction: a comparison between machine learning models using region- and voxel-based morphometric data. Hum. Brain Mapp. **42**(8), 2332–2346 (2021)
16. de Lange, A.M.G., et al.: Multimodal brain-age prediction and cardiovascular risk: the Whitehall II MRI sub-study. Neuroimage **222**, 117292 (2020)
17. Zhu, J.D., et al.: Investigating brain aging trajectory deviations in different brain regions of individuals with schizophrenia using multimodal magnetic resonance imaging and brain-age prediction: a multicenter study. Transl. Psychiatry **13**(1), 82 (2023)
18. Liem, F., et al.: Predicting brain-age from multimodal imaging data captures cognitive impairment. Neuroimage **148**, 179–188 (2017)
19. Mouches, P., Wilms, M., Rajashekar, D., Langner, S., Forkert, N.D.: Multimodal biological brain age prediction using magnetic resonance imaging and angiography with the identification of predictive regions. Hum. Brain Mapp. **43**(8), 2554–2566 (2022)

20. Cai, H., Gao, Y., Liu, M.: Graph transformer geometric learning of brain networks using multimodal MR images for brain age estimation. IEEE Trans. Med. Imaging **42**(2), 456–466 (2023)
21. Ballester, P.L., et al.: Predicting brain age at slice level: convolutional neural networks and consequences for interpretability. Front. Psychiat. **12** (2021)
22. Hwang, I., et al.: Prediction of brain age from routine T2-weighted spin-echo brain magnetic resonance images with a deep convolutional neural network. Neurobiol. Aging **105**, 78–85 (2021)
23. Gupta, U., Lam, P.K., Steeg, G.V., Thompson, P.M.: Improved brain age estimation with slice-based set networks. In: 2021 IEEE 18th International Symposium on Biomedical Imaging (ISBI), pp. 840–844 (2021)
24. Armanious, K., et al.: Age-Net: an MRI-based iterative framework for brain biological age estimation. IEEE Trans. Med. Imaging **40**(7), 1778–1791 (2021)
25. Huang, T.W., et al.: Age estimation from brain MRI images using deep learning. In: 2017 IEEE 14th International Symposium on Biomedical Imaging (ISBI 2017), pp. 849–852 (2017)
26. Han, M., et al.: Segmentation of CT Thoracic Organs by Multi-resolution VB-nets. In: SegTHOR@ ISBI (2019)
27. Di Martino, A., et al.: The autism brain imaging data exchange: towards a large-scale evaluation of the intrinsic brain architecture in autism. Mol. Psychiatry **19**(6), 659–667 (2014)
28. Milham, M., et al.: The ADHD-200 consortium: a model to advance the translational potential of neuroimaging in clinical neuroscience. Front. Syst. Neurosci. **6** (2012)
29. LaMontagne, P.J., et al.: OASIS-3: Longitudinal Neuroimaging, Clinical, and Cognitive Dataset for Normal Aging and Alzheimer Disease. medRxiv (2019)
30. Zuo, X.N., et al.: An open science resource for establishing reliability and reproducibility in functional connectomics. Sci. Data **1**(1), 1–13 (2014)
31. Cole, N.J., et al.: N4ITK: improved N3 Bias correction. IEEE Trans. Med. Imaging **29**(6), 1310–1320 (2010)
32. Jenkinson, M., Bannister, P., Brady, M., Smith, S.: Improved optimization for the robust and accurate linear registration and motion correction of brain images. Neuroimage **17**(2), 825–841 (2002)
33. He, K., Zhang, X., Ren, S., Sun, J.: Deep residual learning for image recognition. In: 2016 IEEE Conference on Computer Vision and Pattern Recognition (CVPR), pp. 770–778 (2016)
34. Shi, Y., et al.: Potential of brain age in identifying early cognitive impairment in subcortical small-vessel disease patients. Front. Aging Neurosci. **14** (2022)

Boundary-Constrained Graph Network for Tooth Segmentation on 3D Dental Surfaces

Yuwen Tan and Xiang Xiang[✉]

Key Lab of Image Processing and Intelligent Control, Ministry of Education,
School of Artificial Intelligence and Automation,
Huazhong University of Science and Technology, Wuhan 430074, China
xex@hust.edu.cn

Abstract. Accurate tooth segmentation on 3D dental models is an important task in computer-aided dentistry. In recent years, several deep learning-based methods have been proposed for automatic tooth segmentation. However, previous tooth segmentation methods often face challenges in accurately delineating boundaries, leading to a decline in overall segmentation performance. In this paper, we propose a boundary-constrained graph-based neural network that establishes the connectivity of mesh cells based on feature distances and utilizes several modules to encode local regions. To enhance segmentation performance in tooth-gingiva boundary regions, we integrate an auxiliary loss to segment the tooth and gingiva. Furthermore, to improve the performance in tooth-tooth boundary regions, we introduce a contrastive boundary-constrained loss that specifically enhances the distinctiveness of features within boundary mesh cells. Following the network prediction, we apply a post-processing step based on the graph cut to refine the boundaries. Experimental results demonstrate that our method achieves state-of-the-art performance in 3D tooth segmentation.

Keywords: Tooth segmentation · 3D dental models · Graph neural network · Boundary refinement

1 Introduction

Computer-aided design (CAD) applications have become an indispensable part of modern dentistry, providing precise diagnosis and efficient treatment planning. Within computer-aided dentistry, tooth segmentation is a significant yet labor-intensive endeavor. The objective of tooth segmentation is to classify each mesh cell of the 3D dental model into distinct categories, following the standards set by the Federation of Dentaire Internationale (FDI) [5]. In various dental diagnoses, including orthodontics and implants, the first step for dentists is to accurately identify individual teeth and gingiva from intra-oral scanned (IOS) dental models obtained from patients. In most cases, a single mesh dental model for either the upper lower jaw contains over 100,000 triangular faces, which could take 30 min for an experienced dentist to annotate. To increase treatment planning

© The Author(s), under exclusive license to Springer Nature Switzerland AG 2024
X. Cao et al. (Eds.): MLMI 2023, LNCS 14349, pp. 94–103, 2024.
https://doi.org/10.1007/978-3-031-45676-3_10

efficiency and alleviate the burden on dentists, there exists a substantial demand for automatic tooth segmentation methods in real-world clinical applications. However, automatic tooth segmentation on dental models is an intricate task, due to the high diversity of tooth shapes, occlusions, and overlaps between teeth.

Several deep-learning-based methods have been proposed [3,9,21,23,25,28], especially the graph-based network [4,9,25,28] show their superiority in the performance for tooth segmentation on 3D dental surfaces. Despite the progress made in tooth segmentation methods, there remain unresolved challenges that must be overcome. When confronted with anomalous dental conditions, such as missing or occlusions, tooth segmentation methods often exhibit suboptimal segmentation performance. Additionally, the inaccurate segmentation of the boundaries results in imperfect segmentation results. Addressing these challenges is crucial for enhancing the accuracy and reliability of automated tooth segmentation in real-world clinical applications.

In this paper, we adopt the graph-based neural network as the backbone that utilizes a one-stream architecture to encode both the coordinates and normal vectors of 3D dental models. The network establishes the connectivity of mesh cells based on feature distances and uses several local modules to encode mesh features. The segmentation of dental models is challenging due to mispredictions that often occur in boundary areas. To enhance the segmentation performance of the tooth-gingiva boundary regions, we apply an auxiliary loss to optimize the prediction results of the tooth and gingiva. We also introduce a contrastive boundary-constrained loss to improve the model performance on tooth-tooth and tooth-gingiva boundaries. After network predictions, a graph-cut post-processing step is employed to further refine the prediction results.

The main contributions of this paper can be summarized as follows: 1) This paper focuses on how to address the inaccurate segmentation of the boundary regions to further improve the segmentation performance; 2) We utilize an auxiliary prediction loss to improve the segmentation performance of boundaries between the tooth and gingiva and a contrastive boundary-constrained loss to enhance the model performance on tooth-tooth and tooth-gingiva boundaries; 3) We conduct extensive experiments on the tooth segmentation dataset and the results demonstrate the superiority of our proposed method.

2   Related Work

2.1   Point Cloud Segmentation

Methods for point cloud segmentation can be classified into two categories, i.e., voxel-based [11,15] and point-based methods [2,12,13,20]. Point-based methods, which operate directly on raw point clouds, are becoming more popular. A pioneering network named PointNet [12] applies the multi-layer perceptron and symmetric function (e.g., max-pooling) to learn and aggregate features, respectively. However, PointNet [12] neglects the local relationships of points since the architecture independently learns on each point. To better encode local areas, PointNet++ [13] builds a hierarchical architecture with several sampling layers, grouping layers, and PointNet layers. PointCNN [7] proposes $\mathcal{X}$-transformation to regularize the irregular points and then uses convolution operations to encode the transformed features. A new operation named Pointconv [20] performs convolution on 3D point clouds with non-uniform sampling. DGCNN [18] uses an EdgeConv operator to capture local contexts by linearly aggregating center point features with edge features. PointTransformer [26] and PCT [2] design more general transformer-based frameworks for 3D point clouds. PointMLP [10] encodes point cloud features through an effective feed-forward residual MLP network. Pointnext [14] revisits Pointnet++ [13] and proposes a more effective learning strategy that gains excellent performance in point cloud analysis tasks.

2.2   3D Dental Segmentation

Recently, numerous deep learning-based methods have been designed for tooth segmentation on 3D dental models. Xu et al. [22] extract hand-crafted features of each cell and then encode these features through a CNN to predict labels. Since dental models can be converted into point cloud models, segmentation methods proposed for point clouds can provide valuable insights for tooth segmentation on dental models. Zanjani et al. [23] propose an end-to-end network that integrates PointCNN with an adversarial discriminator. Tian et al. [17] present an automatic segmentation method via a multi-level hierarchical network based on deep CNNs. MeshSegNet [9] adopts graph-constrained learning modules to encode geometric contents at different stages. Another network named iMeshSegNet [19] based on MeshSegNet, reduces computation cost and enhances segmentation performance. Zhao et al. [27] design a series of graph attention convolution layers to extract local geometric features. Mask-MCNet [24] presents a novel segmentation model that predicts a 3D bounding box of each tooth and then segments each point. TSegNet [1] divides the 3D tooth segmentation framework into two stages, including tooth centroid prediction and individual tooth segmentation. TSGCNet [25] adopts a two-branch structure that encodes coordinates and normal vectors independently and then subsequently fuses them to learn complementary views. Li et al. [8] propose MBESegNet that incorporates several bidirectional enhancement modules. TeethGNN [28] adopts a new two-branch framework to predict label and offset for every graph node respectively. To decrease the burden of the labeling process, Liu et al. [4] apply an unsupervised training method to 3D tooth segmentation, employing a contrastive learning architecture on an unlabeled dataset.

3   Methodolody

In our proposed method, each mesh cell is represented by a 24-dimensional vector consisting of coordinates (12 elements) and normal vectors (12 elements) of four points: three vertices and the central point of the cell. The main segmentation head of the network produces an N × 15 matrix as output, where each row denotes the probabilities of the corresponding cell belonging to one of 15 different classes. The 15 classes represent 14 different classes of teeth and gingiva. The

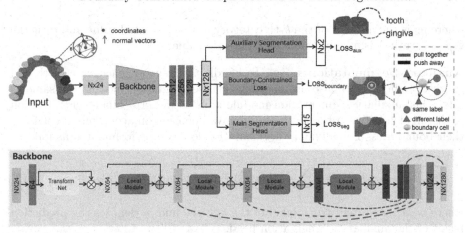

Fig. 1. The architecture of the tooth segmentation network on 3D dental models.

output of the auxiliary segmentation head is an N × 2 matrix, where each row indicates the probabilities belonging to either teeth or gingiva (Fig. 1).

Backbone Architecture. Our network is based on DGCNN [18] and Deep-GCN [6]. We first use an MLP to map the input vector F_0 to a 64-dimension vector F_1. Subsequently, a transform net is utilized to enhance the robustness by transforming the feature F_1 into a canonical space. Then, we sequentially stack four local modules to encode the transferred features. Due to the irregular structure of mesh cells, explicit neighborhood information is not available. Therefore, we employ the k-nearest neighbors algorithm to establish the connectivity relationship, where each mesh cell connects to its k closest neighbors. As we calculate the feature distances instead of the coordinate distances, the connectivity varies across different modules, necessitating the dynamic updating of the corresponding neighborhood graph. The dynamic neighbors can better encode the neighborhood information during the training process.

We concatenate all the local modules' features and one global feature to form a local-global feature vector for segmentation. The local-global feature vector is then passed through several MLP layers to map the high-dimensional feature vectors to a low-dimensional space. The low-dimensional feature vector is then used as input to two segmentation modules to predict the segmentation results and is also used to construct the contrastive boundary loss.

Tooth Segmentation Loss. We use an MLP layer as the segmentation head which outputs the probability of each mesh cell to its corresponding classes. The segmentation loss is formulated as follows

$$L_{seg} = \sum_{k=1}^{15} w_k (1 - \frac{2 * (p_k \cap g_k) + \epsilon}{p_k \cup g_k + \epsilon}) \tag{1}$$

where p_k is the number of the k-th category in the predicted value and g_k is the ground-truth, ϵ is a small value to avoid zero value.

Auxiliary Segmentation Loss. To handle the issue of inaccurate teeth and gingiva prediction, we also use an additional segmentation module the same as [21]. This auxiliary segmentation module aims to perform binary classification to distinguish between the teeth and gingiva, focusing on accurately identifying their boundaries. The auxiliary segmentation loss can be formulated as follows

$$L_{aux} = -\sum_{i}^{N} y_b.log(p_i^b) + (1 - y_b).log(1 - p_i^b) \tag{2}$$

where y_b is the ground truth label of the i-th cell and p_b denotes the prediction result of the auxiliary segmentation head.

Contrastive Boundary-Constrained Loss. A mesh cell is classified to be a boundary cell by comparing its label with the known labels of its neighboring cells. If any neighboring cell has a different label, then the mesh cell is classified as a boundary cell. To formalize the boundary cells, we denote the mesh model as a set $\mathcal{C}$ of mesh cells and each cell as c_i. The neighborhood of each cell is represented as $\mathcal{N}_i = \mathcal{N}(c_i)$ and their corresponding labels are l_i. Boundary cells are further denoted as $\mathcal{B}_g = \{c_k \in \mathcal{C} | \exists c_j \in \mathcal{N}_i, l_j \neq l_k\}$ where we set $\mathcal{N}_i$ to be the n-nearest (n=8) cells of the mesh cell c_i.

For a boundary cell denoted as c_i, the neighbors' features that belong to the same label should be close to the center cell. However, those neighbors with different labels should be pushed apart. We follow the CBL [16] to use the InfoNCE loss and the contrastive boundary-constrained loss can be formulated as

$$L_{\mathcal{B}} = -\frac{1}{|\mathcal{B}_g|} \sum_{c_i \in \mathcal{B}_g} log \frac{\sum_{c_j \in \mathcal{N}_i \wedge l_j = l_i} exp(-d(f_i, f_j)/\tau)}{\sum_{c_k \in \mathcal{N}_i} exp(-d(f_i, f_k)/\tau)} \tag{3}$$

where f_i is the feature of c_i and $d(.,)$ denotes the L_2 distance. The contrastive boundary loss only focuses on the boundary cells. For each boundary cell c_i, the positive pairs are selected from cells with the same label as c_i within its local neighborhood $\mathcal{N}_i$. The negative pairs are those neighbors with different labels. Contrastive learning enhances the distinction of features, improving the performance of segmentation on boundary regions.

Boundary Smoothing Through Post-Processing. To refine the prediction results of the neural network further, we utilize the graph-cut method [3,9] during the post-processing stage. By integrating both the probability term and the smoothness term, this method successfully minimizes an energy function, leading to improved accuracy and a more refined output. We denote the energy function as

$$\mathcal{L}_e = \sum_{i=1}^{N} -log(max(p_i(l_i), \epsilon)) + \alpha \sum_{i}^{N} \sum_{j \in \mathcal{N}_i} S(p_i, p_j, l_i, l_j) \tag{4}$$

where $p_i(l_i)$ denotes the probability of belonging to the class l_i, ϵ is the minimal probability threshold, and α denotes the smoothness term parameter.

4 Experiments

4.1 Implemention Details

Datasets. We have a dataset consisting of 103 dental models, each containing over 100,000 cells. However, we have downsampled them to 24,000 mesh cells. We aim to classify the cells in these models into 15 distinct categories. We have randomly split the dataset into a training set comprising 69 subjects and a testing set with 34 subjects. To improve the generalization ability of our model, we have employed augmentation techniques on the training set, including random rotation, translation, and rescaling of each 3D surface. This has resulted in generating an additional 276 samples from the original dental models. The augmentation process has increased the diversity of our training set, leading to an overall performance improvement.

Competing Methods and Evaluation Metric. Our method is compared with six methods for both 3D part segmentation (i.e., PointNet [12], PointNet++ [13], PCT [2], DGCNN [18]) and tooth segmentation methods on 3D dental models (i.e., MeshSegNet [9] and TSGCNet [25]). For fair comparisons, all networks are trained using the same loss function (i.e., the generalized Dice loss) for 400 epochs except for our proposed method which uses the combination of three losses. The input is 24-dimensional vectors, except for MeshSegNet which only uses a 15-dimensional input. To evaluate the overall segmentation performance across all classes, we use two quantitative metrics: 1) Overall Accuracy (OA) and 2) mean Intersection-over-Union (mIoU). Furthermore, we also calculate the Intersection-over-Union (IoU) for each individual class.

4.2 Performance Comparison

The comprehensive segmentation results are presented in Table 1. Results show that our method achieves the best performance in terms of both OA and mIoU metrics. Compared with PointNet [12], the other two methods (i.e., PointNet++ [13], PCT [20]) and our proposed method yield better results. This finding confirms that the explicit incorporation of local geometric context is crucial for achieving more accurate segmentation results in the tooth segmentation task. Additionally, our method also significantly outperforms the graph-based network DGCNN [18] (83.77% vs. 80.14%), demonstrating the effectiveness of the proposed method which earns more discriminative geometric feature representations on the boundary for accurate tooth segmentation. Besides that, on our dataset, TSGCNet [25] has a poor performance even worse than MeshSegNet [9]. MeshSegNet has heavy computation involved due to the large-scale matrix computations of the large adjacent matrices. Our method is much faster than

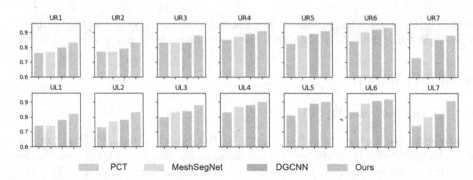

Fig. 2. The refined segmentation results, measured in terms of IoU, are reported for each of the 14 teeth (i.e., UR1-UR7 and UL1-UL7), comparing three state-of-the-art competing methods with our proposed method.

MeshsegNet during the training and performs much better than MeshSegNet (83.77% vs. 77.19%). We also display the IoU for each of the 14 teeth in Fig. 2 and the results further verify the effectiveness of our method in achieving accurate tooth segmentation.

Table 1. The segmentation results for six competing methods and our method on OA and mIoU. **Bold** is the best, *slanted* is 2nd.

Method	Input	OA	mIOU	Gingiva	Teeth	OA(w/p)	mIOU(w/p)
PointNet [12]	4p,4n	79.33	66.21	66.46	66.19	82.15	71.40
PointNet++ [13]	4p,4n	81.72	67.43	70.38	67.22	86.33	76.10
PCT [2]	4p,4n	84.66	73.36	73.03	73.38	87.80	78.98
DGCNN [18]	4p,4n	*88.86*	*80.14*	*76.07*	*80.43*	*91.28*	*84.45*
MeshSegNet [9]	4p,1n	86.98	77.19	73.53	77.45	90.12	82.68
TSGCNet [25]	4p,4n	82.36	68.83	71.13	68.67	86.71	76.53
Ours	4p,4n	**90.98**	**83.77**	**79.62**	**84.06**	**93.00**	**87.61**

4.3   Ablation Study

Table 2 displays the quantitative evaluation of the segmentation results. The parameter k represents the number of nearest neighbors in the local modules. A larger k provides a larger field of view but at the expense of computational efficiency. We evaluate the segmentation results of different k values to choose an appropriate k value. As shown in Table 2, the results of the network improve significantly with the increase of k when k is less than 32. However, when k is set to 40, the performance improvement becomes marginal. To reach the balance between performance and computation cost, we select k as 32 for our experiments. We stack different numbers of local modules in the backbone network. As shown in Table 2, the mIoU and OA reach their peak when the number of local modules increases to 4. However, surpassing this threshold and employing 5 local modules leads to overfitting, resulting in a decline in performance. In our experiments, we stack 4 local modules in the backbone network.

Table 2. Analysis experiments of different k values and the number of local modules.

Neigh.(k)	OA	mIoU	OA(p)	mIoU(p)	Blocks(n)	OA	mIoU	OA(p)	mIoU(p)
k=8	87.07	76.96	90.00	82.32	n=2	87.59	77.61	90.40	82.97
k=16	88.40	79.00	91.13	84.32	n=3	88.43	78.79	91.03	83.85
k=32	89.60	81.31	91.98	85.70	n=4	**89.60**	**81.31**	**91.98**	**85.70**
k=40	89.96	81.83	92.24	86.21	n=5	89.63	81.05	91.88	85.17

Table 3. Ablation study of different boundary refinement techniques.

Seg. Loss	Aux. Loss	Bound. Loss	OA	mIoU	OA(w/p)	mIoU(w/p)
✓			89.60	81.31	91.98	85.70
✓	✓		90.47	82.32	92.67	86.68
✓	✓	✓	**90.98**	**83.77**	**93.00**	**87.61**

To tackle the challenge of mislabeling teeth and gingiva near their boundaries, we propose to adopt an auxiliary segmentation head that discerns whether a given cell belongs to the teeth or gingiva. Our experimental findings, as shown in Table 3, demonstrate that incorporating a loss function for the auxiliary branch results in an enhanced mIoU of 82.32%, compared to the baseline of 81.31%. Considering the intricate nature of irregular tooth shapes and crowding, it becomes significant to address the delineation of boundaries among individual teeth. The introduction of a boundary-constrained loss effectively addresses the boundary concerns in both tooth-gingiva and tooth-tooth scenarios by improving the mIoU to 83.77%. The post-processing step to optimizing the boundary using the graph cut can be employed to enhance the quality of automated segmentations generated by a deep network. We find that mIoU has great improvement through post-processing (87.61% vs. 83.77%).

5   Conclusion

Accurate tooth segmentation on 3D dental models is an important task in computer-aided dentistry. In this paper, we mainly discuss how to improve the segmentation results of tooth-gingiva and tooth-tooth boundaries. The backbone of our proposed method is a graph-based network that establishes connectivity based on feature distances and utilizes several modules to encode local regions. To improve tooth-gingiva boundary segmentation, we incorporate an auxiliary loss to optimize the prediction results for tooth and gingiva. Additionally, we introduce a contrastive boundary-constrained loss to enhance model performance in tooth-tooth and tooth-gingiva boundaries. For further refinement of boundaries, a graph-cut post-processing step is applied after network prediction. The experimental results demonstrate that our approach achieves state-of-the-art performance on the 3D tooth segmentation task.

Acknowledgement. This research was supported by the Natural Science Fund of Hubei Province under Grant 2022CFB823, the HUST Independent Innovation Research Fund under Grant 2021XXJS096, the Alibaba Innovation Research program under Grant CRAQ7WHZ11220001-20978282, and grants from the Key Lab of Image Processing and Intelligent Control, Ministry of Education, China.

References

1. Cui, Z., et al.: TSegNet: an efficient and accurate tooth segmentation network on 3D dental model. Med. Image Anal. **69**, 101949 (2021)
2. Guo, M.H., Cai, J.X., Liu, Z.N., Mu, T.J., Martin, R.R., Hu, S.M.: PCT: point cloud transformer. Comput. Vis. Media **7**(2), 187–199 (2021)
3. Hao, J., et al.: Clinically applicable system for 3D teeth segmentation in intraoral scans using deep learning (2020)
4. He, X., et al.: Unsupervised pre-training improves tooth segmentation in 3-Dimensional intraoral mesh scans. In: International Conference on Medical Imaging with Deep Learning, pp. 493–507 (2021)
5. Herrmann, W.: On the completion of federation dentaire internationale specifications. Zahnarztliche Mitteilungen **57**(23), 1147–1149 (1967)
6. Li, G., Muller, M., Thabet, A., Ghanem, B.: DeepGCNs: can GCNs go as deep as CNNs? In: Proceedings of the IEEE/CVF International Conference on Computer Vision, pp. 9267–9276 (2019)
7. Li, Y., Bu, R., Sun, M., Wu, W., Di, X., Chen, B.: PointCNN: convolution on x-transformed points. In: Advances in neural information processing systems, vol. 31 (2018)
8. Li, Z., Liu, T., Wang, J., Zhang, C., Jia, X.: Multi-scale bidirectional enhancement network for 3D dental model segmentation. In: 2022 IEEE 19th International Symposium on Biomedical Imaging (ISBI), pp. 1–5. IEEE (2022)
9. Lian, C., et al.: Deep multi-scale mesh feature learning for automated labeling of raw dental surfaces from 3D intraoral scanners. IEEE Trans. Med. Imaging **39**(7), 2440–2450 (2020)
10. Ma, X., Qin, C., You, H., Ran, H., Fu, Y.: Rethinking network design and local geometry in point cloud: a simple residual MLP framework. arXiv preprint arXiv:2202.07123 (2022)
11. Maturana, D., Scherer, S.: VoxNet: a 3D convolutional neural network for real-time object recognition. In: 2015 IEEE/RSJ international conference on intelligent robots and systems (IROS), pp. 922–928. IEEE (2015)

12. Qi, C.R., Su, H., Mo, K., Guibas, L.J.: PointNet: deep learning on point sets for 3D classification and segmentation. In: Proceedings of the IEEE Conference on Computer Vision and Pattern Recognition, pp. 652–660 (2017)

13. Qi, C.R., Yi, L., Su, H., Guibas, L.J.: PointNet++: deep hierarchical feature learning on point sets in a metric space. In: Advances in Neural Information Processing Systems, vol. 30 (2017)

14. Qian, G., et al.: PointNeXt: revisiting PointNet++ with improved training and scaling strategies. Adv. Neural. Inf. Process. Syst. **35**, 23192–23204 (2022)

15. Riegler, G., Osman Ulusoy, A., Geiger, A.: OctNet: learning deep 3D representations at high resolutions. In: Proceedings of the IEEE Conference on Computer Vision and Pattern Recognition, pp. 3577–3586 (2017)

16. Tang, L., Zhan, Y., Chen, Z., Yu, B., Tao, D.: Contrastive boundary learning for point cloud segmentation. In: Proceedings of the IEEE/CVF Conference on Computer Vision and Pattern Recognition, pp. 8489–8499 (2022)

17. Tian, S., Dai, N., Zhang, B., Yuan, F., Yu, Q., Cheng, X.: Automatic classification and segmentation of teeth on 3D dental model using hierarchical deep learning networks. IEEE Access **7**, 84817–84828 (2019)

18. Wang, Y., Sun, Y., Liu, Z., Sarma, S.E., Bronstein, M.M., Solomon, J.M.: Dynamic graph CNN for learning on point clouds. ACM Trans. Graph. **38**(5), 1–12 (2019)

19. Wu, T.H., et al.: Two-stage mesh deep learning for automated tooth segmentation and landmark localization on 3D intraoral scans. arXiv preprint arXiv:2109.11941 (2021)

20. Wu, W., Qi, Z., Fuxin, L.: PointConv: deep convolutional networks on 3D point clouds. In: Proceedings of the IEEE/CVF Conference on Computer Vision and Pattern Recognition, pp. 9621–9630 (2019)

21. Xiong, H., et al.: TFormer: 3D tooth segmentation in mesh scans with geometry guided transformer. arXiv preprint arXiv:2210.16627 (2022)

22. Xu, X., Liu, C., Zheng, Y.: 3D tooth segmentation and labeling using deep convolutional neural networks. IEEE Trans. Visual Comput. Graphics **25**(7), 2336–2348 (2018)

23. Zanjani, F.G., et al.: Deep learning approach to semantic segmentation in 3D point cloud intra-oral scans of teeth. In: International Conference on Medical Imaging with Deep Learning, pp. 557–571. PMLR (2019)

24. Zanjani, F.G.: Mask-MCNet: tooth instance segmentation in 3D point clouds of intra-oral scans. Neurocomputing **453**, 286–298 (2021)

25. Zhang, L., et al.: TSGCNet: discriminative geometric feature learning with two-stream graph convolutional network for 3D dental model segmentation. In: Proceedings of the IEEE/CVF Conference on Computer Vision and Pattern Recognition, pp. 6699–6708 (2021)

26. Zhao, H., Jiang, L., Jia, J., Torr, P.H., Koltun, V.: Point transformer. In: Proceedings of the IEEE/CVF International Conference on Computer Vision, pp. 16259–16268 (2021)

27. Zhao, Y., et al.: 3D dental model segmentation with graph attentional convolution network. Pattern Recogn. Lett. **152**, 79–85 (2021)

28. Zheng, Y., Chen, B., Shen, Y., Shen, K.: TeethGNN: semantic 3D teeth segmentation with graph neural networks. IEEE Trans. Visual. Comput. Graphics **29**, 3158–3168 (2022)

FAST-Net: A Coarse-to-fine Pyramid Network for Face-Skull Transformation

Lei Zhao[1,2,3], Lei Ma[3], Zhiming Cui[3], Jie Zheng[1], Zhong Xue[2], Feng Shi[2], and Dinggang Shen[2,3,4(✉)]

[1] School of Information Science and Technology, ShanghaiTech University, Shanghai 201210, China
[2] Shanghai United Imaging Intelligence Co., Ltd., Shanghai 200232, China
dgshen@shanghaitech.edu.cn
[3] School of Biomedical Engineering, ShanghaiTech University, Shanghai 201210, China
[4] Shanghai Clinical Research and Trial Center, Shanghai 201210, China

Abstract. Face-skull transformation, i.e., shape transformation between facial surface and skull structure, has a wide range of applications in various fields such as forensic facial reconstruction and craniomaxillofacial (CMF) surgery planning. However, this transformation is a challenging task due to the significant differences between the geometric topologies of the face and skull shapes. In this paper, we propose a novel coarse-to-fine face-skull transformation network(i.e., FAST-Net) that has a pyramid architecture to gradually improve the transformation level by level. Specifically, using face-to-skull transformation for instance, in the first pyramid level, we use a point displacement sub-network to predict a coarse skull shape of point cloud from a given facial shape of point cloud with a skull template of point cloud as prior information. In the following pyramid levels, we further refine the predicted skull shape by first dividing the skull shape together with the given facial shape into different sub-regions, individually feeding the regions to a new sub-network, and merging the outputs as a refined skull shape. Finally, we generate a surface mesh model for the final predicted skull point cloud by non-rigidly registration with a skull template. Experimental results show that our method achieves the state-of-the-art performance on the task of face-skull transformation.

Keywords: Shape transformation · point cloud learning · face reconstruction · 3D face

1 Introduction

Face-skull transformation refers to the process of transforming shapes between the facial surface and the skull structure of human beings. It involves creating

---

L. Zhao and L. Ma—Equal Contributions.

© The Author(s), under exclusive license to Springer Nature Switzerland AG 2024
X. Cao et al. (Eds.): MLMI 2023, LNCS 14349, pp. 104–113, 2024.
https://doi.org/10.1007/978-3-031-45676-3_11

a facial likeness from a given skull structure and vice versa, which has wide applications in many fields, such as forensic facial reconstruction [10, 11], animation [3, 13] and Craniomaxillofacial (CMF) surgery planning [14]. For example, in forensics, face-skull transformation is a well-known technique used to reconstruct the face of an unknown person from her/his skeletal remains for identification [2, 10]. In the CMF surgery planning, it has been applied to predict normal skull structure from photos of patient with CMF trauma [14]. Despite its potentials, face-skull transformation still remains a challenging task. There are mainly two reasons: (1) both face and skull shapes have complex geometric structure; (2) There are significant differences between geometric topologies of face and skull shapes [7].

Previously, efforts have been made to address this challenging task. For example, Madsen et al. develops a probabilistic joint face-skull model by co-registering the statistical shape models (SSMs) of face and skull shapes [10]. Xiao et al. proposes a sparse-learning based method for face-skull transformation by constructing two dictionaries for normal face and skull models [14]. By sparse learning, a given facial shape can be represented as a weighted sum of normal faces in the face dictionary, and then the learned weights are applied to the skull dictionary, resulting in a new skull model corresponding to the given facial shape. However, these linear methods are based on an unrealistic assumption that people with similar facial shapes have similar skull shapes, which limits prediction accuracy of face and skull. More recently, a learning based-framework has been proposed to model the transformation between face and skull shapes [7]. This method is capable of achieving nonlinear shape transformation between face and skull by directly learning point displacements from a given shape to a target shape. However, it only roughly learns the shape-wise mapping between the entire facial and skull shapes, which unavoidably limits its prediction ability at the local regions.

In this study, we propose a novel FAce-Skull Transformation Network (FAST-Net) that models the face-skull shape relationship at local regions in a coarse-to-fine way. The proposed network has a pyramid structure. Specifically, taking face to bone transformation for example, a coarse skull shape is first predicted using a point displacement sub-network from a facial point cloud with a skull template shape in the first level. In the refinement level of the pyramid structure, the predicted skull shape is refined by dividing it together with the facial point cloud into different regions and predicting new skull sub-regions of point clouds from the divided regions. The predicted skull sub-regions are merged as a refined prediction of the skull shape. The point number and sub-region number of the skull shape are gradually increased from low to high level of the pyramid. By this pyramid design and region-wise transformation, the network can learn more global-to-local relationship between face and skull shapes, thus facilitating in more accurate skull shape prediction.

The contribution of this study is three-fold: (1) We propose a novel pyramid network for precise face-skull shape transformation; (2) In the refinement level, we further improve shape transformation by learning the region-wise transformation between face and skull; (3) Our method achieves state-of-the-art performance on this challenging task.

2   Method

In this section, we will take the skull-to-face transformation for instance to illustrate the details of the proposed FAST-Net. The framework of the proposed FAST-Net is illustrated in Fig. 1. Our FAST-Net adopts a pyramid structure, which gradually learns the global-to-local shape transformation between face and skull shapes. Through the FAST-Net, an accurate and dense skull shape of point cloud can be predicted from an input facial shape. A mesh model of the predicted skull shape is finally generated by non-rigidly registering a template skull model to the predicted skull point cloud [4].

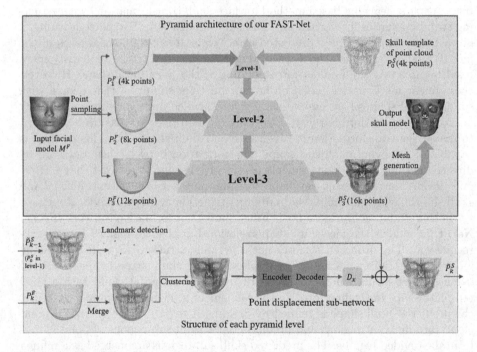

Fig. 1. Architecture of the proposed face-skull transformation network. The network has a pyramid structure, which consists of three levels. In each level of the pyramid, the predicted skull point cloud from the previous level is merged with a new sub-sampled one, and the merged point cloud is divided into sub-regions using KNN. A new skull point cloud can be predicted by separately feeding the sub-regions into network and merging the network outputs as the final result. Particularly, in the first level, the entire facial point cloud together with a skull template point cloud is directly used as input, which can also be regarded as a one-class region. The number of the sub-sampled face point clouds in the three levels are 4K,8K, and 12K, respectively.

2.1   Network Architecture

As shown in Fig. 1, the pyramid architecture of our FAST-Net consists of three levels. The backbone in each level is a PointMLP [9] based encoder-decoder (ED), which is a point displacement sub-network learning point displacements of the input point cloud.

In level-1 of the pyramid network, we coarsely estimate a skull shape of point cloud from a given facial shape. Specifically, given a face surface mesh model M^F, we first sub-sample a facial point cloud P_1^F from it. Then, we merge P_1^F with a skull point cloud P_0^S sub-sampled from a skull template M_T^S, and feed the merged point cloud to the point displacement sub-network in the level-1, outputting an initial skull shape of point cloud $\tilde{P}_1^S$. Due to significant difference between geometric topologies of face and skull shapes, directly transforming face shape to skull shape is difficult for the network. By introducing the prior shape information of the skull template to the input, the network can more easily handle the task of face-to-skull shape transformation, yielding better initial prediction. In this study, the point numbers of the point clouds P_1^F and P_0^S in the first level are both set to 4K. Therefore, the point number of the outputted skull point cloud $\tilde{P}_1^S$ is 8k.

In level-2 and level-3 of the pyramid, we gradually refine the initial skull shape $\tilde{P}_1^S$. These two levels share the same structure. In the refinement level k, we merge the skull point cloud $\tilde{P}_{k-1}^S$ predicted by the previous level with a new face point cloud P_k^F sub-sampled from M^F, resulting in a mixed point cloud P_k^{FS}. We further feed P_k^{FS} to a new point displacement sub-network to predict a refined skull shape of point cloud. However, unlike directly feeding it in level-1, we divide the merged point cloud P_k^{FS} into N_k different sub-regions and individually feed them to the new point displacement sub-network, for the purpose of improving the shape prediction in the local region and reducing the GPU memory consumption in the training. The outputted point clouds of different sub-regions are merged as a new complete skull point cloud $\tilde{P}_k^S$.

In the pyramid structure, we gradually increase the point number of the sub-sampled face point cloud $P_k^F, (k = 1, 2, 3)$ from the first level to the third level, to achieve a dense prediction of skull point cloud. In our implementation, the point numbers of face shape inputs P_1^F, P_2^F and P_3^F are set as 4K, 8K and 12K, respectively. In the meantime, the numbers of the sub-regions N_1, N_2 and N_3 in different levels are set to as 0, 5 and 10, respectively. Moreover, before feeding the sub-regions of point clouds to the sub-network, we sub-sample each sub-region of point cloud to a specific number with two purposes, i.e., simplifying data preparation and improving training performance. In level-2 and level-3, the point numbers of the sub-sampled sub-regions are set to 2.4k and 1.6k, respectively. Therefore, the point numbers of the outputted skull point clouds in level-2 and level-3 are 12k and 16k, respectively.

In this study, we use the K-Nearest Neighbor (KNN) algorithm to divide the face-skull point clouds by skull landmarks. Specifically, we predict skull landmarks from the predicted skull point cloud using a landmark detection network, which is also designed based on PointMLP [9]. The landmark detection network first estimates heatmaps of landmarks on the predicted point cloud, and then regresses landmarks from these heatmaps [12]. By taking the detected landmarks as cluster centers, the merged face-skull point cloud can be divided into different sub-regions.

2.2 Loss Function

The loss function used to train the point displacement sub-networks in each level consists of two components, i.e., shape loss and density loss [8].

Shape loss. Shape loss measures shape similarity between the predicted shape and its ground-truth shape. We employ the Chamfer Distance (CD) as the shape loss L_s.

$$L_s(S^p, S^{gt}) = \frac{1}{2N} \sum_{p \in S^{gt}} \min_{q \in S^p} d(p,q) + \frac{1}{2N} \sum_{q \in S^p} \min_{p \in S^{gt}} d(p,q), \qquad (1)$$

where S^p and S^{gt} denote the predicted shape and the ground-truth shape, respectively. N represents the number of points in the predicted shape. $d(p,q)$ represents the $L2$ distance between points p and q.

Density loss. Density loss is defined as the point density similarity between the predicted shape and its ground-truth shape [15].

$$L_d(S^p, S^{gt}) = \frac{1}{Nm} \sum_{p \in S^{gt}} \sum_{j=1}^{m} \left| d\left(p, O_j[S^{gt}, p]\right) - d\left(p, O_j[S^p, p]\right) \right|, \qquad (2)$$

where $O_j(p, S^{gt})$ denotes the j-th m-nearest points of point p in S^{gt}.

Therefore, the overall loss function for the point displacement sub-networks in each level is the sum of shape loss and density loss:

$$\mathcal{L} = L_s + \lambda L_d, \qquad (3)$$

where λ is the weight of the density loss.

3   Experiments and Results

3.1   Dataset

The dataset used in this study consists of computed tomography (CT) scans of 45 subjects. We first segmented face and skull from the head CT scans. Then, we reconstructed the face and skull mesh surfaces from the segmentation results by using the marching cubes algorithm [5]. We selected the first subject in the dataset as the template, and randomly divided the rest into training set (29 subjects), validation set (8 subjects) and testing set (7 subjects), respectively. In the meantime, we aligned all the subjects in the three sets to the selected template by rigid registration for effective training.

3.2   Implementation Details

We adopted farthest point sampling algorithm [15] as the point sub-sampling approach in this study. Data augmentation is performed by sub-sampling the input facial surface mesh five times following the method proposed in [7]. In the training stage, for each sub-network, we used Adam optimizer. We trained 400 epochs for the network at each level. The initial learning rate was set to 0.001 and then decayed by a factor of 0.8 every 50 epochs. The weight of the density loss λ was set to 0.1.

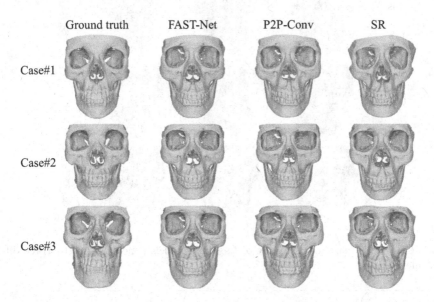

Fig. 2. Visual Comparisons between the skull models generated by the three different methods (i.e., FAST-Net, P2P-Conv and SR) and their ground truth models. Three representative subjects are selected for comparison.

3.3   Experimental Setups

In order to fully evaluate the performance of the proposed method, we performed qualitative and quantitative evaluation on the experimental results. First, we carried out qualitative evaluation by visually comparing the shapes generated by different methods, and presenting their color-coded surface distances from their corresponding ground-truth shapes. Hausdorff distance was used to calculate the surface distances between the predicted shapes and their ground-truth shapes. Then, we performed quantitative evaluation by calculating the shape errors of the predicted shapes with respect to their ground-truth shapes. Chamfer distance (CD) [6] and point mesh face distance (PMFD) [1] were adopted as metrics in the quantitative evaluation. CD is an evaluation metric to measure the bidirectional distance between two point clouds. For each point in each cloud, CD finds the nearest point in the other point set, and sums the square of the distances. PMFD and CD are calculated in the same way, except that PMFD calculates the distance between points and faces.

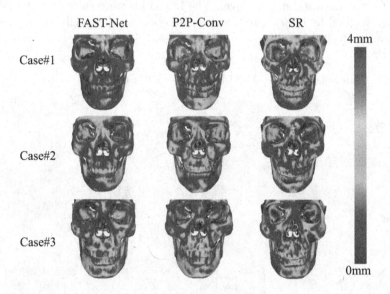

Fig. 3. Color-coded surface distances of the skull models generated by the three different methods.

To demonstrate the effectiveness of the proposed method, we compared our FAST-Net with the sparse learning based-method (SR) [14] and the P2P-Conv method [7]. In the meantime, we carried out ablation studies to evaluate the effectiveness of the important components in the proposed network.

3.4   Results

Qualitative Results: The qualitative and quantitative results of the face-to-skull transformation are as follows. Figure 2 shows visual comparison between the skull models generated by three different methods(i.e., our FAST-Net, P2P-Conv and SR) with respect to their corresponding ground truth skull models. Three representative cases are selected for visual comparison, from which one can easily observe that the skull models generated by our method is more accurate than the other two competing P2P-Conv and SR methods. Results of the color-coded surface distances shown in Fig. 3 also confirms this observation.

Table 1. Quantitative results of the evaluations on our FAST-Net, P2P-Conv and SR. CD and PMFD represents Chamfer distance and point mesh face distance,respectively.

Method	CD (mm)	PMFD (mm)
SR [14]	3.212 ± 0.202	2.992 ± 0.196
P2P-Conv [7]	2.859 ± 0.228	2.632 ± 0.224
FAST-Net	2.701 ± 0.142	2.470 ± 0.148

Quantitative Results: Quantitative results of the three different methods are summarized in Table 1. Compared with P2P-Conv and SR methods, our FAST-Net achieved the best performance in both metrics of CD and PMFD. Moreover, the improvements are statistically significant($P< 0.05$). This result shows the superiority of our method compared with state-of-the-art methods.

Results of Ablation Studies: First, we validated the effectiveness of introducing a skull template as prior information in level-1. Experimental results of level-1 with and without skull template in Table 2 have proved that the skull template input in level-1 can benefit the initial skull prediction. Second, we evaluated the effectiveness of the region division in each pyramid level by training and testing FAST-Net with and without region division. Results of FAST-Net and FAST-Net without region division shown in Table 2 reveal that the proposed region division is effective in improving the face-to-skull transformation. Finally, we conducted experiments to demonstrate the effectiveness of the proposed pyramid architecture in our FAST-Net. According to the results shown in Table 2, as the level increases, the shape error of the skull gradually decreases, which verifies the effectiveness of our pyramid structure.

Table 2. Quantitative results of the ablation studies on introducing prior skull template, region division and pyramid structure.

Method	CD(mm)	PMFD(mm)
level-1 w/o skull template	2.837 ± 0.223	2.599 ± 0.219
FAST-Net w/o region division	2.797 ± 0.183	2.562 ± 0.176
level-1	2.775 ± 0.202	2.546 ± 0.207
level-1+level-2	2.738 ± 0.164	2.511 ± 0.171
level-1+level-2+level-3 (FAST-Net)	2.701 ± 0.142	2.470 ± 0.148

4 Conclusions and Future Works

In this paper, we have proposed a novel face-skull transformation network, FAST-Net. The architecture of the FAST-Net is a pyramid architecture with three levels, where we gradually improve the transformation level by level. Further, we improve the face-skull shape transformation by performing region-wise transformation in each level. Experimental results on real human subjects show that our FAST-Net can achieve better performance compared with the state-of-the-art P2P-Conv and SR methods. Ablation results have also confirmed the effectiveness of important components in our proposal. However, to improve this work, there are still future works to do: (1) we will further evaluate the performance of our FAST-Net on the skull-to-face transformation in future study; (2) We will collect more data to train FAST-Net models for better generalization ability.

Acknowledgement. This work was supported in part by The Key R&D Program of Guangdong Province, China (grant number 2021B0101420006), National Natural Science Foundation of China (grant number 62131015), and Science and Technology Commission of Shanghai Municipality (STCSM) (grant number 21010502600).

References

1. Aspert, N., Santa-Cruz, D., Ebrahimi, T.: MESH: measuring errors between surfaces using the hausdorff distance. In: Proceedings of the IEEE International Conference on Multimedia and Expo, vol. 1, pp. 705–708. IEEE (2002)
2. Guleria, A., Krishan, K., Sharma, V., Kanchan, T.: Methods of forensic facial reconstruction and human identification: historical background, significance, and limitations. Sci. Nat. **110**(2), 8 (2023)
3. Ichim, A.E., Kadleček, P., Kavan, L., Pauly, M.: Phace: physics-based face modeling and animation. ACM Trans. Graph. (TOG) **36**(4), 1–14 (2017)
4. Li, Y., Harada, T.: Non-rigid point cloud registration with neural deformation pyramid. arXiv preprint arXiv:2205.12796 (2022)
5. Lorensen, W., Cline, H.: Marching cubes: a high resolution 3D surface construction algorithm. ACM SIGGRAPH Comput. Graph. **21**, 163 (1987)
6. Ma, L., et al.: Deep simulation of facial appearance changes following craniomaxillofacial bony movements in orthognathic surgical planning. In: de Bruijne, M., et al. (eds.) MICCAI 2021. LNCS, vol. 12904, pp. 459–468. Springer, Cham (2021). https://doi.org/10.1007/978-3-030-87202-1_44
7. Ma, L., et al.: Bidirectional prediction of facial and bony shapes for orthognathic surgical planning. Med. Image Anal., 102644 (2022)
8. Ma, L., et al.: Simulation of postoperative facial appearances via geometric deep learning for efficient orthognathic surgical planning. IEEE Trans. Med. Imaging (2022)
9. Ma, X., Qin, C., You, H., Ran, H., Fu, Y.: Rethinking network design and local geometry in point cloud: a simple residual MLP framework. arXiv preprint arXiv:2202.07123 (2022)
10. Madsen, D., Lüthi, M., Schneider, A., Vetter, T.: Probabilistic joint face-skull modelling for facial reconstruction. In: Proceedings of the IEEE Conference on Computer Vision and Pattern Recognition, pp. 5295–5303 (2018)
11. Valsecchi, A., Damas, S., Cordón, O.: A robust and efficient method for skull-face overlay in computerized craniofacial superimposition. IEEE Trans. Inf. Forensics Secur. **13**(8), 1960–1974 (2018)

12. Wang, Y., Cao, M., Fan, Z., Peng, S.: Learning to detect 3D facial landmarks via heatmap regression with graph convolutional network (2022)
13. Wu, T., Hung, A., Mithraratne, K.: Generating facial expressions using an anatomically accurate biomechanical model. IEEE Trans. Vis. Comput. Graph. **20**(11), 1519–1529 (2014)
14. Xiao, D., et al.: Estimating reference shape model for personalized surgical reconstruction of craniomaxillofacial defects. IEEE Trans. Biomed. Eng. **68**(2), 362–373 (2020)
15. Yin, K., Huang, H., Cohen-Or, D., Zhang, H.: P2P-NET: bidirectional point displacement net for shape transform. ACM Trans. Graph. (TOG) **37**(4), 1–13 (2018)

Mixing Histopathology Prototypes into Robust Slide-Level Representations for Cancer Subtyping

Joshua Butke[1]([✉]), Noriaki Hashimoto[2], Ichiro Takeuchi[2,3], Hiroaki Miyoshi[4], Koichi Ohshima[4], and Jun Sakuma[2,5]

[1] Machine Learning and Data Mining Lab, University of Tsukuba, Tsukuba, Japan
`butkej@mdl.cs.tsukuba.ac.jp`
[2] RIKEN Center for Advanced Intelligence Project, Tokyo, Japan
[3] Department of Mechanical Systems Engineering, Nagoya University, Nagoya, Japan

[4] Department of Pathology, Kurume University, Kurume, Japan
[5] Department of Computer Science, Tokyo Institute of Technology, Tokyo, Japan

Abstract. Whole-slide image analysis via the means of computational pathology often relies on processing tessellated gigapixel images with only slide-level labels available. Applying multiple instance learning-based methods or transformer models is computationally expensive as, for each image, all instances have to be processed simultaneously. The MLP-Mixer is an under-explored alternative model to common vision transformers, especially for large-scale datasets. Due to the lack of a self-attention mechanism, they have linear computational complexity to the number of input patches but achieve comparable performance on natural image datasets. We propose a combination of feature embedding and clustering to preprocess the full whole-slide image into a reduced prototype representation which can then serve as input to a suitable MLP-Mixer architecture. Our experiments on two public benchmarks and one inhouse malignant lymphoma dataset show comparable performance to current state-of-the-art methods, while achieving lower training costs in terms of computational time and memory load. Code is publicly available at https://github.com/butkej/ProtoMixer.

Keywords: Clustering · MLP-Mixer · Computational Pathology

1 Introduction

Whole-slide images (WSIs) are digital files of pathology tissue glass slides, often with gigapixel resolution. The advent of digital slide scanning technology and their increasing ubiquity in many medical facilities, lead to the amount of available WSIs rising drastically. On the basis of data availability there has been a fast advancement in the field of computational pathology (CPATH), applying various computer vision methods for patient specimen (e.g. tissue or individual cells) analysis [2]. Currently, machine learning methods in the CPATH community are

X. Cao et al. (Eds.): MLMI 2023, LNCS 14349, pp. 114–123, 2024.
https://doi.org/10.1007/978-3-031-45676-3_12

often based on multiple instance learning (MIL), in which each individual WSI is regarded as a bag of thousands of patched instances with a single, patient-level, coarse-grained label [4,10]. To improve state-of-the-art performance for subtyping classification problems there are various additions to the base MIL framework of embedding all instances per bag into a feature representation and then classifying a pooled global feature vector, e.g. hard negative mining [1] or instance-level clustering to constrain and refine the feature space during training [15].

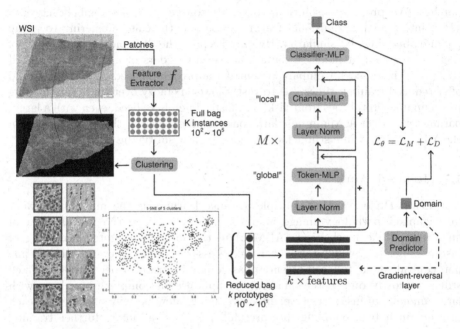

Fig. 1. Overview of our proposed end-to-end framework combining feature embedding and clustering of WSIs into reduced prototypes representations (**left** side) and then feeding the resulting input table into a domain-adversarial Mixer model (**right** side).

Transformer-based models, such as Vision Transformer [5], have quickly gained popularity in most of the computer vision community but their application to CPATH image analysis is still lacking as they are constrained by their fixed input sequence length as well as their prohibitive computational cost due to the self-attention mechanism. The MLP-Mixer model [19] was proposed as a step away from the inductive biases of convolution in CNNs as well as self-attention found in transformer-based models. Instead of those, the model uses only multilayer perceptrons (MLP), the core building block of most neural networks, and applies them either independently to image patches or across patches, effectively operating on a table of "patches × channels" and its transposition. Despite this simplicity, it showed state-of-the-art performance for natural image classification while enabling faster inference and lower computational complexity.

However, the Mixer architecture is originally designed for 16×16 patches from a 224×224 image, similar to the Vision Transformer. Applying Mixer within the MIL framework on bags with tens of thousands of patches is not computationally possible. Thus, we pose the question, if it is possible to apply the Mixer architecture to the CPATH realm of gigapixel images, with the goal of competitive predictive performance while achieving faster training times and lower memory overhead, by reducing a bag down to its essential, describing elements.

Therefore we hypothesize that it is not necessary to use all patches of a given bag, but it is sufficient to use information from the most representative patches. We propose a solution, called ProtoMixer, by first embedding each patch into a high-dimensional feature space and then use clustering to group patch embeddings by similarity. By only keeping the centroid (or prototype) of each cluster we can reduce a dataset by several orders of magnitude per WSI ($10^5 \mapsto 10^1$ instances). Compared to current approaches based on MIL this would offer reduced training times due to vastly lower memory overhead (only loading precomputed prototypes from disk) and faster inference speed, even with a larger parameter budget as the input data space is much smaller than typical bags in the MIL setting, while still achieving good cancer subtyping accuracy.

1.1   Related Works

Attention-Based MIL. Multiple instance learning is the major framework for gigapixel pathology image analysis, especially since the introduction of an attention-like mechanism (ABMIL) in the form of weighted averaging of instances per bag, which takes over the necessary pooling operation [10]. While not equivalent to full self-attention as found in transformer models (which still is to costly due to the quadratic computational complexity on bags with large amounts of instances), weighted averaging offers individual scores for each instance in a bag, enabling interpretability as well as many further training modifications [12].

One of the most prominent, state-of-the-art applications currently is CLAM (Clustering constrained Attention MIL) [15]. CLAM introduced preextracting feature vectors for each patch in a WSI with a pretrained CNN, as well as a multi-class subtyping capable ABMIL. Further enhanced by instance-level clustering within the model to encourage the learning of class specific features, a SVM loss is used together with pseudo-generated instance labels from attention scores.

ReMix Framework. Similar to our proposed method, the ReMix framework [21] first reduces a dataset by embedding patches with a pretrained feature extractor and clustering them and then mixes new, augmented bags from the resulting prototypes. Mixing in this case refers to four different augmentation methods. They found that clustering works as, when learned properly, the representation space is shown to be meaningful and distance metrics can show similarity between related patches. From this setting, the authors subsequently apply various MIL methods, as the framework is model agnostic in this regard and could demonstrate superior performance when compared to their chosen base-lines as well as vastly reduced training costs due the generally low parameter count of MIL method and the reduced dataset size.

2   Method

This work follows standard multiple instance learning conventions, in which a dataset is composed of L bags and formulated as $D = [(B_i, Y_i)]_{i=1}^{L}$, where $B_i = \{X_K\}_{j=1}^{L_K}$ is the i-th bag consisting of K instances (e.g. tessellated patches extracted from a WSI) and a singular bag label Y_i.

Prototype Representation. We reduce a dataset by combining feature embed-ding with clustering. For each bag B_i in our dataset and all K instances within we compute a high-dimensional feature vector $\mathbf{x} \in \mathbb{R}^{1 \times N}$ with a pretrained model, where N is the embedding dimension. Afterwards, any suitable cluster-ing method can be applied to reduce the dataset size. Here we stick with k-means clustering [14]. This enables the construction of a reduced bag, consisting only of the cluster prototypes (or centroids), overall compressing the dataset by several orders of magnitude ($10^5 \mapsto 10^1$).

Intuitions. We argue that the proposed data preprocessing still offers enough information at the prototype level for the Mixer model to build robust whole-slide-level representations for cancer subtyping tasks. Here, we offer several intu-itions on its effectiveness as observed in our experiments:

- **patch-level** By embedding patches to feature vectors we are able to extract information at the lowest level that is often repeated (similar patches).
- **local-level** Clustering all patches and only keeping the prototypes represents local structures such as tissue component classes (e.g. connective tissue). By mixing at the channel-level, the model can learn about each prototype com-ponent.
- **global-level** Finally, there is the token-mixing that acts similar to established self-attention operation and disperses prototype information across different channels to learn a holistic view at the whole-slide level.

WSIs often suffer from tissue imbalance, wherein a large number of simi-lar patches convey redundant information, potentially overshadowing minor-ity patches. Breaking the bag down into prototypes bridging the numerical gap between majority and minority patches, we mitigate this problem to some extent. Additionally, employing the mean of embeddings of similar patch groups enhances tissue representation by providing a less noisy depiction.

MLP-Mixer. Following, we briefly describe the original backbone of MLP-Mixer and how it was adapted to reduced bags of prototypes. MLP-Mixer stacks M isotropic Mixer layers of the same size and with the exact same internal configuration. Each consists of a token-mixing MLP and a channel-mixing MLP. Other classic components include: skip-connections, dropout, and a fully-connected classifier head (cf. Fig. 1). Originally developed for computer vision tasks, input images are tessellated into non-overlapping patches and then unrolled into feature vectors by a per-patch fully connected layer [19].

Since we already obtained feature vectors per patch via the previous embedding step and then reduce our bag of instances to k prototypes we automatically gain the desired input table of $\mathbf{X} = [x_1, ..., x_k] \in \mathbb{R}^{k \times N}$, where k is the chosen amount of clusters to compute and N is the feature dimensionality. The first MLP block is token-mixing and operates on columns of $\mathbf{X}$ (so on the transposed input table $\mathbf{X}^\mathrm{T}$); the second is channel-mixing and operates of rows of $\mathbf{X}$. D_S and D_C are tunable hidden widths in the token-mixing and channel-mixing MLPs. D_S is selected independently of the number of input patches. Therefore, the computational complexity of the model is linear to the number of input patches, unlike ViT whose complexity is quadratic.

$$\mathbf{Y}_{*,i} = \mathbf{X}_{*,i} + \sigma(\mathbf{W}_1 \mathrm{LayerNorm}(\mathbf{X})_{*,i})\mathbf{W}_2 \text{ for } i = 1...C,$$
$$\mathbf{Z}_{j,*} = \mathbf{Y}_{j,*} + \sigma(\mathbf{W}_3 \mathrm{LayerNorm}(\mathbf{Y})_{j,*})\mathbf{W}_4 \text{ for } j = 1...S \tag{1}$$

Thus, each MLP block contains two fully-connected layers ($\mathbf{W}$) with a GELU nonlinearity between them (σ) [9]. After all M Mixer layers, k features are generated and average-pooled into a global vector representation of size N. A final Classifier-MLP computes the class logits.

Domain-Adversarial Training. In pathological image datasets, there is often a large variety of staining conditions observed. When similar staining conditions exist only in a specific class or for a small subset of samples, the trained classification model tends to predict class based on the difference in staining conditions. To decouple the variance in staining intensity from the raw data, stain normalization techniques are often used [17]. However, recently domain-adversarial learning has been used to remove the domain information (which also includes the stain variance) from the model representation [6,11]. Modern works have shown that domain-adversarial training can augment CPATH algorithms further than standard colour augmentation or stain normalization [7]. Methodically, a second branch that predicts the probability of belonging to a certain domain can be added relatively model-agnostic (cf. Fig. 1).

$$\text{Optimization of subtyping classifier } \theta_M \leftarrow \theta_M - \lambda_M \frac{\partial \mathcal{L}_M}{\partial \theta_M} \tag{2}$$

$$\text{Optimization of domain classifier } \theta_D \leftarrow \theta_D - \lambda_D \frac{\partial \mathcal{L}_D}{\partial \theta_D} \tag{3}$$

$$\text{Adversarial update of subtyping classifier } \theta_M \leftarrow \theta_M + \alpha \lambda_D \frac{\partial \mathcal{L}_D}{\partial \theta_M} \tag{4}$$

The parameters θ_M are updated for the subtyping task (by minimizing $\mathcal{L}_M$), and with the adversarial update, the same parameters are updated to prevent the domain of origin to be recovered from the learned representation. We add domain-adversarial training to our framework, expecting that removing the domain information that is still present at the level of patches, embeddings and prototypes, can help to learn robust slide-level representations from just prototype information.

ProtoMixer Framework. Overall, our work proposes a framework we term ProtoMixer (Fig. 1): From a WSI we extract feature embeddings for all segmented patches with output feature dimensionality N and then group these into k clusters, keeping only the prototypes and saving them to disk. Each WSI serves as $k \times N$ input into a Mixer model. Domain-adversarial training with a secondary MLP branch decouples domain information from the learned WSI representation.

3   Experiments

Implementation Details. Each digital slide is fed to a pipeline for automated segmentation and patch tessellation. A binary mask for tissue regions is generated by thresholding the the image after median blurring for edge smoothing, followed by morphological closing to fill artifacts and holes. After segmentation, 224×224 patches are extracted from the foreground contours at a specified magnification level (e.g. 40× or 20×) and saved together with their coordinates.

Similar to many established methods in modern computational pathology algorithms, each patch extracted from a WSI is embedded to a 1024-dimensional feature vector by an ImageNet pretrained, truncated ResNet-50 [3,8]. Patch embedding in this way has become standard practice as it allows to process a WSI much easier on commonly available GPUs, leading to generally faster training times and lower computational costs.

The choice of k and thus the number of prototypes per WSI is a very important hyperparameter in this framework. Similar to the ReMix idea, we sweep over a selection of $\{1, 2, 4, 5, 6, 8, 10, 12, 16\}$ for k, for each dataset and analyze meaningful representation of prototypes and assigned clusters in a two-dimensional t-SNE embedding [16] as well as predictive performance. We choose $k = 5$ for TCGA-RCC and inhouse-Lymphoma datasets and $k = 8$ for CAMELYON.

For the Mixer model, we choose the dimensionality of fully-connected layers to be $D_S = 1024 \, (= N)$, $D_C = 2048$, with the number of stacked Mixer blocks $M = 12$. The domain-adversarial branch running in parallel during training is a three-layer MLP acting as the domain predictor. It transforms the input data space into a domain logit, where the size is equal to the number of individual WSIs in a given dataset. A gradient-reversal layer then modifies the total model loss for each epoch. The domain regularization parameter λ is updated each epoch as $\lambda = \frac{2}{1+\exp(-10r)}$, with $r = \frac{\text{current epoch } e}{\text{total epochs } E} \, \alpha$, inspired by [7].

Table 1. Experimental results of model subtyping performance in terms of macro-F1 score and AUROC on three different datasets. 5-fold cross validation averaged over 5 independent runs. Gray result subset shows available literature results from [20] for CAMELYON16 and TCGA-RCC.

| | Dataset | | | | | |
| | CAMELYON | | TCGA-RCC | | inhouse-Lymphoma | |
	macro-F1	AUROC	macro-F1	AUROC	macro-F1	AUROC
ABMIL [10]	85.00±1.80	90.03±1.30	82.80±1.80	95.90±1.00	N/A	N/A
TransMIL [18]	78.70±3.30	83.80±4.70	85.00±0.70	97.20±0.30	N/A	N/A
DTFD-MIL [22]	85.80±1.70	93.30±0.90	88.40±2.80	97.60±0.90	N/A	N/A
CLAM	**78.72**±0.50	**83.33**±0.69	84.36±0.52	95.90±0.71	70.18±0.38	86.41±0.13
CLAM+prttyps	54.47±0.77	72.30±0.16	23.76±0.00	49.22±0.13	72.29±0.15	86.07±0.02
ReMix+ABMIL	72.26±0.91	78.12±1.04	86.97±0.22	96.34±0.38	78.87±0.25	89.94±0.39
ProtoMixer	72.71±1.01	76.81±1.21	**89.72**±0.47	**97.51**±0.12	**82.67**±0.33	**93.68**±0.18

Datasets. We evaluate our approach on three different datasets. Two are well established benchmark datasets commonly found in current computational pathology research: **CAMELYON** and **TCGA-RCC**. The TCGA-RCC dataset consists of 884 diagnostic WSIs available from different *The Cancer Genome Atlas* programs for renal cell carcinoma. There are three different subtypes made up of 111 slides from 99 cases for chromophobe carcinoma (TCGA-KICH), 489 slides from 483 cases for clear cell carcinoma (TCGA-KIRC) and 284 slides from 264 cases for papillary cell carcinoma. The mean number of extracted patches at 40× magnification is 51496. CAMELYON16 and CAMELYON17 are two competition datasets for breast cancer lymph node metastasis detection [13]. As is convention, these two are often combined into a single dataset for binary prediction, consisting of 899 slides (591 negative and 308 positive) from 370 cases. The mean number of extracted patches per WSI at 40× magnification is 45610.

A third dataset is the **inhouse-Lymphoma** dataset consisting of 1441 WSIs of malignant lymphoma origin with 3 classes (diffuse large B-cell lymphoma (DLBCL); follicular lymphoma (FL); and Reactive subtype). The mean number of extracted patches per WSI at 40× magnification is 14523.

Baselines. We compare against two benchmark baselines in the computational pathology community: CLAM [15] and the ReMix framework [21] with ABMIL as the downstream model. We use the reported settings for these methods. Since the CAMELYON and TCGA-RCC datasets are popular benchmarks in the CPATH community we show additional literature values for different methods to compare against (cf. Table 1, gray subset).

Main Results. Experimental results regarding subtyping performance across the presented datasets are shown in Table 1. We perform 5-fold stratified cross validation for all experiments and report the average and its standard deviation

Table 2. Training costs and run times for different models in comparison to our proposed method on the lymphoma dataset.

	Parameters (M)	Memory Load	Seconds/Epoch
CLAM	0.9	2737.38 MiB	36.47
CLAM+prttyps	0.9	8.43 MiB	4.80
ReMix+ABMIL	0.5	7.46 MiB	4.49
ProtoMixer	14.1	2068.93 MiB	31.02

of macro-F1 score and area under the receiver operating curve (AUROC) across 5 runs. All models are trained for 150 epochs.

From these main results we can see that our proposed ProtoMixer can attain competitive predictive performance, especially on the TCGA-RCC and lymphoma datasets. For the combined CAMELYON dataset there is a drop in performance, which we argue might be related to the low amount of cancer tissue found per WSI. Future research could be directed towards performance for WSIs with low amounts of tumor, e.g. sample multiple adjacent prototypes per cluster instead of only one. An ablation study combining CLAM with the prototypes (CLAM+prttyps) as input fails to achieve good predictive performance, as the CLAM model is not designed to work with only the limited information of prototypes. For CLAM and other ABMIL-related methods larger bags are necessary to make the attention mechanism worthwhile. The Mixer architecture can work on only a select few prototypes and due to their token and channel mixing operation achieve a good whole slide representation in the absence of attention.

Training Costs. Standard MIL methods such as CLAM or ABMIL have a relatively large computational overhead as each bag is processed separately and as a whole. CLAM alleviates this already slightly by precomputing feature vectors ahead of model training. In our proposed case, we achieve an even greater reduction outside of training due to only loading k prototypes per WSI. Regarding the training costs, we report average duration per training epoch and computational memory load in Table 2. To ensure fair comparison we perform all budgeting experiments on a single Nvidia A100 GPU with a batch size of 1 and the same number of prototypes ($k = 5$). Similar to the original ReMix framework [21] we also achieve lowered training costs in terms of runtime per epoch during training as well as memory load by utilizing only prototypes as input to our model, even if the parameter budget of Mixer is substantially larger than CLAM or standard attention-based MIL.

4   Conclusion

This work presents, to our knowledge, the first application of an MLP-Mixer-like architecture to CPATH. Using a combination of feature embedding and clustering for preprocessing enables the use of this model with whole slide data. Although reducing the dataset and discarding much information, Mixer is still able to build good slide-level feature representations for cancer subtyping. Our implementation can compete with current state-of-the-art approaches such as CLAM on major benchmark datasets with competitive predictive performance, 24% memory load and 15% computational time reduction. One drawback of using our proposed method is the lack of interpretability: without attention scores or similar insight available, the decision reached by the model becomes more obtuse.

Acknowledgements. This work was supported by JST CREST JPMJCR21D3 and Grant-in-Aid for Scientific Research (A) 23H00483. J.B. was supported by the Gateway Fellowship program of Research School, Ruhr-University Bochum, Bochum, Germany.

References

1. Butke, J., Frick, T., Roghmann, F., El-Mashtoly, S.F., Gerwert, K., Mosig, A.: End-to-end multiple instance learning for whole-slide cytopathology of urothelial carcinoma. In: MICCAI Workshop on Computational Pathology, pp. 57–68. PMLR (2021)
2. Cui, M., Zhang, D.Y.: Artificial intelligence and computational pathology. Lab. Invest. **101**(4), 412–422 (2021)
3. Deng, J., Dong, W., Socher, R., Li, L.J., Li, K., Fei-Fei, L.: ImageNet: a large-scale hierarchical image database. In: 2009 IEEE Conference on Computer Vision and Pattern Recognition, pp. 248–255. IEEE (2009)
4. Dietterich, T.G., Lathrop, R.H., Lozano-Pérez, T.: Solving the multiple instance problem with axis-parallel rectangles. Artif. Intell. **89**(1–2), 31–71 (1997)
5. Dosovitskiy, A., et al.: An image is worth 16x16 words: transformers for image recognition at scale. arXiv preprint arXiv:2010.11929 (2020)
6. Ganin, Y., et al.: Domain-adversarial training of neural networks. J. Mach. Learn. Res. **17**(1), 2030–2096 (2016)
7. Hashimoto, N., et al.: Multi-scale domain-adversarial multiple-instance CNN for cancer subtype classification with unannotated histopathological images. In: Proceedings of the IEEE/CVF Conference on Computer Vision and Pattern Recognition, pp. 3852–3861 (2020)
8. He, K., Zhang, X., Ren, S., Sun, J.: Deep residual learning for image recognition. In: Proceedings of the IEEE Conference on Computer Vision and Pattern Recognition, pp. 770–778 (2016)
9. Hendrycks, D., Gimpel, K.: Gaussian error linear units (gelus). arXiv preprint arXiv:1606.08415 (2016)
10. Ilse, M., Tomczak, J., Welling, M.: Attention-based deep multiple instance learning. In: International Conference on Machine Learning, pp. 2127–2136. PMLR (2018)
11. Lafarge, Maxime W.., Pluim, Josien P. W.., Eppenhof, Koen A. J.., Moeskops, Pim, Veta, Mitko: Domain-adversarial neural networks to address the appearance variability of histopathology images. In: Cardoso, M.J., et al. (eds.) DLMIA/ML-CDS -2017. LNCS, vol. 10553, pp. 83–91. Springer, Cham (2017). https://doi.org/10.1007/978-3-319-67558-9_10
12. Li, X., et al.: Deep learning attention mechanism in medical image analysis: basics and beyonds. Int. J. Netw. Dyn. Intell. **2**, 93–116 (2023)

13. Litjens, G., et al.: 1399 H&E-stained sentinel lymph node sections of breast cancer patients: the Camelyon dataset. GigaScience **7**(6), giy065 (2018)
14. Lloyd, S.: Least squares quantization in PCM. IEEE Trans. Inf. Theory **28**(2), 129–137 (1982)
15. Lu, M.Y., Williamson, D.F., Chen, T.Y., Chen, R.J., Barbieri, M., Mahmood, F.: Data-efficient and weakly supervised computational pathology on whole-slide images. Nat. Biomed. Eng. **5**(6), 555–570 (2021)
16. Van der Maaten, L., Hinton, G.: Visualizing data using t-SNE. J. Mach. Learn. Res. **9**(11), 2579–2605 (2008)
17. Macenko, M., et al.: A method for normalizing histology slides for quantitative analysis. In: 2009 IEEE international symposium on biomedical imaging: from nano to macro, pp. 1107–1110. IEEE (2009)
18. Shao, Z., Bian, H., Chen, Y., Wang, Y., Zhang, J., Ji, X., et al.: TransMIL: transformer based correlated multiple instance learning for whole slide image classification. Adv. Neural. Inf. Process. Syst. **34**, 2136–2147 (2021)
19. Tolstikhin, I.O., et al.: MLP-mixer: an all-MLP architecture for vision. Adv. Neural. Inf. Process. Syst. **34**, 24261–24272 (2021)
20. Xiong, C., Chen, H., Sung, J., King, I.: Diagnose like a pathologist: transformer-enabled hierarchical attention-guided multiple instance learning for whole slide image classification. arXiv preprint arXiv:2301.08125 (2023)
21. Yang, J., et al.: ReMix: a general and efficient framework for multiple instance learning based whole slide image classification. In: Wang, L., Dou, Q., Fletcher, P.T., Speidel, S., Li, S. (eds.) Medical Image Computing and Computer Assisted Intervention – MICCAI 2022. MICCAI 2022. LNCS, vol. 13432. Springer, Cham (2022). https://doi.org/10.1007/978-3-031-16434-7_4
22. Zhang, H., et al.: DTFD-MIL: double-tier feature distillation multiple instance learning for histopathology whole slide image classification. In: Proceedings of the IEEE/CVF Conference on Computer Vision and Pattern Recognition, pp. 18802–18812 (2022)

Consistency Loss for Improved Colonoscopy Landmark Detection with Vision Transformers

Aniruddha Tamhane, Daniel Dobkin, Ore Shtalrid, Moshe Bouhnik,
Erez Posner[(✉)], and Tse'ela Mida

Intuitive Surgical, Inc., 1020 Kifer Road, Sunnyvale, CA, USA
erez.posner@intusurg.com

Abstract. Colonoscopy is a procedure used to examine the colon and rectum for colorectal cancer or other abnormalities including polyps or diverticula. Apart from the actual diagnosis, manually processing the snapshots taken during the colonoscopy procedure (for medical record keeping) consumes a large amount of the clinician's time. This can be automated through post-procedural machine learning based algorithms which classify anatomical landmarks in the colon. In this work, we have developed a pipeline for training vision-transformers for identifying anatomical landmarks, including appendiceal orifice, ileocecal valve/cecum landmark and rectum retroflection. To increase the accuracy of the model, we utilize a hybrid approach that combines algorithm-level and data-level techniques. We introduce a consistency loss to enhance model immunity to label inconsistencies, as well as a semantic non-landmark sampling technique aimed at increasing focus on colonic findings. For training and testing our pipeline, we have annotated 307 colonoscopy videos and 2363 snapshots with the assistance of several medical experts for enhanced reliability. The algorithm identifies landmarks with an accuracy of 92% on the test dataset.

Keywords: Colonoscopy · Vision Transformer · Landmark Detection · Self-supervised learning · Consistency loss · Data sampling

1 Introduction

Colorectal cancer (CRC) is the third most common cancer worldwide, with an estimated 1.9 million new cases and 930,000 deaths in 2020 [20]. Colonoscopy is an effective diagnostic method for significantly reducing the incidence rate (by up to 52%) and the mortality rate (by up to 62%) of CRC [31]. The American Gastroenterological Institute has defined standards and metrics such as the withdrawal time and the documentation of snapshots of key colon landmarks such as the Appendiceal Orifice (AO), Ileocecal valve (ICV), Cecum landmark (Cec) or Rectum retroflexion (RecRF), for assessing the quality of the colonoscopy [7]. Specifically, the snapshots serve as a visual summary of the patient's colonic

X. Cao et al. (Eds.): MLMI 2023, LNCS 14349, pp. 124–133, 2024.
https://doi.org/10.1007/978-3-031-45676-3_13

health, as well as a measure of the colonoscopy's extent [19]. The snapshots captured during the colonoscopy are presently subjected to a manual annotation process by medical experts subsequent to the procedure's completion. This can potentially cause inefficiency to the process [18] and introduce subjective bias when evaluating the procedure.

Landmark detection in the colorectal or the gastro-intestinal tract is a relatively newer area of research, compared to anomaly detection tasks such as polyp detection [17,22,24]. Detecting the landmarks used for colonoscopy quality indicators with vision transformers was previously explored in [28] as an image classification task. In [16], the same task was approached as ego-motion classification. The relative scarcity of existing research in this area may be due to a lack of annotated data and an inherent subjectivity in the task itself.

In our work, we formulate landmark detection as an image classification task and use a vision transformer as the feature extractor. We incorporate a self-supervised consistency loss pipeline [11] for training the vision transformer to detect colon landmarks from video snapshots, as described in Sect. 4.3. We have also devised a class-wise uniform sampling mechanism described in Sect. 4.2 that effectively tackles two types of class imbalances: inter-class imbalance, which refers to the difference in the prior distribution of frames containing anatomical and non-anatomical landmarks (such as AO, Cec, RecRF vs. non-landmarks), and intra-class imbalance, which stems from the presence of abnormalities (primarily in the non-landmark images), such as polyps and diverticula. We demonstrate the efficacy of both self-supervised training and the data sampling methodology in terms of recall, precision and accuracy.

2   Related Work

In this section we briefly review key prior research that informed us in our work.

Landmark Detection: Landmark detection is a medical task performed on modalities other than colonoscopy as well, such as ultrasound, CT scan and endoscopy videos [6,13,32,33]. Detecting landmarks specifically in colonoscopy has gained lesser academic attention in comparison. Classical machine learning techniques such as K-Means clustering in combination with hand-crafted feature extraction have been used by [3] for detecting certain colon landmarks. In [1,5], CNN- based deep-learning networks were used, while [28] used a vision-transformer based network. In [16], colon landmark detection was formulated as an ego-motion classification task based on depth estimation.

Consistency Regularization: Consistency Regularization is a method which employs training data to produce consistent predictions under different perturbations which can be done through image augmentation [11,25] [2] or model perturbation based methods [29]. Observing an overlap between feature vectors of different landmark classes is observed when images of different classes are visually similar. We employ a consistency regularization in our work to maximize the separation of different classes in the feature space, leading to better classification performance, as explained in Sect. 4.3.

Imbalanced data sampling: Data class imbalance is a well studied problem. Methods for handling class imbalance in machine learning can be grouped into three categories: data-level techniques, algorithm-level methods, and hybrid approaches [15]. Previous works show various methods for handling data imbalance problem through embeddings space re-sampling, specifically oversampling of minority classes which either creates synthetic data points from embeddings distribution or clusters ([4, 14] [21]). Another method is majority class under sampling which is composed of random, neighbourhood and cluster based sampling ([10, 30]). In our work, we employ a probabilistic data-level sampling technique to counter intra-class (random majority under sampling) as well as inter-class imbalance (random over sampling), which is further explained in Sect. 4.2.

3   Data Collection

We collected 307 colonoscopy videos and annotated them as containing one of the three landmarks (AO, Cec/ICV, RecRF) or the Other class (containing no landmark). We also curated 2363 snapshot images to serve as the test set for evaluating the effectiveness of our algorithm. We found that due to strong visual similarities between images of different classes, the annotation task is inherently challenging. To mitigate possible annotation errors, we used the cross-validation based annotation process described in [28] for annotating both videos and snapshots. An example of challenges in differentiating between frames of the different classes can be observed in Fig. 1, as we see only fine grained details differ between different classes. In the first row we can see an example of the resemblance between the first two images from left to right (AO) and the last image which is labeled as non-landmark (Other). Similar examples can be seen in the second row where the first two images from the left to right are Cec examples while the last one is Other.

Training Dataset: Our training dataset has 307 videos constituting of 6.27M frames. Most of our frames ($> 95\%$) belong to a non-landmark (Other) class, and the minority contain a landmark of interest (AO, Cec, RecRF). We balance the dataset as part of our training and evaluation (to get a distribution similar to the snapshots dataset) as described in Sect. 4.

Test Dataset: The test dataset consists of 2363 snapshots frames extracted from 900 different videos, where every frame is cross-validated by two medical experts. The dataset used to assess the classifier comprises of 487 AO, 276 Cecum, 660 RecRF, and 940 non-landmark (Other) images.

4   Training

At a high level, our training procedure consists of three main parts: *1)* data filtering, *2)* data sampling and *3)* ViT based classification network with consistency regularization loss.

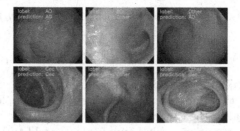

Fig. 1. Examples of intra-class variability (first 2 columns) and inter-class variability (first and last column)

4.1   Data Filtering

Typical colonoscopy videos contain a large amount of non-informative images that are unsuitable for facilitating our classification training. Various factors may render a video frame inappropriate for our downstream task, including a camera positioned too closely to the colon mucosa, frames occluded by colon ridges, moisture upon the lens and rapid motion blurriness. Figure 2 shows examples of such images. In order to be consistent with valid input frames and to increase data efficiency, we trained a binary classifier using a 2D convolutional neural network, and use it to automatically exclude non-viable frames in our training pipeline.

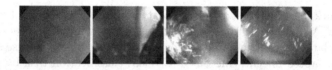

Fig. 2. Non informative frames samples

4.2   Data Sampling

Our training data presents high inter-class imbalance as it is comprised of many observations that do not represent snapshots of anatomic landmarks. Due to this abundance of non-representative frames, random images selection in the training phase may result in a model skewed towards images corresponding to the class *Other*, leading to poor performance on the minority classes. To circumvent this problem, we use a *class-wise uniform sampling* mechanism that effectively balances our dataset by uniformly oversampling frames based on their corresponding label.

Let us denote by $\{(F_i, y_i)\}_{i \in \mathcal{S}}$ the set of training images $F_i \in \mathbb{R}^{H \times W \times 3}$ and their corresponding labels $y_i \in C = \{AO, Cec/ICV, RecRF, Other\}$. The probability $Pr\,(i \in S_c, c \in C)$ of sampling a data item (F_i, y_i) labeled as $c \in C$ is

$$Pr\,(i \in S_c, c \in C) = Pr\,(i \in S_c) \cdot Pr\,(c \in C) = Unif\{S_c\} \cdot Unif\{C\} \tag{1}$$

where $S_c = \{\forall j \in S \mid y_j = c\}$ is the set of all training samples labeled as c. This implies that the probability of drawing a data sample classified as c is inversely proportional to the cardinality of C. Consequently, ensuring that rare classes receive more attention during training, can yield to higher accuracy in classifying them.

Snapshot images are distinguished by the presence of a relatively high number of images featuring gastrointestinal findings, including polyps, diverticuli, and hemorrhoids. These are visually distinct from the rest of the non-landmark frames. As randomly chosen images during training may not accurately reflect this population, we developed an *intra-class abnormalities sampling* (ICAS) strategy that selects images with findings with a predefined probability p, which is a training hyperparameter. If an image F_i is chosen at random from the *Other* class, the probability that the outcome image contains a finding is p. Mathematically speaking, this can be represented as:

$$Pr\,(F_i \text{ contains finding}) = p \cdot Pr\,(y_i = Other) = p\frac{1}{|C|} \tag{2}$$

4.3 ViT-Based Classification Network with Consistency Regularization

The landmark classifier is comprised of two primary components: a Vision Transformer (ViT-B/16) [9] based encoder that maps input images to a lower-dimensional latent space, and a Fully Connected Network (FCN) based projector specifically adapted for carrying out a $|C|$-class classification task by predicting a probability vector for each class. The architecture is described in detail in Sect. 5.

To improve model generalization and to prevent the model from overfitting, we enforce consistency loss by introducing a regularization term to the overall loss. For each pair of training image F_i and its corresponding label y_i, we generate two data samples $(\tilde{F}_A, \tilde{F}_B)$ by applying random augmentation operations on the original image. A supervised cross-entropy (CE) loss is applied on the probability vector obtained from the first image and the ground truth, while the self-supervised consistency loss is applied on the probability vectors for the two distorted views.

For a classification model $f_\theta : \mathbb{R}^{H \times W \times 3} \to (0, 1)^{|C|}$ with softmax layer at the output, we wish to find the optimal weights θ^* by minimizing:

$$\theta^* = \arg\min_\theta \left(\mathcal{L}_{cls} + \lambda\mathcal{L}_{cr}\right) \tag{3}$$

where $\mathcal{L}_{cls} = CE\left(f_\theta\left(\tilde{F}^A\right), \mathbf{y}\right)$ and $\mathcal{L}_{cr} = CE\left(f_\theta\left(\tilde{F}^A\right), f_\theta\left(\tilde{F}^B\right)\right)$ denote the classification and consistency regularization losses, respectively. Here, $f\,(\cdot; \theta)$ is abbreviated as $f_\theta(\cdot)$, and $\mathbf{y} \in \{0,1\}^{|C|}$ denotes the one-hot encoded label. $\tilde{F}^A$ and $\tilde{F}^B$ represent the generated augmented images.

The overall loss can be interpreted as a combination of the cross entropy loss commonly used for classification, and a regularization term that enforces the consistency of the two variations of the input data.

5   Implementation Details

The encoder is composed of a Vision Transformer (ViT-B/16) [9] network pre-trained on the ImageNet [8] dataset, where the output of the fully-connected layer serves as the encoding. The projector comprised a cascade of fully-connected and non-linear layers that are applied to the embeddings generated by the original ViT-B/16 model. Specifically, the FCN projector is composed of 128 dimensional output fully-connected (FC) layer, followed by ReLU nonlinearity, 64 dimensional FC layer, dropout [27], $|C|$ dimensional FC layer and softmax.

Each image F is randomly transformed twice to generate the two distorted views $\tilde{F}^A$ and $\tilde{F}^B$. The image is initially cropped and resized to a resolution of 384×384 pixels, and then subjected to a series of random operations that include grayscale conversion, channel shuffle, Gaussian blur, color jitter, vertical and horizontal flips, and 90° rotation. During testing, we only standardize the image by resizing it to 384×384.

Our training framework was implemented using PyTorch [23]. The model was trained for 75 epochs on an NVIDIA RTX A6000 GPU, where 20,000 images are randomly sampled at each epoch, as described in Sect. 4.2, in batches of size 20. We use the SAM [12] optimizer for minimizing the objective function defined in Eq. 3 and set the training hyperparameters $\lambda = 2$ and $p = 0.2$. The initial learning rate is set to 5×10^{-3}, where it decreases by a multiplicative factor (γ) of 0.5 every 10 epochs.

6   Results

Here, we summarize the performance of the model on the snapshots (test) dataset. We present the performance summary statistics in Table 1 and the class-level statistics in Table 2. As the test data is skewed towards snapshots that are not labelled with any landmark, we prefer using macro recall, macro precision and macro-f1-score [26] over the micro-level metrics. This ensures that the performance evaluation is based on the classifier's ability to correctly identify instances from all classes, regardless of their frequency.

Out of a total of 2363 frames, the model accurately predicted 2180, leading to an overall accuracy rate of 92.26%. To provide a more comprehensive breakdown of the classifier's performance, the experiments outlined in Table 1 are accompanied by their respective confusion matrices in Fig. 3. We see that there is a small level of misclassification between the AO and Cec classes. This may be due to their visual and proximal similarities, causing an overlap of features.

Table 1. Performance evaluation on the test set

$\mathcal{L}_{cr}$	ICAS	Accuracy	macro-Recall	macro-Precision	macro-f1
baseline		0.8993	0.8938	0.8736	0.8800
✓		0.8836	0.8966	0.8521	0.8667
	✓	0.9022	0.8957	0.8773	0.8833
✓	✓	**0.9226**	**0.9207**	**0.8954**	**0.9059**

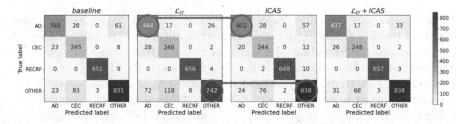

Fig. 3. Confusion matrices for the experiments described in Table 1. The results suggest that the combination of both consistency loss and ICAS produces the optimal results.

To gain deeper insights about the contribution of the semi-supervised loss $\mathcal{L}_{cr}$ and the proposed ICAS sampling strategy to the final results, we conduct an ablation study described in Table 1. It is observed that including either $\mathcal{L}_{cr}$ or ICAS did not yield a considerable improvement in the model's ultimate outcome. Nonetheless, *the combination of both features* produced a remarkable enhancement in the overall level of performance.

A class level analysis is reported in Table 3. It is observed that incorporating $\mathcal{L}_{cr}$ loss increases the *AO* recall to 0.9117 while reducing *Other* recall by more than 9% to 0.7894. However, when ICAS is used in conjunction with $\mathcal{L}_{cr}$, it results in an improvement in *AO* without negatively affecting the other classes. It can also be noted that incorporating ICAS alone have no impact on the model's performance.

We report a high macro-recall of 0.9207. The per-class recall evaluation is provided in Table 2. The recall rates for both *AO* and *Cec/ICV* landmarks exceed 89%, and the recall value obtained for *RecRF* is particularly high - 99.55% - with only 3 misses out of 658 frames. This can be explained by the presence of a dark tube in almost every *RecRF* frame, rendering it easy to identify even in optically challenging circumstances. A high recall rate, implies a low landmark missing-rate during procedures, which in turn reduces the burden of manually annotating snapshot images.

We also report a high macro-precision of 0.8954, where *RecRF* again exhibits a significantly high value, with only three frames being incorrectly predicted as belonging to this class. It is noteworthy that the number of frames categorized as *Other* is notably higher than *AO* and *Cec/ICV*. As a consequence, mislabeled frames have a proportionally greater impact on the precision results for the remaining classes.

Table 2. Per-class performance evaluation

	Recall	Precision	f1	Support
AO	0.8973	0.8846	0.8909	487
CEC/ICV	0.8986	0.7447	0.8144	276
RecRF	0.9955	0.9955	0.9955	660
Other	0.8915	0.9566	0.9229	940

Table 3. Per-class Recall evaluation for $\mathcal{L}_{cr}$ loss and ICAS Sampling

$\mathcal{L}_{cr}$	ICAS	AO	CEC/ICV	RecRF	Other
		0.8172	0.8877	0.9864	0.8840
✓		0.9117 (↑ 9.45%)	0.8913	0.9939	0.7894 (↓ 9.46%)
	✓	0.8254	0.8841	0.9818	0.8915
✓	✓	0.8973 (↑ 8.01%)	0.8986	0.9955	0.8915

7   Conclusion and Inference

Our paper presents a consistency loss based semi-supervised training framework for classifying landmarks in colonoscopy snapshots with vision transformers. Additionally, we achieve performance gain by introducing both the self-supervised consistency loss and a sampling strategy of rare landmarks and colon findings. The classifier's performance was evaluated on a test set comprising 2363 frames annotated by multiple expert gastroenterologists, achieving an accuracy of 92%. We observe that based on the summary statistics reported in Table 1, our algorithm can generalize across multiple colonoscopy videos and identify the key colon landmarks with high accuracy. We also observe that the algorithmic enhancements that we have proposed (i.e. the consistency loss and ICAS) directly contribute to the performance, as evident from Tables 1 and 3.

References

1. Adewole, S., et al.: Deep learning methods for anatomical landmark detection in video capsule endoscopy images. In: Arai, K., Kapoor, S., Bhatia, R. (eds.) FTC 2020. AISC, vol. 1288, pp. 426–434. Springer, Cham (2021). https://doi.org/10.1007/978-3-030-63128-4_32

2. Berthelot, D., Carlini, N., Goodfellow, I., Papernot, N., Oliver, A., Raffel, C.A.: Mixmatch: a holistic approach to semi-supervised learning. In: Advances in Neural Information Processing Systems 32 (2019)

3. Cao, Y., Liu, D., Tavanapong, W., Wong, J., Oh, J., De Groen, P.C.: Automatic classification of images with appendiceal orifice in colonoscopy videos. In: 2006 International Conference of the IEEE Engineering in Medicine and Biology Society, pp. 2349–2352. IEEE (2006)

4. Chawla, N.V., Bowyer, K.W., Hall, L.O., Kegelmeyer, W.P.: SMOTE: synthetic minority over-sampling technique. J. Artif. Intell. Res. **16**, 321–357 (2002)

5. Che, K., et al.: Deep learning-based biological anatomical landmark detection in colonoscopy videos. arXiv preprint arXiv:2108.02948 (2021)

6. Chowdhury, A.S., Yao, J., VanUitert, R., Linguraru, M.G., Summers, R.M.: Detection of anatomical landmarks in human colon from computed tomographic colonography images. In: 2008 19th International Conference on Pattern Recognition, pp. 1–4. IEEE (2008)

7. Cooper, J.A., Ryan, R., Parsons, N., Stinton, C., Marshall, T., Taylor-Phillips, S.: The use of electronic healthcare records for colorectal cancer screening referral decisions and risk prediction model development. BMC Gastroenterol. **20**(1), 1–16 (2020)

8. Deng, J., Dong, W., Socher, R., Li, L.J., Li, K., Fei-Fei, L.: ImageNet: a large-scale hierarchical image database. In: 2009 IEEE Conference on Computer Vision and Pattern Recognition, pp. 248–255. IEEE (2009)

9. Dosovitskiy, A., et al.: an image is worth 16×16 words: transformers for image recognition at scale. arXiv preprint arXiv:2010.11929 (2020)

10. Estabrooks, A., Japkowicz, N.: A mixture-of-experts framework for learning from imbalanced data sets. In: Hoffmann, F., Hand, D.J., Adams, N., Fisher, D., Guimaraes, G. (eds.) IDA 2001. LNCS, vol. 2189, pp. 34–43. Springer, Heidelberg (2001). https://doi.org/10.1007/3-540-44816-0_4

11. Fan, Y., Kukleva, A., Dai, D., Schiele, B.: Revisiting consistency regularization for semi-supervised learning. Int. J. Comput. Vis. **131**, 1–18 (2022). https://doi.org/10.1007/s11263-022-01723-4

12. Foret, P., Kleiner, A., Mobahi, H., Neyshabur, B.: Sharpness-aware minimization for efficiently improving generalization. arXiv preprint arXiv:2010.01412 (2020)

13. Ghesu, F.C., Georgescu, B., Mansi, T., Neumann, D., Hornegger, J., Comaniciu, D.: An artificial agent for anatomical landmark detection in medical images. In: Ourselin, S., Joskowicz, L., Sabuncu, M.R., Unal, G., Wells, W. (eds.) MICCAI 2016. LNCS, vol. 9902, pp. 229–237. Springer, Cham (2016). https://doi.org/10.1007/978-3-319-46726-9_27

14. Jo, T., Japkowicz, N.: Class imbalances versus small disjuncts. ACM SIGKDD Explor. Newsl. **6**(1), 40–49 (2004)

15. Johnson, J.M., Khoshgoftaar, T.M.: Survey on deep learning with class imbalance. J. Big Data **6**(1), 1–54 (2019)

16. Katzir, L., et al.: Estimating withdrawal time in colonoscopies. In: Computer Vision-ECCV 2022 Workshops: Tel Aviv, Israel, October 23–27, 2022, Proceedings, Part III, pp. 495–512. Springer (2023). https://doi.org/10.1007/978-3-031-25066-8_28

17. Mamonov, A.V., Figueiredo, I.N., Figueiredo, P.N., Tsai, Y.H.R.: Automated polyp detection in colon capsule endoscopy. IEEE Trans. Med. Imaging **33**(7), 1488–1502 (2014)

18. McDonald, C.J., Callaghan, F.M., Weissman, A., Goodwin, R.M., Mundkur, M., Kuhn, T.: Use of internist's free time by ambulatory care electronic medical record systems. JAMA Intern. Med. **174**(11), 1860–1863 (2014)
19. Morelli, M.S., Miller, J.S., Imperiale, T.F.: Colonoscopy performance in a large private practice: a comparison to quality benchmarks. J. Clin. Gastroenterol. **44**(2), 152–153 (2010)
20. Morgan, E., et al.: Global burden of colorectal cancer in 2020 and 2040: incidence and mortality estimates from GLOBOCAN. Gut **72**(2), 338–344 (2023)
21. Mullick, S.S., Datta, S., Das, S.: Generative adversarial minority oversampling. In: Proceedings of the IEEE/CVF International Conference on Computer Vision, pp. 1695–1704 (2019)
22. Park, S.Y., Sargent, D., Spofford, I., Vosburgh, K.G., Yousif, A., et al.: A colon video analysis framework for polyp detection. IEEE Trans. Biomed. Eng. **59**(5), 1408–1418 (2012)
23. Paszke, A., et al.: PyTorch: an imperative style, high-performance deep learning library. In: Advances in Neural Information Processing Systems 32, pp. 8024–8035. Curran Associates, Inc. (2019). http://papers.neurips.cc/paper/9015-pytorch-an-imperative-style-high-performance-deep-learning-library.pdf
24. Qadir, H.A., Shin, Y., Solhusvik, J., Bergsland, J., Aabakken, L., Balasingham, I.: Toward real-time polyp detection using fully CNNs for 2D gaussian shapes prediction. Med. Image Anal. **68**, 101897 (2021)
25. Sohn, K., et al.: FixMatch: simplifying semi-supervised learning with consistency and confidence. Adv. Neural. Inf. Process. Syst. **33**, 596–608 (2020)
26. Sokolova, M., Lapalme, G.: A systematic analysis of performance measures for classification tasks. Inf. Process. Manage. **45**(4), 427–437 (2009)
27. Srivastava, N., Hinton, G., Krizhevsky, A., Sutskever, I., Salakhutdinov, R.: Dropout: a simple way to prevent neural networks from overfitting. J. Mach. Learn. Res. **15**(1), 1929–1958 (2014)
28. Tamhane, A., Mida, T., Posner, E., Bouhnik, M.: Colonoscopy landmark detection using vision transformers. In: Imaging Systems for GI Endoscopy, and Graphs in Biomedical Image Analysis: First MICCAI Workshop, ISGIE 2022, and Fourth MICCAI Workshop, GRAIL 2022, Held in Conjunction with MICCAI 2022, Singapore, September 18, 2022, Proceedings, pp. 24–34. Springer (2022). https://doi.org/10.1007/978-3-031-21083-9_3
29. Tarvainen, A., Valpola, H.: Mean teachers are better role models: weight-averaged consistency targets improve semi-supervised deep learning results. In: Advances in Neural Information Processing Systems 30 (2017)
30. Vuttipittayamongkol, P., Elyan, E.: Neighbourhood-based undersampling approach for handling imbalanced and overlapped data. Inf. Sci. **509**, 47–70 (2020)
31. Zhang, J., et al.: Colonoscopic screening is associated with reduced colorectal cancer incidence and mortality: a systematic review and meta-analysis. J. Cancer **11**(20), 5953 (2020)
32. Zhou, S.K., et al.: A review of deep learning in medical imaging: imaging traits, technology trends, case studies with progress highlights, and future promises. In: Proceedings of the IEEE (2021)
33. Zhou, S.K., Xu, Z.: Landmark detection and multiorgan segmentation: representations and supervised approaches. In: Handbook of Medical Image Computing and Computer Assisted Intervention, pp. 205–229. Elsevier (2020)

Radiomics Boosts Deep Learning Model for IPMN Classification

Lanhong Yao[1]([✉]), Zheyuan Zhang[1], Ugur Demir[1], Elif Keles[1],
Camila Vendrami[1], Emil Agarunov[2], Candice Bolan[3], Ivo Schoots[4],
Marc Bruno[4], Rajesh Keswani[1], Frank Miller[1], Tamas Gonda[2], Cemal Yazici[5],
Temel Tirkes[6], Michael Wallace[7], Concetto Spampinato[8], and Ulas Bagci[1]

[1] Department of Radiology, Northwestern University, Chicago, IL 60611, USA
LanhongYao2022@u.northwestern.edu
[2] NYU Langone Health, New York, NY 10016, USA
[3] Mayo Clinic, Rochester, MN 55905, USA
[4] Erasmus Medical Center, 3015 GD Rotterdam, The Netherlands
[5] University of Illinois Chicago, Chicago, IL 60607, USA
[6] Indiana University-Purdue University Indianapolis, Indianapolis, IN 46202, USA
[7] Sheikh Shakhbout Medical City, 11001 Abu Dhabi, United Arab Emirates
[8] University of Catania, 95124 Catania, CT, Italy

Abstract. Intraductal Papillary Mucinous Neoplasm (IPMN) cysts are pre-malignant pancreas lesions, and they can progress into pancreatic cancer. Therefore, detecting and stratifying their risk level is of ultimate importance for effective treatment planning and disease control. However, this is a highly challenging task because of the diverse and irregular shape, texture, and size of the IPMN cysts as well as the pancreas. In this study, we propose a novel computer-aided diagnosis pipeline for IPMN risk classification from multi-contrast MRI scans. Our proposed analysis framework includes an efficient volumetric self-adapting segmentation strategy for pancreas delineation, followed by a newly designed deep learning-based classification scheme with a radiomics-based predictive approach. We test our proposed decision-fusion model in multi-center data sets of 246 multi-contrast MRI scans and obtain superior performance to the state of the art (SOTA) in this field. Our ablation studies demonstrate the significance of both radiomics and deep learning modules for achieving the new SOTA performance compared to international guidelines and published studies (81.9% vs 61.3% in accuracy). Our findings have important implications for clinical decision-making. In a series of rigorous experiments on multi-center data sets (246 MRI scans from five centers), we achieved unprecedented performance (81.9% accuracy). The code is available upon publication.

Keywords: Radiomics · IPMN Classification · Pancreatic Cysts · MRI · Pancreas Segmentation

This project is supported by the NIH funding: NIH/NCI R01-CA246704 and NIH/NIDDK U01-DK127384-02S1.

X. Cao et al. (Eds.): MLMI 2023, LNCS 14349, pp. 134–143, 2024.
https://doi.org/10.1007/978-3-031-45676-3_14

1   Introduction

Pancreatic cancer is a deadly disease with a low 5-year survival rate, primarily because it is often diagnosed at a late stage [1,10]. Early detection is crucial for improving survival rates and gaining a better understanding of tumor patho-physiology. Therefore, research on pancreatic cysts is significant, since some types, such as intraductal papillary mucinous neoplasms (IPMN), can potentially develop into pancreatic cancer [12]. Hence, the diagnosis of IPMN cysts and the prediction of their likelihood of transforming into pancreatic cancer are essential for early detection and disease management. Our study is aligned with this objective and aims to contribute to aiding the early detection of pancreatic cancer.

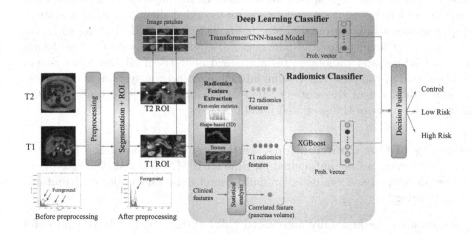

Fig. 1. An overview of our proposed CAD system is shown. Multi-contrast MRI (T1 and T2) are preprocessed with inhomogeneity correction, denoising, and intensity standardization. Clean images are then used to segment the pancreas region. ROI enclosing the segmented pancreases are fed into a **deep learning classifier**. Clinical features selected through statistical analysis are fed into **radiomics classifier**. Decision vectors (probability) from both classifiers are combined via a weighted averaging-based decision fusion strategy for final IPMN cyst stratification.

Diagnosis of IPMN involves a combination of imaging studies, laboratory tests, and sometimes biopsy. Imaging studies utilize CT, MRI, and EUS scans to visualize the pancreas and detect any cystic lesions. The size, location, shape, texture, and other characteristics of the lesions are used for radiographical evaluations. Current international guidelines (AGA, ACG, and IAP) [3,11,13] state that IPMNs should be classified into low-risk or high-risk based on their size, morphology, and presence of high-risk features such as main pancreatic duct involvement, mural nodules, or elevated cyst fluid CEA levels. While high-risk IPMNs should be considered for surgical resection, low-risk IPMNs may be managed with surveillance.

Prior Art. Radiographical identification of IPMNs is important but falls short of diagnostic accuracy, thus there is a need for improving the current standards for IPMN risk stratifications [9]. Several studies have demonstrated the potential of deep learning in IPMN diagnosis, including the use of convolutional neural networks (CNN) [6], inflated neural networks (INN) [10], and neural transformers [15]. However, these studies analyzed MRIs from a single center with a small number of patients and did not include the pancreas segmentation step, only using directly cropped pancreas regions or whole images for classification. Still, these models showed promising results compared to the international guidelines. While deep learning techniques have shown the potential to improve the accuracy and efficiency of IPMN diagnosis and risk stratification [2], further research and validation are needed to determine the clinical utility and their potential impact on patient outcomes. Our work addresses the limitations of current deep learning models by designing a new computer-aided diagnosis (CAD) system, which includes a fully automated pipeline including (1) MRI cleaning with preprocessing, (2) segmentation of pancreas, (3) classification with decision fusion of deep learning and radiomics, (4) statistical analysis of clinical features and incorporation of the pancreas volume into clinical decision system, and (5) testing and validation of the whole system in the multi-center settings. Figure 1 shows an overview of the proposed novel CAD system.

Summary of Our Contributions. To the best of our knowledge, this is the first study having a fully automated pipeline for IPMN diagnosis and risk stratification, developed and evaluated on multi-center data. Our major contributions are as follows:

1. We develop the first fully automated CAD system that utilizes a powerful combination of deep learning, radiomics, and clinical features - all integrated into a single decision support system via a weighted averaging-based decision fusion strategy.
2. Unlike existing IPMN CAD systems, which do not include MRI segmentation of the pancreas and require cumbersome manual annotations on the pancreas, we incorporate the volumetric self-adapting segmentation network (nnUNet) to effectively segment the pancreas from MRI scans with high accuracy and ease.
3. We present a simple yet convenient approach to fusing radiomics that can significantly improve the DL model's performance by up to 20%. This powerful tool allows us to deliver more accurate diagnoses for IPMN.
4. Through rigorous statistical analysis of 8 clinical features, we identify pancreas volume as a potential predictor of risk levels of IPMN cysts. By leveraging this vital information, we enhance the decision fusion mechanism to achieve exceptional overall model accuracy.

2   Materials and Methods

2.1   Dataset

In compliance with ethical standards, our study is approved by the Institutional Review Board (IRB), and necessary privacy considerations are taken into account: all images are de-identified before usage. We obtain 246 MRI scans (both T1 and T2) from five centers: Mayo Clinic in Florida (MCF), Mayo Clinic in Arizona (MCA), Allegheny Health Network (AHN), Northwestern Memorial Hospital (NMH), and New York University Langone Hospital (NYU). All T1 and T2 images are registered and segmentation masks are generated using a fast and reliable segmentation network (see below for the segmentation section). Segmentations are examined by radiologists case by case to ensure their correctness. The ground truth labels of IPMN risk classifications are determined based on either biopsy exams or surveillance information with radiographical evaluation, and overall three balanced classes are considered for risk stratification experiments: healthy (70 cases), low-grade risk (85 cases), and high-grade risk (91 cases).

2.2   Preprocessing

MRI presents unique challenges, including intensity inhomogeneities, noise, non-standardization, and other artifacts. These challenges often arise from variations in acquisition parameters and hardware, even when using the same scanner and operators at different times of the day or with different patients. Therefore, preprocessing MRI scans across different acquisitions, scanners, and patient populations is necessary. In our study, we perform the following preprocessing steps on the images before feeding them into the segmentor and classifiers: Initially, images are reoriented in accordance with the RAS axes convention. Subsequent steps involve the application of bias correction and denoising methodologies, designed to mitigate artifacts and augment image fidelity. Further, we employ Nyul's method [14] for intensity standardization, harmonizing the intensity values of each image with a designated reference distribution. Figure 1 illustrates the image histograms pre- and post-preprocessing, underscoring the efficacy of our standardization procedure. These preprocessing steps can help improve the robustness and reliability of deep learning models.

2.3   Pancreas Segmentation

Pancreas volumetry is a prerequisite for the diagnosis and prognosis of several pancreatic diseases, requiring radiology scans to be segmented automatically as manual annotation is highly costly and inefficient. In this module of the CAD system, our aim is to develop a clinically stable and accurate deep learning-based segmentation algorithm for the pancreas from MRI scans in multi-center settings to prove its generalization efficacy. Among 246 scans, we randomly select 131 MRI images (T2) from multi-center settings: 61 cases from NMH, 15 cases from NYU, and 55 cases from MCF. Annotations are obtained from three centers' data. The segmentation masks are used for pancreas region of interest (ROI) boundary extraction in radiomics and deep learning-based classification rather than exact pixel analysis. We present a robust and accurate deep learning algorithm based on the 3D nnUNet architecture with SGD optimization [7].

2.4 Model Building for Risk Stratification

Radiomics Classifier. Radiomics involves extracting a large number of quantitative features from medical images, offering valuable insights into disease characterization, notably in IPMN where shape and texture are vital to classifications. For this study, 107 distinct features are extracted from both T1 and T2 images, within the ROI enclosing pancreas segmentation mask. They capture characteristics such as texture, shape, and intensity. To mitigate disparities across scales in the radiomics data, we employ the $ln(x + 1)$ transformation and unit variance scaling. Further, analyze 8 clinical features using an OLS regression model to evaluate their predictive efficacy for IPMN risk: diabetes mellitus, pancreas volume, pancreas diagonal, volume over diagonal ratio, age, gender, BMI, and chronic pancreatitis. Through stepwise regression, we refine the model's focus to key features. T-tests reveal significant differences in pancreas volume across IPMN risk groups, consistent with prior medical knowledge. Notably, pancreas volume shows predictive efficacy for IPMN risk, leading to its inclusion as a vital clinical feature in our risk prediction model.

Deep Learning Classifier. We utilize one Transformer-based and four CNN-based architectures to compare and evaluate the IPMN risk assessment. Neural transformers [15] is notably the first application of vision transformers (ViT) in pancreas risk stratification and has obtained promising results but with a limited size of data from a single center. DenseNet [5], ResNet18 [4], AlexNet [8], and MobileNet [16] are all well-known CNN-based architectures that have been developed in recent years and have shown to perform well on various computer vision tasks. Herein, we benchmark these models to create baselines and compare/contrast their benefits and limitations.

Weighted Averaging Based Decision Fusion. The sparse feature vectors learned by DL models pose a challenge for feature fusion with the radiomics model. To address this, we implement weighted averaging-based decision fusion, where we combine the weighted probabilities of the DL classifier (shown below in Eq. 1) and radiomics classifier:

$$
P_c = \begin{cases} P_r, & \text{if } max(P_r) \geq t \\ k * P_d + (1 - k) * P_r, & \text{otherwise} \end{cases} \tag{1}
$$

where k and t are parameters to adjust the decision fusion of two models, and they are selected via grid search during cross-validation. P_d and P_r are probability vectors, predicted by the DL classifier and radiomics classifier, respectively.

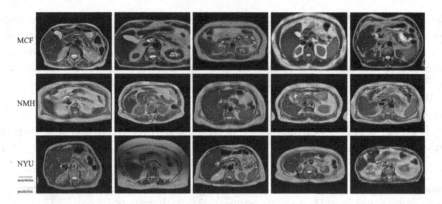

Fig. 2. Qualitative pancreas segmentation visualization on multi-center data. Predicted segmentation maps (yellow line) are highly similar to the ground truth annotations (red line) in anatomical structure regardless of image variance. (Color figure online)

P_c refers to the combined probabilities, based on which we get the final fused predictions. For each case, P_c for three classes add up to one. This P_c is used in the cases when the maximum probability of the radiomics classifier is less than a threshold t, indicating the radiomics classifier is not confident in its decision and could use extra information.

2.5   Training Details

The dataset used in this study consists of 246 patients from five medical centers. Out of these, 49 cases are randomly selected for blind testing (i.e., independent test), and are unseen by any of the models. The remaining 197 cases are split into training and validation sets. Every set incorporates data across all participating medical centers. This consistent distribution ensures an unbiased evaluation environment. We employ the same evaluation procedure for all the models.

The deep learning (DL) models have been developed utilizing the PyTorch framework and executed on an NVIDIA RTX A6000 GPU. The training was conducted with a batch size of 16 across a maximum of 1500 epochs. The radiomics classifier employs XGBOOST with grid search to identify the best parameters. The optimal parameters are determined as follows: number of estimators = 140 and maximum depth = 4.

3   Results

3.1   Segmentation

We employ standard 5-fold cross-validation for the training and use Dice score (higher is better), Hausdorff distance at 95% (HD95, lower is better), Precision, and Recall for quantitative evaluations. Dice score for CT-based segmentation in the literature reaches a plateau value of 85% but MRI segmentations hardly reach 50–60% with a limited number of research papers [17]. Herein, our segmentation results reach 70% for multi-center data, showing a significant increase from the current standards. Figure 2 shows qualitative visualization of the predicted segmentation results compared with the reference standards provided by the radiologists, demonstrating highly accurate segmentations.

We conduct a comprehensive quantitative evaluation. Table 1 shows the quantitative evaluation results on multi-domain, and particularly, we reach one average Dice of 70.11% which has never been achieved before in the literature in multi-center settings.

Table 1. Multi-center pancreas MRI segmentation performance comparison. The model reaches a sufficiently accurate segmentation with average Dice of 70.11.

Data Center	Dice	HD95(mm)	Precision	Recall	Case
MCF	69.51±2.96	28.36±23.59	77.18±3.50	66.11±4.36	61
NMH	66.02±1.84	14.91±9.95	63.39±4.30	71.19±6.23	15
NYU	71.90±3.20	26.59±12.00	75.9±1.59	71.73±4.56	55
Average	70.11±2.96	26.08±18.19	75.06±2.98	69.05±4.70	Sum:131

3.2 Classification

Metrics for evaluating the models' clasification performance include Accuracy (ACC), the Area Under the receiver operating characteristic Curve (AUC), Precision (PR), and Recall (RC). Higher values in these metrics indicate better performance. We boost the classification results with the combination of radiomics and deep learning classifiers and obtain new SOTA results (Table 2). Despite the success of our proposed ensemble, we also identify some challenging cases where our classifiers fail to identify the existence and type of the IPMNs in Fig. 3 (the second row for the failure cases compared to the first row for successfully predicted cases).

Table 2. Quantitative comparison (%) for the influence of combining radiomics with deep learning. We can observe that regardless of network structure, combining radiomics with deep learning can impressively leverage the IPMN classification performance. Similarly, combining the deep learning extracted features can also leverage the performance of radiomics for the IPMN classification.

Network	w/o Radiomics				w/ Radiomics			
Metrics	ACC	AUC	PR	RC	ACC	AUC	PR	RC
w/o DL	-	-	-	-	71.6	89.7	74.7	75.2
DenseNet	57.4	68.4	55.9	56.2	75.8	87.4	74.7	76.0
ResNet18	48.9	66.6	46.3	48.0	73.7	89.3	76.1	76.7
AlexNet	57.0	64.8	60.3	54.7	77.7	**89.8**	76.5	75.2
MobileNet	57.0	64.4	57.7	54.6	71.6	89.3	74.7	75.2
ViT	61.3	71.9	56.2	56.6	**81.9**	89.3	**82.4**	**82.7**

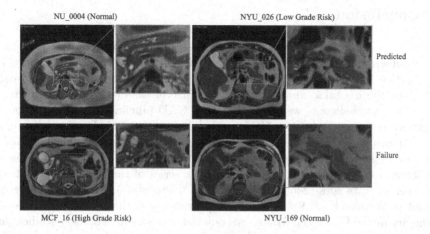

Fig. 3. First and second rows show MRI cases where IPMN diagnosis and stratification are done correctly and incorrectly, respectively. Zoomed versions of the pancreas regions demonstrate the shape and appearance diversity.

We assess the performance of a deep learning (DL) model and a radiomics model individually and in combination, for the classification of IPMN using multi-center multi-contrast MRI data. On a single center scenario, the DL model performs comparably to previous literature [15]. However, when the models are tested on multi-center data, the performance of the DL model decreases, likely due to the heterogeneity of multi-center data compared to a single institution. We also observe that the decision fusion of the radiomics predictions to the DL model prediction improves its performance on multi-center data. This improvement can be attributed to the domain knowledge contained in the handcrafted radiomics features more than the deep features which are high-dimensional and sparse. Radiomics features are designed to capture important characteristics of IPMNs in dense vectors with more controlled variations, indicating a better generalization over multi-center data. The fusion method can be also viewed as a trainable linear layer on probability vectors. Lastly, our findings suggest that the information captured by the DL layers and radiomics features is complementary to a certain degree, and combining them can yield better performance than using either approach alone.

To understand this further, we run experiments of radiomics classifiers with different combinations of features: T1 radiomics, T2 radiomics, combined T1+T2 radiomics, and T1+T2+clinical features. We achieve the following accuracies: 0.573, 0.650, 0.666, and 0.674, respectively, indicating that T1 and T2 features are complementary to each other, and the clinical feature (pancreas volumetry) increases the prediction performance.

4   Conclusion

IPMN cysts are a ticking time bomb that can progress into pancreatic cancer. Early detection and risk stratification of these precancerous lesions is crucial for effective treatment planning and disease control. However, this is no easy feat given the irregular shape, texture, and size of the cysts and the pancreas. To tackle this challenge, we propose a novel CAD pipeline for IPMN risk classification from multi-contrast MRI scans. The proposed CAD system includes a self-adapting volumetric segmentation strategy for pancreas delineation and a newly designed deep learning-based classification scheme with a radiomics-based predictive approach at the decision level. In a series of rigorous experiments on multi-center data sets (246 MRI scans from five centers), we achieve unprecedented performance (81.9% accuracy with radiomics) that surpasses the state of the art in the field (ViT 61.3% without radiomics). Our ablation studies further underscore the pivotal role of both radiomics and deep learning modules for attaining superior performance compared to international guidelines and published studies, and highlight the importance of pancreas volume as a clinical feature.

References

1. Chen, P.T., et al.: Pancreatic cancer detection on CT scans with deep learning: a nationwide population-based study. Radiology **306**(1), 172–182 (2023)
2. Corral, J.E., Hussein, S., Kandel, P., Bolan, C.W., Bagci, U., Wallace, M.B.: Deep learning to classify intraductal papillary mucinous neoplasms using magnetic resonance imaging. Pancreas **48**(6), 805–810 (2019)
3. Elta, G.H., Enestvedt, B.K., Sauer, B.G., Lennon, A.M.: ACG clinical guideline: diagnosis and management of pancreatic cysts. Official J. Am. Coll. Gastroenterol.| ACG **113**(4), 464–479 (2018)
4. He, K., Zhang, X., Ren, S., Sun, J.: Deep residual learning for image recognition. In: Proceedings of the IEEE Conference on Computer Vision and Pattern Recognition, pp. 770–778 (2016)
5. Huang, G., Liu, Z., Van Der Maaten, L., Weinberger, K.Q.: Densely connected convolutional networks. In: Proceedings of the IEEE Conference on Computer Vision and Pattern Recognition, pp. 4700–4708 (2017)
6. Hussein, S., Kandel, P., Bolan, C.W., Wallace, M.B., Bagci, U.: Lung and pancreatic tumor characterization in the deep learning era: novel supervised and unsupervised learning approaches. IEEE Trans. Med. Imaging **38**(8), 1777–1787 (2019)
7. Isensee, F., Jaeger, P.F., Kohl, S.A., Petersen, J., Maier-Hein, K.H.: nnU-Net: a self-configuring method for deep learning-based biomedical image segmentation. Nat. Methods **18**(2), 203–211 (2021)
8. Krizhevsky, A., Sutskever, I., Hinton, G.E.: ImageNet classification with deep convolutional neural networks. Commun. ACM **60**(6), 84–90 (2017)
9. Kuwahara, T., et al.: Usefulness of deep learning analysis for the diagnosis of malignancy in intraductal papillary mucinous neoplasms of the pancreas. Clin. Transl. Gastroenterol. **10**(5), e00045 (2019)

10. LaLonde, R., et al.: INN: inflated neural networks for IPMN diagnosis. In: Shen, D., et al. (eds.) MICCAI 2019. LNCS, vol. 11768, pp. 101–109. Springer, Cham (2019). https://doi.org/10.1007/978-3-030-32254-0_12
11. Lennon, A.M., Ahuja, N., Wolfgang, C.L.: AGA guidelines for the management of pancreatic cysts. Gastroenterology **149**(3), 825 (2015)
12. Luo, G., et al.: Characteristics and outcomes of pancreatic cancer by histological subtypes. Pancreas **48**(6), 817–822 (2019)
13. Marchegiani, G., et al.: Systematic review, meta-analysis, and a high-volume center experience supporting the new role of mural nodules proposed by the updated 2017 international guidelines on IPMN of the pancreas. Surgery **163**(6), 1272–1279 (2018)
14. Nyúl, L.G., Udupa, J.K., Zhang, X.: New variants of a method of MRI scale standardization. IEEE Trans. Med. Imaging **19**(2), 143–150 (2000)
15. Salanitri, F.P., et al.: Neural transformers for intraductal papillary mucosal neoplasms (IPMN) classification in MRI images. In: 2022 44th Annual International Conference of the IEEE Engineering in Medicine & Biology Society (EMBC), pp. 475–479. IEEE (2022)
16. Sandler, M., Howard, A., Zhu, M., Zhmoginov, A., Chen, L.C.: MobileNetV2: inverted residuals and linear bottlenecks. In: Proceedings of the IEEE Conference on Computer Vision and Pattern Recognition, pp. 4510–4520 (2018)
17. Zhang, Z., Bagci, U.: Dynamic linear transformer for 3D biomedical image segmentation. In: Machine Learning in Medical Imaging: 13th International Workshop, MLMI 2022, Held in Conjunction with MICCAI 2022, Singapore, 18 September 2022, Proceedings, pp. 171–180. Springer (2022). https://doi.org/10.1007/978-3-031-21014-3_18

Class-Balanced Deep Learning
with Adaptive Vector Scaling Loss
for Dementia Stage Detection

Boning Tong[1], Zhuoping Zhou[1], Davoud Ataee Tarzanagh[1], Bojian Hou[1],
Andrew J. Saykin[2], Jason Moore[3], Marylyn Ritchie[1], and Li Shen[1(✉)]

[1] University of Pennsylvania, Philadelphia, PA 19104, USA
li.shen@pennmedicine.upenn.edu
[2] Indiana University, Indianapolis, IN 46202, USA
[3] Cedars-Sinai Medical Center, Los Angels, CA 90069, USA

Abstract. Alzheimer's disease (AD) leads to irreversible cognitive
decline, with Mild Cognitive Impairment (MCI) as its prodromal stage.
Early detection of AD and related dementia is crucial for timely treatment and slowing disease progression. However, classifying cognitive normal (CN), MCI, and AD subjects using machine learning models faces
class imbalance, necessitating the use of balanced accuracy as a suitable metric. To enhance model performance and balanced accuracy, we
introduce a novel method called VS-Opt-Net. This approach incorporates the recently developed vector-scaling (VS) loss into a machine
learning pipeline named STREAMLINE. Moreover, it employs Bayesian
optimization for hyperparameter learning of both the model and loss
function. VS-Opt-Net not only amplifies the contribution of minority
examples in proportion to the imbalance level but also addresses the
challenge of generalization in training deep networks. In our empirical
study, we use MRI-based brain regional measurements as features to
conduct the CN vs MCI and AD vs MCI binary classifications. We compare the balanced accuracy of our model with other machine learning
models and deep neural network loss functions that also employ class-
balanced strategies. Our findings demonstrate that after hyperparameter
optimization, the deep neural network using the VS loss function substantially improves balanced accuracy. It also surpasses other models in
performance on the AD dataset. Moreover, our feature importance analysis highlights VS-Opt-Net's ability to elucidate biomarker differences
across dementia stages.

Keywords: Class-Balanced Deep Learning · Hyperparameter
Optimization · Neuroimaging · Mild Cognitive Impairment ·
Alzheimer's Disease

This work was supported in part by the NIH grants U01 AG066833, R01 LM013463,
U01 AG068057, P30 AG073105, and R01 AG071470, and the NSF grant IIS 1837964.
Data used in this study were obtained from the Alzheimer's Disease Neuroimaging
Initiative database (adni.loni.usc.edu), which was funded by NIH U01 AG024904.

X. Cao et al. (Eds.): MLMI 2023, LNCS 14349, pp. 144–154, 2024.
https://doi.org/10.1007/978-3-031-45676-3_15

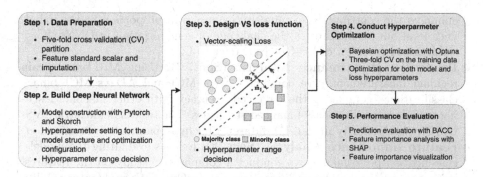

Fig. 1. VS-Opt-Net integrates the VS loss [10] into the STREAMLINE [23] pipeline and employs Bayesian optimization to adaptively learn hyperparameters for both the model and loss function. In Step 3, the VS enlarges the margin of the minority class (m_1) relative to the majority class's margin (m_2).

1   Introduction

Alzheimer's disease (AD) is a degenerative neurological disorder, ranked as the fifth-leading cause of death among Americans aged 65 and older [2]. It leads to irreversible cognitive decline, characterized by gradual cognitive and behavioral impairments [24]. Mild Cognitive Impairment (MCI) is a significant precursor to AD, emphasizing the need for early detection for prompt treatment and disease management [18]. However, distinguishing MCI from cognitively normal (CN) or AD subjects is challenging due to subtle brain changes observed in MCI.

Numerous machine learning algorithms excel in detecting MCI [9,11,20]. However, health datasets, including MCI detection, commonly face imbalanced class distribution [8]. For instance, the MRI data set in Alzheimer's Disease Neuroimaging Initiative (ADNI) [26,27] contains approximately twice as many MCI subjects as CN or AD subjects. Class imbalance can lead to the underrepresentation of minorities, even with highly accurate models. Techniques like data resampling, data augmentation, and class re-weighting have been used to address class imbalance in MCI classification tasks [6,8,16,17,19,29]. However, these approaches may not be as effective for overparameterized models, such as deep neural networks (DNNs), which can suffer from poor generalization [3,10,14,21]. Consequently, such models may overfit the training data, leading to discrepancies in performance when applied to unseen test data.

In light of the challenges faced by existing AD-related classification methods in overparameterized models, we present a novel Bayesian framework that achieves informative predictions for imbalanced data and minimizes generalization error. Our contributions can be summarized as follows:

- **A New Method: VS-Opt-Net (Sect. 2).** We propose VS-Opt-Net, which integrates the vector-scaling loss [10] into the STREAMLINE machine learning pipeline [22,23,25]. Utilizing Bayesian optimization, we adaptively learn hyperparameters for both the model and loss function. VS-Opt-Net not only enhances the contribution of minority examples in proportion to the imbalance level but also addresses the challenge of generalization in DNNs. For a summarized overview of VS-Opt-Net, refer to Fig. 1.

- **Prediction Performance Analysis (Sect. 3).** Using MRI-based brain regional measurements, we conduct CN vs MCI and AD vs MCI binary classifications, comparing the balanced accuracy with other machine learning models employing class-balanced strategies. The results demonstrate VS-Opt-Net's superiority in the AD dataset after hyperparameter optimization.
- **Feature Importance Analysis (Sect. 3).** Besides evaluating the models' classification performance, we conduct a comparative study on the features' impact on prediction. Our findings showcase VS-Opt-Net's explanatory ability in detecting biomarker differences at various dementia stages.

2 Proposed Method

In this section, we cover classification basics, outline the VS loss, explore the STREAMLINE pipeline, and introduce our method, VS-Opt-Net.

Balanced Accuracy and VS Loss. Let (X, Y) be a joint random variable following an underlying distribution $\mathcal{P}(X, Y)$, where $X \in \mathcal{X} \subset \mathbb{R}^d$ is the input, and $Y \in \mathcal{Y} = \{1, \ldots, K\}$ is the label. Suppose we have a dataset $\mathcal{S} = (\mathbf{x}_i, y_i)_{i=1}^n$ sampled i.i.d. from a distribution $\mathcal{P}$ with input space $\mathcal{X}$ and K classes. Let $f : \mathcal{X} \to \mathbb{R}^K$ be a model that outputs a distribution over classes and let $\hat{y}_f = \arg\max_{k \in [K]} f_k(\mathbf{x})$ denote the predicted output. The balanced accuracy (BACC) is the average of the class-conditional classification accuracy:

$$\text{BACC} := \frac{1}{K} \sum_{k=1}^K \mathbb{P}_{\mathcal{P}_k} \left[y = \hat{y}_f(\mathbf{x}) \right]. \tag{BACC}$$

Our approach initially focuses on the VS loss, but it can accommodate other loss functions as well. We provide a detailed description of the VS loss and refer readers to Table 1 for SOTA re-weighting methods designed for training on imbalanced data with distribution shifts. The VS loss [10] unifies multiplicative shift [28], additive shift [14], and loss re-weighting to enhance BACC. For any $(\mathbf{x}, y) \in \mathcal{X} \times \mathcal{Y}$, it has the following form:

$$\ell_{\text{VS}}(y, f(\mathbf{x})) := -w_y \log \left(\frac{e^{l_y f(\mathbf{x})_y + \Delta_y}}{\sum_{j=1}^k e^{l_j f(\mathbf{x})_j + \Delta_j}} \right). \tag{VS}$$

Here, w_j represents the classical weighting term, and l_j and Δ_j are additive and multiplicative logit adjustments. We work with $K = 2$ and aim to find logit parameters (Δ_j, l_j) that optimize BACC. When variables l and Δ are completely unknown and require adaptive optimization based on the model and datasets, we refer to VS as the VS-Opt loss function. The impact of the VS loss on improving balanced accuracy is well-studied in [10,12,21].

STREAMLINE. Simple transparent end-to-end automated machine learning (STREAMLINE) [23] is a pipeline that analyzes datasets with various models through hyperparameter optimization. It serves the specific purpose of comparing performance across datasets, machine learning algorithms, and other Automated machine learning (AutoML) tools. It stands out from other AutoML tools due to its fully transparent and consistent baseline for comparison. This is achieved through a well-designed series of pipeline elements, encompassing exploratory analysis, basic data cleaning, cross-validation partitioning, data scaling and imputation, filter-based feature importance estimation, collective feature selection, ML modeling with hyperparameter optimization over 15 established algorithms, evaluation across 16 classification metrics, model feature importance estimation, statistical significance comparisons, and automatic exporting of all results, plots, a summary report, and models. These features allow for easy application to replication data and enable users to make informed decisions based on the generated results.

VS-Opt-Net: Vector Scaling Loss Optimized for Deep Networks. The following steps introduce VS-Opt-Net, a Bayesian approach for optimizing (VS) loss for DNNs. Figure 1 provides a summary of VS-Opt-Net.

- **Step 1.** We use STREAMLINE for data preprocessing, including train-test split with stratified sampling for five-fold CV. We also impute missing values and scale features to the standard normal distribution for each dataset.

- **Step 2.** We integrate feedforward DNNs and class-balanced DNNs models into STREAMLINE using `skorch`, a PyTorch integration tool with `sklearn`. For DNNs models, we search for optimal structures by setting hyperparameters, including the number of layers and units, activation function, dropout rate, batch normalization usage, as well as optimization configurations such as the function, learning rate, batch size, and epochs. The hyperparameter ranges remain consistent across different models and classification tasks.

- **Step 3.** We integrate VS loss into STREAMLINE for DNNs adaptation. We establish decision boundaries for hyperparameters (l, Δ) in VS loss, as well as for model parameters. We optimize $\tau \in [-1, 2]$ and $\gamma \in [0, 0.5]$ for SOTA losses (see, Table 1), and $l \in [-2, 2]$ and $\Delta \in [0, 1.5]$ for VS-Opt-Net.

- **Step 4.** We employ `Optuna` [1], an open-source Python library used for hyperparameter optimization and built on top of the TPE (Tree-structured Parzen Estimator) algorithm, which is a Bayesian optimization method. We conduct a three-fold CV on the training set, performing a 100-trial Bayesian sweep to optimize both model and loss hyperparameters[1]. For the existing models, we set 'class_weight' to 'None' and 'balanced' to control the use of weights.

---

[1] For existing machine learning models, optimized parameters can be found in https://github.com/UrbsLab/STREAMLINE.

- **Step 5.** We report BACC for evaluating imbalanced classification performance. We use SHAP (SHapley Additive exPlanations) [13] with KernelExplainer to assess feature importance across different models, and top features are visualized by the bar plots and brain region plots.

Table 1. Fixed and tunable hyperparameters for parametric CE losses. N_k denotes sample numbers for class k, and $N_{\min}$ and $N_{\max}$ represent the minimum and maximum sample numbers across all classes; π_k indicates the prior probability of class k.

Loss	Additive	Multiplicative	Optimized Hyperparameter
LDAM [3]	$l = -\frac{1}{2}(N_{\min}/N_k)^{1/4}$	–	–
LA [14]	$l = \tau \log(\pi_k)$	–	τ
CDT [28]	–	$\Delta = (N_k/N_{\max})^{\gamma}$	γ
VS [10]	$l = \tau \log(\pi_k)$	$\Delta = (N_k/N_{\max})^{\gamma}$	τ, γ
l-Opt	l	–	l
Δ-Opt	–	Δ	Δ
VS-Opt	l	Δ	l, Δ

3 Experiments

Datasets. Data for this study were sourced from the ADNI database [26, 27], which aims to comprehensively assess the progression of MCI and AD through a combination of serial MRI, PET, other biological markers, and clinical evaluations. Participants granted written consent, and study protocols gained approval from respective Institutional Review Boards (IRBs)[2]. We collected cross-sectional Freesurfer MRI data from the ADNI site, merging ADNI-1/GO/2 datasets. From the total 1,470 participants (365 CN, 800 MCI, and 305 AD subjects), we selected 317 regional MRI metrics as features. These encompass cortical volume (CV), white matter volume (WMV), surface area (SA), average cortical thickness (TA), and cortical thickness variability (TSD). With these MRI measures as predictors, we performed two binary classifications with noticeable class imbalance: CN vs MCI and AD vs MCI.

---

[2] For the latest information, visit www.adni-info.org.

Baselines. In addition to the deep neural network, we have chosen five commonly used classification models from STREAMLINE to serve as baseline models. These include elastic net, logistic regression, decision tree, random forest, and support vector machine. For all six models, the weight for the k–th class is calculated as $w_k = \pi_k^{-1} = N/(K \cdot N_k)$, where N is the total number of samples, K represents the number of classes (2 for binary classification), N_k is the number of samples for class k, and π_k is the prior probability of class k. We conducted a comparison of DNNs models employing various class-balanced loss functions. Besides the traditional cross-entropy (CE) and weighted cross-entropy (wCE) losses, we evaluated our model against state-of-the-art (SOTA) losses listed in Table 1. These losses incorporate at least one logit adjustment based on class distributions. For models with LA, CDT, and VS loss, we utilized Bayesian optimization to select optimal τ and γ values, which determined l and Δ. Additionally, we introduced two novel approaches: l-Opt and Δ-Opt loss, where we directly optimized logit adjustments l and Δ through Bayesian optimization without class distribution constraints. Furthermore, our proposed method, VS-Opt-Net, optimizes l and Δ together in the VS-Opt loss function.

Prediction Performance Results. We evaluated the prediction performance of various machine learning models for the CN vs MCI and AD vs MCI classification tasks (Table 2). BACC was used to calculate the mean and standard deviation. When comparing models with and without class-balanced weights, we observed that all models showed improvement in BACC after incorporating the weight. However, it is worth noting that the weighted deep neural network underperformed compared to the weighted logistic regression in the CN vs MCI classification, and the weighted SVM in the AD vs MCI classification.

Table 2. Comparison of BACC (mean ± std) for two binary classification tasks using wCE losses with optimized weights w_y and default weight ($w_y = 1$). The table emphasizes that re-weighting alone is ineffective for deep neural networks.

Model	CN VS MCI		AD VS MCI	
	$w_y = 1$	Optimzied w_y	$w_y = 1$	Optimzied w_y
Elastic Net	0.580 ± 0.011	0.650 ± 0.026	0.652 ± 0.040	0.732 ± 0.019
Logistic Regression	0.592 ± 0.037	**0.657 ± 0.042**	**0.738 ± 0.034**	0.742 ± 0.038
Decision Tree	0.569 ± 0.020	0.612 ± 0.027	0.628 ± 0.018	0.679 ± 0.032
Random Forest	0.555 ± 0.018	0.639 ± 0.015	0.657 ± 0.023	0.724 ± 0.024
Support Vector Machine	0.569 ± 0.010	0.650 ± 0.035	0.641 ± 0.014	**0.744 ± 0.042**
Deep Neural Network	**0.606 ± 0.009**	0.633 ± 0.032	0.700 ± 0.055	0.709 ± 0.023

Table 3 compares DNNs models using different class-balanced loss functions. Our numerical analysis shows that models incorporating both additive and multiplicative logit adjustments achieve higher BACC scores than those with only one

Table 3. Balanced accuracy BACC for classification tasks using DNNs models with different loss functions. Cross-validation results are shown as mean ± std in each cell. We tuned the τ and γ in LA, CDT, and VS losses to find l and Δ, following parameters in Table 1. For l-Opt, Δ-Opt, and VS-Opt losses, we adaptively optimized l and Δ.

Loss	CN vs MCI	AD vs MCI
CE	0.606 ± 0.009	0.700 ± 0.055
wCE (w_y)	0.633 ± 0.032	0.709 ± 0.023
LDAM (l)	0.625 ± 0.033	0.726 ± 0.046
LA (l)	0.611 ± 0.037	0.733 ± 0.028
CDT (Δ)	0.608 ± 0.022	0.715 ± 0.033
VS ($l + \Delta$)	0.646 ± 0.035	0.745 ± 0.039
l-Opt	0.641 ± 0.029	0.738 ± 0.037
Δ-Opt	0.608 ± 0.017	0.727 ± 0.043
VS-Opt	**0.669 ± 0.048**	**0.754 ± 0.026**

adjustment, consistent with previous findings in image recognition [10]. Additionally, directly optimizing adjustment parameters with VS-Opt leads to improved prediction performance compared to baselines, enabling our approach to outperform all baseline models.

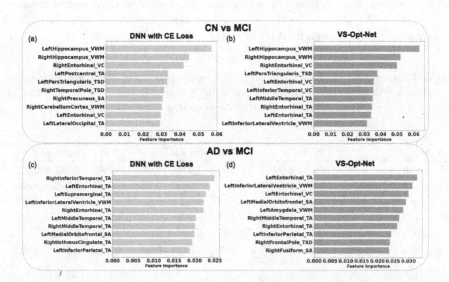

Fig. 2. SHAP feature importance for DNNs with cross-entropy loss (a,c) and VS-Opt-Net (b,d). (a-b) Top regions for CN vs MCI classification. (c-d) Top regions for AD vs MCI classification. Each figure displays the top 10 features for each case.

Feature Importance and Top-Ranked Regions. We analyze feature contributions and assess model classification performance. Figure 2 depicts SHAP feature importance for DNNs with CE and VS-Opt-Net, while Fig. 3 reveals significant brain regions by volume (cortical/white matter), cortical thickness (average/standard deviation), and surface area for VS-Opt-Net. Notably, top-ranking brain regions exhibit similarity between the models, with some regions notably more influential in our model. Cortical/white matter volume and average cortical thickness hold prominent predictive power. Noteworthy features distinguishing CN and MCI encompass hippocampus and right entorhinal cortex volumes. Our model emphasizes the volume of the left entorhinal and left inferior temporal gyri, along with the average thickness of the left middle temporal gyrus-features given less priority by traditional DNNs. For AD vs MCI, key contributors are average thickness of the left entorhinal and volume of the left inferior lateral ventricle. Additionally, contributions from the left entorhinal area and amygdala volumes increase.

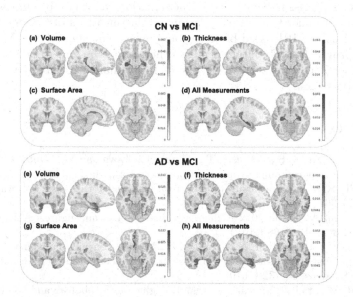

Fig. 3. Brain visualization of the leading 40 features for VS-Opt-Net. Colormap indicates SHAP feature importance; darker shades signify higher significance. Panels (a-d) reveal top features for CN vs MCI classification, while panels (e-h) showcase prime features for AD vs MCI classification. Notably, (a-c) and (e-g) spotlight regions of heightened importance in terms of volume, thickness, and surface area measures for both prediction categories. Panel (d) consolidates (a-c), while panel (h) amalgamates (e-g), displaying the highest importance value when a region encompasses multiple measurements.

The volume reductions of the entorhinal cortex and hippocampus are biomarkers of early Alzheimer's disease. According to prior studies, CN and MCI can be differentiated more accurately using hippocampal volume than lateral neocortical measures [4], which aligns with our feature importance analysis. Additionally, studies have found a significant brain atrophy and thickness decrease in the inferior and middle temporal gyri for MCI patients compared with healthy control [7]. Other studies have reported that AD vs MCI identification is improved by using the entorhinal cortex rather than the hippocampus [5] and the outward deformation of the lateral ventricles. Besides, there is a significant atrophy for the left amygdala when comparing MCI and AD subjects, which is related to the AD severity [15]. The above findings demonstrate the explanatory ability for our model to differentiate between different stages of dementia.

4   Conclusion

We introduced VS-Opt-Net, a novel model integrating the VS loss into STREAMLINE with Bayesian optimization for hyperparameter tuning. It effectively addressed class imbalance and generalization challenges by enhancing the contribution of minority examples. In binary classifications of CN vs MCI and AD vs MCI using MRI-based brain regional measurements, VS-Opt-Net significantly improved BACC, outperforming other models in the AD dataset. Our feature importance analysis revealed successful biomarker explanation at different dementia stages.

References

1. Akiba, T., Sano, S., Yanase, T., Ohta, T., Koyama, M.: Optuna: a next-generation hyperparameter optimization framework. In: Proceedings of the 25th ACM SIGKDD International Conference on Knowledge Discovery & Data Mining, pp. 2623–2631 (2019)
2. Association, A., et al.: 2012 Alzheimer's disease facts and figures. Alzheimer's & Dement. 8(2), 131–168 (2012)
3. Cao, K., Wei, C., Gaidon, A., Arechiga, N., Ma, T.: Learning imbalanced datasets with label-distribution-aware margin loss. In: Advances in Neural Information Processing Systems, vol. 32 (2019)
4. De Santi, S., et al.: Hippocampal formation glucose metabolism and volume losses in MCI and AD. Neurobiol. Aging 22(4), 529–539 (2001)
5. Du, A.T.: Magnetic resonance imaging of the entorhinal cortex and hippocampus in mild cognitive impairment and Alzheimer's disease. J. Neurol. Neurosurg. Psychiatry 71(4), 441–447 (2001)
6. Dubey, R., Zhou, J., Wang, Y., Thompson, P.M., Ye, J.: Analysis of sampling techniques for imbalanced data: an n = 648 ADNI study. Neuroimage 87, 220–241 (2014)
7. Fan, Y., Batmanghelich, N., Clark, C.M., Davatzikos, C., Initiative, A.D.N., et al.: Spatial patterns of brain atrophy in mci patients, identified via high-dimensional pattern classification, predict subsequent cognitive decline. Neuroimage 39(4), 1731–1743 (2008)

8. Hu, S., Yu, W., Chen, Z., Wang, S.: Medical image reconstruction using generative adversarial network for Alzheimer disease assessment with class-imbalance problem. In: 2020 IEEE 6th International Conference on Computer and Communications (ICCC), pp. 1323–1327. IEEE (2020)

9. Kim, D., et al.: A graph-based integration of multimodal brain imaging data for the detection of early mild cognitive impairment (E-MCI). In: Shen, L., Liu, T., Yap, P.-T., Huang, H., Shen, D., Westin, C.-F. (eds.) MBIA 2013. LNCS, vol. 8159, pp. 159–169. Springer, Cham (2013). https://doi.org/10.1007/978-3-319-02126-3_16

10. Kini, G.R., Paraskevas, O., Oymak, S., Thrampoulidis, C.: Label-imbalanced and group-sensitive classification under overparameterization. In: Advances in Neural Information Processing Systems, vol. 34, pp. 18970–18983 (2021)

11. Li, J., et al.: Persistent feature analysis of multimodal brain networks using generalized fused lasso for EMCI identification. In: Martel, A.L., et al. (eds.) MICCAI 2020. LNCS, vol. 12267, pp. 44–52. Springer, Cham (2020). https://doi.org/10.1007/978-3-030-59728-3_5

12. Li, M., Zhang, X., Thrampoulidis, C., Chen, J., Oymak, S.: Autobalance: Optimized loss functions for imbalanced data. In: Advances in Neural Information Processing Systems, vol. 34, pp. 3163–3177 (2021)

13. Lundberg, S.M., Lee, S.I.: A unified approach to interpreting model predictions. In: Advances in Neural Information Processing Systems, vol. 30 (2017)

14. Menon, A.K., Jayasumana, S., Rawat, A.S., Jain, H., Veit, A., Kumar, S.: Long-tail learning via logit adjustment. arXiv preprint arXiv:2007.07314 (2020)

15. Miller, M.I., et al.: Amygdala atrophy in MCI/Alzheimer's disease in the BIO-CARD cohort based on diffeomorphic morphometry. In: Medical Image Computing and Computer-Assisted Intervention: MICCAI... International Conference on Medical Image Computing and Computer-Assisted Intervention, vol. 2012, p. 155. NIH Public Access (2012)

16. Miller, M.I., et al.: Amygdala atrophy in MCI/Alzheimer's disease in the BIO-CARD cohort based on diffeomorphic morphometry. In: Medical Image Computing and Computer-Assisted Intervention: MICCAI. International Conference on Medical Image Computing and Computer-Assisted Intervention, vol. 2012, p. 155. NIH Public Access (2012)

17. Puspaningrum, E.Y., Wahid, R.R., Amaliyah, R.P., et al.: Alzheimer's disease stage classification using deep convolutional neural networks on oversampled imbalance data. In: 2020 6th Information Technology International Seminar (ITIS), pp. 57–62. IEEE (2020)

18. Rasmussen, J., Langerman, H.: Alzheimer's disease – why we need early diagnosis. Degenerative Neurol. Neuromuscul. Dis. Volume 9, 123–130 (2019)

19. Sadegh-Zadeh, S.A., et al.: An approach toward artificial intelligence Alzheimer's disease diagnosis using brain signals. Diagn. 13(3), 477 (2023)

20. Shen, L., et al.: Identifying neuroimaging and proteomic biomarkers for MCI and AD via the elastic net. In: Liu, T., Shen, D., Ibanez, L., Tao, X. (eds.) MBIA 2011. LNCS, vol. 7012, pp. 27–34. Springer, Heidelberg (2011). https://doi.org/10.1007/978-3-642-24446-9_4

21. Tarzanagh, D.A., Hou, B., Tong, B., Long, Q., Shen, L.: Fairness-aware class imbalanced learning on multiple subgroups. In: Uncertainty in Artificial Intelligence, pp. 2123–2133. PMLR (2023)

22. Tong, B., et al.: Comparing amyloid imaging normalization strategies for Alzheimer's disease classification using an automated machine learning pipeline. AMIA Jt. Summits Transl. Sci. Proc. 2023, 525–533 (2023)

23. Urbanowicz, R., Zhang, R., Cui, Y., Suri, P.: Streamline: a simple, transparent, end-to-end automated machine learning pipeline facilitating data analysis and algorithm comparison. In: Genetic Programming Theory and Practice XIX, pp. 201–231. Springer (2023). https://doi.org/10.1007/978-981-19-8460-0_9

24. Uwishema, O., et al.: Is Alzheimer's disease an infectious neurological disease? a review of the literature. Brain Behav. **12**(8), e2728 (2022)

25. Wang, X., et al.: Exploring automated machine learning for cognitive outcome prediction from multimodal brain imaging using streamline. AMIA Jt. Summits Transl. Sci. Proc. **2023**, 544–553 (2023)

26. Weiner, M.W., Veitch, D.P., Aisen, P.S., et al.: The Alzheimer's disease neuroimaging initiative: a review of papers published since its inception. Alzheimers Dement. **9**(5), e111-94 (2013)

27. Weiner, M.W., Veitch, D.P., Aisen, P.S., et al.: Recent publications from the Alzheimer's disease neuroimaging initiative: reviewing progress toward improved AD clinical trials. Alzheimer's Dement. **13**(4), e1–e85 (2017)

28. Ye, H.J., Chen, H.Y., Zhan, D.C., Chao, W.L.: Identifying and compensating for feature deviation in imbalanced deep learning. arXiv preprint arXiv:2001.01385 (2020)

29. Zeng, L., Li, H., Xiao, T., Shen, F., Zhong, Z.: Graph convolutional network with sample and feature weights for Alzheimer's disease diagnosis. Inf. Process. Manage. **59**(4), 102952 (2022)

Enhancing Anomaly Detection in Melanoma Diagnosis Through Self-Supervised Training and Lesion Comparison

Jules Collenne(✉)[iD], Rabah Iguernaissi[iD], Séverine Dubuisson[iD], and Djamal Merad[iD]

Aix Marseille University, Université de Toulon, CNRS, LIS, Marseille, France
jules.collenne@lis-lab.fr

Abstract. Melanoma, a highly aggressive form of skin cancer notorious for its rapid metastasis, necessitates early detection to mitigate complex treatment requirements. While considerable research has addressed melanoma diagnosis using convolutional neural networks (CNNs) on individual dermatological images, a deeper exploration of lesion comparison within a patient is warranted for enhanced anomaly detection, which often signifies malignancy. In this study, we present a novel approach founded on an automated, self-supervised framework for comparing skin lesions, working entirely without access to ground truth labels. Our methodology involves encoding lesion images into feature vectors using a state-of-the-art representation learner, and subsequently leveraging an anomaly detection algorithm to identify atypical lesions. Remarkably, our model achieves robust anomaly detection performance on ISIC 2020 without needing annotations, highlighting the efficacy of the representation learner in discerning salient image features. These findings pave the way for future research endeavors aimed at developing better predictive models as well as interpretable tools that enhance dermatologists' efficacy in scrutinizing skin lesions.

Keywords: Melanoma · Skin Cancer · Lesion Comparison · Representation Learning · Anomaly Detection · Early Detection

1 Introduction

The concept of comparing skin lesions within a single patient, known as the "ugly duckling sign", was introduced by Grob et al. [3] in 1998. According to this concept, visually distinct skin lesions within a patient are more likely to be indicative of cancers. Despite the extensive research on skin lesion classification, the exploration of lesion comparison within patients remains relatively limited, hampering the detection of anomalies that may signify malignancy. Concurrently, convolutional neural networks (CNNs) applied to dermoscopic images have exhibited

X. Cao et al. (Eds.): MLMI 2023, LNCS 14349, pp. 155–163, 2024.
https://doi.org/10.1007/978-3-031-45676-3_16

remarkable success in diagnosing skin diseases [5,13], yet the incorporation of lesion comparison remains underutilized. We posit that integrating the "ugly duckling sign" concept can significantly enhance these diagnostic outcomes.

Previous studies investigating the comparison of skin lesions for melanoma detection [2,4,12] have shown promising results, demonstrating increased detection rates when leveraging the "ugly duckling sign", whether analyzed by dermatologists or artificial intelligence algorithms. Automating this process commonly involves extracting features from a base model trained on the dataset, followed by comparing these features—an approach shared by our work. However, several crucial parameters must be considered in pursuit of this goal.

Soenksen et al. [10] utilize a CNN for classification, extracting features reduced to two dimensions using PCA and compared with Euclidean distance. Yu et al. [14] propose a hybrid model combining classification CNNs and transformers for lesion comparison. However, classification CNNs rely exclusively on labelled data and exhibit bias towards the primary task. Surprisingly, representation learners, purpose-built models for generating comparable feature vectors, have not been explored in prior work. These models employ a modified CNN architecture with a distinct training setup, comparing vector representations of two images. By bringing similar images closer and pushing dissimilar images apart, representation learners offer a promising approach for lesion comparison, distinct from the traditional use of classification CNNs in existing skin lesion studies.

In our approach, we take a significant stride by conducting our analysis without relying on any annotations. The underlying concept is rooted in the understanding that melanoma, like any cancer, entails the uncontrolled proliferation of malignant cells. Given a sufficient amount of data, the aberrant growth patterns exhibited by cancerous cells can be viewed as anomalies, especially when compared to other lesions of a patient. Thus, we posit that with a substantial volume of medical images, cancer detection could potentially be achieved without the need for annotations, as exemplified by our current methodology. To carry out our experiments, we employed the SimSiam [1] siamese neural network, leveraging the ResNet-50 [6] architecture as the backbone. This choice offers the advantage of reduced computational complexity compared to alternative models, while still yielding a robust representation of the skin lesion images.

By addressing the research gaps outlined above, we strive to develop more intelligent and interpretable tools that augment dermatologists' capabilities in analyzing skin lesions, facilitating early melanoma detection and improving patient outcomes.

2   Methods

Our study comprises two modules, as illustrated in Fig. 1. Firstly, we train an encoder, employing a representation learner, to generate vector representations of images in the latent space. Subsequently, for each patient, we compare the generated vectors of their lesions using a k-nearest neighbors-inspired algorithm.

This anomaly detection process yields an "Ugly-Duckling Index" (UDI) which quantifies the degree of atypicality of a lesion compared to others.

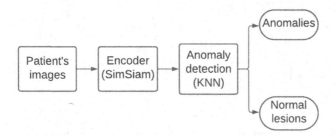

Fig. 1. Pipeline of our anomaly detection

2.1   Features Extraction

For feature extraction, we utilized the SimSiam architecture [1] in conjunction with a ResNet-50 model [6]. This siamese architecture consists of two identical branches, both accepting the same input image that has undergone distinct augmentations. Notably, it does not necessitate the inclusion of negative pairs. During training, one branch undergoes a stop-gradient operation, preventing the model from producing identical vectors for all inputs. We maintained the original model and architecture, making only one adjustment after feature extraction: reducing the size of the vectors from 2048 to 100 by using the UMAP algorithm [7]. This modification was made to ensure that our feature vectors were represented in a dimension that strikes a balance between capturing pertinent information and avoiding excessive dimensionality, which could produce the curse of dimensionality. We also normalize the values per feature using the MinMax scaler to facilitate anomaly detection. Our model is trained on 150 epochs, the loss function is the cosine similarity, the optimizer is SGD and the learning rate is 0.05. Before entering the model, the images are resized from their original size to 224 × 224, which is the expected size for ResNet-50. We kept the original data augmentation techniques of the original SimSiam paper which is a composition of random operations such as resize, crop, random grayscale, gaussian blur, horizontal flip and color jitter. The model seems to have converged after the 150 epochs (Fig. 2).

Following the completion of training on the designated training set, we extracted the features from all images encompassing the dataset, which encompassed both the train and test sets. The acquired feature vectors served as the foundation for the subsequent stage of anomaly detection, facilitating the progression of our analysis.

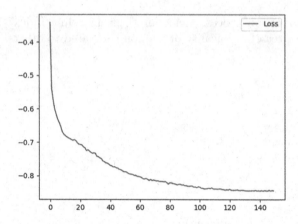

Fig. 2. Loss of ResNet-50 per epoch during SimSiam training.

2.2 Deriving the Ugly Duckling Index Through k-Nearest Neighbors Computation

The process of feature extraction yielded a collection of vector representations for each patient, providing the means to compare and analyze these vectors. For this, we devise a straightforward algorithm inspired by the k-nearest neighbors approach, wherein the relative distances between all lesions are compared. Lesions that exhibit greater distance from their respective k-nearest neighbors than other lesions are assigned a higher Ugly-Duckling Index (UDI). This index, computed per patient, quantifies the distinctiveness of a given lesion in relation to its k-neighbors. The higher the index, the more different the lesion is from its neighbors. First, we compute the distances between all vectors corresponding to the patient's lesions, and subsequently, for each vector, we compute the average distance to its k-nearest neighbors (Eq. 1).

$$D_i = \frac{\sum_{j \in K_i} d(v_i, v_j)}{k} \tag{1}$$

with v_i the features' vector of image i, k the number of neighbours, K_i the ensemble of k neighbours of i and d the distance function (in our case the cosine distance). We divide the resulting values by the average of these values (denoted as $\overline{D}$) in order to obtain the UDI (Eq. 2).

$$UDI_i = \frac{D_i}{\overline{D}} \tag{2}$$

In order to facilitate comparison, we also normalize the resulting values by removing the mean and scaling to unit variance, considering their range in $\mathbb{R}_+^*$. Among various distance metrics available, we opt for the cosine distance, aligning with the cosine similarity utilized by our siamese network for vector comparison. By leveraging this choice, we effectively capitalize on the learned representations,

as the cosine distance and cosine similarity are related by the simple equation
$cosine\_distance = 1 - cosine\_similarity$. Figure 3 presents an example of a two-
dimensional visualization of the results.

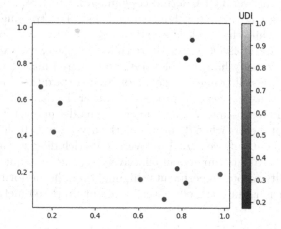

Fig. 3. Visualisation of our anomaly detection algorithm when using $k = 1$ on a 2
dimensional cloud of dots. Each point represents a lesion's features' vector while its
colour represents its UDI.

Our algorithm was intentionally designed to mirror the approach employed by
dermatologists when comparing lesions. Unlike a singular, ideal representation
of normal lesions, typical lesions exhibit diversity and can be characterized by
multiple clusters. Therefore, it is important to note that lesions that deviate
significantly from the average appearance of a normal lesion may not necessarily
be considered "ugly ducklings". By analyzing the k nearest vectors, our algorithm
effectively identifies vectors that are distant from any other cluster, enabling the
accurate prediction of true "ugly ducklings".

We manage cases where patients have a lower number of lesions than the k
parameter by computing the distance to the centroid of the lesions (similar to
prior approaches [10]). To further evaluate the effectiveness of this approach, we
conducted a comparative analysis, contrasting the outcomes obtained using this
method with those achieved through our current algorithm. The results of this
comparative assessment will be presented in the subsequent section.

3   Results

The evaluation of our model on the test set involves comparing the computed
Ugly-Duckling Index (UDI) with the ground truth annotations. In this section,
we present details about the dataset utilized in our experiments, analyze the performance of our model while varying the k parameter, and subsequently examine
the results obtained using the optimal value of k.

3.1   Dataset

We conducted our experiments using the ISIC 2020 dataset [8], which is the largest available dataset for skin lesions (examples of images are presented Fig. 4). It consists of over 32,000 dermoscopy images, which we randomly divided into three subsets for training, validation and test. The training set comprises 70% of patients, while the validation and test sets contain 15% each. Although the ISIC 2020 dataset is extensive, it contains images with various artifacts, including excessive body hair, bubbles, ink, and colored marks. In a traditional classification task, a well-trained model learns to disregard these artifacts as they are irrelevant to the classification process. However, in the context of detecting atypical images in a self-supervised manner, the model may perceive these artifacts as significant factors when determining the atypicality of a lesion. By utilizing the ISIC 2020 dataset, we aimed to leverage its rich diversity and substantial size to train and evaluate our model effectively. In the subsequent sections, we present the results of our experiments, highlighting the performance achieved and providing insights into the effectiveness of our proposed methodology.

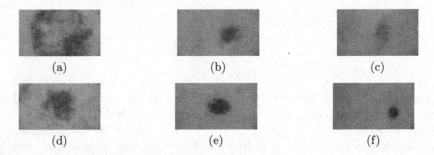

Fig. 4. Examples of images from the ISIC 2020 dataset from a single patient are shown above. All images except for (a) depict nevus.

3.2   Anomaly Detection

We conducted a thorough analysis of our anomaly detection results, comparing them to the ground truth annotations. The evaluation focused on determining the optimal value for the parameter k, and the summarized findings are presented in Fig. 5. This analysis was exclusively performed on the validation data. Our study revealed a clear trend where increasing the value of k resulted in improved performances until reaching a peak at $k = 3$, after which the performance gradually declined. Notably, when utilizing the distance to the centroid of all patient's lesions, the Area Under the ROC Curve (AUC) reached its lowest value, indicating the limited relevance of this approach to our problem. We hypothesize that "ugly duckling" lesions, being typically isolated, can be adequately assessed by comparing them to their three nearest neighbors to determine their deviation

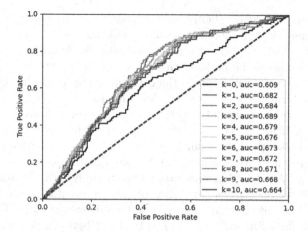

Fig. 5. AUC comparison on the validation data with varying values of the parameter k

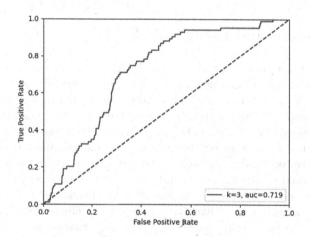

Fig. 6. ROC curve of the model on the test set for $k = 3$.

from other lesions. As lesions are represented by multiple clusters, increasing the number of neighbors may lead to the averaging out of distances, resulting in similar values for all lesions. Using a high value of k leads to the inclusion of the distance to the centroid, which may inadvertently diminish the potential information from patients with a low number of lesions. Using the average of coordinates is comparable to using a high k value, which implies the assumption of a unique normality for lesions, located at a single point in space, which may not accurately reflect reality. Skin lesions present a diverse spectrum of patterns that can still be benign, suggesting that representing them as multiple distinct clusters in the latent space would be more appropriate. Additionally, images that do not belong to any cluster (being too far apart from any cluster) are classified as 'ugly duckling' and are more likely to be malignant.

Following the identification of the optimal value for k on the validation data, we proceeded to assess the performance of our model on the test set using the selected value of $k = 3$. Surprisingly, the corresponding AUC is 0.719, indicating even higher performance compared to the validation data. The peak balanced accuracy achieved by threshold selection is 0.70, aligning with a sensitivity of 0.77 and a specificity of 0.62, thereby illustrating the optimal discriminative performance. Although these results remain below the performance of models that leverage ground-truth annotations, they are notably promising, particularly when taking into account that annotations are entirely omitted from the entire pipeline. It also demonstrates capability of the representation learner to learn visual features in a way that can be leveraged for subsequent tasks. The threshold also affords the flexibility to fine-tune the trade-off between detecting a greater number of melanomas or upholding a higher level of specificity in our model. For a visual depiction of the model's performance, see the ROC curve in Fig. 6.

4   Conclusions and Future Work

Our study showcases compelling outcomes by leveraging representation learners in a fully self-supervised manner on dermoscopic images. Despite the inherent difficulty of this task, our developed architecture demonstrates proficiency in melanoma detection, avoiding the reliance on labeled data and relying solely on inter-lesion comparisons within patients. Subsequent investigations can investigate performance gains achieved by integrating this information into state-of-the-art classification models, as well as exploring novel representation learning techniques focused on the interpretability of the learnt vectors. Recent advancements in one-class classification models [9] and specialized loss functions for siamese networks [11] have diligently addressed the challenge of vector representation, thereby facilitating more seamless image comparisons. Although our approach presently hinges on the k-nearest neighbors algorithm, necessitating diligent parameter tuning, we acknowledge its potential for refinement, making it an enticing avenue for future research endeavors aimed at automating this anomaly detection paradigm. It is noteworthy that this approach could have a substantial impact on broader applications, particularly in the realm of diagnosing various types of cancers and illnesses.

References

1. Chen, X., He, K.: Exploring simple Siamese representation learning. In: Proceedings of the IEEE/CVF Conference on Computer Vision and Pattern Recognition (CVPR), pp. 15750–15758 (2021)
2. Gaudy-Marqueste, C., et al.: Ugly duckling sign as a major factor of efficiency in Melanoma detection. JAMA Dermatol. **153**(4), 279–284 (2017). https://doi.org/10.1001/jamadermatol.2016.5500

3. Grob, J.J., Bonerandi, J.J.: The 'ugly duckling' sign: identification of the common characteristics of Nevi in an individual as a basis for melanoma screening. Arch. Dermatol. **134**(1), 103–104 (1998)
4. Grob, J.J., et al.: Diagnosis of Melanoma: importance of comparative analysis and "ugly duckling" sign. J. Clin. Oncol. **30**(15_suppl), 8578 (2012). https://doi.org/10.1200/jco.2012.30.15_suppl.8578
5. Ha, Q., Liu, B., Liu, F.: Identifying melanoma images using efficientNet ensemble: winning solution to the SIIM-ISIC Melanoma classification challenge. arXiv:2010.05351 (2020). https://doi.org/10.48550/ARXIV.2010.05351. https://arxiv.org/abs/2010.05351
6. He, K., Zhang, X., Ren, S., Sun, J.: Deep residual learning for image recognition (2015). https://doi.org/10.48550/ARXIV.1512.03385. https://arxiv.org/abs/1512.03385
7. McInnes, L., Healy, J., Saul, N., Großberger, L.: UMAP: uniform manifold approximation and projection. J. Open Source Softw. **3**(29), 861 (2018). https://doi.org/10.21105/joss.00861
8. Rotemberg, V., et al.: A patient-centric dataset of images and metadata for identifying melanomas using clinical context. Sci. Data **8**, 34 (2021). https://doi.org/10.1038/s41597-021-00815-z
9. Ruff, L., et al.: Deep one-class classification. In: Dy, J., Krause, A. (eds.) Proceedings of the 35th International Conference on Machine Learning. Proceedings of Machine Learning Research, vol. 80, pp. 4393–4402. PMLR (2018). https://proceedings.mlr.press/v80/ruff18a.html
10. Soenksen, L.R., et al.: Using deep learning for dermatologist-level detection of suspicious pigmented skin lesions from wide-field images. Sci. Transl. Med. **13**(581), eabb3652 (2021). https://doi.org/10.1126/scitranslmed.abb3652. https://www.science.org/doi/abs/10.1126/scitranslmed.abb3652
11. Wang, T., Isola, P.: Understanding contrastive representation learning through alignment and uniformity on the hypersphere. In: III, H.D., Singh, A. (eds.) Proceedings of the 37th International Conference on Machine Learning. Proceedings of Machine Learning Research, vol. 119, pp. 9929–9939. PMLR (2020). https://proceedings.mlr.press/v119/wang20k.html
12. Wazaefi, Y., et al.: Evidence of a limited intra-individual diversity of Nevi: intuitive perception of dominant clusters is a crucial step in the analysis of Nevi by dermatologists. J. Invest. Dermatol. **133**(10), 2355–2361 (2013). https://doi.org/10.1038/jid.2013.183. https://www.sciencedirect.com/science/article/pii/S0022202X15359911
13. Winkler, J.K., et al.: Melanoma recognition by a deep learning convolutional neural network-performance in different Melanoma subtypes and localisations. Eur. J. Cancer **127**, 21–29 (2020). https://doi.org/10.1016/j.ejca.2019.11.020. https://www.sciencedirect.com/science/article/pii/S0959804919308640
14. Yu, Z., et al.: End-to-end ugly duckling sign detection for melanoma identification with transformers. In: de Bruijne, M., et al. (eds.) MICCAI 2021. LNCS, vol. 12907, pp. 176–184. Springer, Cham (2021). https://doi.org/10.1007/978-3-030-87234-2_17

DynBrainGNN: Towards Spatio-Temporal Interpretable Graph Neural Network Based on Dynamic Brain Connectome for Psychiatric Diagnosis

Kaizhong Zheng, Bin Ma, and Badong Chen[✉]

National Key Laboratory of Human-Machine Hybrid Augmented Intelligence,
National Engineering Research Center for Visual Information and Applications,
and Institute of Artificial Intelligence and Robotics, Xi'an Jiaotong University,
Xi'an, China
chenbd@mail.xjtu.edu.cn

Abstract. Mounting evidence has highlighted the involvement of altered functional connectivity (FC) within resting-state functional networks in psychiatric disorder. Considering the fact that the FCs of the brain can be viewed as a network, graph neural networks (GNNs) have recently been applied to develop useful diagnostic tools and analyze the brain connectome, providing new insights into the functional mechanisms of the psychiatric disorders. Despite promising results, existing GNN-based diagnostic models are usually unable to incorporate the dynamic properties of the FC network, which fluctuates over time. Furthermore, it is difficult to produce temporal interpretability and obtain temporally attended brain markers elucidating the underlying neural mechanisms and diagnostic decisions. These issues hinder their possible clinical applications for the diagnosis and intervention of psychiatric disorder. In this study, we propose DynBrainGNN, a novel GNN architecture to analysis dynamic brain connectome, by leveraging dynamic variational autoencoders (DVAE) and spatio-temporal attention. DynBrainGNN is capable of obtaining disease-specific dynamic brain network patterns and quantifying the temporal properties of brain. We conduct experiments on three distinct real-world psychiatric datasets, and our results indicate that DynBrainGNN achieves exceptional performance. Moreover, DynBrainGNN effectively identifies clinically meaningful brain markers that align with current neuro-scientific knowledge.

1 Introduction

Psychiatric disorders (e.g., depression and autism) are associated with abnormal mood, cognitive impairments, insomnia, which significantly impact the well-being of those affected [8]. Despite decades of research, the underlying pathophysiology of psychiatric disorders remains largely elusive. Functional magnetic resonance imaging (fMRI) [15]

K. Zheng and B. Ma—Contribute equally to this work.

Supplementary Information The online version contains supplementary material available at https://doi.org/10.1007/978-3-031-45676-3_17.

has emerged as a promising noninvasive neuroimaging modality to investigate the psychiatric neural mechanism and identify neuroimaging biomarkers. In particular, fMRI can quantify functional connectivity (FC) between different brain regions by measuring pairwise correlations of fMRI time series. Previous studies have identified abnormal brain function biomarkers in diverse patient populations with psychiatric disorders by comparing FC between patients and healthy controls [1, 28].

In recent years, graph neural networks [12] have shown great potential in representing graph-structured data, especially in brain analysis. In these studies, brain is modeled as a complex graph consisting of nodes representing brain regions of interest (ROIs) and edges representing the functional connectivity (FC) between these ROIs. So far, researchers have leveraged GNNs to develop end-to-end diagnostic models for psychiatric disorder which identify important nodes/edges for decision-making [13, 25]. However, existing GNN-based diagnostic models still face challenges in incorporating the dynamic features of FC network and providing temporal interpretability (e.g., dwell time, fractional windows and number of transitions).

To address these issues, we develop a novel GNN-based psychiatric diagnostic framework based on dynamic brain connectom via dynamic variational autoencoders (DVAE) and spatio-temporal attention. Our proposed model incorporates temporal features of dynamic FC to improve the classification performance and obtains dynamic subgraph to uncover neural basis of brain disorders. In addition, we adopt k-means clustering analysis based on spatio-spatially attended graph representations to identify distinct brain temporal states and compare the temporal properties (e.g., dwell time, fractional windows and number of transitions) between patients and healthy controls (HC). We term our model Dynamic Brain Graph Neural Networks (DynBrainGNN) and provides predictions and interpretations simultaneously. To summarize, the major contributions are threefold:

- To our best knowledge, this is the first attempt to develop a novel *built-in* interpretable GNN with dynamic FC for psychiatric disorders analysis.
- We design a spatio-temporal attention module that can incorporate temporal and spatial properties of dynamic FC. More importantly, our spatial attention module could obtain edge-wise attention which is more critical in psychiatric diagnosis [9]. Additionally, we rely on dynamic variational autoencoders to enhance the informativeness and reliability of the spatio-temporal representations obtained.
- Extensive experiments on three psychiatric datasets demonstrates that DynBrainGNN achieves overwhelming performance and obtains disease-specific dynamic brain network connections/patterns coinciding with neuro-scientific knowledge.

2 Proposed Model

2.1 Problem Definition

Psychiatric diagnosis based on dynamic FC can be regarded as dynamic graph classification task. The goal of this task is to train a graph neural network that learns the representation of the dynamic graph for predicting labels. Specifically, given a set of dynamic graphs $\left\{ \mathcal{G}_{dyn}^1, \mathcal{G}_{dyn}^2, ..., \mathcal{G}_{dyn}^N \right\}$, our proposed model learns $\left\{ h_{\mathcal{G}_{dyn}}^1, h_{\mathcal{G}_{dyn}}^2, ..., h_{\mathcal{G}_{dyn}}^N \right\}$ to predict the corresponding labels $\{Y_1, Y_2, ..., Y_N\}$, where $\mathcal{G}_{dyn}^i = \left\{ \mathcal{G}^i(1), ..., \mathcal{G}^i(T) \right\}$ is the sequence of brain functional graphs with T time-points for the i-th participant, $h_{\mathcal{G}_{dyn}}^i$ denotes the representation of $\mathcal{G}_{dyn}^i$ and N is the number of participants.

2.2 Overall Framework of DynBrainGNN

The flowchart of DynBrainGNN is illustrated in Fig. 1 . It consists of four modules: graph encoder, spatial attention module, temporal attention module and DVAE. First, we adopt the graph encoder followed by the spatial attention module using a sequence of dynamic brain functional graphs to produce a sequence of spatial graph representation $h_{G(t)}$. Then we leverage temporal transition followed by temporal attention module to generate a sequence of spatio-temporal graph representation $h_{GT(t)}$. To guarantee the reliability and reconstruction of h_{GT}, the decoder is used over h_{GT} to recover h_G. Finally, a sequence of h_{GT} are summed to generate $\tilde{h}_{G_{dyn}}$, which can predict labels.

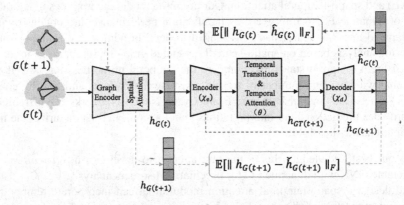

Fig. 1. The overall architecture of our proposed DynBrainGNN. The graph encoder followed by the spatial attention module is used to generate a sequence of spatially attended graph representation vectors $h_{G(t)}$. The temporal transitions and temporal attention module are used to produce a sequence of spatio-temporal graph representations vectors $h_{GT(t+1)}$. $\hat{h}_{G(t)}$ and $\check{h}_{G(t+1)}$ are recovered by the decoder, where $\hat{h}_{G(t)} = \mathcal{X}_d \circ \mathcal{X}_e \left(h_{G(t)} \right)$ and $\check{h}_{G(t+1)} = \mathcal{X}_d \circ \theta \circ \mathcal{X}_e \left(h_{G(t)} \right)$. To ensure reliability of decoder, $\mathbb{E}\left[\left\| h_{G(t)} - \hat{h}_{G(t)} \right\|_F \right]$ and $\mathbb{E}\left[\left\| h_{G(t+1)} - \check{h}_{G(t+1)} \right\|_F \right]$ are computed.

2.3 Construction of Dynamic Functional Graph

A sequence of dynamic brain functional graphs are transformed by the dynamic functional connectivity matrices (dFC). The dFC is constructed by the sliding-window algorithm, where the temporal window of width L is shifted across time series of length T with stride S, resulting in $W = [T - L/S]$ windowed dFC matrices. Each dFC matrix is measured by the Pearson's correlation coefficient of windowed time series between pairwise ROIs.

After obtaining dFC matrices, we further obtain a sequence of dynamic brain functional graphs $\mathcal{G}(t) = (A(t), X(t))$, where the binary adjacency matrix $A(t)$ is quantified by thresholding the top 20-percentile values of all absolute correlation coefficients as connected, and otherwise unconnected. As for the node features $X(t)$, we only use the FC values for simplicity. Specifically, the node feature $X_i(t)$ for node i can be defined as $X_i(t) = [\rho_{i1}, \ldots, \rho_{in}]^{\mathrm{T}}$, where ρ_{ij} is the Pearson's correlation coefficient for node i and node j.

2.4 Graph Encoder

We adopt graph convolutional network (GCN) as the graph encoder. Its propagation rule for the l-th layer is defined as:

$$H^l = \sigma\left(D^{-\frac{1}{2}}\hat{A}D^{-\frac{1}{2}}\Theta^{l-1}\right), \tag{1}$$

where $\hat{A} = A + I$, $D = \sum_j \hat{A}_{ij}$ is a diagonal degree matrix, Θ^{l-1} are trainable parameters and $\sigma(\cdot)$ is the sigmoid activation function.

2.5 Spatio-Temporal Attention-Based READOUT Module

We design two novel attention-based READOUT modules, namely the Spatial Attention READOUT (SAR) and Temporal Attention READOUT (TAR). The modules are inspired by the stochastic attention mechanism introduced in [16]. The attention mask $\mathcal{Z} \in [0,1]^N$ is computed by an attention function $\mathcal{S}$ that maps $\mathbb{R}^{D \times N} \rightarrow [0,1]^N$, taking H as a prior. Formally, this procedure is defined as: $\mathcal{Z} = \mathcal{S}(H)$.

In SAR, H represents edge embeddings transformed by node embeddings, i.e., $H_{\text{space}} = [x_i; x_j]$, where x_i and x_j are node embeddings for node i and node j, and $[\cdot; \cdot]$ denotes the concatenation operation. For TAR, H represents temporal embeddings, concatenating a sequence of graph representations at different times.

For $\mathcal{S}$, both modules utilize a multi-layer perceptron (MLP) based attention. Specifically, an MLP layer followed by a sigmoid function is used to map H into $\mathcal{Z} \in [0,1]$. Then, attention masks are sampled from Bernoulli distributions, and the gumbel-softmax reparameterization trick [10] is applied to update $\mathcal{S}$, ensuring that the gradient of the attention mask is computable. Formally, this procedure is defined as:

$$\mathcal{Z} = \text{Gumbel\_Softmax}\left(\text{Sigmoid}\left(\text{MLP}\left(H\right)\right)\right). \tag{2}$$

After obtaining $\mathcal{Z}_{\text{space}}$, we will further extract specially attended graph representation h_G through graph encoder: $h_G = \text{GCN}(\mathcal{Z}_{\text{space}} \odot G)$, where $\odot$ element-wise multiplication. For TAR, temporally attended graph representation is obtained: $h_{GT} = \mathcal{Z}_{\text{time}} \otimes h_G$ where $\otimes$ refers to the Kronecker product (Fig. 2).

2.6 Dynamic Variational Autoencoders (DVAE)

Dynamic Variational Autoencoders (DVAE) consists of encoder, temporal transition and decoder. The encoder and decoder are MLPs, which are used to reduce dimension and reconstruct the graph representation, respectively. In addition, the temporal

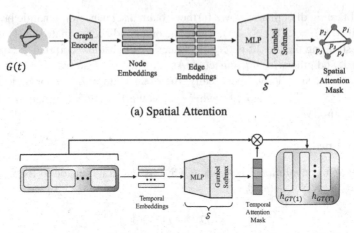

(a) Spatial Attention

(b) Temporal Attention

Fig. 2. (a) Spatial attention-based READOUT module (SAR). With the attention function $\mathcal{S}$, SAR compute spatial attention mask (edge distribution). (b) Temporal attention-based READOUT module (TAR). With the attention function $\mathcal{S}$, TAR calculate temporal attention mask which represents the weights of different time.

transition is a long short-term memory (LSTM) which encodes the sequence of graph representation vectors at the next time: $h_{G(t+1)} = \text{LSTM}\left(h_{G(t)}\right)$.

The objective of DVAE can be defined as:

$$
\mathcal{L}_{\text{DVAE}} = \alpha \left(\sum_{t=1}^{T} \mathbb{E}\left[\left\|h_{G(t)} - \hat{h}_{G(t)}\right\|_F\right] + \sum_{t=1}^{T-1} \mathbb{E}\left[\left\|h_{G(t+1)} - \check{h}_{G(t+1)}\right\|_F\right] \right)
$$
$$
- \beta \left(\sum_{t=1}^{T} \mathbb{E}\left[D_{\text{KL}}\left[q\left(Z|h_{G(t)}\right) \| p\left(Z\right)\right]\right] \right),
$$
(3)

in which $\hat{h}_{G(t)} = \mathcal{X}_d \circ \mathcal{X}_e\left(h_{G(t)}\right)$, $\check{h}_{G(t+1)} = \mathcal{X}_d \circ \theta \circ \mathcal{X}_e\left(h_{G(t)}\right)$, $q\left(Z|h_G\right)$ denotes the encoder model, $\|\cdot\|_F$ is the Frobenius norm and $p\left(Z\right)$ represents a prior distribution, which is assumed to follow an isotropic Gaussian, α and β are the scaling coefficients of the regularization term.

Besides, in order to guarantee the compactness and informativeness of spatio-temporally attended graph representation h_{GT}, we further adopt one regularization term:

$$
\mathcal{L}_{\text{MI}} = \gamma \left(\sum_{t=1}^{T-1} I\left(h_{G(t+1)}, h_{GT(t+1)}\right) \right),
$$
(4)

where $I\left(\cdot\right)$ represents the mutual information and γ is the scaling coefficient. Here, we directly estimate $I\left(h_G, h_{GT}\right)$ with the recently proposed matrix-based Rényi's α-order mutual information [5, 26], which is computationally efficient and mathematically well-defined. For more details, we refer interested readers to supplementary material.

Thus, the objective of DynBrainGNN can be defined as:

$$\mathcal{L} = \mathcal{L}_{\text{CE}} + \mathcal{L}_{\text{DVAE}} + \mathcal{L}_{\text{MI}}, \tag{5}$$

where $\mathcal{L}_{\text{CE}}$ denodes the cross entropy loss.

3 Experiments

3.1 Dataset

Publicly available fMRI data from ABIDE, REST-meta-MDD and SRPBS are used for our experiments. Preprocessing is provided in the supplementary material.

- ABIDE [4]: This dataset[1] openly shares more than 1000 fMRI data. A total of 289 autism patients (ASD) and 293 HCs are provided by ABIDE.
- REST-meta-MDD [24]: This dataset[2] is collected from twenty-five independent research groups in China. In this study, we utilize the fMRI data of 397 MDD patients and 427 HCs.
- SRPBS [21]: This is a multi-disorder MRI dataset[3]. In this work, we consider fMRI data of 234 patients with schizophrenia and 92 HCs.

3.2 Baselines

To evaluate the performance of DynBrainGNN, we compare it against three static shallow models, including support vector machines (SVM) [17], random forest (RF) [20], LASSO [19], three static GNN baselines (GCN [12], GAT [22], GIN [23]) and two GNNs designed for brain networks: BrainGNN [13] and IBGNN [3]. We also include three dynamic models including LSTM [7], DynAERNN [6], EvolveGCN [18] and two dynamic models designed for brain networks: STAGIN-SERO [11], STAGIN-GARO [11].

Table 1. Range of hyper-parameters and final specification. S(), A() and R() represent SRPBS, ABIDE and REST-meta-MDD.

Hyper-parameter	Range Examined	Final Specification
#GNN Layers	[1,2,3,4]	2
#MLP Layers	[1,2,3,4]	2
#Temporal Transitions Layers	[1,2,3]	1
#Hidden Dimensions	[64,128,256,512]	128, 256,128
Learning Rate	[1e−2,1e−3,1e−4]	1e−3
Batch Size	[32,64,128]	64
Weight Decay	[1e−3, 1e−4, 1e−5]	1e−4
α	[1e−4, 5e−4, 1e−3, 5e−3, 1e−2]	S(5e−4), A(5e−4), R(1e−3)
β	[5e−5, 1e−4, 5e−4,1e−3, 5e−3]	S(1e−4), A(1e−3), R(5e−4)
γ	[0.01,0.03,0.05,0.1,0.3,0.5,1]	S(0.1), A(0.05), R(0.3)

---

[1] http://fcon_1000.projects.nitrc.org/indi/abide/.

[2] http://rfmri.org/REST-meta-MDD/.

[3] https://bicr-resource.atr.jp/srpbsfc/.

3.3 Experimental Settings

The hyper-parameters are set through a grid search or based on the recommended settings of related work. To assess the performance of DynBrainGNN and baselines, we employ a five-fold cross-validation approach. Table 1 shows range of hyper-parameters and final specification of DynBrainGNN.

3.4 Evaluation on Classification Performance

Table 2 demonstrates the classification performances in terms of Accuracy, F1-score and matthew's correlation coefficient (MCC). As can be seen, our framework yields impressive improvements over all static and dynamic baselines. DynBrainGNN outperforms the previous traditional static classifiers (e.g., SVM) and static GNN (e.g., GCN), suggesting that incorporating dynamic features of the FC network can improve classification performance. In addition, compared with dynamic deep models such as DynAERNN, DynBrainGNN achieves more than 7% absolute improvements on ABIDE, indicating the efficacy of our brain dynamic network-oriented design.

Table 2. The classification performance, standard deviations of DynBrainnGNN and the baselines on three datasets. * denotes a significant improvement based on paired t-test with $p < 0.05$ compared with baselines. The best performances are in bold.

Method	ABIDE			REST-meta-MDD			SRPBS			Type of FC
	ACC	F1	MCC	ACC	F1	MCC	ACC	F1	MCC	
SVM	0.72 ± 0.04	0.73 ± 0.03	0.44 ± 0.07	0.73 ± 0.02	0.72 ± 0.02	0.45 ± 0.05	0.83 ± 0.05	0.89 ± 0.03	0.59 ± 0.12	Static
LASSO	0.71 ± 0.03	0.71 ± 0.03	0.42 ± 0.06	0.71 ± 0.01	0.70 ± 0.02	0.41 ± 0.03	0.83 ± 0.04	0.88 ± 0.03	0.57 ± 0.10	Static
RF	0.70 ± 0.03	0.70 ± 0.04	0.41 ± 0.06	0.70 ± 0.04	0.66 ± 0.04	0.40 ± 0.09	0.79 ± 0.04	0.87 ± 0.02	0.42 ± 0.13	Static
GCN	0.71 ± 0.02	0.66 ± 0.02	0.42 ± 0.04	0.70 ± 0.02	0.67 ± 0.02	0.41 ± 0.03	0.84 ± 0.01	0.90 ± 0.01	0.60 ± 0.03	Static
GAT	0.72 ± 0.02	0.72 ± 0.03	0.46 ± 0.04	0.71 ± 0.03	0.70 ± 0.04	0.43 ± 0.07	0.83 ± 0.02	0.89 ± 0.01	0.54 ± 0.06	Static
GIN	0.70 ± 0.03	0.72 ± 0.03	0.40 ± 0.07	0.70 ± 0.02	0.69 ± 0.04	0.41 ± 0.04	0.85 ± 0.04	0.90 ± 0.02	0.62 ± 0.10	Static
BrainGNN	0.71 ± 0.05	0.70 ± 0.05	0.39 ± 0.09	0.70 ± 0.03	0.68 ± 0.03	0.39 ± 0.06	0.84 ± 0.02	0.89 ± 0.01	0.58 ± 0.05	Static
IBGNN	0.68 ± 0.04	0.66 ± 0.04	0.34 ± 0.08	0.69 ± 0.03	0.67 ± 0.06	0.38 ± 0.05	0.84 ± 0.02	0.90 ± 0.01	0.59 ± 0.05	Static
LSTM	0.69 ± 0.04	0.69 ± 0.04	0.38 ± 0.08	0.70 ± 0.03	0.66 ± 0.04	0.38 ± 0.07	0.84 ± 0.02	0.90 ± 0.02	0.59 ± 0.04	Dynamic
DynAERNN	0.68 ± 0.03	0.68 ± 0.05	0.35 ± 0.05	0.68 ± 0.01	0.65 ± 0.01	0.36 ± 0.03	0.81 ± 0.02	0.87 ± 0.02	0.51 ± 0.06	Dynamic
EvolveGCN	0.72 ± 0.06	0.71 ± 0.03	0.44 ± 0.11	0.70 ± 0.06	0.68 ± 0.03	0.39 ± 0.11	0.85 ± 0.03	0.90 ± 0.02	0.60 ± 0.05	Dynamic
STAGIN-SERO	0.72 ± 0.02	0.70 ± 0.04	0.44 ± 0.04	0.71 ± 0.01	0.69 ± 0.03	0.41 ± 0.02	0.86 ± 0.03	0.90 ± 0.02	0.64 ± 0.12	Dynamic
STAGIN-GARO	0.71 ± 0.02	0.72 ± 0.03	0.42 ± 0.03	0.72 ± 0.01	0.70 ± 0.04	0.43 ± 0.03	0.85 ± 0.01	0.89 ± 0.02	0.60 ± 0.05	Dynamic
DynBrainGNN	**0.75 ± 0.02***	**0.76 ± 0.02***	**0.51 ± 0.05***	**0.74 ± 0.02***	**0.73 ± 0.02***	**0.49 ± 0.04***	**0.87 ± 0.02***	**0.91 ± 0.01***	**0.67 ± 0.04***	Dynamic

4   Interpretation Analysis

4.1   Disease-Specific Brain Dynamic Network Connections

The spatial attention mask $\mathcal{Z}_{space}$ and temporal attention mask $\mathcal{Z}_{time}$ provide interpretations of dynamically dominant and fluctuant connections. For each participant, dynamically dominant subgraph $\mathcal{G}_{dsub}$ and fluctuant subgraph $\mathcal{G}_{fsub}$ can be defined as:

$$\mathcal{G}_{dsub} = \frac{1}{T}\sum_{t=1}^{T}\left(\mathcal{Z}(t)\right), \mathcal{G}_{fsub} = \sqrt{\frac{1}{T}\sum_{t=1}^{T}\left(\mathcal{Z}(t) - \bar{\mathcal{Z}}\right)^2}, \quad (6)$$

where $\mathcal{Z}(t) = \mathcal{Z}_{space}(t) \odot \mathcal{Z}_{time}(t)$ and $\bar{\mathcal{Z}}$ denotes the average of the sequence of $\mathcal{Z}(t)$.

To assess the differences in dynamic subgraphs between healthy controls (HCs) and patients with autism spectrum disorder (ASD), we compute the average dynamically dominant and fluctuant subgraphs and selected the top 50 edges. Figure 3 shows the results, indicating that the dynamically dominant subgraph of ASD patients exhibits rich interactions between the limbic and subcortical brain networks, while connections within the sensorimotor network are significantly less than those of HCs. These findings are consistent with previous studies [2] that have reported abnormal functional connectivity within and between the limbic and sensorimotor networks in ASD, suggesting that these patterns may contribute to the pathophysiology of ASD. Furthermore, we observe that connections within the fronto-parietal network in patients are sparser than HCs, indicating dynamic functional alterations in this neural system. This observation aligns with a previous finding [14] showing that dynamic brain configurations within the lateral and medial frontoparietal networks occur less frequently in ASD than in HCs.

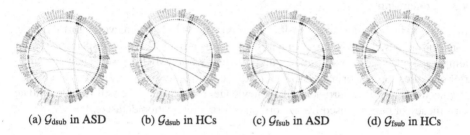

(a) $\mathcal{G}_{\text{dsub}}$ in ASD   (b) $\mathcal{G}_{\text{dsub}}$ in HCs   (c) $\mathcal{G}_{\text{fsub}}$ in ASD   (d) $\mathcal{G}_{\text{fsub}}$ in HCs

Fig. 3. The obtained dynamic subgraph on ABIDE. The colors of brain neural systems are described as: visual network (VN), somatomotor network (SMN), dorsal attention network (DAN), ventral attention network (VAN), limbic network (LIN), frontoparietal network (FPN), default mode network (DMN), cerebellum (CBL) and subcortial network (SBN), respectively. (Color figure online)

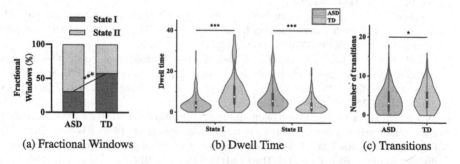

(a) Fractional Windows   (b) Dwell Time   (c) Transitions

Fig. 4. Temporal properties of brain states including (a) Fractional windows (b) Dwell time and (c) number of transitions on ABIDE. Note that the optimal number of clusters is 2 on ABIDE. $^*P < 0.05,^{***}P < 0.0001$. The interpretation analyses of REST-meta-MDD and SRPBS are provided in the supplementary material.

4.2 Temporal Properties

We provide the interpretations of temporal properties to understand brain flexibility and adaptability in psychiatric disorders. Specifically, We first apply a k-means clustering algorithm on windowed spatio-temporally attended graph representations h_{GT} to assess dynamic brain patterns (states). The optimal number of clusters is determined using a cluster validity analysis based on silhouette scores. We then quantify group differences in the temporal properties of these states, including dwell time (i.e., the duration of consecutive windows belonging to one state), fractional windows (i.e., the fraction of total windows belonging to one state), and number of transitions (i.e., the number of transitions between states), using a two-sample t-test with false discovery rate (FDR) correction (Fig. 4). Our analysis reveals that ASD patients have higher fractional windows and mean dwell time in state II, which is consistent with a recent neuroimaging study [27].

4.3 Conclusion

In this study, we develop a novel interpretable GNN framework for dynamic connectome-based psychiatric diagnosis, which provides disorder-specific spatio-temporal explanations. Experiments on three real-world fMRI datasets show the superior classification performance and validate the rationality of our discovered biomarkers/mechanisms with neuro-scientific knowledge. In the future, we will further validate the efficacy of our proposed framework on external replication datasets.

Acknowledgements. This work was supported by the National Natural Science Foundation of China with grant numbers (62088102, U21A20485, 61976175).

References

1. Biswal, B.B., et al.: Toward discovery science of human brain function. Proc. Natl. Acad. Sci. **107**(10), 4734–4739 (2010)
2. Boedhoe, P.S., et al.: Subcortical brain volume, regional cortical thickness, and cortical surface area across disorders: findings from the enigma ADHD, ASD, and OCD working groups. Am. J. Psychiatry **177**(9), 834–843 (2020)
3. Cui, H., Dai, W., Zhu, Y., Li, X., He, L., Yang, C.: Interpretable graph neural networks for connectome-based brain disorder analysis. In: Wang, L., Dou, Q., Fletcher, P.T., Speidel, S., Li, S. (eds.) MICCAI 2022. LNCS, vol. 13438, pp. 375–385. Springer, Cham (2022). https://doi.org/10.1007/978-3-031-16452-1_36
4. Di Martino, A., et al.: The autism brain imaging data exchange: towards a large-scale evaluation of the intrinsic brain architecture in autism. Mol. Psychiatry **19**(6), 659–667 (2014)
5. Giraldo, L.G.S., Rao, M., Principe, J.C.: Measures of entropy from data using infinitely divisible kernels. IEEE Trans. Inf. Theory **61**(1), 535–548 (2014)
6. Goyal, P., Chhetri, S.R., Canedo, A.: dyngraph2vec: Capturing network dynamics using dynamic graph representation learning. Knowl.-Based Syst. **187**, 104816 (2020)
7. Greff, K., Srivastava, R.K., Koutník, J., Steunebrink, B.R., Schmidhuber, J.: LSTM: a search space odyssey. IEEE Trans. Neural Netw. Learn. Syst. **28**(10), 2222–2232 (2016)
8. Hyman, S.E.: A glimmer of light for neuropsychiatric disorders. Nature **455**(7215), 890 (2008)
9. Insel, T.R., Cuthbert, B.N.: Brain disorders? Precisely. Science **348**(6234), 499–500 (2015)
10. Jang, E., Gu, S., Poole, B.: Categorical reparameterization with gumbel-softmax. In: International Conference on Learning Representations (2016)
11. Kim, B.H., Ye, J.C., Kim, J.J.: Learning dynamic graph representation of brain connectome with spatio-temporal attention. In: Advances in Neural Information Processing Systems, vol. 34, pp. 4314–4327 (2021)

12. Kipf, T.N., Welling, M.: Semi-supervised classification with graph convolutional networks. In: International Conference on Learning Representations (2016)
13. Li, X., et al.: BrainGNN: interpretable brain graph neural network for fMRI analysis. Med. Image Anal. **74**, 102233 (2021)
14. Marshall, E., et al.: Coactivation pattern analysis reveals altered salience network dynamics in children with autism spectrum disorder. Netw. Neurosci. **4**(4), 1219–1234 (2020)
15. Matthews, P.M., Jezzard, P.: Functional magnetic resonance imaging. J. Neurol. Neurosurg. Psychiatr. **75**(1), 6–12 (2004)
16. Miao, S., Liu, M., Li, P.: Interpretable and generalizable graph learning via stochastic attention mechanism. In: International Conference on Machine Learning, pp. 15524–15543. PMLR (2022)
17. Pan, X., Xu, Y.: A novel and safe two-stage screening method for support vector machine. IEEE Trans. Neural Netw. Learn. Syst. **30**(8), 2263–2274 (2018)
18. Pareja, A., et al.: EvolveGCN: evolving graph convolutional networks for dynamic graphs. In: Proceedings of the AAAI Conference on Artificial Intelligence, vol. 34, pp. 5363–5370 (2020)
19. Ranstam, J., Cook, J.: Lasso regression. J. Br. Surg. **105**(10), 1348–1348 (2018)
20. Rigatti, S.J.: Random forest. J. Insurance Med. **47**(1), 31–39 (2017)
21. Tanaka, S.C., et al.: A multi-site, multi-disorder resting-state magnetic resonance image database. Sci. Data **8**(1), 227 (2021)
22. Velickovic, P., Cucurull, G., Casanova, A., Romero, A., Lio, P., Bengio, Y., et al.: Graph attention networks. Stat **1050**(20), 10–48550 (2017)
23. Xu, K., Hu, W., Leskovec, J., Jegelka, S.: How powerful are graph neural networks? In: International Conference on Learning Representations (2018)
24. Yan, C.G., et al.: Reduced default mode network functional connectivity in patients with recurrent major depressive disorder. Proc. Natl. Acad. Sci. **116**(18), 9078–9083 (2019)
25. Ying, Z., Bourgeois, D., You, J., Zitnik, M., Leskovec, J.: Gnnexplainer: generating explanations for graph neural networks. In: Advances in Neural Information Processing Systems, vol. 32 (2019)
26. Yu, S., Giraldo, L.G.S., Jenssen, R., Principe, J.C.: Multivariate extension of matrix-based rényi's α-order entropy functional. IEEE Trans. Pattern Anal. Mach. Intell. **42**(11), 2960–2966 (2019)
27. Yue, X., et al.: Abnormal dynamic functional network connectivity in adults with autism spectrum disorder. Clin. Neuroradiol. **32**(4), 1087–1096 (2022)
28. Zhang, Y., et al.: Identification of psychiatric disorder subtypes from functional connectivity patterns in resting-state electroencephalography. Nat. Biomed. Eng. **5**(4), 309–323 (2021)

Precise Localization Within the GI Tract by Combining Classification of CNNs and Time-Series Analysis of HMMs

Julia Werner[1]($\boxtimes$), Christoph Gerum[1], Moritz Reiber[1], Jörg Nick[2], and Oliver Bringmann[1]

[1] Department of Computer Science, University of Tübingen, Tübingen, Germany
julia-helga.werner@uni-tuebingen.de
[2] Department of Mathematics, ETH Zürich, Zürich, Switzerland

Abstract. This paper presents a method to efficiently classify the gastroenterologic section of images derived from Video Capsule Endoscopy (VCE) studies by exploring the combination of a Convolutional Neural Network (CNN) for classification with the time-series analysis properties of a Hidden Markov Model (HMM). It is demonstrated that successive time-series analysis identifies and corrects errors in the CNN output. Our approach achieves an accuracy of 98.04% on the Rhode Island (RI) Gastroenterology dataset. This allows for precise localization within the gastrointestinal (GI) tract while requiring only approximately 1M parameters and thus, provides a method suitable for low power devices.

Keywords: Medical Image Analysis · Wireless Capsule Endoscopy · GI Tract Localization

1 Introduction

The capsule endoscopy is a medical procedure that has been used for investigating the midsection of the GI tract since early 2000 [3,12]. This minimally invasive method allows to visualize the small intestine, which is in most part not accessible through standard techniques using flexible endoscopes [22]. The procedure starts by swallowing a pill-sized capsule. While it moves through the GI tract by peristalsis, it sends captured images from an integrated camera with either an adaptive or a defined frame rate to an electronic device. The overall aim of this procedure is to detect diseases affecting the small intestine such as tumors and its preliminary stages, angiectasias as well as chronic diseases [17,22,24]. Since the esophagus, stomach and colon can be more easily assessed by standard techniques, the small intestine section is of main interest in VCE studies.

This work has been partly funded by the German Federal Ministry of Education and Research (BMBF) in the project MEDGE (16ME0530).

All images of the small intestine should be transmitted for further evaluation by medical experts who are qualified to check for anomalies. The frame rate of the most prominent capsules ranges from 1 to 30 frames per second with a varying resolution between 256×256 and 512×512 depending on the platform [22]. For example, the PillCam® SB3 by Medtronic lasts up to 12 h with an adaptive frame rate of 2 to 6 frames per second [18]. This should ensure passing through the whole GI tract before the energy of the capsule's battery is depleted. However, a capsule can also require more than one day to pass through the whole GI tract leading to an incomplete record of images due to depletion of the capsule's battery after maximal 12 h. In this procedure, the energy is the bottleneck and small changes of the architecture can increase the overall energy requirement leading to a shorter battery lifetime with the risk of running out of energy without covering the small intestine. However, modifications such as capturing images with a higher resolution might improve the recognition ability of clinicians and thus, it is desirable to increase the limited resolution or add more functions (e.g. zooming in or out, anomaly detection on-site) helping to successfully scan the GI tract for anomalies at the cost of increasing energy demands. The images taken before the small intestine are not of interest but demand their share of energy for capturing and transmitting the images.

This paper presents a method for very accurately determining the location of the capsule by on-site evaluation using a combination of neural network classification and time-series analysis by a HMM. This neglects the necessity to consume electric energy for transmitting images of no interest. If this approach is integrated into the capsule it can perform precise self-localization and the transition from the stomach to the small intestine is verified with high confidence. From this moment onwards, all frames should be send out for further evaluation. A major part of the energy can be saved since the data transmission only starts after the capsule enters the small intestine and therefore can be used for other valuable tasks. For example, the frame rate or resolution could be increased while in the small intestine or additionally, a more complex network for detecting anomalies on-site could be employed.

1.1   Related Work

In the field of gastroenterology, there have been different approaches to perform localization of a capsule within the GI tract [16] including but not limited to magnetic tracking [19,26], video-based [15,28] and electromagnetic wave techniques [7,27]. However, the datasets are typically not publicly available which prevents the reproduction of the results. Charoen et al. [2] were the first to publish a dataset with millions of images classified into the different sections of the GI tract. They achieved an accuracy of 97.1% with an Inception ResNet V2 [23] architecture on the RI dataset and therefore successfully demonstrated precise localization without aiming for an efficient realization on hardware. To the best of our knowledge, there is no superior result than the baseline with this dataset. However, a large network with 56M parameters as the Inception ResNet V2 is not suitable for low-power embedded systems since the accompanied high energy demand results in a short battery lifetime. Thus, we present a new approach for this problem setting using the same dataset and the same split resulting in a higher accuracy while requiring a much smaller network and less parameters.

2   Methodology

2.1   Inference

To improve the diagnosis within the GI tract, a tool for accurate self-localization of the capsule is presented. Since the energy limitation of such a small device needs to be considered, it is crucial to limit the size of the typical large deep neural network. The presented approach achieves this by improving the classification results of a relatively small CNN with subsequent time-series analysis.

CNNs have been successfully used for many different domains such as computer vision, speech and pattern recognition tasks [1,8,11] and thus, were employed for the classification task in this work. MobileNets [10] can be categorized as leight weight CNNs, which have been used for recognition tasks while being efficiently employed on mobile devices. Since a low model complexity is essential for the capsule application, the MobileNetV3-Small [9] was utilized and is in the following interchangeably referred to as CNN. For subsequent time-series analysis, a HMM was chosen, since the statistical model is well established in the context of time series data [20]. Due to the natural structure of the GI tract in humans, the order of states within a VCE is known. The capsule traverses the esophagus, stomach, small intestine and colon sequentially in every study. This inherent structure of visited locations can be directly encoded into the transition probabilities of the HMM. Finally, the predictions from the CNN are interpreted as the emissions of a HMM and the Viterbi algorithm [5] is used to compute the most likely sequence of exact locations, given the classifications of the CNN.

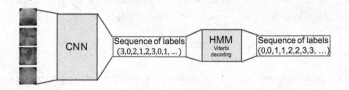

Fig. 1. Illustration of the presented approach (GI tract images from [2]).

Hence, the presented method for localizing the four different gastroenterology sections consists of two phases as depicted in Fig. 1. For each patient, the CNN classifies chronologically received input data from the RI gastroenterology VCE dataset [2] into the four given classes. The respective output labels/predictions of the CNN are fed into a HMM, which uses the Viterbi algorithm for determining the most likely sequence of states given the observations from the CNN. With four hidden states and often much more than 10000 observations per patient, the size of the matrix storing the likelihood values for each hidden state and each observation has usually more than 40000 entries. However, the less storage is required, the more useful this method becomes for low power devices. Furthermore, as the decoding is performed backwards a larger matrix leads to an increasing delay in classification. Thus, to limit the size of the matrix, a sliding window of size n was used to build the matrix with a predefined shape. After succeeding the nth observation, for each new addition to the matrix the first column is removed, ensuring the predefined shape. The designated route the capsule moves along is known by the given anatomy of the GI tract. Therefore, specific assumptions can be made confidentially, e.g. the capsule cannot simply skip an organ, nor does the capsule typically move backwards to an already passed organ. To exploit this advantage of prior knowledge, the Viterbi decoding is used to detect the transitions for each organ by limiting the possible transition to the subsequent organ until a transition is detected.

2.2 HMM and Viterbi Decoding

HMMs are popular tools in the context of time-dependent data, e.g. in pattern recognition tasks [13,21,25], characterized by low complexity compared to other models. The probabilistic modeling technique of a HMM assumes an underlying Markov chain, which describes the dynamic of the hidden states $\{S_1, \ldots, S_n\}$. In the present setting, the exact location of the capsule is interpreted as the hidden state, which gives $\{S_1 = \text{Esophagus}, S_2 = \text{Stomach}, S_3 = \text{Small intestine}, S_4 = \text{Colon}\}$. At any time point t, an emission $X_t \in \{K_1, \ldots, K_m\}$, corresponding to one of the locations as classified by the neural network, is observed and assumed to be sampled from the emission probabilities, which only depend on the hidden state S_t. The model is thus completely determined by the transition probabilities a_{ij} of the Markov chain S_t (as well as its initial distribution $\pi_i = P(X_1 = S_i)$ for $i \leq 4$) and the emission probabilities $b_j(k)$, which are given by the probabilities

$$a_{ij} = P(X_{t+1} = S_j | X_t = S_i), \quad b_j(k) = P(O_t = K_k | X_t = S_j),$$

for $i, j \leq 4$ and $k \leq 4$, where O_t denotes the observation at time t [4,14]. The objective of the model is then to infer the most likely hidden states, given observations $O = (O_1, \ldots, O_t)$, which is effectively realized by the Viterbi algorithm [5]. The Viterbi algorithm then efficiently computes the most likely sequence of gastroenterologic states $X = (X_1, \ldots, X_t)$, given the evaluations of the neural network $(O_1, \ldots, O_t)$, namely

$$\arg \max_{X_1, \ldots, X_t} P(X_1, \ldots, X_t, O_1, \ldots O_t).$$

To determine good approximations for the transition and emission probabilities in this problem setting, a grid search was performed. For the diagonal and superdiagonal entries of the probability matrix, a defined number of values was tested as different combinations and for each variation the average accuracy computed for all patients (all other values were set to zero). The final probabilities were then chosen based on the obtained accuracies from the grid search and implemented for all following experiments. An additional metric is employed to evaluate the time lag between the first detection of an image originating from the small intestine and the actual passing of the capsule at this position. This delay arises due to the required backtrace of the matrix storing the log-likelihoods during the Viterbi decoding before the final classification can be performed.

3   Results and Discussion

To determine the window size used during Viterbi decoding for subsequent experiments, the average accuracy as well as the average delays of the Viterbi decoding after classification with the CNN+HMM combination are plotted over different window sizes (see Fig. 2). A larger window size results in a higher accuracy and a larger delay, while a window size reduction leads to an accuracy loss but also a decrease of the delay. A reasonable tradeoff seems to be given by a window size of 300 samples, which was chosen for further experiments.

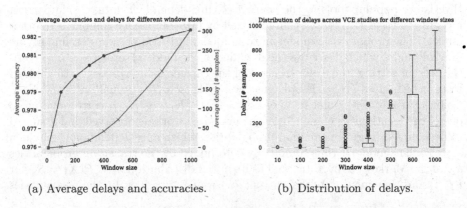

(a) Average delays and accuracies.   (b) Distribution of delays.

Fig. 2. Delays and accuracies for different window sizes sliding over the log-likelihood matrix of the Viterbi decoding.

Table 1 displays the results of the CNN+HMM combination in comparison to only using the CNN on different input image sizes (averaged describes the average over the values per patient). The $n \times n$ center of the input image was cropped and used as an input to explore the reduced complexity in terms of accuracy. This demonstrates that combining the CNN with time-series analysis can compensate a proportion of false classifications of the CNN and enhances its overall classification abilities notably. However, for all subsequent experiments an input image size of 320×320 was used for better comparability with the results of the original authors.

In all experiments, the MobileNetV3 was trained for 10 epochs on the RI training set with the AdamW optimizer and a learning rate of 0.001 with the HANNAH framework [6]. The presented approach for localization within the

Table 1. Results of the CNN+HMM approach with different input image sizes.

Input size	64×64		120×120		320×320	
Metric	CNN	CNN+HMM	CNN	CNN+HMM	CNN	CNN+HMM
Accuracy [%]	90.60	96.16	93.94	97.37	96.95	98.04
Averaged MAE	0.1178	0.0560	0.07895	0.0406	0.0463	0.0350
Averaged R2-Score	0.1764	0.6889	0.4454	0.7819	0.7216	0.8077
Average Delay	–	19.11	–	17.87	–	19.19

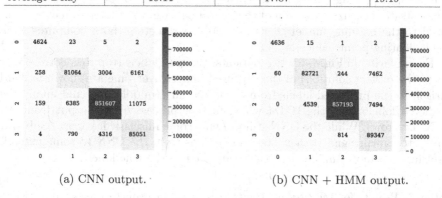

(a) CNN output.   (b) CNN + HMM output.

Fig. 3. Confusion matrices of the CNN output (a) and the CNN+HMM combination (b) (classes: esophagus (0), stomach (1), small intestine (2) and colon (3)).

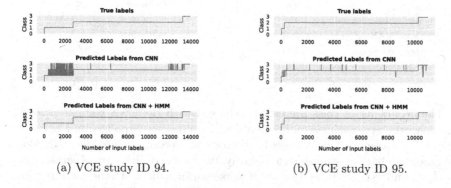

(a) VCE study ID 94.   (b) VCE study ID 95.

Fig. 4. Comparison of class predictions, examplarily shown for two VCE studies.

GI tract by combining classification with the CNN and the time-series analysis of the HMM achieved an accuracy of 98.04% with a window size of $w = 300$. This is an improvement compared to only applying the MobileNetV3 on its own (96.95%). The corresponding confusion matrices are shown in Fig. 3, displaying the improved classification per class of the combination CNN+HMM (b) and compared to solely using the CNN (a). It becomes apparent that particularly the classification of stomach and colon images was notably enhanced (3004 vs.

244 images misclassified as small intestine and 4316 vs. 814 images misclassified as colon). Hence, even with low resolutions high accuracies were achieved (Table 1) and an overall improved self-localization was demonstrated (Fig. 3). Importantly, this allows to use low resolutions within the sections outside of the small intestine while still providing a precise classification. Therefore, less energy is required within these sections and the energy can be either used for other tasks or simply leads to a longer battery lifetime. Table 2 extends the results by displaying additional metrics in comparison to the baseline [2]. This demonstrates the size reduction of the neural network to $\approx$ 1M parameters compared to the baseline model [2] with $\approx$ 56M parameters (both computed with 32-bit floating-point values).

Examplarily, in Fig. 4, for two patients, the true labels (top row, corresponding to the perfect solution) of the captured images over time are shown in comparison to the predicted labels from the CNN only (middle row) and finally the predicted labels from the HMM which further processed the output from the CNN (last row). While the CNN still presents some misclassifications, the HMM is able to capture and correct false predictions from the CNN to some extend, resulting in a more similar depiction compared to the true labels.

Table 2. Results for the presented approach in comparison to the baseline results (Mean values over all 85 tested VCE studies).

Metric	MobilenetV3 + HMM ($w = 300$)	MobilenetV3	Baseline [2]
Accuracy [%]	98.04	96.95	97.1
Number of Parameters	$\approx$ 1M	$\approx$ 1M	$\approx$ 56M
Averaged MAE	0.0350	0.0463	–
Averaged R2-Score	0.8077	0.7216	–
Average Delay (# Frames)	19.19	–	–

The accuracies of class prediction achieved with the CNN compared to the combinatorial approach over all patient VCE studies are visualized in Fig. 5. The CNN+HMM combination achieved superior results compared to the CNN alone for almost all patient studies. Examplarily, two of the outliers are observed more closely to understand the incidences of worse performance with the combinatorial approach (Fig. 6). Presented in Fig. 6a, one VCE study shows poor results for both approaches. The CNN misclassifies the majority of the images achieving an accuracy of only 25.20%. Subsequently, as the preceding classification of the CNN is mostly incorrect, the Viterbi decoding cannot classify the error-prone labels correctly received from the CNN resulting in a very early misclassification as colon. Since the HMM cannot take a step back and the CNN provides such an error-prone output, the remaining images are also classified as colon resulting in an accuracy of 5.93%. For one VCE study (Fig. 6b) overall good results can be achieved, but classification with the combination of CNN+HMM showed a slightly worse result than classifying with the CNN only. It becomes apparent that the CNN has trouble classifying images near the transition of small intestine and colon leading to a delayed transition detection by the HMM. It is expected that a large proportion of misclassifications can be avoided with time-dependent transition probabilities.

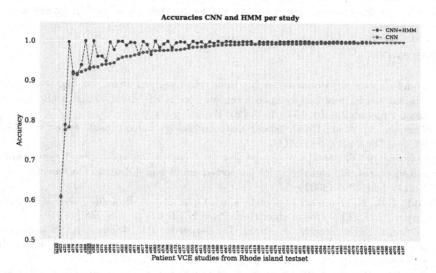

Fig. 5. Accuracies of the CNN compared to the combinatorial approach CNN+HMM. Marked in red are two studies with worse results if the combination is used, more details can be found in Fig. 6a and Fig. 6b. (Color figure online)

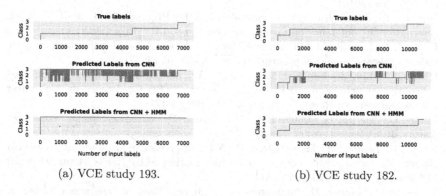

(a) VCE study 193.   (b) VCE study 182.

Fig. 6. Comparison of class predictions for two outlier VCE studies.

4   Conclusion

A pipeline to accurately classify the current gastroenterologic section of VCE images was proposed. The combination of a CNN followed by time-series analysis can automatically classify the present section of the capsule achieving an accuracy of 98.04%. Limited by the small size of the embedded device in practical use, considering the given énergy constraints is crucial. The presented approach requires only $\approx$ 1M parameters while providing a higher accuracy than the current baseline. The resulting energy reduction provides more options on when and how relevant images are captured and will be further addressed in future work.

References

1. Abdel-Hamid, O., Mohamed, A.R., Jiang, H., Deng, L., Penn, G., Yu, D.: Convolutional neural networks for speech recognition. IEEE/ACM Trans. Audio Speech Lang. Process. **22**(10), 1533–1545 (2014)
2. Charoen, A., et al.: Rhode island gastroenterology video capsule endoscopy data set. Sci. Data **9**(1), 602 (2022)
3. Costamagna, G., et al.: A prospective trial comparing small bowel radiographs and video capsule endoscopy for suspected small bowel disease. Gastroenterology **123**(4), 999–1005 (2002)
4. Eddy, S.R.: Hidden markov models. Curr. Opin. Struct. Biol. **6**(3), 361–365 (1996)
5. Forney, G.D.: The Viterbi algorithm. Proc. IEEE **61**(3), 268–278 (1973)
6. Gerum, C., Frischknecht, A., Hald, T., Bernardo, P.P., Lübeck, K., Bringmann, O.: Hardware accelerator and neural network co-optimization for ultra-low-power audio processing devices. arXiv preprint arXiv:2209.03807 (2022)
7. Goh, S.T., Zekavat, S.A., Pahlavan, K.: DOA-based endoscopy capsule localization and orientation estimation via unscented Kalman filter. IEEE Sens. J. **14**(11), 3819–3829 (2014)
8. He, K., Zhang, X., Ren, S., Sun, J.: Deep residual learning for image recognition. In: Proceedings of the IEEE Conference on Computer Vision and Pattern Recognition, pp. 770–778 (2016)
9. Howard, A., et al.: Searching for mobilenetv3. In: Proceedings of the IEEE/CVF International Conference on Computer Vision, pp. 1314–1324 (2019)
10. Howard, A.G., et al.: Mobilenets: efficient convolutional neural networks for mobile vision applications. arXiv preprint arXiv:1704.04861 (2017)
11. Huang, G., Liu, Z., Van Der Maaten, L., Weinberger, K.Q.: Densely connected convolutional networks. In: Proceedings of the IEEE Conference on Computer Vision and Pattern Recognition, pp. 4700–4708 (2017)
12. Iddan, G., Meron, G., Glukhovsky, A., Swain, P.: Wireless capsule endoscopy. Nature **405**(6785), 417–417 (2000)
13. Kenny, P., Lennig, M., Mermelstein, P.: A linear predictive hmm for vector-valued observations with applications to speech recognition. IEEE Trans. Acoust. Speech Sig. Process. **38**(2), 220–225 (1990)
14. Manning, C., Schutze, H.: Foundations of Statistical Natural Language Processing. MIT Press, Cambridge (1999)
15. Marya, N., Karellas, A., Foley, A., Roychowdhury, A., Cave, D.: Computerized 3-dimensional localization of a video capsule in the abdominal cavity: validation by digital radiography. Gastrointest. Endosc. **79**(4), 669–674 (2014)
16. Mateen, H., Basar, R., Ahmed, A.U., Ahmad, M.Y.: Localization of wireless capsule endoscope: a systematic review. IEEE Sens. J. **17**(5), 1197–1206 (2017)
17. McLaughlin, P.D., Maher, M.M.: Primary malignant diseases of the small intestine. Am. J. Roentgenol. **201**(1), W9–W14 (2013)

18. Monteiro, S., de Castro, F.D., Carvalho, P.B., Moreira, M.J., Rosa, B., Cotter, J.: Pillcam® sb3 capsule: Does the increased frame rate eliminate the risk of missing lesions? World J. Gastroenterol. **22**(10), 3066 (2016)
19. Pham, D.M., Aziz, S.M.: A real-time localization system for an endoscopic capsule. In: 2014 IEEE Ninth International Conference on Intelligent Sensors, Sensor Networks and Information Processing (ISSNIP), pp. 1–6. IEEE (2014)
20. Rabiner, L., Juang, B.: An introduction to hidden Markov models. IEEE ASSP Mag. **3**(1), 4–16 (1986)
21. Rabiner, L.R.: A tutorial on hidden Markov models and selected applications in speech recognition. Proc. IEEE **77**(2), 257–286 (1989)
22. Smedsrud, P.H., et al.: Kvasir-capsule, a video capsule endoscopy dataset. Sci. Data **8**(1), 142 (2021)
23. Szegedy, C., Ioffe, S., Vanhoucke, V., Alemi, A.: Inception-v4, inception-resnet and the impact of residual connections on learning. In: Proceedings of the AAAI Conference on Artificial Intelligence, vol. 31 (2017)
24. Thomson, A., Keelan, M., Thiesen, A., Clandinin, M., Ropeleski, M., Wild, G.: Small bowel review: diseases of the small intestine. Dig. Dis. Sci. **46**, 2555–2566 (2001)
25. Trentin, E., Gori, M.: Robust combination of neural networks and hidden Markov models for speech recognition. IEEE Trans. Neural Netw. **14**(6), 1519–1531 (2003)
26. Yim, S., Sitti, M.: 3-D localization method for a magnetically actuated soft capsule endoscope and its applications. IEEE Trans. Rob. **29**(5), 1139–1151 (2013)
27. Zhang, L., Zhu, Y., Mo, T., Hou, J., Rong, G.: Design and implementation of 3D positioning algorithms based on RF signal radiation patterns for in vivo micro-robot. In: 2010 International Conference on Body Sensor Networks, pp. 255–260. IEEE (2010)
28. Zhou, M., Bao, G., Pahlavan, K.: Measurement of motion detection of wireless capsule endoscope inside large intestine. In: 2014 36th Annual International Conference of the IEEE Engineering in Medicine and Biology Society, pp. 5591–5594. IEEE (2014)

Towards Unified Modality Understanding for Alzheimer's Disease Diagnosis Using Incomplete Multi-modality Data

Kangfu Han[1,4], Fenqiang Zhao[4], Dajiang Zhu[2], Tianming Liu[3], Feng Yang[1(✉)], and Gang Li[4(✉)]

[1] School of Biomedical Engineering, Southern Medical University, Guangzhou, China
yangf@smu.edu.cn
[2] Department of Computer Science and Engineering, The University of Texas at Arlington, Arlington, USA
[3] School of Computing, The University of Georgia, Athens, USA
[4] Department of Radiology and BRIC, University of North Carolina at Chapel Hill, Chapel Hill, USA
gang_li@med.unc.edu

Abstract. Multi-modal neuroimaging data, e.g., magnetic resonance imaging (MRI) and positron emission tomography (PET), has greatly advanced computer-aided diagnosis of Alzheimer's disease (AD) and its prodromal stage, i.e., mild cognitive impairment (MCI). However, incomplete multi-modality data often limits the diagnostic performance of deep learning-based methods, as only partial data can be used for training neural networks, and meanwhile it is challenging to synthesize missing scans (e.g., PET) with meaningful patterns associated with AD. To this end, we propose a novel unified modality understanding network to directly extract discriminative features from incomplete multi-modal data for AD diagnosis. Specifically, the incomplete multi-modal neuroimages are first branched into the corresponding encoders to extract modality-specific features and a Transformer is then applied to adaptively fuse the incomplete multi-modal features for AD diagnosis. To alleviate the potential problem of domain shift due to incomplete multi-modal input, the cross-modality contrastive learning strategy is further leveraged to align the incomplete multi-modal features into a unified embedding space. On the other hand, the proposed network also employs inter-modality and intra-modality attention weights for achieving local- to-local and local-to-global attention consistency so as to better transfer the diagnostic knowledge from one modality to another. Meanwhile, we leverage multi-instance attention rectification to rectify the localization of AD-related atrophic area. Extensive experiments on ADNI datasets with 1,950 subjects demonstrate the superior performance of the proposed methods for AD diagnosis and MCI conversion prediction.

Keywords: Alzheimer's Disease · Unified Modality Understanding · Cross-modality Contrastive Learning · Attention Consistency · Multi-Instance

© The Author(s), under exclusive license to Springer Nature Switzerland AG 2024
X. Cao et al. (Eds.): MLMI 2023, LNCS 14349, pp. 184–193, 2024.
https://doi.org/10.1007/978-3-031-45676-3_19

1   Introduction

Alzheimer's disease (AD), the most cause of dementia, is a chronic and irreversible neurodegenerative disease characterized by progressive memory loss and cognitive function impairment [1], and it's estimated that 1 out of 85 people will be living with AD by 2050 globally [3]. Early detection of AD is of great significance in clinical treatment and recent decades have witnessed the great clinical value of neuroimaging, including structural magnetic resonance imaging (sMRI) and fluorodeoxyglucose positron emission tomography (PET), on early detection of AD, owing to their strong capability in capturing brain pathological changes associated with AD in vivo [2].

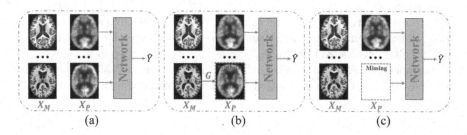

Fig. 1. Illustration of multi-modality neuroimage analysis for Alzheimer's disease Diagnosis using incomplete MRI and PET scans. (a) Neural networks only trained on complete multi-modality data; (b) Neural networks trained on multi-modality data after imputing missing PET scans by a generative model G; (c) Neural network trained on incomplete multi-modality data by learning a unified representation for AD diagnosis.

In the literature, early studies extracted handcrafted features (e.g., brain tissue density map [10], cortical thickness [11], textural or shape features of bilateral hippocampi [19]) from sMRI for AD diagnosis. More recent studies [6,13] have used advanced techniques to extract multiple discriminative patches after landmark localization [13] and multi-level features [6] from sMRI to construct convolutional neural networks (CNNs) for AD diagnosis. Instead of using single-modality neuroimages, multi-modal neuroimage analysis has demonstrated promising diagnostic performance by providing shared and complementary information. Conventional multi-modality classification model applied complete multi-modal neuroimages for AD diagnosis as shown in Fig. 1(a). However, only part of the enrolled datasets had complete multi-modal neuroimages and therefore the diagnostic performance and generalization ability are limited, especially for deep learning-based methods. To tackle the incomplete multi-modality problem, recent studies [5,14] designed a generative model G (e.g., generative adversarial network (GAN)) to synthesize the missing PET scans for AD diagnosis as in Fig. 1(b). Though the synthesis of missing scans has greatly increased the number of training samples and therefore improved diagnostic performance, it is still challenging to synthesize the PET scans with plausible subtle changes related to AD from sMRI scans for AD diagnosis.

To this end, we construct a unified modality understanding network for AD diagnosis (Fig. 1(c)), which can adaptively extract discriminative representations from incomplete multi-modality neuroimages by a 1-layer Transformer [4] after modality-specific feature extraction. The contributions of this work can be summarized as follows:

1. We construct a unified modality understanding network to adaptively extract discriminative features from incomplete multi-modality neuroimages for AD diagnosis.
2. A module of cross-modality contrastive learning and local attention consistency constraint was designed to align the extracted modality-incomplete and modality-complete features from different levels into a unified embedding space.
3. The multi-instance attention rectification was designed to avoid high attention weights on normal control samples so as to enforce the model focus on atrophic regions for AD diagnosis.
4. Experiments on ADNI datasets with 1,950 subjects in the task of AD diagnosis and MCI conversion prediction have demonstrated the superiority of the proposed methods.

2   Methods

Our Unified Modality Understanding network for AD diagnosis using incomplete multi-modality neuroimages (i.e., MRI and PET) is shown in Fig. 2. Specifically, the incomplete multi-modality inputs were firstly branched into two convolutional neural networks (CNN) based encoders that consist of 11 convolutional layers to extract modality-specific features, and a 1-layer Transformer was then applied to extract modality-(in)complete features for AD diagnosis, which can be formulated as:

$$\mathbf{F}_M, \mathbf{F}_C, \mathbf{F}_P, \mathbf{A} = Transformer(MRI(\mathbf{X}_M), PET(\mathbf{X}_P)) \qquad (1)$$

where $\mathbf{X}_M$ and $\mathbf{X}_P$ is the MRI and PET images, respectively, $\mathbf{F}_C$ is the modality-complete features and $\mathbf{A}$ is the attention weights obtained by Transformer. In another, the features $\mathbf{F}_M$ and $\mathbf{F}_P$ from single modality can be extracted by mask operation in Transformer for subjects with complete multi-modal neuroimages. Finally, the diagnostic prediction $\mathbf{P}$ can be obtained by a classifier (It is worth noting that only some PET scans are missing in the enrolled dataset):

$$\mathbf{P} = \begin{cases} classifier(\mathbf{F}_C), \text{if Complete modality} \\ classifier(\mathbf{F}_M), \text{if PET scans missing} \end{cases} \qquad (2)$$

where the classifier contains 2 fully-connected layers with units of 64 and 2, respectively.

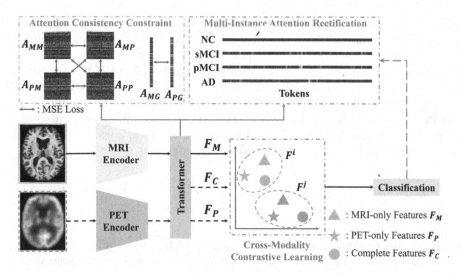

Fig. 2. The proposed Unified Modality Understanding Network using incomplete MRI and PET scans, in which $\mathbf{A}_{MM}$, $\mathbf{A}_{MP}$, $\mathbf{A}_{PM}$ and $\mathbf{A}_{PP}$ is local-to-local attention weights, while $\mathbf{A}_{MG}$ and $\mathbf{A}_{PG}$ is the local-to-global attention weights obtained by Transformer. $\mathbf{F}^i$ and $\mathbf{F}^j$ denote the features extracted from the ith and j^{th} subject, repsectively. The black dashed lines in this figure mean that the corresponding representations among subjects with missing PET scans do not need to be extracted.

Cross-Modality Contrastive Learning (CMC). As domain shift usually exists among multi-modal representations that may harness the diagnostic performance of the proposed methods when using a Transformer to adaptively fuse incomplete multi-modal features for AD diagnosis, We leveraged cross-modality contrastive learning [12] to align the extracted features into a unified embedding space in a mini-batch, the CMC with $\mathbf{F}_M$, $\mathbf{F}_P$ extracted single-modal neuroimages and $\mathbf{F}_C$ extracted from complete multi-modal neuroimages can be formulated as:

$$\mathcal{L}_{cmc} = -\left(log\frac{d(\mathbf{F}_C^+,\mathbf{F}_M^+)}{\sum_{i,j}d(\mathbf{F}_C^+,\mathbf{F}_M^j)} + log\frac{d(\mathbf{F}_C^+,\mathbf{F}_P^+)}{\sum_{i,j}d(\mathbf{F}_C^i,\mathbf{F}_P^j)} + log\frac{d(\mathbf{F}_M^+,\mathbf{F}_P^+)}{\sum_{i,j}d(\mathbf{F}_M^i,\mathbf{F}_P^j)}\right) \quad (3)$$

where the $d(\cdot,\cdot)$ is the scaled dot-product function, upon which larger values indicate smaller distances between positive pairs. Therefore, the CMC loss can enforce the extracted features from incomplete multi-modality neuroimages into a unified embedding space, thus avoiding sub-optimal diagnosis for subjects without PET scans.

Local Attention Consistency Constraint (LAC). Though the cross-modality contrastive learning has aligned the extracted modality-incomplete and modality-complete features from the global level, we further designed a local attention consistency constraint (LAC) to transfer informative knowledge from one modality to another from the local perspective. For subjects with complete multi-modal neuroimages, the attention weights $\mathbf{A}$ obtained by Transformer can be divided into four local-to-local attention parts (i.e., MRI to MRI attention weights $\mathbf{A}_{MM}$, MRI to PET $\mathbf{A}_{MP}$, PET to PET $\mathbf{A}_{PP}$, and PET to MRI $\mathbf{A}_{PM}$) and two local-to-global parts (i.e., MRI to Global attention weights $\mathbf{A}_{MG}$ and PET to Global $\mathbf{A}_{PG}$). Therefore, the LAC aims to minimize the attention weights of corresponding parts, thereby enabling the local consistent attention between different modalities, which can be formulated as:

$$\mathcal{L}_{lac} = MSE(\mathbf{A}_{MG}, \mathbf{A}_{PG}) + \sum_{i,j \in (MM,MP,PM,PP), i \neq j} MSE(\mathbf{A}_i, \mathbf{A}_j) \quad (4)$$

Multi-Instance Attention Rectification (MAR). Generally, the attention mechanism is commonly used in neural networks to increase explainable ability of the obtained model. However, assigning higher attention weights to normal control (NC) subjects is unreasonable, as there are no AD-related features in NC images. With this consideration, we employed the principle of multi-instance learning, that the positive bags include at least one positive instance, while the negative bags do not include positive instances, to rectify the attention weights obtained by Transformer with its prediction. This can be formulated as:

$$\mathcal{L}_{mar} = CE(Max(\mathbf{A}_{MG}), \mathbf{P}) + \mathbb{I}(CE(Max(\mathbf{A}_{PG}), \mathbf{P})) \quad (5)$$

where the CE is the cross-entropy loss, $\mathbf{P}$ is the prediction for the given subjects in a mini-batch, and $\mathbb{I}$ is the indicator to mark whether the given subjects have PET scans.

Implementation. Finally, the total loss of the proposed network, including the cross-entropy loss $\mathcal{L}_{ce}$ for classification, $\mathcal{L}_{cmc}$, $\mathcal{L}_{lac}$ and $\mathcal{L}_{mar}$, was formulated as:

$$\mathcal{L} = \mathcal{L}_{ce} + \alpha \mathcal{L}_{cmc} + \beta \mathcal{L}_{lac} + \gamma \mathcal{L}_{mar} \quad (6)$$

where α, β, and γ are the super-parameters and were set to 1 by default in the proposed network. We trained the proposed network for 80 epochs by randomly selecting 4 subjects with complete multi-modality neuroimages and 2 subjects with incomplete multi-modality neuroimages in a mini-batch iteratively, and applying Adam optimizer with a learning rate of 0.0001 based on Pytorch and a single GPU (i.e., NVIDIA GeForce RTX 3090 24GB).

3   Materials and Results

Datasets and Image Preprocessing. The public dataset, Alzheimer's Disease Neuroimaging Initiative (ADNI) (http://adni.loni.usc.edu) with 1,950 subjects was enrolled in this work for AD diagnosis and MCI conversion prediction, in which subjects were diagnosed as normal control (NC), stable mild cognitive impairment (sMCI), progressive mild cognitive impairment (pMCI) or Alzheimer's disease (AD) in terms of the standard clinical criteria, including mini-mental state examination (MMSE) score and/or clinical dementia rating sum of boxes. To summarize, the baseline ADNI-1 dataset contains 229 NC, 165 sMCI, 175 pMCI, and 188 AD subjects, while 371 subjects have PET scans; the ADNI-2&3 contains 538 NC, 323 sMCI, 109 pMCI, and 223 AD subjects, while 811 subjects have PET scans. The MRI scans were firstly preprocessed to correct their intensity inhomogeneity and remove the skull by FreeSurfer [17,18], which were then together with their corresponding PET scans linearly aligned onto a common Colin27 template [7] using FLIRT [8,9]. Finally, all scans were cropped into the sizes of $146 \times 182 \times 150$.

Table 1. Classification results in terms of ACC, SEN, SPE and AUC obtained by models trained on ADNI-1 and tested on ADNI-2&3. The suffix '-c' means evaluation with complete multi-modality images, '-s' means with additional synthesized images, and '-i' means incomplete images.

Methods	AD Diagnosis				MCI Conversion Prediction			
	ACC	SEN	SPE	AUC	ACC	SEN	SPE	AUC
ROI-c	0.829	0.794	0.866	0.892	0.696	0.514	0.759	0.675
CNN-c	0.820	0.815	0.825	0.882	0.767	0.346	0.911	0.716
LDMIL-c	0.900	0.895	0.905	0.941	0.791	0.495	0.892	0.783
Ours-c	0.931	0.945	0.915	0.966	0.817	0.607	0.889	0.817
ROI-s	0.845	0.824	0.854	0.901	0.734	0.578	0.787	0.730
CNN-s	0.905	0.776	0.959	0.930	0.787	0.514	0.879	0.779
LDMIL-s	0.925	0.874	0.946	0.958	0.815	0.587	0.892	0.810
Ours-s	0.940	0.834	0.983	0.972	0.829	0.624	0.898	0.844
Ours-i	0.945	0.897	0.965	0.974	0.829	0.679	0.879	0.832

Classification Results. In this part, we compared the proposed methods with 1) **ROI**: the ROI-based support vector machine (SVM), 2) **CNN**: multi-modality convolutional neural network, and 3) **LDMIL** [13]: landmark-based deep multi-instance learning. Since the compared methods have no capability in disease diagnosis using incomplete multi-modality neuroimages, we constructed a Hybrid GAN (HGAN) [14] to synthesize the missing PET scans for data imputation. As a result, the classification performance in terms of accuracy (ACC), sensitivity (SEN), specificity (SPE), and area under the receiver operating characteristic curve (AUC) in the task of AD diagnosis and MCI conversion prediction are summarized in Table 1.

From Table 1, we can observe that the methods using incomplete multi-modality neuroimages, where some PET images are synthesized though, achieved better classification performance, suggesting that a larger dataset can greatly improve the generalization ability of learning-based methods. The proposed method (**Ours-c** and **Ours-i**) trained on complete and incomplete multi-modality neuroimages achieve better classification performance in both task of

AD diagnosis and MCI conversion prediction when compared with the conventional machine learning and deep learning methods **LDMIL**. Finally, our model (**Ours-s**) achieved similar classification results on the testing datasets after data imputation by HGAN, which indicates the effectiveness and rationality of the proposed method.

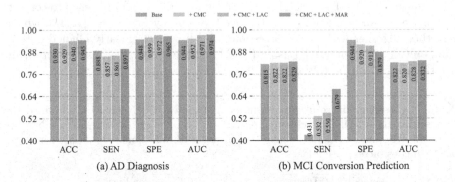

Fig. 3. Results of ablation study in the task of (a) AD diagnosis and (b) MCI conversion prediction in terms of ACC, SEN, SPE, and AUC, obtained by the models trained on ADNI-1 and tested on ADNI-2&3.

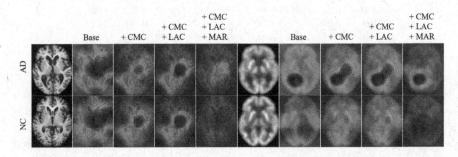

Fig. 4. Visualization of attention weights obtained by different ablated models on two randomly selected AD (first row) and NC (second row) subjects

Ablation Study. Experiments in this part were implemented to validate the effectiveness of the module of cross-modality contrastive learning (CMC), local attention consistency constraint (LAC), and multi-instance attention rectification (MAR). The classification results for the task of AD diagnosis and MCI conversion prediction obtained by the models trained on ADNI-1 were summarized in Fig. 3. It can be observed that with the utilization of CMC, LAC, and MAR step by step, the proposed method achieved better diagnostic performance on both tasks of AD diagnosis and MCI conversion prediction, validating its effectiveness on brain disease diagnosis. Moreover, from the attention figure in Fig. 4, the addition of the CMC, LAC, and MAR gradually help the proposed model rectify the attention in Transformer on AD-related regions, while yielding lower attention on corresponding regions among NC subjects.

Table 2. Comparison with previous studies that conducted experiments on the ADNI dataset in the task of AD diagnosis and MCI conversion prediction. The results are reported in their papers.

Methods	Data	AD Diagnosis				MCI Conversion Prediction			
		ACC	SEN	SPE	AUC	ACC	SEN	SPE	AUC
Liu et al. [13]	MRI	0.911	0.881	0.935	0.959	0.769	0.421	0.824	0.776
Qiao et al. [16]		0.896	–	–	0.949	–	–	–	–
Han et al. [6]		0.913	0.872	0.946	0.965	0.802	0.745	0.821	0.814
Pan et al. [14]	MRI + PET	0.936	0.952	0.915	0.970	0.774	0.772	0.791	0.825
Gao et al. [5]		0.927	0.891	0.940	0.956	0.753	0.773	0.741	0.786
Pan et al. [15]		0.931	0.947	0.909	0.972	0.797	0.813	0.753	0.844
Ours		0.945	0.897	0.965	0.974	0.829	0.679	0.879	0.832

Compared with Previous Studies. To demonstrate the superiority of the proposed methods, we compared them with the state-of-the-art deep learning-based models trained on ADNI-1 for AD diagnosis and MCI conversion prediction, which were summarized in Table 2. From this, we can observe that the models using multi-modality neuroimages achieved better diagnostic performance on both tasks of AD diagnosis and MCI conversion prediction, which indicates the great clinical importance of multi-modality neuroimages on AD diagnosis. On the other hand, the proposed method using incomplete multi-modality neuroimages yields comparable or better classification performances without the need for missing modality data imputation, which suggests its feasibility and effectiveness on brain disease diagnosis.

4   Conclusion

In this work, we construct a unified modality understanding network with cross-modality contrastive learning, local attention consistency constraint, and multi-instance attention rectification to align modality-specific and modality-complete representation in a unified embedding space and better localize AD-related atrophic regions for AD diagnosis using incomplete multi-modality neuroimages. Experiments on ADNI datasets with 1,950 subjects have demonstrated its superiority in AD diagnosis and MCI conversion prediction.

Acknowledgements. Data used in the preparation of this article were obtained from the Alzheimer's Disease Neuroimaging Initiative (ADNI) database (adni.loni.usc.edu). As such, the investigators within the ADNI contributed to the design and implementation of ADNI and/or provided data but did not participate in the analysis or writing of this report. A complete listing of ADNI investigators can be found at:http://adni.loni.usc.edu/wp-content/uploads/how_to_apply/ADNI_Acknowledgement_List.pdf. This work was supported in part by the National Natural Science Foundation of China under Grant 61771233, and Grant 61702182 to Feng Yang; in part by Guangdong Basic and Applied Basic Research Foundation under Grant No. 2023A1515011260, and Science and Technology Program of Guangzhou under Grant No. 202201011672 to Feng Yang.

References

1. Association, A.: 2019 Alzheimer's disease facts and figures. Alzheimer's Dement. **15**(3), 321–387 (2019)
2. Baron, J., et al.: In vivo mapping of gray matter loss with voxel-based morphometry in mild Alzheimer's disease. Neuroimage **14**(2), 298–309 (2001)
3. Brookmeyer, R., Johnson, E., Ziegler-Graham, K., Arrighi, H.M.: Forecasting the global burden of Alzheimer's disease. Alzheimer's Dement. **3**(3), 186–191 (2007)
4. Dosovitskiy, A., et al.: An image is worth 16×16 words: transformers for image recognition at scale. In: International Conference on Learning Representations (2021)
5. Gao, X., Shi, F., Shen, D., Liu, M.: Task-induced pyramid and attention GAN for multimodal brain image imputation and classification in Alzheimer's disease. IEEE J. Biomed. Health Inform. **26**(1), 36–43 (2022)
6. Han, K., He, M., Yang, F., Zhang, Y.: Multi-task multi-level feature adversarial network for joint Alzheimer's disease diagnosis and atrophy localization using sMRI. Phys. Med. Biol. **67**(8), 085002 (2022)
7. Holmes, C.J., Hoge, R., Collins, L., Evans, A.C.: Enhancement of t1 MR images using registration for signal averaging. NeuroImage **3**(3, Supplement), S28 (1996)
8. Jenkinson, M., Bannister, P., Brady, M., Smith, S.: Improved optimization for the robust and accurate linear registration and motion correction of brain images. Neuroimage **17**(2), 825–841 (2002)
9. Jenkinson, M., Smith, S.: A global optimisation method for robust affine registration of brain images. Med. Image Anal. **5**(2), 143–156 (2001)
10. Klöppel, S., et al.: Automatic classification of MR scans in Alzheimer's disease. Brain **131**(3), 681–689 (2008)
11. Li, G., Nie, J., Wu, G., Wang, Y., Shen, D.: Consistent reconstruction of cortical surfaces from longitudinal brain MR images. Neuroimage **59**(4), 3805–3820 (2012)
12. Li, W., et al.: UNIMO: towards unified-modal understanding and generation via cross-modal contrastive learning. In: Proceedings of the 59th Annual Meeting of the Association for Computational Linguistics and the 11th International Joint Conference on Natural Language Processing (Volume 1: Long Papers), pp. 2592–2607 (2021)
13. Liu, M., Zhang, J., Adeli, E., Shen, D.: Landmark-based deep multi-instance learning for brain disease diagnosis. Med. Image Anal. **43**, 157–168 (2018)
14. Pan, Y., Liu, M., Lian, C., Xia, Y., Shen, D.: Spatially-constrained fisher representation for brain disease identification with incomplete multi-modal neuroimages. IEEE Trans. Med. Imaging **39**(9), 2965–2975 (2020)

15. Pan, Y., Liu, M., Xia, Y., Shen, D.: Disease-image-specific learning for diagnosis-oriented neuroimage synthesis with incomplete multi-modality data. IEEE Trans. Pattern Anal. Mach. Intell. **44**(10), 6839–6853 (2022)

16. Qiao, H., Chen, L., Ye, Z., Zhu, F.: Early Alzheimer's disease diagnosis with the contrastive loss using paired structural MRIs. Comput. Methods Programs Biomed. **208**, 106282 (2021)

17. Sled, J.G., Zijdenbos, A.P., Evans, A.C.: A nonparametric method for automatic correction of intensity nonuniformity in MRI data. IEEE Trans. Med. Imaging **17**(1), 87–97 (1998)

18. Ségonne, F., et al.: A hybrid approach to the skull stripping problem in MRI. Neuroimage **22**(3), 1060–1075 (2004)

19. Wang, L., et al.: Large deformation diffeomorphism and momentum based hippocampal shape discrimination in dementia of the Alzheimer type. IEEE Trans. Med. Imaging **26**(4), 462–470 (2007)

COVID-19 Diagnosis Based on Swin Transformer Model with Demographic Information Fusion and Enhanced Multi-head Attention Mechanism

Yunlong Sun, Yiyao Liu, Junlong Qu, Xiang Dong, Xuegang Song, and Baiying Lei[✉]

Guangdong Key Laboratory for Biomedical Measurements and Ultrasound Imaging, National-Regional Key Technology Engineering Laboratory for Medical Ultrasound, School of Biomedical Engineering, Shenzhen University Medical School, Shenzhen 518060, China
leiby@szu.edu.cn

Abstract. Coronavirus disease 2019 (COVID-19) is an acute disease, which can rapidly become severe. Hence, it is of great significance to realize the automatic diagnosis of COVID-19. However, existing models are often inapplicable for fusing patients' demographic information due to its low dimensionality. To address this, we propose a COVID-19 patient diagnosis method with feature fusion and a model based on Swin Transformer. Specifically, two auxiliary tasks are added for fusing computed tomography (CT) images and patients' demographic information, which utilizes the patients' demographic information as the label for the auxiliary tasks. Besides, our approach involves designing a Swin Transformer model with Enhanced Multi-head Self-Attention (EMSA) to capture different features from CT data. Meanwhile, the EMSA module is able to extract and fuse attention information in different representation subspaces, further enhancing the performance of the model. Furthermore, we evaluate our model in COVIDx CT-3 dataset with different tasks to classify Normal Controls (NC), COVID-19 cases and community-acquired pneumonia (CAP) cases and compare the performance of our method with other models, which show the effectiveness of our model.

Keywords: COVID-19 diagnosis · Swin Transformer · Demographic information fusing · Enhanced Multi-head Self-attention

1 Introduction

Coronavirus disease 2019 (COVID-19) is an acute disease with a mortality rate of approximately 2% [1]. COVID-19 has spread rapidly and become a worldwide pandemic. Timely and extensive screening for COVID-19 plays a crucial role in combating the disease. Furthermore, COVID-19 pneumonia can quickly develop into a very serious disease. If we can find the patients with mild symptoms as early as possible, we not only can significantly reduce the mortality rate through effective intervention but also possibly avoid the waste of medical resources by eliminating the suspected patients [2].

© The Author(s), under exclusive license to Springer Nature Switzerland AG 2024
X. Cao et al. (Eds.): MLMI 2023, LNCS 14349, pp. 194–204, 2024.
https://doi.org/10.1007/978-3-031-45676-3_20

Therefore, how to screen patients with COVID-19 efficiently is highly essential while it requires a lot of time to help prevent and control the progress of COVID-19 [3].

Among various solutions, computed tomography (CT) images are used to examine the ground glass opacity (GGO) and the extrapulmonary zone to identify possible clues of COVID-19 [4]. According to the latest guidelines for coronavirus pneumonia in China, CT can be used for clinical diagnosis instead of nucleic acid detection [5]. However, the shortage of medical personnel in many countries and the manual review of a large number of CT images are time-consuming challenges. Figure 1 shows CT images of healthy people, community-acquired pneumonia (CAP) patients and COVID-19 patients. The challenge of mild COVID-19 diagnosis is that the focus of early patients is fuzzy with a low resolution and low gradient feature. It is very difficult to spot features of early COVID-19 through CT images by human eyes. Therefore, it is vital to realize the automatic diagnosis of COVID-19 [6].

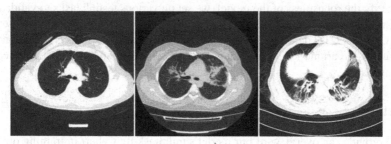

Fig. 1. The CT scans of normal control (left), COVID-19 case (middle), and CAP case (right).

In recent years, since deep learning has achieved promising disease diagnosis results [7], leading many researchers to propose COVID-19 diagnostic models based on CT images and yield good results [8–11]. However, there are still some limitations in current models. The first is that most of the existing research is often based on a single center with a limited number of subjects and high dimension data. Given the global prevalence of COVID-19, utilizing the data from multi-centers can enhance the robustness of models. Studies have shown that the performance of the conventional model in cross-center data is poor [12]. The second is that some studies have shown that fusing the patients' demographic information with the image data can improve the performance of the model [13]. But the patients' demographic information dimension of patients is too small to play a significant role in fusion with CT images. Referring to the idea of transfer learning [14, 15], auxiliary tasks are exploited to include patients' demographic information as auxiliary labels. In order to enhance the ability of handling high-dimensional data like CT images [16], we design two auxiliary tasks. The third is that the current studies did not focus on cases of COVID-19 patients, characterized by low resolutions, low gradient features and global features while Swin Transformer [17] models have a stronger ability to extract global features with high resolution and gradient. In order to further enhance its ability of extracting features, taking inspiration from multi-attention fusion [18, 19], we upgrade the multi-head attention mechanism of Swin Transformer.

To address these issues, we choose Swin Transformer as the backbone of our model for its ability to handle multi-scale data and transfer information in adjacent windows. Meanwhile, compared with other similar models, Swin Transformer reduces computation. Our model aims to improve the performance of cross-center training and testing. The main contributions of our model are as follows:

1) We design two auxiliary tasks to integrate the patients' demographic information and CT image. This approach can effectively improve the fusion effect of patients' demographic information and CT image, reduce the dependence on the amount of data, and improve the robustness of cross-center validation.
2) We introduce an enhanced multi-head self-attention (EMSA) mechanism to assign and fuse weighted importance of representation subspaces in different Swin Transformer layers. In this way, we can make each head of attention focus on as many positions as possible.

To verify the robustness of the model, we set up various testing tasks to validate the model based on datasets from 10 different centers. Our model is validated using several tasks in COVIDx CT-3 datasets, and the experimental results as well as comparison with other state-of-the-art models prove that our model achieves good results in cross-center validation tasks.

2   Method

Figure 2 provides an overview of our model's multitask flow chart in training our proposed model. The whole training process comprises 3 main parts: 1) gender prediction, 2) age prediction and 3) diagnosis prediction. Notably, gender prediction and age prediction are designed as auxiliary tasks. We first train the model to predict patients' gender and age. Subsequently, we add two new fully connected layers to train the model to diagnosis.

2.1   Swin Transformer

The backbone network of our model is Swin Transformer, which is suitable for 3D medical images. Compared with traditional Vision Transformer, Swin Transformer has 2 main differences: 1) Swin Transformer takes a hierarchical method to build up the model, which is similar to traditional convolutional neural networks. It successively downsamples the images 4x, 8x, 16x and 32x in the process of obtaining multi-scale feature maps, while traditional Vision Transformer directly perform downsampling operations at a certain downsampling rate at the beginning, and the subsequent feature map

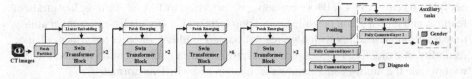

Fig. 2. Overall flow chart of the tasks with our proposed model.

size remains unchanged. 2) Windows Multi-head Self-Attention (W-MSA) is used in Swin Transformer, which divides the feature map into multiple non-overlapping window regions, with each window performing a separate multi-head self-attention operation. This considerably reduces the computation compared with Vision Transformer, in which multi-head self-attention operation is performed to the whole feature map. To implement the information transfer between adjacent windows, Shifted Windows Multi-head Self-Attention (SW-MSA) is added to Swin Transformer.

In order to capture different feature from the images, we proposed EMSA. We replaced all instances of multi-head self-attention (MSA) operation with EMSA. Figure 3 illustrate the detailed structure of Swin Transformer block, in which the W-MSA and SW-MSA module are replaced by our W-EMSA and SW-EMSA module. The detailed model flow chart is as Fig. 3.

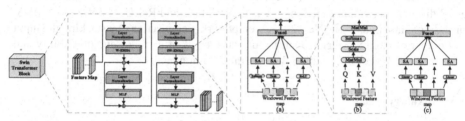

Fig. 3. Detailed structure of Swin Transformer block and difference between our EMSA (a), SA (b) and MSA (c). For each head of EMSA, different functions are applied as activation function in order to capture different features.

2.2 Enhanced Multi-head Self-attention

Figure 3 also illustrates the detailed structure of EMSA module. As an important part of EMSA, the self-attention layers make global interaction between image patches possible. In self-attention layers, the input feature vector $X \in \mathbb{R}^{N \times D}$ is initially transformed into 3 distinct vectors: Query vector $Q = XW_Q$, Key vector $K = XW_K$ and Value vector $V = XW_V$ ($W_Q, W_K, W_V \in \mathbb{R}^{D \times D}$), where W_Q, W_K and W_V are learnable matrices. Here, N represents the number of input vectors, and D denotes the dimensionality of each vector. The implementation of self-attention layers can be represented as follow:

$$\text{Attention}(Q, K, V) = \text{Softmax}\left(\frac{QK^T}{\sqrt{d_k}}\right)V, \tag{1}$$

$$\text{Output} = \text{Attention}(Q, K, V)W^O. \tag{2}$$

where d_k denotes the dimensionality of Key vector and W^O represents a learnable weight matrix.

To improve the performance of self-attention layers, MSA was introduced based on the structure of self-attention. In the implement of MSA with h heads, we first divided Q, K, V into h parts. For the i-th head, Query vector $Q_i = QW_Q^i$, Key vector $K_i = KW_K^i$

and Value vector $V_i = XV_V^i$ are calculated. Finally, the results from each head are concatenated and fused by a learnable matrix $W^O \in \mathbb{R}^{D \times D}$. Generally, the implementation of MSA can be described as follows:

$$head_i = Attention(Q_i, K_i, V_i), \qquad (3)$$

$$\text{MSA}(Q, K, V) = \text{Concate}(head_1, \ldots, head_h)W^O. \qquad (4)$$

However, MSA also has a deficiency that sometimes it may not effectively focus on different positions as we expected, primarily due to the same activation function. As is shown in Fig. 3, only one linear activation function is applied. Our proposed EMSA uses different activation functions to transform the feature map into different non-linear transformations, which ensures that each head pays attention to different subspaces. We use 3 different activation functions (i.e., sigmoid, tanh and ReLU) to form our EMSA. In addition, inspired by ResNet [20], we employ skip connections by stacking the input and output of attention mechanism to further strengthen its representation ability and facilitates the flow of information through the attention layers.

2.3 Auxiliary Tasks

In order to improve the processing ability of cross-center data in small samples and effectively fuse population information of subjects, we employ the population information of the subjects as auxiliary tasks-based tags to design two auxiliary tasks. The first one is to predict the sample gender by CT images, and the second is to predict the sample age.

The specific training process is as follows: first, the proposed model is trained using a classifier to determine the gender of the subjects. Subsequently, the weights of the Swin Transformer layer are retained to reconstruct a regression model that predict the age of the subjects. Finally, the weights of the Swin Transformer layers are preserved and a new classifier is added for disease diagnosis. In this way, the model can fuse population information with CT image during training model and learn more features such as gender and age information hidden in CT images to enhance its robustness. Moreover, compared to fusing population information fused directly, this auxiliary task-based pre-training approach allows for the elimination of the population information of the patients after the model training. This offers convenience in usage and promotes the protection of privacy.

3 Experiments and Results

3.1 Experimental Setup

The CT images come from COVIDx CT-3 dataset [21]. The dataset consists of CT image data collected from 10 centers and is officially divided into a training set, a validation set and a test set. In the training set, there's 516 normal controls, 3558 COVID-19 patients and 777 community-acquired pneumonia (CAP) patients. The validation set includes 220 normal controls, 281 COVID-19 patients and 258 CAP patients. Finally, the test set comprises 217 normal controls, 293 COVID-19 patients and 208 CAP patients. To ensure consistency in the input data, all CT images are upsampled or downsampled to the size of (64,128,128) since the image sizes may vary between patients. To increase the number of samples and reduce the probability of over fitting, we randomly apply data augmentation to a subset of CT images. The specific operations of data augmentation include random flipping, random rotation, random contrast adjustment, and random elastic deformation, which can simulate variations and noise in the real world and enhance generalization of the model.

In order to test the effectiveness of each modules and compare it with other networks, we design two classification tasks: 1) classification between normal controls and COVID-19 patients, 2) Classification among normal controls, COVID-19 patients and CAP patients to further test and compare the ability of classification of our method.

To achieve comprehensive and objective assessment of the classification performance of the proposed method, we select seven classification evaluation metrics, including accuracy (Acc), precision (Pre), recall (Rec), specificity (Spec), F1-score (F1), Kappa, area under the receiver operating characteristic (ROC) curve (AUC). All experiments are implemented by configuring the PyTorch framework on NVIDIA GTX 1080Ti GPU with 11 GB of memory.

For the proposed model, we choose Adam optimizer for optimization and categorical cross-entropy as the loss function. We set the batch size as 8, the number of epochs as 200, the learning rate as 10^{-4}, the first-order exponential decay rates for the moment estimates as 0.9, and the second-order exponential decay rates for the moment estimates as 0.999. In data augmentation, we set augmentation probability rate as 0.5, which indicates that half of the training data is randomly selected and then randomly augmented.

3.2 Comparison with Other Methods

In order to evaluate the classification performance of the proposed method, we choose several state-of-the-art image classification methods for comparison, including several state-of-the-art networks: ResNet34 [20], ResNet50 [22], ShuffleNet [23], EfficientNet [24] and DenseNet121 [25], and ViT [26]. The results of the comparative experiment are illustrated in Tables 1 and 2. It can be observed that our proposed method achieves the best performance on our dataset, and these results demonstrate the superiority of our proposed method in achieving accurate classification results compared to other networks (Fig. 4).

3.3 Ablation Study

In this section, we adopt Swin Transformer as the baseline of our network, and then incorporate auxiliary tasks and EMSA to optimize its performance on the COVIDx CT-3 dataset. We use our proposed method to conduct two classification tasks above. Tables 1 and 2 and Fig. 5 show the results of the ablation experiments for different modules in the proposed method. Among the experimental setups, "Gender" indicates that gender prediction is added as an auxiliary task. "Age" represents that age prediction is added as an auxiliary task. "EMSA" indicates that EMSA is adopted in Swin Transformer.

Table 1. Classification performance comparison of state-of-the-art methods on the test set (Two-classification) (%).

Model	Acc	Pre	Recall	Spec	F1	Kappa	AUC
ResNet50	93.32	98.43	83.33	99.21	88.04	85.22	91.27
ResNet34	96.29	97.20	92.67	98.43	94.44	91.97	95.55
DenseNet121	97.03	95.39	96.67	97.24	96.85	93.66	96.96
EfficientNet	93.81	90.32	93.33	94.09	93.57	86.84	93.71
ShuffleNet	86.39	89.60	72.73	94.80	78.97	70.06	83.76
ViT	95.30	**99.26**	88.16	**99.60**	91.59	89.75	93.88
Proposed	**98.02**	98.67	**96.10**	99.20	**97.05**	**95.78**	**99.86**

Table 2. Classification performance comparison of state-of-the-art methods on the test set (Three-classification) (%).

Model	Acc	Pre	Recall	Spec	F1	Kappa	AUC
ResNet50	91.83	92.86	91.78	95.78	92.01	87.52	93.78
ResNet34	92.16	93.08	92.63	95.90	92.67	88.00	94.27
DenseNet121	92.48	92.82	92.49	96.04	92.64	88.35	94.26
EfficientNet	88.40	89.72	88.20	93.67	88.89	81.79	90.94
ShuffleNet	87.58	90.35	86.74	93.38	87.82	80.91	90.06
ViT	91.50	91.76	91.75	95.49	91.75	86.67	93.62
Proposed	**93.79**	**94.43**	**94.01**	**96.81**	**94.04**	**90.37**	**95.41**

These results presented in Tables 3 and 4 show that the diagnosis accuracy can be effectively improved by demographic information fusion, which also demonstrates that feature extraction ability can be improved through the training of auxiliary tasks. Although slightly inferior in Fig. 5 (c), our method still reaches the highest average AUC.

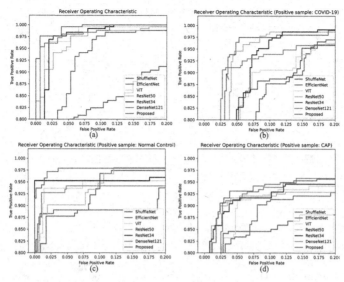

Fig. 4. ROC of comparison between our method and others. Picture (a) shows the ROC curve of two-classification task while (b), (c), and (d) shows three-classification. In picture (b), (c) and (d), the positive sample are set as COVID-19, normal control, and CAP respectively.

Table 3. Ablation experiments of different modules in proposed method (Two-classification) (%).

Gender	Age	EMSA	Acc	Pre	Recall	Spec	F1	Kappa	AUC
			95.54	98.41	88.57	99.24	91.93	89.93	93.91
√			96.53	97.78	92.31	98.85	94.37	92.33	95.58
	√		96.04	**99.19**	89.05	**99.63**	92.41	90.94	94.34
√	√		97.03	99.35	93.29	99.58	95.12	93.82	95.83
		√	96.78	97.76	92.91	98.86	94.81	92.84	95.88
√	√	√	**98.02**	98.67	**96.10**	99.20	**97.05**	**95.78**	**99.86**

Table 4. Ablation experiments of different modules in proposed method (Three-classification) (%).

Gender	Age	EMSA	Acc	Pre	Recall	Spec	F1	Kappa	AUC
			89.38	91.49	89.32	94.44	89.87	83.75	91.88
√			91.01	92.41	89.99	95.14	90.91	85.87	92.56
	√		90.69	91.63	91.60	95.12	91.42	85.78	93.36
√	√		92.32	92.71	91.88	95.90	92.23	87.85	93.89
		√	91.50	92.92	90.17	95.37	91.19	86.82	92.77
√	√	√	**93.79**	**94.43**	**94.01**	**96.81**	**94.04**	**90.37**	**95.41**

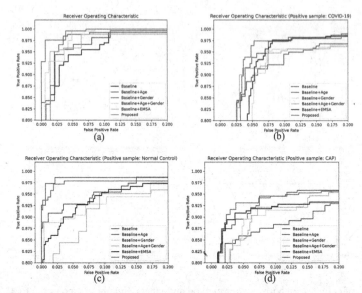

Fig. 5. ROC curves of our ablation study. Picture (a) shows the ROC curve of two-classification task while (b), (c), and (d) shows three-classification. In picture (b), (c) and (d), the positive sample are set as COVID-19, normal control and CAP respectively.

4   Conclusion

In this paper, we develop a new method based on Swin Transformer for rapid prediction of COVID-19 from CT images. We design two auxiliary tasks, gender prediction and age prediction, in order to fuse the feature from patient's population information and CT image, which can further enhance the performance. Besides, we devise an EMSA model to upgrade Swin Transformer to mine and fuse attention information in different representation subspaces, which can improve the performance of the model. The experimental results show that our model can effectively improve the diagnosis accuracy and robustness of the model and it's superior to the related models as well.

Acknowledgements. This work was supported partly by National Natural Science Foundation of China (Nos. U22A2024, U1902209 and 62271328), National Natural Science Foundation of Guangdong Province (Nos. 202020A1515110605, and 2022A1515012326), Shenzhen Science and Technology Program (Nos. JCYJ20220818095809021).

References

1. Xu, Z., Shi, L., Wang, Y., Zhang, J., Huang, L., Zhang, C., et al.: Pathological findings of COVID-19 associated with acute respiratory distress syndrome. Lancet Respir. Med. **8**(4), 420–422 (2020). https://doi.org/10.1016/S2213-2600(20)30076-X
2. Risch, H.A.: Early outpatient treatment of symptomatic, high-risk COVID-19 patients that should be ramped up immediately as key to the pandemic crisis. Am. J. Epidemiol. **189**(11), 1218–1226 (2020). https://doi.org/10.1093/aje/kwaa093

3. Lunz, D., Batt, G., Ruess, J.: To isolate, or not to isolate: a theoretical framework for disease control via contact tracing. medRxiv 1–9 (2020)

4. Kong, W., Agarwal, P.P.: Chest imaging appearance of COVID-19 infection. Radiol. Cardiothorac. Imaging **2**(1), e200028 (2020). https://doi.org/10.1148/ryct.2020200028

5. China NHCotPsRo: New diagnosis and treatment of coronary pneumonia. (2020)

6. Zhang, K., Liu, X., Shen, J., Li, Z., Sang, Y., Wu, X., et al.: Clinically applicable AI system for accurate diagnosis, quantitative measurements, and prognosis of COVID-19 pneumonia using computed tomography. Cell **181**(6), 1423-1433.e11 (2020). https://doi.org/10.1016/j.cell.2020.04.045

7. Suzuki, K.: Overview of deep learning in medical imaging. Radiol. Phys. Technol. **10**(3), 257–273 (2017). https://doi.org/10.1007/s12194-017-0406-5

8. Mei, X., et al.: Artificial intelligence–enabled rapid diagnosis of patients with COVID-19. Nat. Med. **26**(8), 1224–1228 (2020). https://doi.org/10.1038/s41591-020-0931-3

9. Chen, J., Wu, L., Zhang, J., Zhang, L., Gong, D., Zhao, Y., et al.: Deep learning-based model for detecting 2019 novel coronavirus pneumonia on high-resolution computed tomography. Sci. Rep. **10**(1), 19196 (2020). https://doi.org/10.1038/s41598-020-76282-0

10. Hochreiter, S., Schmidhuber, J.: Long short-term memory. Neural Comput. **9**(8), 1735–1780 (1997). https://doi.org/10.1162/neco.1997.9.8.1735

11. Li, L., Qin, L., Xu, Z., Yin, Y., Wang, X., Kong, B., et al.: Using artificial intelligence to detect COVID-19 and community-acquired pneumonia based on pulmonary CT: evaluation of the diagnostic accuracy. Radiology **296**(2), E65–E71 (2020). https://doi.org/10.1148/radiol.2020200905

12. Silva, P., Luz, E., Silva, G., Moreira, G., Silva, R., Lucio, D., et al.: COVID-19 detection in CT images with deep learning: a voting-based scheme and cross-datasets analysis. Inform. Med. Unlocked. **20**, 100427 (2020). https://doi.org/10.1016/j.imu.2020.100427

13. Liu, M., Zhang, J., Adeli, E., Shen, D.: Joint classification and regression via deep multi-task multi-channel learning for Alzheimer's disease diagnosis. IEEE Trans. Biomed. Eng. **66**(5), 1195–1206 (2019). https://doi.org/10.1109/TBME.2018.2869989

14. Hazarika, D., Poria, S., Zimmermann, R., Mihalcea, R.: Conversational transfer learning for emotion recognition. Inf. Fusion. **65**, 1–12 (2021). https://doi.org/10.1016/j.inffus.2020.06.005

15. Jaiswal, A., Gianchandani, N., Singh, D., Kumar, V., Kaur, M.: Classification of the COVID-19 infected patients using DenseNet201 based deep transfer learning. J. Biomol. Struct. Dyn. **39**(15), 5682–5689 (2021). https://doi.org/10.1080/07391102.2020.1788642

16. Ali, F., El-Sappagh, S., Islam, S.M.R., Kwak, D., Ali, A., Imran, M., et al.: A smart healthcare monitoring system for heart disease prediction based on ensemble deep learning and feature fusion. Inf. Fusion. **63**, 208–222 (2020). https://doi.org/10.1016/j.inffus.2020.06.008

17. Liu, Z., Lin, Y., Cao, Y., Hu, H., Wei, Y., Zhang, Z., et al.: Swin transformer: hierarchical vision transformer using shifted windows, pp. 10012–10022

18. Jiang, M.-x, Deng, C., Shan, J.-s, Wang, Y.-y, Jia, Y.-j, Sun, X.: Hierarchical multi-modal fusion FCN with attention model for RGB-D tracking. Inf. Fusion **50**, 1–8 (2019). https://doi.org/10.1016/j.inffus.2018.09.014

19. Vaswani, A., Shazeer, N., Parmar, N., Uszkoreit, J., Jones, L., Gomez, A.N., et al.: Attention is all you need. Adv. Neural. Inf. Process. Syst. **30**, 5998–6008 (2017)

20. He, K., Zhang, X., Ren, S., Sun, J.: Deep residual learning for image recognition. In: Proceedings of the IEEE Conference on Computer Vision and Pattern Recognition, pp. 770–778 (2016)

21. Tuinstra, T., Gunraj, H., Wong, A.: COVIDx CT-3: a large-scale, multinational, open-source benchmark dataset for computer-aided COVID-19 screening from chest CT Images (2022)

22. Xie, S., Girshick, R., Dollár, P., Tu, Z., He, K.: Aggregated residual transformations for deep neural networks. In: Proceedings of the IEEE Conference on Computer Vision and Pattern Recognition, pp. 1492–1500 (2017)

23. Zhang, H., Wu, C., Zhang, Z., Zhu, Y., Lin, H., Zhang, Z., et al.: ResNest: split-attention networks. In: Proceedings of the IEEE/CVF Conference on Computer Vision and Pattern Recognition, pp. 2736–2746 (2022)

24. Liu, Z., Mao, H., Wu, C.-Y., Feichtenhofer, C., Darrell, T., Xie, S.: A convnet for the 2020s. Proceedings of the IEEE/CVF Conference on Computer Vision and Pattern Recognition, pp. 11976–11986 (2022)

25. Huang, G., Liu, Z., Van Der Maaten, L., Weinberger, K.Q.: Densely connected convolutional networks. In: Proceedings of the IEEE Conference on Computer Vision and Pattern Recognition, pp. 4700–4708 (2017)

26. Dosovitskiy, A., Beyer, L., Kolesnikov, A., Weissenborn, D., Zhai, X., Unterthiner, T., et al.: An image is worth 16×16 words: transformers for image recognition at scale. arXiv preprint https://arxiv.org/abs/2010.11929 (2020)

MoViT: Memorizing Vision Transformers for Medical Image Analysis

Yiqing Shen, Pengfei Guo, Jingpu Wu, Qianqi Huang, Nhat Le, Jinyuan Zhou, Shanshan Jiang, and Mathias Unberath[✉]

Johns Hopkins University, Baltimore, USA
{yshen92,unberath}@jhu.edu

Abstract. The synergy of long-range dependencies from transformers and local representations of image content from convolutional neural networks (CNNs) has led to advanced architectures and increased performance for various medical image analysis tasks due to their complementary benefits. However, compared with CNNs, transformers require considerably more training data, due to a larger number of parameters and an absence of inductive bias. The need for increasingly large datasets continues to be problematic, particularly in the context of medical imaging, where both annotation efforts and data protection result in limited data availability. In this work, inspired by the human decision-making process of correlating new "evidence" with previously memorized "experience", we propose a Memorizing Vision Transformer (MoViT) to alleviate the need for large-scale datasets to successfully train and deploy transformer-based architectures. MoViT leverages an external memory structure to cache history attention snapshots during the training stage. To prevent overfitting, we incorporate an innovative memory update scheme, attention temporal moving average, to update the stored external memories with the historical moving average. For inference speedup, we design a prototypical attention learning method to distill the external memory into smaller representative subsets. We evaluate our method on a public histology image dataset and an in-house MRI dataset, demonstrating that MoViT applied to varied medical image analysis tasks, can outperform vanilla transformer models across varied data regimes, especially in cases where only a small amount of annotated data is available. More importantly, MoViT can reach a competitive performance of ViT with only 3.0% of the training data. In conclusion, MoViT provides a simple plug-in for transformer architectures which may contribute to reducing the training data needed to achieve acceptable models for a broad range of medical image analysis tasks.

Keywords: Vision Transformer · External Memory · Prototype Learning · Insufficient Data

1 Introduction

With the advent of Vision Transformer (ViT), transformers have gained increasing popularity in the field of medical image analysis [2], due to the capability of

X. Cao et al. (Eds.): MLMI 2023, LNCS 14349, pp. 205–213, 2024.
https://doi.org/10.1007/978-3-031-45676-3_21

capturing long-range dependencies. However, ViT and its variants require considerably larger dataset sizes to achieve competitive results with convolutional neural networks (CNNs), due to larger model sizes and the absence of convolutional inductive bias [2,10]. Indeed, ViT performs worse than ResNet [4], a model of similar capacity, on the ImageNet benchmark [11], if ViT does not enjoy pretraining on JFT-300M [12], a large-scale dataset with 303 million weakly annotated natural images. The drawback of requiring exceptionally large datasets prevents transformer-based architectures to fully evolve its potential in the medical image analysis context, where data collection and annotation continue to pose considerable challenges. To capitalize on the benefits of transformer-based architectures for medical image analysis, we seek to develop an effective ViT framework capable of performing competitively even when only comparably small data is available.

In the literature, the problematic requirement for large data is partly alleviated by extra supervisory signals. Data-efficient Image Transformer (DeiT), for example, distills hard labels from a strong teacher transformer [14]. Unfortunately, this approach only applies to problems where data-costly, high-capacity teacher transformer can be developed. Moreover, DeiT enables the training on student transformers exclusively for mid-size datasets, between 10k to 100k samples, and the performance dramatically declines when the data scale is small [14]. Concurrently, another line of work attempts to introduce the shift, scale, and distortion invariance properties from CNNs to transformers, resulting in a series of hybrid architecture designs [9,17,19,21]. To give a few examples, Van *et al.* fed the extracted features from CNNs into a transformer for multi-view fusion in COVID diagnosis [15]. Barhoumi *et al.* extended a single CNN to multiple CNNs for feature extraction before the fusion by a tansformer [1]. Importantly, they note that pre-training on ImageNet is still required to fuse convolutional operations with self-attention mechanisms, particularly in the medical context [1]. Yet, pre-training on large-scale medical dataset is practically unaffordable, due to the absence of centralized dataset as well as the privacy regularizations.

Our developments to combat the need for large data to train transformer-models is loosely inspired by the process that clinicians use when learning how to diagnose medical images from a relative very limited number of cases compared to regular data size. To mimic this human decision-making process, where new information or "evidence" is often conceptually correlated with previously memorized facts or "experience", we present the *Memorizing Vision Transformer* (MoViT) for efficient medical image analysis. MoViT introduces external memory, allowing the transformer to access previously memorized experience, *i.e.* keys and values, in the self-attention heads generated during the training. In the inference stage, the external memory then enhances the instance-level attention by looking up the correlated memorized facts. Introducing external memory enables long-range context to be captured through attention similar to language modeling, which provides supplementary attention with the current ViT and variants [3,6,18].

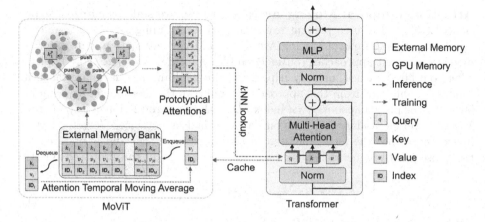

Fig. 1. An overview of the proposed Memorizing Vision Transformer (MoViT).

The contributions of this paper are three-fold, summarized as follows. (1) A novel *Memorizing Vision Transformer* (MoViT), which introduces storage for past attention cues by caching them into external memory, without introducing additional trainable parameters. (2) A new approach to updating the memory using a *Attention Temporal Moving Average* scheme, that accumulates attention snapshots and optimizes data in the external memory dynamically. In contrast, previous work, such as [18], is restricted to a random dropping-out scheme to keep a fixed amount of external memorized events. (3) A new *post hoc* scheme, *Prototypical Attention Learning*, to distill the large-scale cached data into a representative prototypical subset, which accelerates computation during inference. Experiments are carried out across different modalities, *i.e.* Magnetic Resonance (MRI) Images and histopathological images, demonstrating superior performance to vanilla transformer models across all data regimes, especially when only small amounts of training samples are available.

2   Methods

Memorizing Vision Transformer. Memorizing Vision Transformer (MoViT) accumulates snapshots of attention cues *i.e.* key k and value v generated by the attention heads as the memorized "experience", and caches them to an external memory bank in the form of an indexed triplet (ID, k, v) during the training process. The enumerated index ID of the data sample and the generated attention fact (k, v) are prepared for an efficient lookup and update in the subsequent Attention Temporal Moving Average (ATMA) scheme. Importantly, the gradients are not back-propagated into the memory bank, and thus, the caching operation only costs slightly extra training. This approach is practical in that the proposed MoViT can be easily plugged into any Vision Transformer (ViT) or its variants, by replacing one or multiple vanilla transformer blocks with MoViT blocks. An overview of the MoViT framework is presented in Fig. 1.

Attention Temporal Moving Average. To remove stale memorized "experience" [16], and also to prevent overfitting to the training set, we introduce a novel Attention Temporal Moving Average (ATMA) strategy to update external memory. Current approaches routinely employ a fixed capacity to store triplets, where outdated cached memory triplets are dropped and new facts are taken in randomly. Different from this approach, we improve the mechanism by accumulating all the past snapshots $w.r.t$ index ID, with the introduction of Exponential Moving Average update [13]. The outdated cached "experience" $(k_{\text{old}}, v_{\text{old}})$ $w.r.t$ index ID denote the fact generated in the previous epoch, and is updated by the subsequently generated facts $(k_{\text{generated}}, v_{\text{generated}})$ in the proposed ATMA according to:

$$\begin{cases} k_{\text{new}} = \alpha_k \cdot k_{\text{generated}} + (1 - \alpha_k) \cdot k_{\text{old}}, \\ v_{\text{new}} = \alpha_v \cdot v_{\text{generated}} + (1 - \alpha_v) \cdot v_{\text{old}}, \end{cases} \tag{1}$$

where the subscripts "new" denotes the updated facts to the external memory, and α_k, α_v are the friction terms. In the smoothing process, both coefficients uniformly follow the ramp-down scheme [8] for a steady update. Specifically, the coefficients are subject to the current training epoch number t, $i.e.$

$$\alpha = \begin{cases} 1 - \alpha_0 \cdot \exp(-t_0(1 - \frac{t}{t_0})^2, & t \le t_0 \\ 1 - \alpha_0, & t > t_0 \end{cases} \tag{2}$$

with $a_0 = 0.01$ and t_0 set to 10% of the total training epochs as in previous work [13]. The number of stored facts M is exclusively correlated with the dataset scale, and network architecture $i.e.$ $M = \#(\text{training samples}) \times \#(\text{attention heads})$, where the number of attention heads is often empirically set between the range of 3–12 [2], leading to a bounded M.

Prototypical Attention Learning. We write all the cached experience from the training stage of MoViT as $\mathcal{F} = \{(k_i, v_i)\}_{i=1}^M$. Then, prototypical attention facts refer to a small number of representative facts to describe $\mathcal{F}$ $i.e.$ $\mathcal{P} = \{(k_i^p, v_i^p)\}_{i=1}^P$, where P represents the total number of prototypes. To distill the external memorized facts into representative prototypes for efficient inference, we introduce Prototypical Attention Learning (PAL), which is applied $post\ hoc$ to the external memory after model training. To identify the prototype keys from the cached keys $\{k_i\}_{i=1}^M$, we leverage the Maximum Mean Discrepancy (MMD) metric [7] to measure the discrepancy between two distributions. Subsequently, the objective in PAL is steered toward minimizing the MMD metric, $i.e.$

$$MMD^2 = \frac{1}{P^2} \sum_{i,j=1}^P D(k_i^p, k_j^p) - \frac{1}{PM} \sum_{i,j=1}^{P,M} D(k_i^p, k_j) + \frac{1}{M^2} \sum_{i,j=1}^M D(k_i, k_j), \tag{3}$$

where $D(\cdot, \cdot)$ denotes the cosine similarity. We employ a greedy search to find $\{k_i^p\}_{i=1}^P$ from Eq. (3). To integrate all information after deriving prototype keys, we leverage the weighted average to derive the associated $\{v_i^p\}_{i=1}^P$ $i.e.$

$$v_i^p = \sum_{j=1}^{M} w_{j,i} v_j \ \text{ with } w_{j,i} = \frac{\exp(D(v_j, v_i^p)/\tau)}{\sum \exp_{k=1}^{M}(D(v_k, v_i^p)/\tau)}, \tag{4}$$

where the weights $w_{j,i}$ are normalized by the softmax operation, and temperature τ is a hyper-parameter to determine the confidence of normalization.

Inference Stage. To apply the attention facts $\mathcal{F}$ or prototypes $\mathcal{P}$ stored in the external memory during inference, approximate k-nearest-neighbor (kNN) search is employed to look up the top k pairs of (key, value) $w.r.t$ the given the local queries. In this way, the same batch of queries generated from the test sample is used for both the multi-head self attentions and external memory retrievals. With the retrieved keys, the attention matrix is derived by computing the softmax operated dot product with each query. Afterwards, we use the attention matrix to compute a weighted sum over the retrieved values. The results attended to local context and external memories are combined using a learned gate scheme [18].

3   Experiments

Datasets. Evaluations are performed on two datasets curated from different modalities. (1) Histology Image Dataset: *NCT-CRC-HE-100K* is a public Hematoxylin & Eosin (H&E) stained histology image dataset with $100,000$ patches without overlap, curated from $N = 86$ colorectal cancer samples [5]. All RGB images are scaled to 224×224 pixels at the magnification of $20\times$. To simulate various data availability conditions, some experiments use a subset from *NCT-CRC-HE-100K* as the training set. In terms of the test set, an external public dataset *CRC-VAL-HE-7K* with 7180 patches from $N = 50$ patients is employed. This dataset was designed to classify the nine tissue categories from histology image patches, and we use the top-1 test accuracy as the evaluation metric. (2) MRI Dataset: This in-house dataset includes 147 scans with malignant gliomas curated from $N = 92$ patients. All data has been deidentified properly to comply with the Institutional Review Board (IRB). Each scan contains five MRI sequences, namely T1-weighted (T1w), T2-weighted (T2w), fluid-attenuated inversion recovery (FLAIR), gadolinium enhanced T1-weighted (Gd-T1w), and amide proton transfer-weighted (APTw). Labels are generated manually at the slice level by experienced radiologists. The corresponding slices from each scan are concentrated after co-registration and z-score normalization, resulting in an input size of $256 \times 256 \times 5$. A proportion of 80% of the patients are divided into the training set, and the remaining 20% as the test set *i.e.*, 1770 training samples and 435 test samples. This is a binary classification task to distinguish malignant gliomas from normal tissue. We use accuracy, area under the ROC curve (AUC), precision, recall, and F1-score as the evaluation metrics.

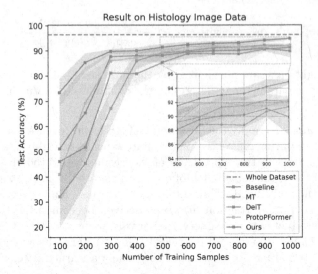

Fig. 2. Performance comparisons of MoViT plugged into the last layer of ViT-Tiny with counterparts, using a varying number of training samples from the Histology image dataset (0.1% to 1% data employed). The solid line and shadow regions represent the average and the standard deviation of test accuracy computed from seven random runs, respectively. The dashed line denotes the performance of the baseline (vanilla ViT) trained with the entire dataset, regarded as the performance upper bound.

Implementations. All experiments are performed on one NVIDIA GeForce RTX 3090 GPU with 24 GB memory. An AdamW optimizer is used with a Cosine Annealing learning rate scheduler, where the initial learning rates are 2×10^{-3} for the MRI dataset, 5×10^{-4} for the Histology image dataset; with the maximum number of training epochs set to 100. We plug the MoViT into the last layer of ViT-Tiny, $i.e.$ 12 transformer layers with 3, and set $k = 32$ for the kNN lookup. In PAL, we set the number of prototypes $P = \#$(number of class) $\times$ 32, and temperature $\tau = 0.5$ for Eq. (4). Comparisons are made to Memorizing Transformer (MT) [18], DeiT [14], and ProtoPFormer $i.e.$ a prototypical part

Table 1. Performance comparison trained on the entire dataset in terms of test accuracy (%). We employ three ViT configurations $i.e.$ ViT-Tiny, ViT-Small, and ViT-Base.

Method	ViT-Tiny	ViT-Small	ViT-Base
Baseline	$96.462_{\pm 0.213}$	$95.850_{\pm 0.503}$	$94.231_{\pm 0.511}$
MT [18]	$96.511_{\pm 0.312}$	$96.621_{\pm 0.108}$	$95.102_{\pm 0.272}$
DeiT [14]	$96.439_{\pm 0.331}$	$96.216_{\pm 0.213}$	$93.246_{\pm 0.259}$
ProtoPFormer [20]	$96.712_{\pm 0.521}$	$96.032_{\pm 0.364}$	$93.002_{\pm 0.752}$
MoViT (Ours)	$\mathbf{97.792}_{\pm 0.293}$	$\mathbf{97.326}_{\pm 0.138}$	$\mathbf{95.989}_{\pm 0.205}$

network framework for ViT [20], where the vanilla ViT is regarded as the baseline. To compute the mean and standard deviation of the metrics, all models are trained from scratch for seven random runs.

Results on Histology Image Dataset. To simulate the case where only small data is available as in many medical image analysis tasks, we use a limited proportion of the training set, across varied data regimes, and use the whole *NCT-CRC-HE-100K* as the test set for a fair comparison. As shown in Fig. 2, MoViT improves over the baseline at any data scale, especially when the number of samples is particularly small, *i.e.* 0.1%, where we can observe a similar trend with a large proportion of the data between 1%–100%. Notably, our method can achieve a close margin to the entire-dataset-trained model (96.462% ± 0.213%) using only 1.0% data (94.927% ± 0.378%), and a competitive performance (96.341% ± 0.201%) with 3.0% data. Additionally, our approach also significantly reduces the performance fluctuations *i.e.* standard deviation, leading to a more stable performance. For example, vanilla ViT is 20.901% when trained with 0.1% data and ours is 5.452% *i.e.* approximately four times smaller. Moreover, our method can consistently outperform state-of-the-art data-efficient transformer DeiT [14] and pure prototype learning method ProtoPFormer [20]. We notice that Memorizing Transformer (MT) [18] performs worse than the baseline although achieving almost 100% training accuracy, where the gap becomes significant with 0.1%–0.4% data, which we attribute to the overfitting issue. The large margin between the performance of MT and MoViT implies that ATMA and PAL can alleviate the overfitting issues during the memorization of the facts. Performance comparison is also performed on the entire training set *i.e.* using 100% *NCT-CRC-HE-100K* as the training set, with different ViT configurations. In Table 1, our method can consistently outperform its counterparts with a large margin, which demonstrates its applicability and scalability to large datasets. This suggests that MoViT scales well to a wide range of data scales as a by-product. The averaged training times per epoch on ViT-Tiny are 162.61(s) for baseline ViT, 172.22(s) for MT, 109.8(s) for DeiT, 639.4(s) for ProtoPFormer, and 171.49(s) for our approach. Our method can boost performance with a reduced training data scale.

Table 2. Quantitative comparison on the MRI dataset. MoViT achieves the highest performance across all metrics, further suggesting its ability to perform well in applications where limited data is available.

Method	Accuracy($\uparrow$)	AUC($\uparrow$)	Precision($\uparrow$)	Recall($\uparrow$)	F1-score($\uparrow$)
Baseline	$74.01_{\pm0.40}$	$80.62_{\pm0.40}$	$57.64_{\pm0.66}$	$79.01_{\pm0.68}$	$66.64_{\pm0.54}$
MT [18]	$77.92_{\pm0.36}$	$84.83_{\pm0.40}$	$62.54_{\pm0.68}$	$81.84_{\pm0.64}$	$70.95_{\pm0.56}$
DeiT [14]	$78.63_{\pm0.40}$	$85.65_{\pm0.39}$	$63.07_{\pm0.67}$	$84.68_{\pm0.56}$	$72.20_{\pm0.54}$
ProtoPFormer [20]	$77.53_{\pm0.40}$	$85.74_{\pm0.38}$	$61.12_{\pm0.67}$	$86.07_{\pm0.53}$	$71.54_{\pm0.55}$
Ours	$\mathbf{82.05_{\pm0.36}}$	$\mathbf{88.38_{\pm0.30}}$	$\mathbf{65.94_{\pm0.66}}$	$\mathbf{94.43_{\pm0.36}}$	$\mathbf{77.67_{\pm0.47}}$

Table 3. Ablations on MRI dataset with MoViT-Tiny as the backbone. We can observe that the exclusion of either ATMA or PAL results in decreased performance with varying degrees.

ATMA	PAL	Accuracy(↑)	AUC(↑)	Precision(↑)	Recall(↑)	F1-score(↑)
		$78.63_{\pm0.39}$	$86.14_{\pm0.38}$	$62.75_{\pm0.68}$	$86.07_{\pm0.60}$	$72.62_{\pm0.54}$
	✓	$79.35_{\pm0.34}$	$87.43_{\pm0.31}$	$61.92_{\pm0.63}$	$96.55_{\pm0.31}$	$75.57_{\pm0.49}$
✓		$79.54_{\pm0.38}$	$87.13_{\pm0.32}$	$62.27_{\pm0.66}$	$95.85_{\pm0.31}$	$75.49_{\pm0.48}$
✓	✓	$82.05_{\pm0.36}$	$88.38_{\pm0.30}$	$65.94_{\pm0.66}$	$94.43_{\pm0.36}$	$77.67_{\pm0.47}$

Results on MRI Dataset. As depicted in Table 2, our proposed MoViT achieves the highest performance in terms of all metrics on the MRI dataset, where the dataset scale is relatively small, by nature. Specifically, MoViT can improve the AUC by a margin of 0.026 to the state-of-the-art transformer *i.e.* 0.857 achieved by ProtoPFromer; and can achieve better performance (AUC of 0.821) than baseline with 30% training data. Empirically, our method is superior to other modalities in the generalization ability.

Ablation Study. To investigate the contribution of each functional block, ablation studies are performed on the MRI dataset. As shown in Table 3, the proposed MoViT benefits from both ATMA and PAL. Although each module brings a similar AUC improvement from 0.010 to 0.013, the exclusion of the two modules suffers an AUC decline of 0.022. Conclusively, the reported results suggest the effectiveness and indispensability of ATMA and PAL.

4   Conclusion

In conclusion, we show that using memory in transformer architectures is beneficial for reducing the amount of training data needed to train generalizable transformer models. The reduction in data needs is particularly appealing in medical image analysis, where large-scale data continues to pose challenges. Our model, the Memorizing Vision Transformer (MoViT) for medical image analysis, caches and updates relevant key and value pairs during training. It then uses them to enrich the attention context for the inference stage. MoViT's implementation is straightforward and can easily be plugged into various transformer models to achieve performance competitive to vanilla ViT with much less training data. Consequently, our method has the potential to benefit a broad range of applications in the medical image analysis context. Future work includes a hybrid of MoViT with convolutional neural networks for more comprehensive feature extraction.

Acknowledgments. This work was supported in part by grants from the National Institutes of Health (R37CA248077, R01CA228188). The MRI equipment in this study was funded by the NIH grant: 1S10OD021648.

References

1. Barhoumi, Y., et al.: Scopeformer: n-CNN-ViT hybrid model for intracranial hemorrhage classification. arXiv preprint arXiv:2107.04575 (2021)
2. Dosovitskiy, A., et al.: An image is worth 16×16 words: transformers for image recognition at scale. arXiv preprint arXiv:2010.11929 (2020)
3. Guo, P., et al.: Learning-based analysis of amide proton transfer-weighted MRI to identify true progression in glioma patients. NeuroImage: Clin. **35**, 103121 (2022)
4. He, K., et al.: Deep residual learning for image recognition. In: Proceedings of the IEEE conference on CVPR, pp. 770–778 (2016)
5. Kather, J.N., et al.: 100,000 histological images of human colorectal cancer and healthy tissue. Zenodo (2018). https://doi.org/10.5281/zenodo.1214456
6. Khandelwal, U., et al.: Generalization through memorization: nearest neighbor language models. arXiv preprint arXiv:1911.00172 (2019)
7. Kim, B., et al.: Examples are not enough, learn to criticize! criticism for interpretability. In: Advances in Neural Information Processing Systems 29 (2016)
8. Laine, S., et al.: Temporal ensembling for semi-supervised learning. arXiv preprint arXiv:1610.02242 (2016)
9. Li, Y., et al.: LocalViT: bringing locality to vision transformers. arXiv preprint arXiv:2104.05707 (2021)
10. Liu, Y., et al.: Efficient training of visual transformers with small datasets. Adv. Neural. Inf. Process. Syst. **34**, 23818–23830 (2021)
11. Russakovsky, O., et al.: ImageNet large scale visual recognition challenge. Int. J. Comput. Vis. **115**(3), 211–252 (2015)
12. Sun, C., et al.: Revisiting unreasonable effectiveness of data in deep learning era. In: Proceedings of the ICCV, pp. 843–852 (2017)
13. Tarvainen, A., et al.: Mean teachers are better role models: weight-averaged consistency targets improve semi-supervised deep learning results. In: Advances in Neural Information Processing Systems 30 (2017)
14. Touvron, H., et al.: Training data-efficient image transformers & distillation through attention. In: International Conference on Machine Learning, pp. 10347–10357. PMLR (2021)
15. van Tulder, G., Tong, Y., Marchiori, E.: Multi-view analysis of unregistered medical images using cross-view transformers. In: de Bruijne, M., et al. (eds.) MICCAI 2021. LNCS, vol. 12903, pp. 104–113. Springer, Cham (2021). https://doi.org/10.1007/978-3-030-87199-4_10
16. Wang, X., et al.: Cross-batch memory for embedding learning. In: Proceedings of the CVPR, pp. 6388–6397 (2020)
17. Wu, H., et al.: CvT: Introducing convolutions to vision transformers. In: Proceedings of the ICCV, pp. 22–31 (2021)
18. Wu, Y., Rabe, M.N., Hutchins, D., Szegedy, C.: Memorizing transformers. In: International Conference on Learning Representations (2022). https://openreview.net/forum?id=TrjbxzRcnf-
19. Xu, W., et al.: Co-scale conv-attentional image transformers. In: Proceedings of the ICCV, pp. 9981–9990 (2021)
20. Xue, M., et al.: ProtoPFormer: concentrating on prototypical parts in vision transformers for interpretable image recognition. arXiv preprint arXiv:2208.10431 (2022)
21. Yuan, K., et al.: Incorporating convolution designs into visual transformers. In: Proceedings of the ICCV, pp. 579–588 (2021)

Fact-Checking of AI-Generated Reports

Razi Mahmood[1(✉)], Ge Wang[1], Mannudeep Kalra[2], and Pingkun Yan[1]

[1] Department of Biomedical Engineering and Center for Biotechnology and Interdisciplinary Studies, Rensselaer Polytechnic Institute, Troy, NY 12180, USA
mahmor@rpi.edu
[2] Department of Radiology, Massachusetts General Hospital, Harvard Medical School, Boston, MA 02114, USA

Abstract. With advances in generative artificial intelligence (AI), it is now possible to produce realistic-looking automated reports for preliminary reads of radiology images. However, it is also well-known that such models often hallucinate, leading to false findings in the generated reports. In this paper, we propose a new method of fact-checking of AI-generated reports using their associated images. Specifically, the developed examiner differentiates real and fake sentences in reports by learning the association between an image and sentences describing real or potentially fake findings. To train such an examiner, we first created a new dataset of fake reports by perturbing the findings in the original ground truth radiology reports associated with images. Text encodings of real and fake sentences drawn from these reports are then paired with image encodings to learn the mapping to real/fake labels. The examiner is then demonstrated for verifying automatically generated reports.

Keywords: Generative AI · Chest X-rays · Fact-checking · Radiology Reports

1 Introduction

With the developments in radiology artificial intelligence (AI), many researchers have turned to the problem of automated reporting of imaging studies [3,5,11,13, 15,16,22,26]. This can significantly reduce the dictation workload of radiologists, leading to more consistent reports with improved accuracy and lower overall costs. While the previous work has largely used image captioning [23,27] or image-to-text generation methods for report generation, more recent works have been using large language models (LLMs) such as GPT-4 [6,12]. These newly emerged LLMs can generate longer and more natural sentences when prompted with good radiology-specific linguistic cues [4,7].

However, with powerful language generation capabilities, hallucinations or false sentences are prevalent as it is difficult for those methods to identify their own errors. This has led to fact-checking methods for output generated by LLMs and large vision models (LVMs) [10,18,21]. Those methods detect errors either through patterns of phrases found repeatedly in text or by consulting other

© The Author(s), under exclusive license to Springer Nature Switzerland AG 2024
X. Cao et al. (Eds.): MLMI 2023, LNCS 14349, pp. 214–223, 2024.
https://doi.org/10.1007/978-3-031-45676-3_22

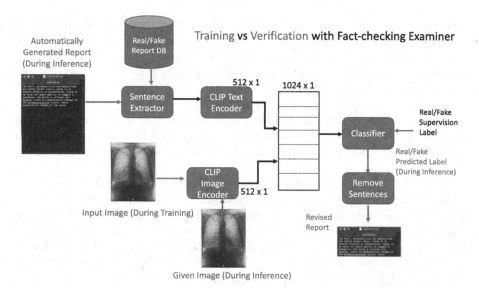

Fig. 1. Illustration of the training (blue) and inference (red) phases of the image-driven fact-checking examiner. The operations common to both phases are in black. (Color figure online)

external textual sources for the veracity of information [10,18,21]. In radiology report generation, however, we have a potentially good source for fact checking, namely, the associated images, as findings reported in textual data must be verifiable through visual detection in the associated imaging. Since most methods of report generation already examine the images in order to detect findings and generate the sentences, bootstrapping them with an independent source of verification is needed in order to identify their own errors.

In this paper, we propose a new imaging-driven method of fact-checking of AI-generated reports. Specifically, we develop a fact-checking examiner to differentiate between real and fake sentences in reports by learning the association between an image and sentences describing real or potentially fake findings. To train such an examiner, we first create a new dataset of fake reports by perturbing the findings in the original ground truth radiology reports associated with images. Text encodings of real and fake sentences drawn from these reports are then paired with image encodings to learn the mapping to real or fake labels via a classifier. The utility of such an examiner is demonstrated for verifying automatically generated reports by detecting and removing fake sentences. Future generative AI approaches can use the examiner to bootstrap their report generation leading to potentially more reliable reports.

Our overall approach to training and inference using the examiner is illustrated in Fig. 1. To create a robust examiner that is not attuned for any particular automated reporting software, it is critical to create a dataset for training that encompasses a wide array of authentic and fabricated samples. Hence we first synthesize a dataset of real and fake reports using a carefully controlled process of perturbation of actual radiology reports associated with the images. We then pair each image with sentences from its corresponding actual report as real sentences with real label, and the perturbed sentences from fake reports as fake sentences with fake label. Both textual sentence and images are then encoded by projecting in a joint image-text embedding space using the CLIP model [19]. The encoded vectors of image and the paired sentence are then concatenated to form the feature vector for classification. A binary classifier is then trained on this dataset to produce a discriminator for real/fake sentences associated with a given image.

The fact-checker can be used for report verification in inference mode. Given an automatically produced radiology report, and the corresponding input imaging study, the examiner extracts sentences from the report, and the image-sentence pair is then subjected to the same encoding process as used in training. The combined feature vector is then given to the classifier for determination of the sentence as real or fake. A revised report is assembled by removing those sentences that are deemed fake by the classifier to produce the new report.

2   Generation of a Synthetic Report Dataset

The key idea in synthetic report generation is to center the perturbation operations around findings described in the finding sections of reports, as these are critical to preliminary reads of imaging studies.

2.1   Modeling Finding-Related Errors in Automated Reports

The typical errors seen in the finding sections of reports can be due to (a) addition of incorrect findings not seen in the accompanying image, (b) exchange errors, where certain findings are missed and others added, (c) reverse findings reported i.e. positive instance reported when negative instances of them are seen in image and vice versa, (d) spurious or unnecessary findings not relevant for reporting, and finally (e) incorrect description of findings in terms of fine-grained appearance, such as extent of severity, location correctness, etc.

From the point of real/fake detection, we focus on the first 3 classes of errors for synthesis as they are the most common. Let $R = \{S_i\}$ be a ground-truthed report corresponding to an image I consisting of sentences $\{S_i\}$ describing corresponding findings $\{F_i\}$. Then we can simulate a random addition of a new finding by extending the report R as $R_a = \{S_i\} \cup \{S_a\}$ where S_a describes a new finding $F_a \notin \{F_i\}$. Similarly, we simulate condition (b) through an exchange of finding where one finding sentence S_r is removed to be replaced by another finding sentence S_a as $R_e = \{S_i\} - \{S_r\} \cup \{S_a\}$. Finally, we can simulate the replacement of positive with negative findings and vice versa to form a revised report $R_r = \{S_i\} - \{S_p\} \cup \{S_{p'}\}$ where S_p is a sentence corresponding to a finding F_p and $S_{p'}$ is a sentence corresponding to the finding $F_{p'}$ which is in opposite sense of the meaning. For example, a sentence "There is pneumothorax', could be replaced by "There is no pneumothorax" to represent a reversal of polarity of the finding. Figure 2 shows examples of each of the type of operations of add, exchange and reverse findings respectively.

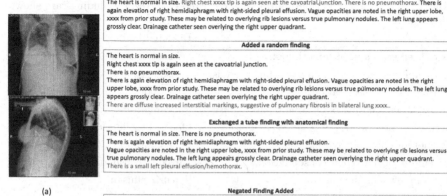

Fig. 2. Illustration of the fake reports drawn from actual reports. (a) Frontal and lateral views of a chest X-ray. (b) Corresponding original and fake radiology reports. The affected sentences during the synthesis operation are shown in red. (Color figure online)

2.2 Detecting Findings in Sentences

Since detecting findings is key to our approach, our synthetic dataset generation focused on chest X-ray datasets, as finding detectors are well-developed for these datasets. Further, the majority of work on automated reporting has been done on chest X-rays and finding-labeled datasets are publicly available [8,9,24]. However, most of the existing approaches summarize findings at the report level. To locate findings at the sentence level, we used NLP tools such as Spacy [1] to separate sentences. We then combined ChexPert [8] labeler and NegSpacy [1] parser to extract positive and negative findings from sentences. Table 1 shows examples of findings detected in sentences. The detected findings were then validated against the ground truth labels provided at the report level in the datasets. All unique findings across reports were then aggregated into a pool $\{F_{pool}\}$ and all unique sentences in the original reports were aggregated and mapped to their findings (positive or negative) to create the pool of sentences $\{S_{pool}\}$.

2.3 Fake Report Creation

For each original report R associated with an image I, we create three instances of fake reports R_a, R_e, and R_r corresponding to the operations of addition, exchange and reversal of findings, respectively. Specifically, for creating R_a type of reports, we randomly draw from S_{pool} a sentence that contains a randomly

Table 1. Illustration of extracting findings from reports. Negated findings are shown within square brackets.

Sentences	Detected findings
There is effusion and pneumothorax.	'effusion', 'pneumothorax'
No pneumothorax, pleural effusion, but there is lobar air space consolidation.	'consolidation', ['pneumothorax', 'pleural effusion']
No visible pneumothorax or large pleural effusion.	['pneumothorax', 'pleural effusion']
Specifically, no evidence of focal consolidation, pneumothorax, or pleural effusion.	['focal consolidation'], ['pneumothorax', 'pleural effusion']
No definite focal alveolar consolidation, no pleural effusion demonstrated	['alveolar consolidation'], ['pleural effusion']

Table 2. Details of the fake report dataset distribution. 2557 frontal views were retained for images. 64 negative findings were retained and 114 positive findings.

Dataset	Patients	Images/Views	Reports	Pos/Neg Findings	Unique Sentences	Image-Sent Pairs	Fake Reports
Original	1786	7470/2557	2557	119/64	3850	25535	7671
Training	1071	2037	2037	68	2661	20326	4074
Testing	357	254	254	68	919	2550	508

selected finding $F_a \notin \{F_i\}$, where $\{F_i\}$ are the set of findings in R (positive or negative). Similarly, to create R_e, we randomly select a finding pair (F_{ei}, F_{eo}), where $F_{ei} \in \{F_i\}$ and $F_{eo} \in \{F_{pool}\} - \{F_i\}$. We then remove the associated sentence with F_{ei} in R and replace it with a randomly chosen sentence associated with F_{eo} in $\{S_{pool}\}$. Finally, to create the reversed findings reports, R_r, we randomly select a positive or negative finding $F_p \in \{F_i\}$, remove its corresponding sentence and swap it with a randomly chosen sentence $S_{p'} \in \{S_{pool}\}$, containing findings $F_{p'}$ that is reversed in polarity. The images, their perturbed finding, and associated sentences were recorded in each case of fake reports so that they could be used to form the pairing dataset for training the fact-checking examiner described next.

3 Fact-Checking of AI-Generated Reports

3.1 Fact-Checking Examiner

The fact-checking examiner is a classifier using deep-learned features derived from joint image-text encodings. Specifically, since we combine images with textual sentences, we chose a feature encoding that is already trained on joint image and text pairs. In particular, we chose the CLIP joint image-text embedding model [19] to project the image and textual sentences into a common 512-length encoding space. While other joint image-text encoders could potentially work, we chose CLIP as our encoder because it was pre-trained on natural image-text pair and subsequently tuned on radiology report-image pairs [3]. We then concatenate the image and textual embedding into a 1024-length feature vector to train a binary classifier. In our splits, the real/fake incidence distribution was relatively balanced (2:1) so that the accuracy could be used as a reliable measure of performance. We experimented with several classifiers ranging from support vector machines (SVM) to neural net classifiers. As we observed similar performance, we retained a simple linear SVM as sufficient for the task.

3.2   Improving the Quality of Reports Through Verification

We apply the fact-checking examiner to filter our incorrect/irrelevant sentences in automatically produced reports as shown in Fig. 1 (colored in red). Specifically, given an automatically generated report for an image, we pair the image with each sentence of the report. We then use the same CLIP encoder used in training the examiner, to encode each pair of image and sentence to form a concatenated feature vector. The examiner predicted fake sentences are then removed to produce the revised report.

We develop a new measure to judge the improvement in the quality of the automatic report after applying the fact-checking examiner. Unlike popular report comparison measures such as BLEU [17], ROUGE [14] scores that perform lexical comparisons, we use a semantic similarity measure formed from encoding the reports through large language models such as SentenceBERT [20]. Specifically, let $R = \{S_i\}$, $R_{auto} = \{S_{auto}\}$, $R_{corrected} = \{S_{corrected}\}$ be the original, automated, and corrected reports with their sentences respectively. To judge the improvement in quality of the report, we adopt SentenceBERT [20] to encode the individual sentences of the respective reports to produce an average encoding per report as $E_R, E_{auto}, E_{corrected}$ respectively. Then the quality improvement score, $QI(R)$ per triple of reports $(R, R_{auto}, R_{corrected})$ is given by the difference in the cosine similarity between the pairwise encodings as

$$QI(R, R_{auto}, R_{corrected}) = d(E_R, E_{corrected}) - d(E_R, R_{auto}), \qquad (1)$$

where d is the cosine similarity between the average encodings. This measure allows for unequal lengths of reports. A positive value indicates an improvement while a negative value indicates a worsening of the performance. The overall improvement in the quality of automatically generated reports is then given by

$$QI = (n_{positive} + n_{same} - n_{negative})/n_R \qquad (2)$$

where

$$n_{positive} = |arg_R(d(E_R, E_{corrected}) > d(E_R, R_{auto}))|$$
$$n_{same} = |arg_R(d(E_R, E_{corrected}) = d(E_R, R_{auto}))| \qquad (3)$$
$$n_{negative} = |arg_R(d(E_R, E_{corrected}) < d(E_R, R_{auto}))|$$

are the number of times the corrected reports are closer to original reports, same similarity as the AI report, or worse than the AI report by applying the examiner respectively, and n_R is the total number of automated reports evaluated.

Table 3. Performance of real/fake discriminator.

	Prec.	Recall	F1	Sup.
Real class	0.86	0.93	0.90	3648
Accuracy	–	–	0.84	5044
Macro Avg	0.82	0.77	0.79	5044
Weighted Avg	0.84	0.84	0.84	5044

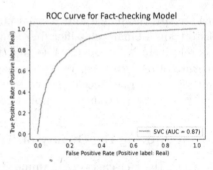

Fig. 3. performance of real/fake report sentence differentiation.

4   Results

To test our approach for fact-checking of radiology reports, we selected an open access dataset of chest X-rays from Indiana University [2] provided on Kaggle, which contains 7,470 chest X-Ray (frontal and lateral views) images with corresponding 2,557 non-duplicate reports from 1786 patients. The dataset also came with annotations documenting important findings at the report level. Of the 1786 patients, we used a (60-20-20)% patient split for training the examiner, testing the examiner, and evaluating its effectiveness in report correction respectively, thus ensuring no patient overlap between the partitions (Table 3).

4.1   Fake Report Dataset Created

By applying NLP methods of sentence extraction, we extracted 3850 unique sentences from radiology reports. By applying the finding extractor at the sentence level as described in Sect. 2.2, we catalogued a total of 119 distinct positive and 64 negative findings as shown Table 2. Using these findings and their sentences in the 2557 unique reports, and the 3 types of single perturbation operations described in Sect. 2.1, we generated 7,671 fake reports as shown in Table 2.

The training and test dataset for the fact-checking examiner was generated by randomly drawing sentences from sentence pool $\{S_{pool}\}$. Each image was first paired with each sentence from its original report and the pair was given the "Real" label. The perturbed sentence drawn from $\{S_{pool}\}$ from the fake reports was then retrieved from each fake report and paired with the image and given the "Fake" label. The list of pairs produced were processed to remove duplicate pairings. By this process, we generated 20,326 pairs of images with real/fake sentences for training, and 2,550 pairs for testing as shown in Table 2 using 80% of the 1,786 patients.

4.2   Fact-Checking Examiner Accuracy

Using the train-test splits shown in Table 2, we trained fact-checking examiner with encodings of image-sentence pairs shown in Table 2. The resulting classifier

Table 4. Report quality evaluation on two automatically generated report datasets.

Dataset	Patients	Reports	$n_{positive}$	n_{same}	$n_{negative}$	QI score
Synthetic Reports	358	3661	1105	1008	1548	15.63%
NIH Reports	198	198	60	55	83	16.1%

achieved an average accuracy of 84.2% and the AUC was 0.87 as shown in Fig. 3b. The precision, recall, F-score, macro accuracy by average and weighted average methods are shown in Fig. 3a. By using 10 fold cross-validation in the generation of the (60-20-20) splits for the image-report dataset, and using different classifiers provided in the Sklearn library (decision tree, logistic regression, etc.) the average accuracy lay in the range 0.84 ± 0.02. The errors seen in the classification were primarily for complex description of findings containing modifiers and negations.

4.3 Overall Report Quality Improvement Evaluation

We evaluated the efficacy of the fact-checking examiner on two report datasets, one synthetic with controlled "fakeness" and another dataset generated by a published algorithm described in [22]. Specifically, using the 20% partition of patients from the Indiana reports that was not used to train or test the examiner, we selected 3089 of the fake reports shown in Table 2. We evaluated the improvement in report quality using the method described in Sect. 3.2. These results are summarized in Table 4. Since our fake reports had only one fake sentence added, the performance improvement while still present, is modest around 15.63%.

To test the performance on automated reports generated by existing algorithms, we obtained a reference dataset consisting of freshly created reports on the NIH image dataset [24] created by radiologists as described in [25]. We retained the output of an automated report generation algorithm for the same images described in [22]. A total of 198 pairs of original and automatically created reports along with their associated imaging from the NIH dataset was used for this experiment. The results of quality improvement is shown in Table 4. As it can be seen, the quality improvement is slightly greater for reports produced by automated report extraction methods.

5   Conclusion

In this paper, we have proposed for the first time, an image-driven verification of automatically produced radiology reports. A dataset was carefully constructed to elicit the different types of errors produced by such methods. A Novel fact-checking examiner was developed using pairs of real and fake sentences with their corresponding imaging. The work will be extended in future to cover larger variety of defects and extended evaluation on a larger number automated reports.

References

1. NegSpacy Parser. https://spacy.io/universe/project/negspacy
2. Demmer-Fushma, D., et al.: Preparing a collection of radiology examinations for distribution and retrieval. J. Am. Med. Inf. Assoc. (JAMIA) **23**(2), 304–310 (2014)
3. Endo, M., Krishnan, R., Krishna, V., Ng, A.Y., Rajpurkar, P.: Retrieval-based chest x-ray report generation using a pre-trained contrastive language-image model. Proc. Mach. Learn. Res. **158**, 209–219 (2021)
4. Krause, J., Johnson, J., Krishna, R., Fei-Fei, L.: A hierarchical approach for generating descriptive image paragraphs. In: IEEE CVPR (2017)
5. Gale, W., Oakden-Rayner, L., Carneiro, G., Bradley, A.P., Palmer, L.J.: Producing radiologist-quality reports for interpretable artificial intelligence. arXiv preprint arXiv:1806.00340 (2018)
6. Grewal, H., et al.: Radiology gets chatty: the chatgpt saga unfolds. Cureus **15** (2023). https://doi.org/10.7759/CUREUS.40135. https://www.cureus.com/articles/161200-radiology-gets-chatty-the-chatgpt-saga-unfolds
7. Guo, J., Lu, S., Cai, H., Zhang, W., Yu, Y., Wang, J.: Long text generation via adversarial training with leaked information. In: AAAI-2018, pp. 5141–5148 (2018)
8. Irvin, J., et al.: Chexpert: a large chest radiograph dataset with uncertainty labels and expert comparison. In: Thirty-Third AAAI Conference on Artificial Intelligence (2019)
9. Johnson, A.E.W., et al.: Mimic-cxr: a large publicly available database of labeled chest radiographs. arXiv preprint arXiv:1901.07042 (2019)
10. Lab, N.J.: Ai will start fact-checking. we may not like the results. https://www.niemanlab.org/2022/12/ai-will-start-fact-checking-we-may-not-like-the-results/
11. Li, C.Y., Liang, X., Hu, Z., Xing, E.P.: Knowledge-driven encode, retrieve, paraphrase for medical image report generation. arXiv preprint arXiv:1903.10122 (2019)
12. Li, X., et al.: Artificial general intelligence for medical imaging. arXiv preprint arXiv:2306.05480 (2023)
13. Li, Y., Liang, X., Hu, Z., Xing, E.P.: Hybrid retrieval-generation reinforced agent for medical image report generation. In: Advances in Neural Information Processing Systems, pp. 1530–1540 (2018)
14. Lin, C.Y.: Rouge: a package for automatic evaluation of summaries. In: Workshop on Text Summarization Branches Out (2004)
15. Liu, G., et al.: Clinically accurate chest x-ray report generation. arXiv:1904.02633v (2019)
16. Pang, T., Li, P., Zhao, L.: A survey on automatic generation of medical imaging reports based on deep learning. BioMed. Eng. OnLine **22**(1), 48 (2023). https://doi.org/10.1186/s12938-023-01113-y
17. Papineni, K., Roukos, S., Ward, T., Zhu, W.J.: Bleu: a method for automatic evaluation of machine translation. In: Proceedings of the 40th Annual Meeting of the Association for omputational Linguistics, pp. 311–318. Association for Computational Linguistics, Philadelphia (2002). https://doi.org/10.3115/1073083.1073135. https://aclanthology.org/P02-1040
18. Passi, K., Shah, A.: Distinguishing fake and real news of twitter data with the help of machine learning techniques. In: ACM International Conference Proceeding Series, pp. 1–8 (2022). https://doi.org/10.1145/3548785.3548811
19. Radford, A., et al.: Learning transferable visual models from natural language supervision. Proc. Mach. Learn. Res. **139**, 8748–8763 (2021). https://arxiv.org/abs/2103.00020v1

20. Reimers, N., Gurevych, I.: Sentence-bert: sentence embeddings using siamese bert-networks. CoRR abs/1908.10084 (2019). http://arxiv.org/abs/1908.10084
21. Suprem, A., Pu, C.: Midas: multi-integrated domain adaptive supervision for fake news detection (2022). https://arxiv.org/pdf/2205.09817.pdf
22. Syeda-Mahmood, T., et al.: Chest X-Ray report generation through fine-grained label learning. In: Martel, A.L., et al. (eds.) MICCAI 2020. LNCS, vol. 12262, pp. 561–571. Springer, Cham (2020). https://doi.org/10.1007/978-3-030-59713-9_54
23. Vinyals, O., Toshev, A., Bengio, S., Erhan, D.: Show and tell: a neural image caption generator. In: Proceedings of the IEEE Conference on Computer Vision and Pattern Recognition, pp. 3156–3164 (2015)
24. Wang, X., Peng, Y., Lu, L., Lu, Z., Bagheri, M., Summers, R.M.: Chestx-ray8: hospital-scale chest x-ray database and benchmarks on weakly-supervised classification and localization of common thorax diseases (2017). https://uts.nlm.nih.gov/metathesaurus.html
25. Wu, J.T., et al.: Comparison of chest radiograph interpretations by artificial intelligence algorithm vs radiology residents. JAMA Netw. Open **3**, e2022779–e2022779 (2020). https://doi.org/10.1001/JAMANETWORKOPEN.2020.22779. https://jamanetwork.com/journals/jamanetworkopen/fullarticle/2771528
26. Xiong, Y., Du, B., Yan, P.: Reinforced transformer for medical image captioning. In: Suk, H.-I., Liu, M., Yan, P., Lian, C. (eds.) MLMI 2019. LNCS, vol. 11861, pp. 673–680. Springer, Cham (2019). https://doi.org/10.1007/978-3-030-32692-0_77
27. Xu, K., et al.: Show, attend and tell: neural image caption generation with visual attention. In: International Conference on Machine Learning, pp. 2048–2057 (2015)

Is Visual Explanation with Grad-CAM More Reliable for Deeper Neural Networks? A Case Study with Automatic Pneumothorax Diagnosis

Zirui Qiu[1]([✉]), Hassan Rivaz[2], and Yiming Xiao[1]

[1] Department of Computer Science and Software Engineering, Concordia University, Montreal, Canada
leoqiuzirui@gmail.com
[2] Department of Electrical and Computer Engineering, Concordia University, Montreal, Canada

Abstract. While deep learning techniques have provided the state-of-the-art performance in various clinical tasks, explainability regarding their decision-making process can greatly enhance the credence of these methods for safer and quicker clinical adoption. With high flexibility, Gradient-weighted Class Activation Mapping (Grad-CAM) has been widely adopted to offer intuitive visual interpretation of various deep learning models' reasoning processes in computer-assisted diagnosis. However, despite the popularity of the technique, there is still a lack of systematic study on Grad-CAM's performance on different deep learning architectures. In this study, we investigate its robustness and effectiveness across different popular deep learning models, with a focus on the impact of the networks' depths and architecture types, by using a case study of automatic pneumothorax diagnosis in X-ray scans. Our results show that deeper neural networks do not necessarily contribute to a strong improvement of pneumothorax diagnosis accuracy, and the effectiveness of GradCAM also varies among different network architectures.

Keywords: Grad-CAM · Deep learning · Interpretability

1 Introduction

With rapid development, deep learning (DL) techniques have become the state-of-the-art in many vision applications, such as computer-assisted diagnosis. Although initially proposed for natural image processing, staple Convolutional Neural Networks (CNNs), such as VGG and ResNet architectures, have become ubiquitous backbones in computer-assisted radiological applications due to their robustness and flexibility to capture task-specific, complex image features. Furthermore, the more recent Vision Transformer (ViT), a new class of DL architecture that leverage the self-attention mechanism to encode long-range contextual information is attracting great attention in medical image processing, with

X. Cao et al. (Eds.): MLMI 2023, LNCS 14349, pp. 224–233, 2024.
https://doi.org/10.1007/978-3-031-45676-3_23

evidence showing superior performance than the more traditional CNNs [3]. Although these DL algorithms can offer excellent accuracy, one major challenge that hinders their wide adoption in clinical practice is the lack of transparency and interpretability in their decision-making process. So far, various explainable AI (XAI) methods have been proposed [1], and among these, direct visualization of the saliency/activation maps has gained high popularity, likely due to their intuitiveness for fast uptake in clinical applications. With high flexibility and ease of implementation for different DL architectures, the Gradient-weighted Class Activation Mapping (Grad-CAM) technique [9], which provides visual explanation as heatmaps with respect to class-wise decision has been applied widely in many computer-assisted diagnostic and surgical vision applications. However, in almost all previous investigations, Grad-CAM outcomes are only demonstrated qualitatively. To the best of our knowledge, the impacts of different deep learning architectures and sizes on the robustness and effectiveness of Grad-CAM have not been investigated, but are important for the research community.

To address the mentioned knowledge gap, we benchmarked the performance of Grad-CAM on DL models of three types of popular architectures, including VGG, ResNet, and ViT, with varying network depths/sizes of each one. We explored the impacts of DL architectures on GradCAM by using pneumothorax diagnosis from chest X-ray images as a case study. Pneumothorax is a condition that is characterized by the accumulation of air in the pleural space, and can lead to lung collapse, posing a significant risk to patient health if not promptly diagnosed and treated. As the radiological features of pneumothorax (i.e., air invasion) can be subtle to spot in X-ray scans, the task provides an excellent case to examine the characteristics of different DL models and Grad-CAM visualization. In summary, our work has two main contributions. **First**, we conducted a comprehensive evaluation of popular DL models including CNNs and Transformers for pneumothorax diagnosis. **Second**, we systematically compared the effectiveness of visual explanation using Grad-CAM across these staple DL models both qualitatively and quantitatively. Here, we analyzed the impact of network architecture choices on diagnostic accuracy and effectiveness of Grad-CAM results.

2   Related Works

The great accessibility of public chest X-ray datasets has allowed a large amount works [7] on the diagnosis and segmentation of lung diseases using deep learning algorithms, with a comprehensive review provided by Calli et al. [2]. So far, many previous reports adopted popular DL models that were first designed for natural image processing. For example, Tian et al. [14] leveraged ResNet and VGG models with multi-instance transfer learning for pneumothorax classification. Wollek at al. [15] employed the Vision Transformer to perform automatic diagnosis for multiple lung conditions on chest X-rays. To incorporate visual explanation for DL-based pneumothorax diagnosis, many have adopted Grad-CAM and its variants [15,16] for both CNNs and ViTs. Yuan et al. [16] proposed a human-guided design to enhance the performance of Saliency Map, Grad-CAM, and

Integrated Gradients in visual interpretability of pneumothorax diagnosis using CNNs. Most recently, Sun et al. [13] proposed the Attri-Net, which employed Residual blocks and multi-label explanations that align with clinical knowledge for improved visual explanation in chest X-ray classification.

3   Material and Methodology

3.1   Deep Learning Model Architectures

Our study explored a selection of staple deep learning architectures, including VGG, ResNet and ViT, which are widely used in both natural and medical images. Specifically, the VGG models [11] are characterized by their multiple 3×3 convolution layers, and for the study, we included VGG-16 and VGG-19. The ResNet models [5] leverage skip connections to enhance residual learning and training stability. Here, we incorporated ResNet18, ResNet34, ResNet50, and ResNet101, which comprise 18, 34, 50, and 101 layers, respectively. Lastly, the Vision Transformers initially proposed by Dosovitskiy et al. [4] treat images as sequences of patches/tokens to model their long-range dependencies without the use of convolution operations. To test the influence of network sizes, we adopted the ViT_small and ViT_base variants, both with 12 layers but differing in input features, and the ViT_large variant with 24 layers. All these models were pretrained on ImageNet-1K and subsequently fine-tuned for the task of Pneumothorax vs. Healthy classification using the curated public dataset.

3.2   Grad-CAM Visualization

To infer the decision-making process of DL models, the Grad-CAM technique [9] creates a heatmap by computing the gradients of the target class score with respect to the feature maps of the last convolutional layer. Specifically, for VGG16 and VGG19, we applied Grad-CAM to the last convolution layer. For ResNet models, we targeted the final bottleneck layer and in Vision Transformer variants, the technique was applied to the final block layer before the classification token is processed. Ideally, an effective visual guidance should provide high accuracy (i.e., correct identification of the region of interest by high values in the heatmap) and specificity (i.e., tight bound around the region of interest). Note that for each Grad-CAM heatmap, the value is normalized to [0,1].

3.3   Dataset Preprocessing and Experimental Setup

For this study, we used the SIIM-ACR Pneumothorax Segmentation dataset from Kaggle[1]. It contains chest X-Ray images of 9,000 healthy controls and 3,600 patients with pneumothorax. In addition, regions related to pneumothorax were

---

[1] SIIM-ACR Pneumothorax Segmentation: https://www.kaggle.com/competitions/siim-acr-pneumothorax-segmentation/data.

manually segmented for 3576 patients. For our experiments, we created a balanced subset of 7,200 cases (50% with pneumothorax) from the original dataset. From the curated data collection, we divided the cases into 7,000 for training, 1,000 for validation, and 1,000 for testing while balancing the health vs. pneumothorax ratio in each set. For DL model training and testing, each image was processed using Contrast Limited Adaptive Histogram Equalization (CLAHE) and normlaized with z-transform. In addition, all images were re-scaled to the common dimension of 224×224 pixels. In terms of training, the VGG and ResNet models utilized the cross-entropy loss function with the Adam optimizer, and were trained at a learning rate of 1e-4 for 50 epochs. The ViT models employed a cross-entropy loss function and the Stochastic Gradient Descent (SGD) method for optimization with a learning rate of 1e-4. They were trained for 300 epochs. Finally, to boost our model's performance and mitigate overfitting, we used data augmentations in training, including the addition of random Gaussian noise, rotations (up to $10\,^{\circ}C$), horizontal flips, and brightness/contrast shifts.

3.4 Evaluation Metrics

To evaluate the performance of different DL models in pneumothorax diagnosis, we assessed the accuracy, precision, recall, and area under the curve (AUC) metrics. In terms of assessing the effectiveness of the Grad-CAM results for each model, we propose to use two different measures. First, we compute the difference between the means of the Grad-CAM heatmap values within and outside the ground truth pneumothorax segmentation, and refer to this metric as $Diff_{GradCAM}$. We hypothesize that an effective Grad-CAM visualization should generate a high positive $Diff_{GradCAM}$ because the ideal heatmap should accumulate high values primarily within the pathological region (i.e., air invasion in the pleural space). The scores from the models were further compared by a one-way ANOVA test and Tukey's post-hoc analysis, and a p-value < 0.05 indicated a statistically significant difference. Second, we compute the Effective Heat Ratio (EHR) [15], which is the ratio between the thresholded area of the Grad-CAM heatmap within the ground truth segmentation and the total threshold area. The thresholds were computed in equidistant steps, and the Area Under the Curve (AUC) is calculated over all EHRs and the associated threshold values to assess the quality of Grad-CAM results. Both metrics reflect the accuracy and specificity of visual explanation for the networks.

4 Results

4.1 Pneumothorax Diagnosis Performance

The performance of pneumothorax diagnosis for all DL models is listed in Table 1. In terms of accuracy, ResNet models offered the best results, particularly with ResNet50 at 88.20%, and the ViT and VGG ranked the second and the last. The similar trend held for precision and AUC. However, as for recall, the obtained scores were similar across different architecture types. When looking into different network sizes for each architecture type, the results showed that deeper neural networks did not necessarily produce superior diagnostic performance. Specifically, the two popular VGG models didn't result in large discrepancy in accuracy, recall, and AUC while VGG16 has better precision. For the ResNet models, ResNet18, ResNet34 and ResNet101 had similar performance, with the accuracy, recall, and AUC peaked slightly at ResNet50. This likely means that ResNet18 has enough representation power to perform the classification, and therefore deeper networks do not improve the results. Finally, for the ViTs, ViT_base resulted in better performance than the small and large versions, with slight performance deterioration for ViT_large.

Table 1. Pneumothorax diagnosis performance across all DL models

Model	Accuracy	Precision	Recall	AUC
VGG16	83.70%	0.8224	0.8800	0.9069
VGG19	83.20%	0.7747	0.9080	0.9098
ResNet18	87.60%	0.8821	0.8680	0.9420
ResNet34	87.20%	0.8758	0.8600	0.9417
ResNet50	88.20%	0.8802	0.8820	0.9450
ResNet101	87.60%	0.8798	0.8780	0.9434
ViT Small	85.80%	0.8137	0.8820	0.9302
ViT Base	86.20%	0.8602	0.8740	0.9356
ViT Large	84.40%	0.8356	0.8640	0.9197

4.2   Qualitative Grad-CAM Evaluation

We present the Grad-CAM heatmaps for three patient cases across all tested DL models in Fig. 1, with the ground truth pneumothorax segmentation overlaid in the X-ray scans. In the presented cases, all the heatmaps correctly indicated the side of the lungs affected by the disease. However, the overlap with the pneumothorax segmentation varied. The VGG16 and VGG19 models pinpointed the pneumothorax areas for the first two cases while VGG19 failed to do so for the third case. Note that both models presented a secondary region. The ResNet18, 34, and 50 models successfully highlighted the problematic area, with the heatmap of ResNet18 slightly off-center, while the ResNet101 model showed activation in two regions. In comparison, ViT models exhibited more dispersed Grad-CAM patterns than the CNNs, and the amount of unrelated areas increased with the model size.

4.3   Quantitative Grad-CAM Evaluation

For the Grad-CAM heatmap of each tested model, we computed the $Diff_{GradCAM}$ and EHR AUC metrics across all test cases to gauge their effectiveness of visually interpreting the decision-making process of DL algorithms. Here, $Diff_{GradCAM}$ and EHR AUC are reported in Table 2 and Fig. 2, respectively. For $Diff_{GradCAM}$, the ANOVA test showed a group-wise difference ($p<0.001$). In general, the CNNs offered better results than the ViT models, despite ViTs' good pneumothorax diagnosis accuracy. VGG16 and ResNet101 ranked the best ($p<0.05$), but the associated standard deviations of CNN models were also higher. Within each architecture type, the scores for VGG16 and VGG19 were similar ($p>0.05$), the ViT models also don't differ significantly ($p>0.05$), and ResNet101's score was significantly higher than ResNet34 ($p<0.05$). Between VGG and ResNet models, comparisons between ResNet18, ResNet50, VGG16, and VGG19 did not yield any significant differences ($p>0.05$). Among all the tested models, ResNet50 achieved the highest max score and ResNet34 had the lowest min score. As for EHR AUC, similar to the case of $Diff_{GradCAM}$, the CNN models also outperformed the ViT ones, with ResNet101 leading the scores at 0.0319 and VGG16 ranking the second at 0.0243. Compared with VGG16, the EHR AUC score of VGG19 was very similar.

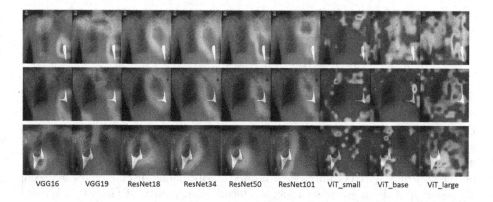

Fig. 1. Demonstration of Grad-CAM results across different deep learning models for three patients X-ray images (one patient per row), with the manual pneumothorax segmentation overlaid in white color.

Table 2. Difference of mean Grad-CAM values within and outside the manual pneumothorax segmentation ($Diff_{GradCAM}$).

	VGG16	VGG19	ResNet18	ResNet34	ResNet50	ResNet101	ViT_small	ViT_base	ViT_large
mean	0.186	0.166	0.162	0.133	0.142	0.183	0.052	0.081	0.051
std	0.212	0.220	0.262	0.273	0.251	0.265	0.143	0.166	0.164
min	−0.213	−0.365	−0.366	−0.452	−0.287	−0.320	−0.251	−0.233	−0.319
max	0.773	0.715	0.787	0.782	0.801	0.780	0.652	0.786	0.685

Among the ResNet variants, ResNet18, ResNet34, and ResNet50 scored similary in the range of 0.21–0.23, generally lower than those of VGG models. With the deepest architecture among the ResNet models, ResNet101 had a large increase of the EHR AUC metric even though its pneumothorax diagnosis accuracy was similar to the rest. For the ViT models, the EHR AUC improved gradually with the increasing size of the architecture, ranging from 0.0145 to 0.0171.

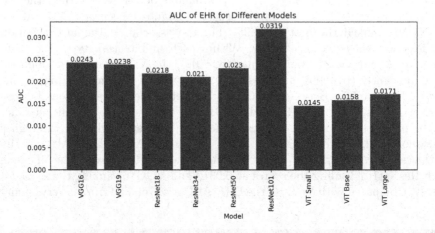

Fig. 2. EHR AUC results of different models

5   Discussion

Deep learning architectures like VGG, ResNet, and ViT all achieved commendable diagnostic performance in the range of 0.84–0.88%. As far as the experiments are concerned, deeper networks and the use of ViT do not contribute to better accuracy. Furthermore, with limited performance discrepancies between all tested models in pneumothorax diagnosis, the quality of Grad-CAM visualizations didn't necessarily correlate with the model's accuracy. The observed distinct behaviors of the Grad-CAM heatmaps may largely be due to the DL model's respective architecture types (pure CNN vs. Residual blocks vs. Transformer) and varying network sizes/depths. In our case study, while the Grad-CAM heatmaps of VGG16 and VGG19 accurately pinpointed areas of interest, VGG19 sometimes missed the pathological region, indicating that depth alone doesn't guarantee perfect feature localization. In addition, their activation maps often captured the upper chest areas, which may not be directly relevant for clinical interpretability. On the other hand, ResNet models, particularly ResNet18, 34, and 50, consistently highlighted relevant regions with a single cluster, albeit with slight deviations from the pneumothorax region in some cases. This could be attributed to the network's ability to focus on the most critical features through its residual connections. However, the much deeper ResNet101 model was prone to have two distinct areas in the Grad-CAM heatmaps. This was also noted by Seo et al. [10] in their vertebral maturation stage classification with ResNets, likely due to the fact that deeper architectures can capture more intricate representations [11]. In contrast, ViT models produced more dispersed Grad-CAM patterns compared to the included CNN ones. The inherent global contextual perception ability of Transformers might be responsible for this observation. As we only employed a single Chest X-ray dataset with a limited size, it's plausible that the ViTs could benefit from a larger dataset, potentially leading to better visualization outcomes [12]. Moreover, as the size of the ViT model increases, the proportion of irrelevant areas in the Grad-CAM visualizations also appears to increase, implying the interplay between model architecture and dataset size and the adverse cascading effects in deep Transformer models [17]. In previous investigations for DL-based computer-assisted diagnosis [6,10], multiple network models of different designs and natures were often benchmarked together. However, to the best of our knowledge, a systematic investigation for the impact of network architecture types and network depths on Grad-CAM visualization, especially with quantitative assessments has not been conducted to date. As transparency is becoming increasingly important for the safety and adoptiblity of DL algorithms, the relevant insights are of great importance to the XAI community.

The presented study has a few limitations. First, a typical issue in medical deep learning is the lack of large, well-annotated datasets. To facilitate the training process, we employed DL models that were pre-trained using natural images and then fine-tuned using domain-specific data. As noted by Lee et al. [6], in comparison to training from scratch, model fine-tuning also has a better advantage to provide more sparse and specific visualization for the target regions in Grad-CAM heatmaps. Furthermore, additional data augmentation was also implemented to mitigate overfitting issues. Second, as visual explanation of the DL algorithms is intended to allow easier incorporation of human participation, user studies to validate the quantitative metrics would be beneficial [8]. However, this requires more elaborate experimental design and inclusion of clinical experts, and will be included in our future studies. Lastly, we utilized pneumothorax diagnosis in chest X-ray as a case study to investigate the impact of DL model architectures on Grad-CAM visualization. It is possible that the observed trend may be application-specific. To confirm this, we will explore different datasets with varying disease types and imaging contrasts in the future.

6   Conclusion

In this study, we performed a comprehensive assessment of popular DL architectures, including VGG, ResNet and ViT, and their variants of different sizes for pneumothorax diagnosis in chest X-ray, and investigated the impacts of network depths and architecture types on visual explanation provided by Grad-CAM. While the accuracy for pneumothorax vs. healthy classification is similar across different models, the CNN models offer better specificity and accuracy than the ViTs when comparing the resulting heatmaps from Grad-CAM. Furthermore, the network size can affect both the model accuracy and Grad-CAM outcomes, with the two factors not necessarily in synch with each other. We hope the insights from our study can help better inform future explainable AI research, and we will further confirm the observations with more extensive studies involving more diverse datasets and DL models in the near future.

References

1. Arrieta, A.B., et al.: Explainable artificial intelligence (XAI): concepts, taxonomies, opportunities and challenges toward responsible AI. Inf. Fusion **58**, 82–115 (2020)
2. Çallı, E., Sogancioglu, E., van Ginneken, B., van Leeuwen, K.G., Murphy, K.: Deep learning for chest x-ray analysis: a survey. Med. Image Anal. **72**, 102125 (2021)
3. Chen, J., et al.: Transunet: transformers make strong encoders for medical image segmentation. arXiv preprint arXiv:2102.04306 (2021)
4. Dosovitskiy, A., et al.: An image is worth 16x16 words: transformers for image recognition at scale. arXiv preprint arXiv:2010.11929 (2020)
5. He, K., Zhang, X., Ren, S., Sun, J.: Deep residual learning for image recognition. In: Proceedings of the IEEE Conference on Computer Vision and Pattern Recognition, pp. 770–778 (2016)
6. Lee, Y.-H., Won, J.H., Kim, S., Auh, Q.-S., Noh, Y.-K.: Advantages of deep learning with convolutional neural network in detecting disc displacement of the temporomandibular joint in magnetic resonance imaging. Sci. Rep. **12**(1), 11352 (2022)
7. Mijwil, M.M.: Implementation of machine learning techniques for the classification of lung x-ray images used to detect covid-19 in humans. Iraqi J. Sci. 2099–2109 (2021)
8. Rong, Y., et al.: Towards human-centered explainable AI: user studies for model explanations. arXiv preprint arXiv:2210.11584 (2022)
9. Selvaraju, R.R., Cogswell, M., Das, A., Vedantam, R., Parikh, D., Batra, D.: Gradcam: visual explanations from deep networks via gradient-based localization. In: Proceedings of the IEEE International Conference on Computer Vision, pp. 618–626 (2017)
10. Seo, H., Hwang, J.J., Jeong, T., Shin, J.: Comparison of deep learning models for cervical vertebral maturation stage classification on lateral cephalometric radiographs. J. Clin. Med. **10**(16), 3591 (2021)
11. Simonyan, K., Zisserman, A.: Very deep convolutional networks for large-scale image recognition. arXiv preprint arXiv:1409.1556 (2014)
12. Steiner, A., Kolesnikov, A., Zhai, X., Wightman, R., Uszkoreit, J., Beyer, L.: How to train your vit? data, augmentation, and regularization in vision transformers. arXiv preprint arXiv:2106.10270 (2021)
13. Sun, S., Woerner, S., Maier, A., Koch, L.M., Baumgartner, C.F.: Inherently interpretable multi-label classification using class-specific counterfactuals. arXiv preprint arXiv:2303.00500 (2023)
14. Tian, Y., Wang, J., Yang, W., Wang, J., Qian, D.: Deep multi-instance transfer learning for pneumothorax classification in chest x-ray images. Med. Phys. **49**(1), 231–243 (2022)
15. Wollek, A., et al.: Attention-based saliency maps improve interpretability of pneumothorax classification. Radiol. Artif. Intell. **5**(2), e220187 (2022)

16. Yuan, H., Jiang, P.-T., Zhao, G.: Human-guided design to explain deep learning-based pneumothorax classifier. In: Medical Imaging with Deep Learning, Short Paper Track (2023)

17. Zhou, D., et al.: Deepvit: towards deeper vision transformer. arXiv preprint arXiv:2103.11886 (2021)

Group Distributionally Robust Knowledge Distillation

Konstantinos Vilouras[1]([✉]), Xiao Liu[1,2], Pedro Sanchez[1], Alison Q. O'Neil[1,2], and Sotirios A. Tsaftaris[1]

[1] School of Engineering, University of Edinburgh, Edinburgh EH9 3FB, UK
{konstantinos.vilouras,xiao.liu,pedro.sanchez,s.tsaftaris}@ed.ac.uk
[2] Canon Medical Research Europe Ltd., Edinburgh EH6 5NP, UK
alison.oneil@mre.medical.canon

Abstract. Knowledge distillation enables fast and effective transfer of features learned from a bigger model to a smaller one. However, distillation objectives are susceptible to *sub-population shifts*, a common scenario in medical imaging analysis which refers to groups/domains of data that are underrepresented in the training set. For instance, training models on health data acquired from multiple scanners or hospitals can yield subpar performance for minority groups. In this paper, inspired by distributionally robust optimization (DRO) techniques, we address this shortcoming by proposing a group-aware distillation loss. During optimization, a set of weights is updated based on the per-group losses at a given iteration. This way, our method can dynamically focus on groups that have low performance during training. We empirically validate our method, *GroupDistil* on two benchmark datasets (natural images and cardiac MRIs) and show consistent improvement in terms of worst-group accuracy.

Keywords: Invariance · Knowledge Distillation · Sub-population Shift · Classification

1 Introduction

The rapid success of deep learning can be largely attributed to the availability of both large-scale training datasets and high-capacity networks able to learn arbitrarily complex features to solve the task at hand. Recent practices, however, pose additional challenges in terms of real-world deployment due to increased complexity and computational demands. Therefore, developing lightweight versions of deep models without compromising performance remains an active area of research.

Knowledge Distillation. To this end, knowledge distillation is a promising technique aiming to guide the learning process of a small model (*student*) using a larger pre-trained model (*teacher*). This guidance can be exhibited in various forms; for example, Hinton et al. [3] propose to match the soft class probability

X. Cao et al. (Eds.): MLMI 2023, LNCS 14349, pp. 234–242, 2024.
https://doi.org/10.1007/978-3-031-45676-3_24

distributions predicted by the teacher and the student. Tian et al. [10] augment this framework with a contrastive objective to retain the dependencies between different output dimensions (since the loss introduced in [3] operates on each dimension independently). On the contrary, Romero et al. [7] consider the case of matching intermediate teacher and student features (after projection to a shared space since the original dimensions might be mismatched). Another interesting application involves distilling an ensemble of teachers into a single student model, which has been previously explored in the case of histopathology [9] and retinal [4] images.

Distillation May Lead to Poor Performance. Although knowledge distillation has been widely used as a standard approach to model compression, our understanding of its underlying mechanisms remains underexplored. Recently, Ojha et al. [6] showed, through a series of experiments, that a distilled student exhibits both useful properties such as invariance to data transformations, as well as biases, e.g., poor performance on rare subgroups inherited from the teacher model. Here, we attempt to tackle the latter issue which can be especially problematic, for instance, for medical tasks where we have access to only a few data for certain groups in the training set. Note that there also exist concurrent works [11] similar to ours that consider the case of distillation on long-tailed[1] datasets.

In this paper, we consider the task of knowledge distillation under a common type of distribution shift called *sub-population shift* where both training and test sets overlap, yet per-group proportions differ between sets. In sum, our contributions are the following:

- We propose a simple, yet effective, method that incorporates both the original distillation objective and also group-specific weights that allow the student to achieve high accuracy even on minority groups.
- We evaluate our method on two publicly available datasets, i.e., on natural images (Waterbirds) and cardiac MRI data (M&Ms) and show improvements in terms of worst-group accuracy over the commonly used knowledge distillation loss of [3].

2   Methodology

Preliminaries. Let D denote the data distribution from which triplets of data instances $x \in \mathcal{X}$, labels $y \in \mathcal{Y}$ and domains $d \in \mathcal{D}$ are sampled, respectively. Let also the teacher $f_T : \mathcal{X} \to \mathcal{Z}_T$ and student $f_S : \mathcal{X} \to \mathcal{Z}_S$ map input samples to class logits. The vanilla knowledge distillation loss introduced in Hinton et al. [3] is defined as

$$\mathcal{L}_{KD} = (1 - \alpha) \cdot H(y, \sigma(z_S)) + \alpha\tau^2 \cdot D_{KL}(\sigma(z_T/\tau), \sigma(z_S/\tau)), \qquad (1)$$

where $H(p, q) = -\mathbb{E}_p[\log q]$ refers to cross-entropy, $D_{KL}(p, q) = H(p, q) - H(p)$ is the Kullback-Leibler divergence metric, σ is the softmax function, τ is the temperature hyperparameter and α the weight that balances the loss terms.

---

[1] i.e., on sets where the majority of data comes from only a few classes.

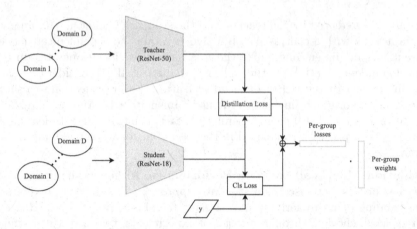

Fig. 1. Overview of the proposed method *GroupDistil*. Data from multiple domains $d = 1, ..., |\mathcal{D}|$ are fed to both the teacher and student to produce class logits. A soft version of the probability distribution over classes induced by both models is used to calculate the distillation loss, whereas the hard version ($\tau = 1$) of the student's probability distribution is used for the classification loss. The final loss is calculated as the dot product between the per-group losses and the group weights. This way, more emphasis is placed on domains where the student has low performance.

In our case, inspired by the groupDRO algorithm of Sagawa et al. [8], we propose an alternative objective that incorporates group-specific weights (termed *GroupDistil*). An overview of our method is presented in Fig. 1. The final form of the proposed loss function is shown in Eq. 2, where w_d and $\mathcal{L}_{KD}^d$ refer to the weight and the dedicated distillation loss for group d, respectively. This allows the student to emphasize domains where performance remains low (e.g., rare subgroups within the training set) by upweighting their contribution to the overall loss. In that sense, the final optimization objective can be divided into two steps, i.e., first performing exponentiated gradient ascent on group weights and then minibatch stochastic gradient descent on student's weights. Algorithm 1 provides a full description of the proposed method. The main difference with the groupDRO algorithm [8] is the choice of loss function, i.e., we use the distillation loss instead of cross-entropy, and it is highlighted in Algorithm 1 in blue.

$$\mathcal{L}_{GroupDistil} = \sum_{d=1}^{|\mathcal{D}|} w_d \cdot \mathcal{L}_{KD}^d, \tag{2}$$

Algorithm 1: *GroupDistil* method

Input: Data distribution for a given domain P_d; Learning rate η_θ; Group
    weight step size η_w

Initialize (uniform) group weights $w^{(0)}$ and student weights $\theta_s^{(0)}$

for $t = 1, ..., T$ **do**

$\quad$ $d \sim \mathcal{U}(1, ..., |\mathcal{D}|)$ `// Draw a random domain`

$\quad$ $(x, y) \sim P_d$ `// Sample (data, labels) from domain d`

$\quad$ $\ell \leftarrow \mathcal{L}_{KD}^d(x, y)$ `// Calculate distillation loss (Eq. 1)`

$\quad$ $w' \leftarrow w^{(t-1)}; w_d' \leftarrow w_d' \cdot \exp(\eta_w \ell)$ `// Update weight for domain d`

$\quad$ $w^{(t)} \leftarrow w' / \sum_{d'} w_{d'}'$ `// Normalize new weights`

$\quad$ $\theta_s^{(t)} \leftarrow \theta_s^{(t-1)} - \eta_\theta w_d^{(t)} \nabla \ell$ `// Optimize student weights via SGD`

end

---

3 Datasets

3.1 Waterbirds

Waterbirds [5,8] is a popular benchmark used to study the effect of sub-population shifts. More specifically, it contains instances from all possible combinations of $\mathcal{Y} = \{$landbird, waterbird$\}$ labels and $\mathcal{D} = \{$land, water$\}$ backgrounds (4 domains in total), respectively. However, it is collected in a way such that uncommon pairs (i.e., landbirds on water and waterbirds on land) occur less frequently, thus creating an imbalance. Also, note that the level of imbalance in the training set is different than that of the test set. Training, validation and test sets consist of 4795, 1199 and 5794 samples, respectively. In our setup, we first resize images to fixed spatial dimensions (256×256), then extract the 224×224 center crop and normalize using Imagenet's mean and standard deviation. A few representative samples of this dataset are depicted in Fig. 2.

Fig. 2. Examples of images from Waterbirds dataset. Left: waterbird on water. Middle left: waterbird on land. Middle right: landbird on land. Right: landbird on water.

3.2 M&Ms

The multi-centre, multi-vendor and multi-disease cardiac image segmentation (M&Ms) dataset [1] contains 320 subjects. Subjects were scanned at 6 clinical centres in 3 different countries using 4 different magnetic resonance scanner vendors (Siemens, Philips, GE, and Canon) i.e., domains A, B, C and D. For each subject, only the end-systole and end-diastole phases are annotated. Voxel resolutions range from $0.85 \times 0.85 \times 10$ mm to $1.45 \times 1.45 \times 9.9$ mm. Domain A contains 95 subjects. Domain B contains 125 subjects. Both domains C and D contain 50 subjects. We show example images of the M&Ms data in Fig. 3. Note that, while the dataset was originally collected for the task of segmentation, we instead use it here for classification.

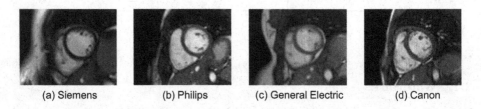

(a) Siemens   (b) Philips   (c) General Electric   (d) Canon

Fig. 3. Examples of cardiac MRI images from M&Ms dataset [1].

4 Experimental Results

We now discuss our results. First, we briefly discuss hyperparameter choices made on a per-dataset basis. In each case, we compare the performance of three types of *students*: an individual student that was trained with a group robustness method [8] from scratch (*groupDRO* hereafter), a distilled student trained with the objective of Eq. 1 (*KD* hereafter) and also a distilled student trained with our proposed *GroupDistil* method as described in Eq. 2 and Algorithm 1 (*GroupDistil* hereafter).

Waterbirds. We consider the case of distilling the knowledge from a ResNet-50 teacher model pre-trained with groupDRO to a ResNet-18 student (shorthanded as R50 → R18). Note that our proposed method does not impose any restrictions on either the choice of model architectures or the teacher's pre-training strategy (groupDRO was merely chosen to ensure that the teacher learns robust features on the given training set). For our objective, we use the following hyperparameters: batch size = 128, total number of epochs = 30, Adam optimizer with learning rate $\eta_\theta = 10^{-4}$, temperature $T = 4$, $\alpha = 0.9$ and group weight step size $\eta_w = 0.01$. Final results on the official test set are depicted in Table 1.

Our *GroupDistil* method shows consistent improvement ($> 2\%$) compared to the knowledge distillation objective (KD) of [3] in terms of worst-group accuracy.

Table 1. Distillation results on Waterbirds dataset. First two rows show performance of each model trained from scratch. The last two rows refer to vanilla knowledge distillation (KD) and our proposed method (GroupDistil), respectively. In those cases, we fix the teacher (1st row) and initialize each R18 model with ImageNet-pretrained weights. We report both the average (2nd column) and worst-group (3rd column) accuracy on the official test set. Final results have been averaged across 3 random seeds.

Setup	Adjusted avg acc. (%)	Worst-group acc. (%)
groupDRO (R50)	90.9	86.8
groupDRO (R18)	86.0	79.6
KD (R50 → R18)	95.2 ± 0.1	78.0 ± 1.5
GroupDistil (R50 → R18)	86.6 ± 0.8	**81.7 ± 0.8**

This implies that our distilled student does not rely on spurious attributes (background information) to classify each type of bird. This is also evident from the fact that worst-group accuracy in our case does not deviate much (approx. 5%) from the average accuracy in the test set. Note that our distilled student even outperforms a ResNet-18 model trained from scratch using groupDRO, showing that smaller models can largely benefit from larger ones in this setup.

M&Ms. In the case of M&Ms, since there is no official benchmark for studying sub-population shifts, we divide the available data in training and test sets as follows: First, we consider a binary classification task using the two most common patient state classes[2], i.e., hypertrophic cardiomyopathy (HCM) and healthy (NOR). Then, we conduct our experiments using two types of splits as presented in Table 2. Splits are defined in such a way that allows us to measure performance on a specific type of scanner (domain) that is underrepresented in the training set. Thus, for the first (resp. second) split, we keep only 2 patients per class from Domain C (resp. D) in the train set, and the rest in the test set. Note that, for fair comparison, we ensure that the test set is always balanced.

Table 2. M&Ms dataset. For each type of experiment, we show the total number of patients per dataset (train or test), domain (A, C, or D) and class (HCM or NOR), respectively. Note that we extract 20 2D frames from each patient for both training and testing.

Experiment	Class	Train Domains (A/C/D)	Test Domains (A/C/D)
Split 1	HCM	25/2/10	-/3/-
	NOR	21/2/14	-/3/-
Split 2	HCM	25/5/2	-/-/8
	NOR	21/11/2	-/-/8

---

[2] This choice eliminates Domain B due to lack of available data.

Table 3. Distillation results on M&Ms dataset, 1st split (testing on Domain C). First two rows show performance of each model trained from scratch. The last two rows refer to vanilla knowledge distillation (KD) and our proposed method (GroupDistil), respectively. We report the average accuracy on the test set. Final results have been averaged across 5 random seeds.

Setup	Accuracy (%)
groupDRO (R18)	53.3
groupDRO (D121)	46.7
KD (R18 → D121)	55.8 ± 7.0
GroupDistil (R18 → D121)	**62.8** ± 4.9

Table 4. Distillation results on M&Ms dataset, 2nd split (testing on Domain D). First two rows show performance of each model trained from scratch. The last two rows refer to vanilla knowledge distillation (KD) and our proposed method (GroupDistil), respectively. We report the average accuracy on the test set. Final results have been averaged across 5 random seeds.

Setup	Accuracy (%)
groupDRO (R18)	62.2
groupDRO (D121)	56.2
KD (R18 → D121)	64.2 ± 7.4
GroupDistil (R18 → D121)	**66.5** ± 2.8

For this dataset, we use a ResNet-18 (pre-trained with groupDRO) as teacher and a DenseNet-121 as a student model (R18 → D121 setup). As a pre-processing step, we first extract a random 2D frame from each 4D input (in total, we extract 20 frames from each patient per epoch) and then apply data augmentations such as intensity cropping, random rotation and flip. The input for each model is a 224×224 crop from each frame. We also used the following hyperparameters: batch size = 128, total number of epochs = 5, Adam optimizer with learning rate $\eta_\theta = 5 \cdot 10^{-4}$, temperature $T = 4$, $\alpha = 0.9$ and group weight step size $\eta_w = 0.01$. Results for the first split are shown in Table 3, whereas for the second split in Table 4.

As in Waterbirds, similar observations can be made for the M&Ms dataset. It is clear that in both types of splits, our method reaches the highest accuracy. Also note that the teacher and the student trained from scratch have low performance on the test set, indicating the challenging nature of this dataset; yet, distilled students can significantly outperform them. The results for KD method exhibit high variance (possibly due to the limited number of available data), indicating that it could be unstable in this setup. On the contrary, our method remains fairly robust and shows consistent performance improvements over the rest of the methods.

5   Conclusion and Future Work

In this paper we consider the task of knowledge distillation under a challenging type of distribution shift, i.e., sub-population shift. We showed that adding group-specific weights to a popular distillation objective provides a significant boost in performance, which even outperforms the same student architecture trained from scratch with a group robustness method in terms of worst-group accuracy. We also made sure that our proposed method remains fairly general, allowing arbitrary combinations of teacher-student models.

A limitation of our work is the fact that we assume access to fully labeled data, i.e., with both label and domain annotations, which is restrictive in practice. Therefore, for future work, we plan to investigate methods that infer domains directly from data as in [2,12].

Acknowledgements. This work was supported by the University of Edinburgh, the Royal Academy of Engineering and Canon Medical Research Europe by a PhD studentship to Konstantinos Vilouras. S.A. Tsaftaris also acknowledges the support of Canon Medical and the Royal Academy of Engineering and the Research Chairs and Senior Research Fellowships scheme (grant RCSRF1819\ 8\ 25), and the UK's Engineering and Physical Sciences Research Council (EPSRC) support via grant EP/X017680/1.

References

1. Campello, V.M., et al.: Multi-centre, multi-vendor and multi-disease cardiac segmentation: The m&ms challenge. IEEE TMI (2021)
2. Creager, E., Jacobsen, J.H., Zemel, R.: Environment inference for invariant learning. In: International Conference on Machine Learning, pp. 2189–2200. PMLR (2021)
3. Hinton, G., Vinyals, O., Dean, J.: Distilling the knowledge in a neural network. arXiv preprint arXiv:1503.02531 (2015)
4. Ju, L., et al.: Relational subsets knowledge distillation for long-tailed retinal diseases recognition. In: de Bruijne, M., et al. (eds.) MICCAI 2021. LNCS, vol. 12908, pp. 3–12. Springer, Cham (2021). https://doi.org/10.1007/978-3-030-87237-3_1
5. Koh, P.W., et al.: WILDS: a benchmark of in-the-wild distribution shifts. In: International Conference on Machine Learning, pp. 5637–5664. PMLR (2021)
6. Ojha, U., Li, Y., Lee, Y.J.: What knowledge gets distilled in knowledge distillation? arXiv preprint arXiv:2205.16004 (2022)
7. Romero, A., Ballas, N., Kahou, S.E., Chassang, A., Gatta, C., Bengio, Y.: FitNets: hints for thin deep nets. arXiv preprint arXiv:1412.6550 (2014)
8. Sagawa, S., Koh, P.W., Hashimoto, T.B., Liang, P.: Distributionally robust neural networks for group shifts: on the importance of regularization for worst-case generalization. arXiv preprint arXiv:1911.08731 (2019)
9. Tellez, D., et al.: Whole-slide mitosis detection in H&E breast histology using PHH3 as a reference to train distilled stain-invariant convolutional networks. IEEE Trans. Med. Imaging **37**(9), 2126–2136 (2018)
10. Tian, Y., Krishnan, D., Isola, P.: Contrastive representation distillation. arXiv preprint arXiv:1910.10699 (2019)

11. Wang, S., Narasimhan, H., Zhou, Y., Hooker, S., Lukasik, M., Menon, A.K.: Robust distillation for worst-class performance. arXiv preprint arXiv:2206.06479 (2022)
12. Zhang, M., Sohoni, N.S., Zhang, H.R., Finn, C., Ré, C.: Correct-N-Contrast: a contrastive approach for improving robustness to spurious correlations. arXiv preprint arXiv:2203.01517 (2022)

A Bone Lesion Identification Network (BLIN) in CT Images with Weakly Supervised Learning

Kehao Deng[1,2], Bin Wang[1], Shanshan Ma[1], Zhong Xue[1],
and Xiaohuan Cao[1(✉)]

[1] Shanghai United Imaging Intelligence Co., Ltd, Shanghai, China
xiaohuan.cao@uii-ai.com
[2] ShanghaiTech University, Shanghai, China

Abstract. Malignant bone lesions often lead to poor prognosis if not detected and treated in time. It also influences the treatment plan for primary tumor. However, diagnosing these lesions can be challenging due to their subtle appearance resemblances to other pathological conditions. Precise segmentation can help identify lesion types but the regions of interest (ROIs) are often difficult to delineate, particularly for bone lesions. We propose a bone lesion identification network (BLIN) in whole body non-contrast CT scans based on weakly supervised learning through class activation map (CAM). In the algorithm, location of the focal box of each lesion is used to supervise network training through CAM. Compared with precise segmentation, focal boxes are relatively easy to be obtained either by manual annotation or automatic detection algorithms. Additionally, to deal with uneven distribution of training samples of different lesion types, a new sampling strategy is employed to reduce overfitting of the majority classes. Instead of using complicated network structures such as grouping and ensemble for long-tailed data classification, we use a single-branch structure with CBAM attention to prove the effectiveness of the weakly supervised method. Experiments were carried out using bone lesion dataset, and the results showed that the proposed method outperformed the state-of-the-art algorithms for bone lesion classification.

Keywords: weakly supervision · class activation map · attention · sampling

1 Introduction

Bone lesion is a group of pathological conditions where abnormal tissues grow or replace normal tissues inside the bone. It can include over a hundred specific diseases, but most of them are rare in clinical practice. Bone tumor is one of the major lesions and can be malignant or benign, where malignant tumors are also classified as primary or metastasis. Malignant primary bone tumors are less common than benign primary tumors, but metastatic tumors are more common.

X. Cao et al. (Eds.): MLMI 2023, LNCS 14349, pp. 243–252, 2024.
https://doi.org/10.1007/978-3-031-45676-3_25

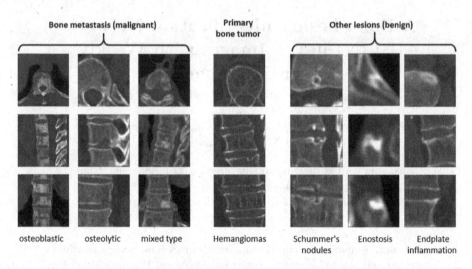

Fig. 1. Some common types of bone lesions in CT image.

Metastasis is spread from other cancers and may indicate a higher degree of cancer and is life-threatening. Therefore, it is important to detect bone abnormalities from any CT scans and also essential to identify the types of lesions.

There are many types of bone lesions, and common bone lesions can be divided into seven categories, as shown in Fig1. Some bone lesions have similar appearance in CT image and are not clinically obvious, which poses considerable difficulties in the clinical diagnosis, especially for distinguishing malignant and benign lesions. Therefore, it is quite necessary to classify bone lesions by artificial intelligence. In this paper, we propose a bone lesion identification network (BLIN) in non-contrast CT scans based on weakly supervised learning through class attention maps (CAM)using bounding boxes. The reason is that bounding boxes are relatively easy to be obtained either by manual annotation or automatic detection algorithms.

There are two challenges to identify different bone lesions accurately. *First*, the lesion number of different types often subject to a long-tailed distribution [8, 9] in reality, and the tailed category may have very low sampling probability, as the blue bar shown in Fig. 3. During the training process, the loss of the head category could dominate those of the tails, causing the classification of the minority category less effective. Thus, it is difficult for the network to learn useful information for the tailed category compared to the head category. In order to deal with this problem, Zhou et al. proposes a BBN [22] architecture, which uses two branches: one branch uses original data distribution and one branch uses balanced data distribution by sampling strategy. Xiang et al. proposes a LFME [20] method. It distills multiple teacher models into a unified model. However, this network structure is complicated and consumes more computational resources. Besides, some loss functions such as focal loss [10], CB loss [4] have also been proposed to classify long-tailed data.

Second, the appearance of different bone lesions in CT image is difficult to distinguish, and the lesions are hard to delineate accurately since the boundary is not clear in bone structures. To build an effective identification model, the model should focus on the lesion rather than the normal bone structures. Ouyang et al. [12] proposed an attention network for classification problem. In this work, an additional segmentation network is introduced and the similarity between the segmentation and CAM of classification network is applied as an additional loss to guide the classification network training. However, the segmentation label is difficult to obtain for bone lesions.

To tackle the two challenges, the proposed BLIN framework aims to use bounding boxes and class labels (from clinical results) to train a lesion identification network in a weakly supervised fashion. Our contribution can be summarized as follows: (1) an improved sampling strategy is designed to alleviate the over-fitting problem of long-tailed data; (2) a class activation map is used to design a new CAM loss by using lesion bounding box rather than segmentation, since the bounding box is more easy to obtain clinically and the boundary of bone lesions is hard to segment.

2   Method

2.1   Bone Lesion Identification Network

The structure of the proposed BLIN model is shown in Fig. 2. The main backbone is a 3D ResNet-18 [6] with several residual blocks. After many residual blocks, a fully connected layer is employed to generate the classification output. The input of the network is a cropped 3D patch based on the expanded bounding box of the lesion. The lesion bounding box can be determined by manual label. In our work, an FPN-based [9] lesion detection module is used to detect the lesion automatically. The output of BLIN is the predicted lesion type, and herein seven common types (as shown in Fig. 1) are used.

Many methods have been proposed to either identify lesion types through the segmented shapes and appearances or using a deep learning classifier without any shape information. Notice that the bounding boxes of detected lesions not only reflect the location but also the size of them, they can be used to constrain deep learning algorithms so that lesion identification can effectively focus on lesion and its surrounding tissues, not other structures. Thus, in BLIN, we incorporate a Convolutional Block Attention Module (CBAM) [19] to improve the classification performance. We introduce the bounding box as a prior knowledge so that the network can focus on the lesion regions to well distinguish different bone lesions.

As shown in Fig. 2, we combine the CAM and the bounding box attention for loss computation, and the implementation of CAM loss is shown in Fig. 4. .

2.2   Sampling Strategy

Class balance is used for dealing with long-tailed datasets via two different strategies: sampling [1–3,11,14]and re-weighting [4,7,13,17]. For the former,

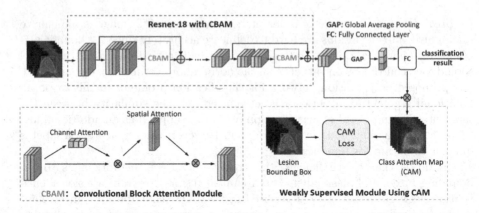

Fig. 2. The proposed BLIN framework using CBAM as the attention mechanism. Bounding boxes of lesions are used to guide classifier training using CAM loss.

over-sampling is normally applied for small sample size (increase sampling probability), and under-sampling is used for the large sample size (decrease sampling probability), but under-sampling will not use the full data sufficiently, so we choose the over-sampling strategy. Generally, sampling probability can be inversely proportional to the number of samples, but can bring serious overfitting problem for long-tailed data, particularly for the groups with less samples.

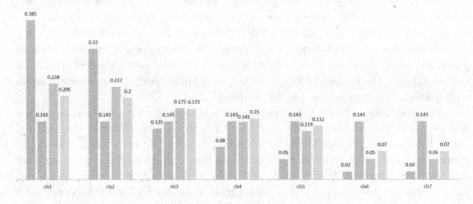

Fig. 3. Sample rate during training using different sampling strategies. Blue: original distribution of sample rate. Orange: sample rate using the traditional method. Green: sample rate using our proposed sampling strategy with n=5. Yellow: sample rate using our proposed sampling strategy with n=10. (Color figure online)

In this work, we propose a new data sampling strategy. First, we set a threshold n as the ratio of the maximum and the minimum sampling probability, where the the maximum of samples is m, and the relationship between our sampling probability and the number of samples is $1/(x+m/n)$. This can ensure that, for the large sample size, the sampling probability is relatively lower. For the tailed class with small sample size, it can also avoid the overfitting problem caused by repeated sampling. The results of different sampling strategies are shown in Fig. 3. we have balanced the long tail of the data distribution, meanwhile the tail category is not sampled too much. This can effectively reduce the overfitting caused by repeated sampling of the small sample size by using our proposed sampling strategy.

2.3   Identification Model with CBAM

The different bone lesions in CT image is difficult to distinguish, and the boundary is not clear in bone structures. In order to make our class activation map focus on the lesion area more accurately, we introduce CBAM [19], a mixture of spatial attention and channel attention to improve the accuracy of the class activation map. Channel attention is the result of pooling in spatial dimensions, and spatial attention is pooling in channel dimensions. Moreover, since the added average pooling and maximum pooling cannot carry out normal gradient calculation during the back propagation process, we use 1×1 convolution on the obtained features and combine with the average pooling and maximum pooling layers to ensure the network trained normally [23].

2.4   Cam Loss

We use the bounding box as the weak supervised information instead of using segmentation result. The bounding box can incorporate more context information and can avoid the difficulty and effort of accurately delineating the lesions. As shown in Fig. 4, the weakly supervised information is derived from the bounding box, where we set 1 in the bounding box and 0 out of the bounding box, and we denote this as gt. For the CAM, we set a threshold T to get the binarized CAM. The CAM is obtained by

$$CAM(x) = w_c' F(x), \tag{1}$$

where $F(x)$ is the feature map before global average pooling (GAP) and w_c is the fully connected weight for class c. The CAM_b is the final binary CAM using threshold T:

$$CAM_b(x) = \begin{cases} 1 & CAM(x) \geq T \\ 0 & \text{otherwise} \end{cases}. \tag{2}$$

We use CAM_b and gt to calculate our cam loss, N represents the total number of samples:

$$L_{cam} = \frac{1}{N} \sum_{i=1}^{i=N} (1 - \sum CAM_b(x_i) \odot gt(x_i) / \sum gt(x_i)). \tag{3}$$

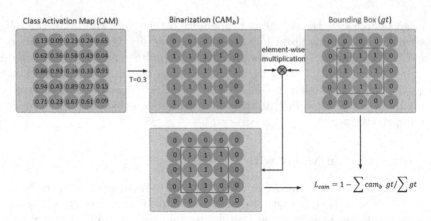

Fig. 4. An illustration for calculating CAM loss.

The final loss is consisted of two parts: classification loss L_{CE} using cross entropy and Cam loss:

$$L = L_{CE} + \lambda L_{cam}, \tag{4}$$

where $\lambda = (\frac{epoch}{Epoch})^2$ is a weight parameter and will increase during training.

2.5   Implementation Details

The proposed BLIN algorithm is implemented using Pytorch, and RTX 2080 GPU was used for training and testing. Each input image was cropped to $112 \times 112 \times 112$ according to the longest edge based on the bounding box of lesions. Random rotation and translation are used for data augmentation. We apply the Adam optimizer, the learning rate is set to 0.001 initially and will decrease after each 10 epochs, the decrease weight is 0.5. Since the parameters of the dilated convolution [18] may affect the dispersion of the class activation map, we set all the parameters of the dilated convolution to 1. A L2 regularization with a weight of 0.002 is also applied to avoid overfitting problem. The batch size is set to 64. In sampling, threshold T=0.3. For the dataset, we use 70% for training and 30% for testing.

3   Experiments

3.1   Datasets and Evaluation Metrics

The bone lesion dataset has 955 CT scans from 955 patients. Totally, we have 9422 lesions. The common bone lesions can be divided into seven categories, three of which are malignant. They are osteoblastic (cls1, 3678 lesions), osteolytic (cls2,2991 lesions), mixed type (cls4,722 lesions). Benign lesions mainly include four categories: Enostosis(cls3,1182 lesions), Schummer's nodules(cls5,516 lesions), Endplate inflammation(cls6,167 lesions), Heman-giomas(cls7,166 lesions). The average precision is used to evaluate the final performance.

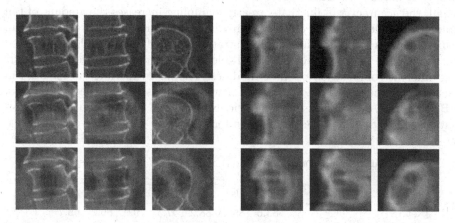

Fig. 5. CAM of bone lesions. The first row is original CT images. The second row and the third row are the CAM maps without using CAM Loss and using proposed CAM Loss, respectively.

3.2   Performance

We compare the experimental result with other SOTA methods, including classification using Focal loss [10]; using CB loss [4] and EQLv2 [15] to tackle the long-tailed problem. Besides, a common CE loss is also applied as the classification baseline. To ensure the fairness of the comparison, we use the same network backbone to train the model. From table 1we can see that the proposed method achieves the best performance for bone lesion classification.

Table 1. The performance(precision) of ours compared with other methods in our dataset

Method	cls1	cls2	cls3	cls4	cls5	cls6	cls7	mAP
CE	83.6	89.5	88.2	47.3	79.0	69.2	66.7	74.8
Focal loss[5]	81.9	88.6	86.3	44.6	85.4	50.0	**90.0**	75.3
EQLv2[21]	81.6	90.5	87.9	47.3	88.6	64.8	71.6	76.0
CB Loss[11]	83.7	86.6	88.6	45.6	**89.4**	71.4	85.0	78.6
Ours	**85.5**	**90.7**	**90.0**	49.2	83.5	**78.6**	78.3	**79.5**

Our results not only achieve the best performance on the head categories, but also have a significant improvement for the tail category (cls6, cls7), compared with traditional CE Loss.

3.3   Ablation Study

We conduct a series of ablation experiments to evaluate the performance of the proposed modules of BLIN. Since the number of the tail category is relatively small, the accuracy will fluctuate greatly. Sampling increases the amount of tail training samples, thus effectively improving the accuracy of tail classification. Three modules including the sampling strategy, CBAM and CAM loss are tested with the same backbone ResNet-18. Table 2 lists the performance by comparing different combination of these modules. The post-experimental results can be further improved. Three parts: the new sampling strategy, the single use of CBAM and CAM loss has improved the experimental results, and their combination will further improve the final performance.

Table 2. The results(precision) of the ablation experiment

Model	cls1	cls2	cls3	cls4	cls5	cls6	cls7	mAP
ResNet-18	83.6	89.5	88.2	47.3	79.0	69.2	66.7	74.8
ResNet-18+Sampling	81.9	88.6	86.3	44.6	85.4	50.0	90.0	75.3
ResNet-18+CBAM	85.3	88.9	86.3	47.8	87.5	64.3	81.3	77.4
ResNet+Cam loss	84.4	89.1	88.2	48.3	**88.5**	55.6	85.7	77.1
ResNet-18+CBAM+Cam loss	85.7	89.5	89.4	47.9	87.1	73.3	75.4	78.3
ResNet-18+All	**85.5**	**90.7**	**90.0**	**49.2**	83.5	**78.6**	**78.3**	**79.5**

4   Conclusion

We proposed a weakly supervised classification method using CAM and new sampling strategy for classifying bone lesions, which is a typical long-tailed problem. Bounding boxes of bone lesions and cam were used jointly as weakly supervised information to assist in the classification of imbalanced datasets. Comparative and ablation studies showed that our method outperformed others and the individual proposed modules are valid.

References

1. Buda, M., Maki, A., Mazurowski, M.A.: A systematic study of the class imbalance problem in convolutional neural networks. Neural Netw. **106**, 249–259 (2018)
2. Byrd, J., Lipton, Z.: What is the effect of importance weighting in deep learning? In: International Conference on Machine Learning, pp. 872–881. PMLR (2019)
3. Chawla, N.V., Bowyer, K.W., Hall, L.O., Kegelmeyer, W.P.: SMOTE: synthetic minority over-sampling technique. J. Artif. Intell. Res. **16**, 321–357 (2002)
4. Cui, Y., Jia, M., Lin, T.-Y., Song, Y., Belongie, S.: Class-balanced loss based on effective number of samples. In: Proceedings of the IEEE/CVF Conference on Computer Vision and Pattern Recognition, pp. 9268–9277 (2019)

5. Drummond, C., Holte, R.C., et al.: C4. 5, class imbalance, and cost sensitivity: why under-sampling beats over-sampling. In: Workshop on Learning from Imbalanced Datasets II, vol. 11, pp. 1–8 (2003)
6. He, K., Zhang, X., Ren, S., Sun, J.: Deep residual learning for image recognition. In Proceedings of the IEEE Conference on Computer Vision and Pattern Recognition, pp. 770–778 (2016)
7. Huang, C., Li, Y., Change Loy, C., Tang, X.: Learning deep representation for imbalanced classification. In: Proceedings of the IEEE Conference on Computer Vision and Pattern Recognition, pp. 5375–5384 (2016)
8. Kendall, M.G., et al.: The advanced theory of statistics. vols. 1. The advanced theory of statistics, vols. 1, 1(Ed. 4) (1948)
9. Lin, T.-Y., Dollár, P., Girshick, R., He, K., Hariharan, B., Belongie, S.: Feature pyramid networks for object detection. In: Proceedings of the IEEE Conference on Computer Vision and Pattern Recognition, pp. 2117–2125 (2017)
10. Lin, T.-Y., Goyal, P., Girshick, R., He, K., Dollár, P.: Focal loss for dense object detection. In: Proceedings of the IEEE International Conference on Computer Vision, pp. 2980–2988 (2017)
11. More, A.: Survey of resampling techniques for improving classification performance in unbalanced datasets. arXiv preprint arXiv:1608.06048 (2016)
12. Ouyang, X., et al.: Dual-sampling attention network for diagnosis of Covid-19 from community acquired pneumonia. IEEE Trans. Med. Imaging **39**(8), 2595–2605 (2020)
13. Ren, M., Zeng, W., Yang, B., Urtasun, R.: Learning to reweight examples for robust deep learning. In: International Conference on Machine Learning, pp. 4334–4343. PMLR (2018)
14. Shen, L., Lin, Z., Huang, Q.: Relay backpropagation for effective learning of deep convolutional neural networks. In: Leibe, B., Matas, J., Sebe, N., Welling, M. (eds.) ECCV 2016. LNCS, vol. 9911, pp. 467–482. Springer, Cham (2016). https://doi.org/10.1007/978-3-319-46478-7_29
15. Tan, J., Lu, X., Zhang, G., Yin, C., Li, Q.: Equalization loss v2: a new gradient balance approach for long-tailed object detection. In Proceedings of the IEEE/CVF Conference on Computer Vision and Pattern Recognition, pp. 1685–1694 (2021)
16. Van Horn, G., Perona, P.: The devil is in the tails: fine-grained classification in the wild. arXiv preprint arXiv:1709.01450 (2017)
17. Wang, Y.-X., Ramanan, D., Hebert, M.: Learning to model the tail. In: Advances in Neural Information Processing Systems 30 (2017)
18. Wei, Y., Xiao, H., Shi, H., Jie, Z., Feng, J., Huang, T.S.: Revisiting dilated convolution: a simple approach for weakly-and semi-supervised semantic segmentation. In: Proceedings of the IEEE Conference on Computer Vision and Pattern Recognition, pp. 7268–7277 (2018)
19. Woo, S., Park, J., Lee, J.-Y., Kweon, I.S.: CBAM: convolutional block attention module. In: Ferrari, V., Hebert, M., Sminchisescu, C., Weiss, Y. (eds.) ECCV 2018. LNCS, vol. 11211, pp. 3–19. Springer, Cham (2018). https://doi.org/10.1007/978-3-030-01234-2_1
20. Xiang, L., Ding, G., Han, J.: Learning from multiple experts: self-paced knowledge distillation for long-tailed classification. In: Vedaldi, A., Bischof, H., Brox, T., Frahm, J.-M. (eds.) ECCV 2020. LNCS, vol. 12350, pp. 247–263. Springer, Cham (2020). https://doi.org/10.1007/978-3-030-58558-7_15
21. Zhou, B., Khosla, A., Lapedriza, A., Oliva, A., Torralba, A.: Learning deep features for discriminative localization. In: Proceedings of the IEEE Conference on Computer Vision and Pattern Recognition, pp. 2921–2929 (2016)

22. Zhou, B., Cui, Q., Wei, X.-S., Chen, Z.-M.: BBN: bilateral-branch network with cumulative learning for long-tailed visual recognition. In: Proceedings of the IEEE/CVF Conference on Computer Vision and Pattern Recognition, pp. 9719–9728 (2020)
23. Zhou, Q., Zou, H., Wang, Z.: Long-tailed multi-label retinal diseases recognition via relational learning and knowledge distillation. In: Medical Image Computing and Computer Assisted Intervention-MICCAI 2022: 25th International Conference, Singapore, September 18–22, 2022, Proceedings, Part II, pp. 709–718. Springer, Cham (2022). https://doi.org/10.1007/978-3-031-16434-7_68

Post-Deployment Adaptation with Access to Source Data via Federated Learning and Source-Target Remote Gradient Alignment

Felix Wagner[1]([✉]), Zeju Li[2], Pramit Saha[1], and Konstantinos Kamnitsas[1,3,4]

[1] Department of Engineering Science, University of Oxford, Oxford, UK
felix.wagner@eng.ox.ac.uk
[2] Nuffield Department of Clinical Neurosciences, University of Oxford, Oxford, UK
[3] Department of Computing, Imperial College London, London, UK
[4] School of Computer Science, University of Birmingham, Birmingham, UK

Abstract. Deployment of Deep Neural Networks in medical imaging is hindered by distribution shift between training data and data processed after deployment, causing performance degradation. Post-Deployment Adaptation (PDA) addresses this by tailoring a pre-trained, deployed model to the *target* data distribution using limited labelled or entirely unlabelled target data, while assuming no access to *source* training data as they cannot be deployed with the model due to privacy concerns and their large size. This makes reliable adaptation challenging due to limited learning signal. This paper challenges this assumption and introduces **FedPDA**, a novel adaptation framework that brings the utility of learning from remote data from Federated Learning into PDA. FedPDA enables a deployed model to obtain information from source data via remote gradient exchange, while aiming to optimize the model specifically for the target domain. Tailored for FedPDA, we introduce a novel optimization method `StarAlign` (**S**ource-**T**arget **R**emote Gradient **Align**ment) that aligns gradients between source-target domain pairs by maximizing their inner product, to facilitate learning a target-specific model. We demonstrate the method's effectiveness using multi-center databases for the tasks of cancer metastases detection and skin lesion classification, where our method compares favourably to previous work. Code is available at: https://github.com/FelixWag/StarAlign.

Keywords: adaptation · domain shift · federated learning

1 Introduction

Effectiveness of Deep Neural Networks (DNNs) relies on the assumption that training (source) and testing (target) data are drawn from the same distribution (domain). When DNNs are applied to target data from a distribution (target domain) that differs from the training distribution (source domain), i.e. there is distribution shift, DNNs' performance degrades [13,16]. For instance, such shift

© The Author(s), under exclusive license to Springer Nature Switzerland AG 2024
X. Cao et al. (Eds.): MLMI 2023, LNCS 14349, pp. 253–263, 2024.
https://doi.org/10.1007/978-3-031-45676-3_26

can occur, when a pre-trained DNN model is deployed to a medical institution with data acquired from a different scanner or patient population than the training data. This hinders reliable deployment of DNNs in clinical workflows.

Related Work: Variety of approaches have been investigated to alleviate this issue. Domain Generalization (DG) approaches assume access to data from multiple source domains during training. They aim to learn representations that are invariant to distribution shift, enabling better generalization to any unseen domain [10,20,26]. Due to the great heterogeneity in medical imaging, achieving universal generalization may be too optimistic. Instead, Domain Adaptation (DA) methods [4,12,16] aim to learn a model that performs well on a specific target domain. They assume that (commonly unlabelled) target data and labelled source data is collected in advance and centrally aggregated to train a model *from scratch*. Privacy concerns in healthcare limit the scalability of these approaches.

Federated Learning (FL) [25] enables training a single model on multiple decentralised databases. The original Federated Averaging algorithm (`FedAvg`) [25] and its extensions target scenarios where different "nodes" in the federated *consortium* have data from different source domains [18,21,23,33]. They train a single model to generalize well across unseen domains. In contrast, approaches for *Personalization* in FL [1,15,23,24] learn multiple models, one for each source domain but do not support adaptation to unseen target domains.

Federated Domain Adaptation (FDA) methods [11,22,27] enable DA in a federated setting. These methods assume abundant unlabelled data on the target domain and perform distribution matching to enforce domain invariance. They train a model from scratch for each target which limits their practicality, while the enforced invariance can lead to loss of target-specific discriminative features.

Post-Deployment Adaptation (PDA) methods (or Test-Time Adaptation or Source-free DA) seek to enable a pre-trained/deployed model to optimize itself for the specific target domain by learning only from limited labelled or unlabelled target data processed after deployment [3,6,7,17,30,31]. These methods assume that deployed models have no access to the source data due to privacy, licensing, or data volume constraints. In practice, however, adaptation using solely limited labelled or unlabelled data can lead to overfitting or unreliable results.

Contribution: This work presents a novel PDA framework to *optimize a pre-trained, deployed model* for a *specific target domain* of deployment, assuming *limited labelled data at the deployment node*, while *remotely obtaining information from source data* to facilitate target-specific adaptation.

- The framework enables a deployed DNN to obtain information from source data without data exchange, using remote gradient exchange. This overcomes PDA's restriction of unavailable access to source data. This combines FL with PDA into a new framework, **FedPDA** (Fig. 1). Unlike FL, FedPDA optimizes a target-specific model instead of a model that generalises to any distribution (FL) or source distributions (personalized FL). While DA requires central aggregation of data, FedPDA does not transfer data. FedPDA also differs from FDA by adapting a *pre-trained deployed* model at the user's endpoint, rather than training a model *pre-deployment from scratch* with abundant unlabelled

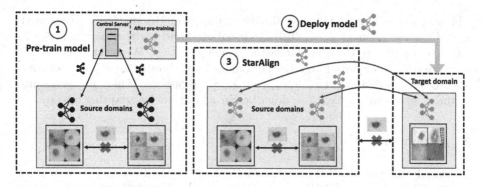

Fig. 1. The **FedPDA** framework consists of three steps: (1) The model is trained on source domains (here via FL); (2) The model is deployed to the target domain; (3) Adaptation is done via gradient alignment between all source-target domain pairs with `StarAlign` without cross-domain data exchange.

target data and ML developer's oversight. This addresses the practical need for reliable adaptation with limited data and technical challenge of optimizing from a source-specific initial optimum to a target-specific optimum.
- Tailored specifically for this setting, we introduce the `StarAlign` optimization algorithm (**S**ource-**T**arget **R**emote Gradient **Align**ment) for decentralised multi-domain training. It extracts gradients from source data and aligns them with target data gradients by maximising the gradient inner product, to regularize adaptation of a target-specific model.

We evaluate the method through extensive experiments on two medical datasets: cancer metastases classification in histology images using Camelyon17 with 5 target domains [2], and a very challenging setting for skin lesion classification using a collection of 4 databases (HAM [29], BCN [9], MSK [8], D7P [19]). In numerous settings, our method achieves favourable results compared to state-of-the-art PDA, (personalized) FL, FDA and gradient alignment methods.

2 Background

Problem Setting: We consider the general case with a set of S source domains $\{\mathcal{D}_1, \ldots, \mathcal{D}_S\}$ and one target domain $\mathcal{D}_T$, where $T = S+1$. Each domain's dataset $\mathcal{D}_k := \{(x_i^k, y_i^k)\}_{i=1}^{n_k}$ is drawn from a data distribution p_k, where x_i is an image and y_i the corresponding label. We assume there are little available labels in the target domain, $|\mathcal{D}_T| \ll |\mathcal{D}_k|, \forall k \in [1, S]$.

In a setting with multiple source domains, optimizing a DNN's parameters θ is commonly done with Empirical Risk Minimization (ERM), by minimizing a cost R_K for each domain $\mathcal{D}_k$, which is the expectation of a loss $\mathcal{L}$ over $\mathcal{D}_k$:

$$\min_{\theta} R_{ERM}(\theta) := \frac{1}{S} \sum_{k=1}^{S} R_k(\theta) = \frac{1}{S} \sum_{k=1}^{S} \mathbb{E}_{(x,y) \sim \mathcal{D}_k}[\mathcal{L}(\theta; x, y)]. \tag{1}$$

This can be performed via centralised training (e.g. DG methods) or approximated via FL when $\{\mathcal{D}_1, \ldots, \mathcal{D}_S\}$ are distributed and without data sharing. In our experiments we focus on the latter challenging case. We assume model θ is pre-trained and 'deployed' to a target domain where limited labelled $\mathcal{D}_T$ is available. Due to domain shift $p_T \neq p_k, \forall k \in [1, S]$, we assume θ may not generalize to the target domain. Next, we describe our algorithm for adapting θ to p_T.

3   Method

Using only limited labelled target data $\mathcal{D}_T$ for PDA is likely to give unreliable results. Therefore, our FedPDA framework, enables a deployed model to obtain information from source data using remote gradient exchange via FL (Fig. 1). To achieve this, our StarAlign algorithm for FedPDA aligns gradients of all domains with target domain gradients by maximising their inner product. This ensures that source-derived gradients promote learning features that are relevant to the target domain.

3.1   Theoretical Derivation of Source-Target Alignment

First, we derive our algorithm in the *theoretical* setting where source and target data $\{\mathcal{D}_1, \ldots, \mathcal{D}_S, \mathcal{D}_T\}$ are available for centralised training. The subsequent section presents our *distributed algorithm* for remote domains, developed to approximate the optimization (Eq. 2) in the theoretical setting. We form pairs between each domain and target $\mathcal{D}_T$ to extract target-relevant information from each source domain. For the k-th pair we minimise the combined cost $R_{kT}(\theta) = R_k(\theta) + R_T(\theta)$. To align gradients $G_k = \mathbb{E}_{\mathcal{D}_k}[\nabla_\theta \mathcal{L}_k(\theta; x, y)]$ of the k-th domain and the target domain $G_T = \mathbb{E}_{\mathcal{D}_T}[\nabla_\theta \mathcal{L}_T(\theta; x, y)]$, we maximize their inner product $G_T \cdot G_k$. Hence over all pairs, we minimize the total cost:

$$R_{total}^{\text{StarAlign}} = \frac{1}{S+1} \sum_{k=1}^{S+1} R_{kT}^{\text{Align}}(\theta) = \frac{1}{S+1} \sum_{k=1}^{S+1} (R_{kT}(\theta) - \delta G_T \cdot G_k), \quad (2)$$

with δ a hyperparameter. Using dot product's distributive property we get:

$$R_{total}^{\text{StarAlign}} = R_T(\theta) + \frac{1}{S+1} \sum_{k=1}^{S+1} R_k(\theta) - \delta G_T \cdot \sum_{k=1}^{S+1} \frac{G_k}{S+1} \quad (3)$$

This allows us to interpret the effect of optimizing Eq. 2. We minimize the target domain's cost, regularized by the average cost over all domains, which mitigates overfitting the limited data $\mathcal{D}_T$. The third term forces the average gradient over all domain data, $\sum_{k=1}^{S+1} \frac{G_k}{S+1}$, to align with the gradients from the target domain G_T. We clarify that domain pairs summed in Eq. 2 include the pair target-to-target with R_{TT} and $G_T \cdot G_T$ terms. This ensures that the average gradient of all domains $\sum_{k=1}^{S+1} \frac{G_k}{S+1}$ in Eq. 3 always has a component along the direction of G_T, avoiding zero or very low dot products, for instance due to

---

Algorithm 1. Centralised gradient alignment for source-target domain pairs

1: θ_T: Model pre-trained on source data $\{\mathcal{D}_1, \ldots, \mathcal{D}_S\}$, deployed to Target
2: **for** $j = 1, 2 \ldots$ **do**
3:     **for** $\mathcal{D}_k \in \{\mathcal{D}_1, \ldots, \mathcal{D}_S, \mathcal{D}_T\}$ **do**
4:         $\hat{\theta}_k \leftarrow \theta_T; \theta_k \leftarrow \theta_T$ ▷ Create copies of deployed model θ_T
5:         $d_T \sim \mathcal{D}_T$ ▷ Sample batch from domain T
6:         $\hat{\theta}_k \leftarrow \hat{\theta}_k - \alpha \nabla_{\hat{\theta}_k} \mathcal{L}(\hat{\theta}_k; d_T)$ ▷ Compute gradient of d_T and update model
7:         $d_k \sim \mathcal{D}_k$ ▷ Sample batch from domain k
8:         $\hat{\theta}_k \leftarrow \hat{\theta}_k - \alpha \nabla_{\hat{\theta}_k} \mathcal{L}(\hat{\theta}_k; d_k)$ ▷ Compute gradient of d_k and update model
9:         $\theta_k \leftarrow \theta_k + \beta(\hat{\theta}_k - \theta_k)$ ▷ 1st order approximation update
10:     $\theta_T \leftarrow \frac{1}{S+1} \sum_{k=1}^{S+1} \theta_k$ ▷ Update model with avg. over domains

---

almost perpendicular gradients in high dimensional spaces, leading to smoother optimization trajectory. We found this effective in preliminary experiments.

Directly optimising dot-products in Eq. 2 is computationally expensive as it requires second-order derivatives. Instead, minimizing Eq. 2 is approximated via Algorithm 1 and first-order derivatives in *L4-L9*. This is based on the result in [28] that the update step $\beta(\hat{\theta} - \theta)$ (*L9*) approximates optimisation of Eq. 2 for a pair of domains. β is a scaling hyperparameter. This was used in [28] for *centralised* multi-source DG, aligning gradients of *source-source* pairs to learn a general model (*no target*). Here, we minimize Eq. 2 to align *source-target* pairs of gradients and learn a *target-specific* model.

3.2   Source-Target Remote Gradient Alignment for FedPDA

We assume a model pre-trained on source data $\{\mathcal{D}_1, \ldots, \mathcal{D}_S\}$ is deployed on a computational node with access to target data D_T but not source data. Below, we describe the case when each source dataset $\{\mathcal{D}_1, \ldots, \mathcal{D}_S\}$ is held in a separate compute node (e.g. federation). We now derive the distributed StarAlign Algorithm 2, approximating minimization of Eq. 2 without data exchange between nodes.

The aim is to approximate execution of Algorithm 1 on the target node. Lack of access to source data, however, prevents calculation of source gradients in *L8* of Algorithm 1. Instead, we approximate the gradient of each source domain $\mathcal{D}_s$ separately, by computing the average gradient direction $\bar{g}_s$ on the specific source node. For this, we perform τ local optimisation steps on source node s and average their gradients, obtaining $\bar{g}_s$. By transferring $\bar{g}_s$ from each source node to the target node, we can approximate the updates in *L8* of Algorithm 1 on the target node.

Algorithm 2 presents the distributed StarAlign method that aligns gradients between all source-target domain pairs. First, target and source node models get initialised with the pre-trained model θ_T (*L1*). Each source node s computes its average gradient direction $\bar{g}_s$ and communicates it to the target node (*L3–L9*). After receiving the average gradient directions of source domains, the target node performs the interleaving approximated updates (*L14-15*) for source and

Algorithm 2. Distributed `StarAlign` algorithm for FedPDA

1: θ_T: Model pre-trained on source data $\{\mathcal{D}_1, \ldots, \mathcal{D}_S\}$, deployed to Target node
2: **for** $j = 1$ to E **do**   ▷ `where E total communication rounds`
3:    **for** $\mathcal{D}_s \in \{\mathcal{D}_1, \ldots, \mathcal{D}_S\}$ **in parallel do**   ▷ *Performed at each Source node*
4:      $\theta_s \leftarrow \theta_T$ ▷ `Obtain latest model from Target node`
5:      **for** $i = 1$ to τ **do**   ▷ `Local iterations`
6:        $d_s \sim \mathcal{D}_s$ ▷ `Sample batch`
7:        $g_s^i \leftarrow \nabla_{\theta_s} \mathcal{L}(\theta_s; d_s)$ ▷ `Compute gradient`
8:        $\theta_s \leftarrow \theta_s - \alpha g_s^i$ ▷ `Update model`
9:      $\overline{g}_s = \frac{1}{\tau} \sum_{i=1}^{\tau} g_s^i$ ▷ *Compute avg. gradient and* **SEND** *to Target node*
10:   **for** $k \in \{1, \ldots, S, T\}$ **do**   ▷ *Code below is performed at Target node*
11:     $\theta_k \leftarrow \theta_T$; $\hat{\theta}_k \leftarrow \theta_T$
12:     **for** $i = 1$ to τ **do**   ▷ `Local iterations`
13:       $d_T \sim \mathcal{D}_T$ ▷ `Sample batch`
14:       $\hat{\theta}_k \leftarrow \hat{\theta}_k - \alpha \nabla_{\hat{\theta}_k} \mathcal{L}(\hat{\theta}_k; d_T)$ ▷ `Compute gradient on` d_T `and update`
15:       **if** $k \neq T$ **then** $\hat{\theta}_k \leftarrow \hat{\theta}_k - \alpha \overline{g}_k$ ▷ `Update with average gradient`
16:       **else** Repeat L13-14: Resample $d_T' \sim \mathcal{D}_T$, update $\hat{\theta}_k \leftarrow \hat{\theta}_k - \alpha \nabla_{\hat{\theta}_k} \mathcal{L}(\hat{\theta}_k; d_T')$
17:     $\theta_k \leftarrow \theta_k - \beta(\hat{\theta}_k - \theta_k)$ ▷ `1st order approximation update`
18: $\theta_T = \frac{1}{S+1} \sum_{k=1}^{S+1} \theta_k$ ▷ *Update* θ_T *and* **SEND** *to Source nodes*

---

target domain gradients for each domain pair for τ steps. *L16* implements the case of the target-target pair (Sect. 3.1) with a second actual update rather than approximation. Finally, *L17* performs the first-order approximation update. Note, that we perform this update step after τ steps, which we found empirically to give better results than performing it after each interleaving update. This process is repeated for E communication rounds.

4 Experiments

Cancer Detection: We use Camelyon17 [2] dataset. The goal is to predict whether a histology image contains cancer (binary classification). The data was collected from 5 different hospitals, which we treat as 5 domains.
Skin Lesion Diagnosis: We train a model to classify 7 skin lesion types in dermatoscopic images. To create a multi-domain benchmark, we downloaded four public datasets acquired at 4 medical centers: HAM [29], BCN [9], MSK [8] and D7P [19]. Due to missing classes in MSK and D7P, we combine them in one that contains all classes. Therefore we experiment with 3 domains.

4.1 Experimental Setup

Setup: For all experiments we use DenseNet-121 [14]. We perform 'leave-one-domain-out' evaluations, iterating over each domain. One domain is held-out as target and the rest are used as source domains. Each domain dataset is divided into train, validation and test sets (60,20,20% respectively). The training sets

Table 1. Test accuracy (%) on tumour detection (3 seeds averaged).

Method		D_T		Target domain					
Pre-Train	Adapt	Lab.	Unlab.	Hospital 1	Hospital 2	Hospital 3	Hospital 4	Hospital 5	Average
FedAvg	None	×	×	88.0	80.5	76.2	85.4	75.2	81.1
None	From scratch	✓	×	92.7	88.4	92.9	91.3	94.7	92.0
FedBN	None (avg. $\mathcal{D}_S$ BN stats)	×	×	50.3	53.8	70.7	50.0	74.3	59.8
FedBN	None ($\mathcal{D}_T$ BN stats)	×	✓	89.4	84.5	89.6	91.4	84.8	87.9
FedBN	Fine-tuning	✓	×	93.1	88.2	93.7	95.8	**97.2**	93.6
FedBN	Supervised TENT	✓	×	92.8	86.8	91.6	94.1	93.1	91.7
FedBN	FedPDA: FedBN	✓	×	91.7	75.7	94.7	83.3	55.2	80.1
FedBN	FedPDA: PCGrad	✓	×	93.8	90.0	94.9	95.1	95.0	93.8
FedBN	FedPDA: StarAlign	✓	×	**95.0**	**91.4**	**96.8**	**96.8**	94.5	**94.9**
None	FDA-KD3A	×	✓	94.3	**91.4**	90.4	93.6	94.3	92.8

are used for pre-training and in FedPDA for source domains. As we investigate how to adapt with limited labelled data in the target domain, we only use 1.8% of a target domain's dataset as labelled for PDA on Camelyon17, and 6% for skin-lesion. The process is then repeated for other target domains held-out.

Metrics: Each setting is repeated for 3 seeds. From each experiment we select 5 model snapshots with best validation performance and report their average performance on the test set (averaging 15 models per setting). The two classes in Camelyon17 are balanced, therefore we report *Accuracy* (Table 1). Skin-lesion datasets have high class-imbalance and hence *Accuracy* is inappropriate because methods can increase it by collapsing and predicting solely the majority classes. Instead, we report *Weighted Accuracy* (Table 2) defined as: $\sum_{c=1}^{C} \frac{1}{C} \text{acc}_c$, where C is the number of classes and acc_c represents accuracy for class c.

Methods: The 'Pre-Train' column in Tables 1 and 2 indicates the pre-deployment training method, employing FedAvg [25] or FedBN [23] with $\tau = 100$ local iterations per communication round. FedBN pre-training is used thereafter as it outperformed FedAvg when BN statistics were adapted (D_T **BN stats** below). Column 'Adapt' indicates the post-deployment adaptation method. 'Lab' and 'Unlab' indicate whether labelled or unlabelled target data are used for adaptation. We compare StarAlign with: no adaptation (**None**), training the model just on target data (**from scratch**), FedBN without adaptation using average Batch Normalization (BN) statistics from source domains (**avg. D_S BN stats**) and when estimating BN stats on unlabelled D_T data (D_T **BN stats**) [23], and **fine-tuning** the model using only target labelled data. We also compare with the PDA method **TENT** [31], adapted to use labelled data (supervised) via cross entropy instead of unsupervised entropy for fair comparison, and the state-of-the-art FDA method **KD3A** [11] (using whole training set as unlabelled). We also attempt to perform FedPDA by simply integrating the target domain node into the federated system and resuming FedBN starting from the pre-trained model (**FedPDA: FedBN**), and with another gradient alignment method, a variant of **PCGrad** [32], which we made applicable for FedPDA. PCGrad projects gradients that point away from each other onto the normal plane of each other.

Table 2. Test weighted accuracy (%) on skin lesion diagnosis (3 seeds averaged).

Methods		D_T		Target domain			
Pre-Train	Adapt	Lab.	Unlab.	BCN	HAM	MSK & D7P	Average
FedAvg	None	×	×	26.0	37.6	25.2	29.6
None	From scratch	✓	×	34.8	40.0	24.5	33.1
FedBN	None (avg. $\mathcal{D}_S$ BN stats)	×	×	26.6	34.8	24.3	28.6
FedBN	None ($\mathcal{D}_T$ BN stats)	×	✓	35.4	39.7	29.5	34.9
FedBN	Fine-tuning	✓	×	37.6	46.2	25.5	37.8
FedBN	Supervised TENT	✓	×	40.1	47.9	29.2	39.0
FedBN	FedPDA: FedBN	✓	×	39.6	49.9	31.2	40.2
FedBN	FedPDA: PCGrad	✓	×	38.4	48.3	22.4	36.4
FedBN	FedPDA: StarAlign	✓	×	**44.4**	**53.6**	**33.5**	**43.8**
None	FDA-KD3A	×	✓	41.4	47.4	31.8	40.2

StarAlign uses $\tau = 100$, and $\beta = 0.01$ or $\beta = 0.2$ for Camelyon17 and skin lesion tasks respectively, configured on validation set.

4.2 Results and Discussion

Camelyon17 Results - Table 1: StarAlign achieves the highest average accuracy over 5 hospitals among all methods, only outperformed in 1 out of 5 settings. In our FedPDA framework, replacing StarAlign with the prominent gradient alignment method PCGrad, degrades the performance. This shows the potential of our proposed gradient alignment method.

Skin Lesion Diagnosis Results - Table 2: We observe that performance of all methods in this task is lower than in Camelyon17. This is a much more challenging task due to very high class-imbalance and strong domain shift[1]. It is extremely challenging to learn to predict the rarest minority classes under the influence of domain shift from very limited target data, which influences weighted-accuracy greatly. Accomplishing improvements in this challenging setting demonstrates promising capabilities of our method even in highly imbalanced datasets.

StarAlign consistently outperforms all compared methods. This includes PDA methods that cannot use source data (target fine-tuning and Supervised TENT) and the state-of-the-art FDA method KD3A, showing the potential of Fed-PDA. When FedPDA is performed with StarAlign, it outperforms FedPDA performed simply with a second round of FedBN training when the target node is connected to the FL system along with source nodes. FedBN can be viewed as Personalised FL method, as it learns one model per client via client-specific BN layers. Results demonstrate that learning one target-specific model with StarAlign yields better results. It also outperforms FedPDA with PCGrad, showing our

---

[1] Our baselines achieve Accuracy 75–90% on source domains of the skin lesion task, comparable to existing literature [5], indicating they are well configured.

gradient alignment method's effectiveness. This shows `StarAlign`'s potential to adapt a model to a distribution with high class imbalance and domain shift.

5   Conclusion

This work presents FedPDA, a framework unifying FL and PDA without data exchange. FedPDA enables a pre-trained, deployed model to obtain information from source data via remote gradient communication to facilitate post-deployment adaptation. We introduce `StarAlign` which aligns gradients from source domains with target domain gradients, distilling useful information from source data without data exchange to improve adaptation. We evaluated the method on two multi-center imaging databases and showed that `StarAlign` surpasses previous methods and improves performance of deployed models on new domains.

Acknowledgment. Felix Wagner is supported by the EPSRC Centre for Doctoral Training in Health Data Science (EP/S02428X/1), by the Anglo-Austrian Society, and by an Oxford-Reuben scholarship. Pramit Saha is supported in part by the UK EPSRC Programme Grant EP/T028572/1 (VisualAI) and a UK EPSRC Doctoral Training Partnership award. The authors also acknowledge the use of the University of Oxford Advanced Research Computing (ARC) facility in carrying out this work (http://dx.doi.org/10.5281/zenodo.22558).

References

1. Arivazhagan, M.G., Aggarwal, V., Singh, A.K., Choudhary, S.: Federated learning with personalization layers. arXiv preprint arXiv:1912.00818 (2019)
2. Bandi, P., et al.: From detection of individual metastases to classification of lymph node status at the patient level: the camelyon17 challenge. IEEE Trans. Med. Imaging (2018)
3. Bateson, M., Kervadec, H., Dolz, J., Lombaert, H., Ayed, I.B.: Source-free domain adaptation for image segmentation. Med. Image Anal. (2022)
4. Ben-David, S., Blitzer, J., Crammer, K., Kulesza, A., Pereira, F., Vaughan, J.W.: A theory of learning from different domains. Mach. Learn. (2010)
5. Cassidy, B., Kendrick, C., Brodzicki, A., Jaworek-Korjakowska, J., Yap, M.H.: Analysis of the ISIC image datasets: usage, benchmarks and recommendations. Med. Image Anal. (2022)
6. Chen, C., Liu, Q., Jin, Y., Dou, Q., Heng, P.-A.: Source-free domain adaptive fundus image segmentation with denoised pseudo-labeling. In: de Bruijne, M., (eds.) MICCAI 2021. LNCS, vol. 12905, pp. 225–235. Springer, Cham (2021). https://doi.org/10.1007/978-3-030-87240-3_22
7. Chidlovskii, B., Clinchant, S., Csurka, G.: Domain adaptation in the absence of source domain data. In: SIGKDD (2016)
8. Codella, N.C., et al.: Skin lesion analysis toward melanoma detection: a challenge at the 2017 international symposium on biomedical imaging (ISBI), hosted by the international skin imaging collaboration (ISIC). In: ISBI (2018)

9. Combalia, M., et al.: Bcn20000: dermoscopic lesions in the wild. arXiv preprint arXiv:1908.02288 (2019)

10. Dou, Q., Coelho de Castro, D., Kamnitsas, K., Glocker, B.: Domain generalization via model-agnostic learning of semantic features. In: NeurIPS **32** (2019)

11. Feng, H., et al.: KD3A: Unsupervised multi-source decentralized domain adaptation via knowledge distillation. In: ICML (2021)

12. Ganin, Y., et al.: Domain-adversarial training of neural networks. J. Mach. Learn. Res. (2016)

13. Geirhos, R., Temme, C.R., Rauber, J., Schütt, H.H., Bethge, M., Wichmann, F.A.: Generalisation in humans and deep neural networks. In: NeurIPS (2018)

14. Huang, G., Liu, Z., Van Der Maaten, L., Weinberger, K.Q.: Densely connected convolutional networks. In: CVPR (2017)

15. Jiang, M., Yang, H., Cheng, C., Dou, Q.: IOP-FL: inside-outside personalization for federated medical image segmentation. In: IEEE TMI (2023)

16. Kamnitsas, K., et al.: Unsupervised domain adaptation in brain lesion segmentation with adversarial networks. In: IPMI (2017)

17. Karani, N., Erdil, E., Chaitanya, K., Konukoglu, E.: Test-time adaptable neural networks for robust medical image segmentation. Med. Image Anal. (2021)

18. Karimireddy, S.P., Kale, S., Mohri, M., Reddi, S., Stich, S., Suresh, A.T.: SCAFFOLD: stochastic controlled averaging for federated learning. In: ICML (2020)

19. Kawahara, J., Daneshvar, S., Argenziano, G., Hamarneh, G.: Seven-point checklist and skin lesion classification using multitask multimodal neural nets. IEEE J. Biomed. Health Inf. (2018)

20. Li, D., Yang, Y., Song, Y.Z., Hospedales, T.: Learning to generalize: meta-learning for domain generalization. In: AAAI (2018)

21. Li, T., Sahu, A.K., Zaheer, M., Sanjabi, M., Talwalkar, A., Smith, V.: Federated optimization in heterogeneous networks. In: Proceedings of Machine Learning and Systems (2020)

22. Li, X., Gu, Y., Dvornek, N., Staib, L.H., Ventola, P., Duncan, J.S.: Multi-site FMRI analysis using privacy-preserving federated learning and domain adaptation: abide results. Med. Image Anal. (2020)

23. Li, X., Jiang, M., Zhang, X., Kamp, M., Dou, Q.: FedBN: Federated learning on non-iid features via local batch normalization. In: ICLR (2021)

24. Liu, Q., Chen, C., Qin, J., Dou, Q., Heng, P.A.: FedDG: Federated domain generalization on medical image segmentation via episodic learning in continuous frequency space. In: CVPR (2021)

25. McMahan, B., Moore, E., Ramage, D., Hampson, S., Arcas, B.A.Y.: Communication-efficient learning of deep networks from decentralized data. In: AISTATS (2017)

26. Muandet, K., Balduzzi, D., Schölkopf, B.: Domain generalization via invariant feature representation. In: ICML (2013)

27. Peng, X., Huang, Z., Zhu, Y., Saenko, K.: Federated adversarial domain adaptation. arXiv preprint arXiv:1911.02054 (2019)

28. Shi, Y., et al.: Gradient matching for domain generalization. In: ICLR (2022)

29. Tschandl, P., Rosendahl, C., Kittler, H.: The ham10000 dataset, a large collection of multi-source dermatoscopic images of common pigmented skin lesions. Sci. Data (2018)

30. Valvano, G., Leo, A., Tsaftaris, S.A.: Stop throwing away discriminators! Re-using adversaries for test-time training. In: Albarqouni, S., et al. (eds.) DART/FAIR -2021. LNCS, vol. 12968, pp. 68–78. Springer, Cham (2021). https://doi.org/10.1007/978-3-030-87722-4_7

31. Wang, D., Shelhamer, E., Liu, S., Olshausen, B., Darrell, T.: Tent: fully test-time adaptation by entropy minimization. In: ICLR (2021)
32. Yu, T., Kumar, S., Gupta, A., Levine, S., Hausman, K., Finn, C.: Gradient surgery for multi-task learning. In: NeurIPS (2020)
33. Zhao, Y., Li, M., Lai, L., Suda, N., Civin, D., Chandra, V.: Federated learning with non-iid data. arXiv preprint arXiv:1806.00582 (2018)

Data-Driven Classification of Fatty Liver From 3D Unenhanced Abdominal CT Scans

Jacob S. Leiby[1], Matthew E. Lee[1], Eun Kyung Choe[2], and Dokyoon Kim[1,3(✉)]

[1] Department of Biostatistics, Epidemiology, and Informatics, Perelman School of Medicine, University of Pennsylvania, Philadelphia, PA 19104, USA
Dokyoon.kim@pennmedicine.upenn.edu

[2] Department of Surgery, Seoul National University Hospital Healthcare System Gangnam Center, Seoul 06236, South Korea

[3] Institute for Biomedical Informatics, University of Pennsylvania, Philadelphia, PA 19104, USA

Abstract. Fatty liver disease is a prevalent condition with significant health implications and early detection may prevent adverse outcomes. In this study, we developed a data-driven classification framework using deep learning to classify fatty liver disease from unenhanced abdominal CT scans. The framework consisted of a two-stage pipeline: 3D liver segmentation and feature extraction, followed by a deep learning classifier. We compared the performance of different deep learning feature representations with volumetric liver attenuation, a hand-crafted radiomic feature. Additionally, we assessed the predictive capability of our classifier for the future occurrence of fatty liver disease. The deep learning models outperformed the liver attenuation model for baseline fatty liver classification, with an AUC of 0.90 versus 0.86, respectively. Furthermore, our classifier was better able to detect mild degrees of steatosis and demonstrated the ability to predict future occurrence of fatty liver disease.

Keywords: Fatty Liver Disease · Deep Learning · Computed Tomography

1 Introduction

Fatty liver disease is an increasingly prevalent condition that is estimated to affect roughly 25% of the global population [1]. It can lead to more severe manifestations including non-alcoholic steatohepatitis, cirrhosis, and liver failure [2, 3]. Additionally, fatty liver is associated with an increased risk of type II diabetes, cardiovascular disease, chronic kidney disease, and cancer [3]. Early and accurate detection of fatty liver could allow for clinical management and help prevent progression to adverse outcomes.

Liver biopsy is the gold standard for assessment of fatty liver disease; however, it is invasive and not suitable for screening large numbers of individuals [4]. Non-invasive imaging techniques including ultrasonography and magnetic resonance imaging (MRI) have proven to be effective tools in quantifying hepatic fat and are often performed for population screening and initial examination of individuals with suspected fatty liver disease [4]. MRI quantifies the presence of hepatic fat through measuring the proton density fat fraction (PDFF) and is considered to be one of the most accurate and reproducible non-invasive diagnostic tools [4–7]. Studies have compared the utility of unenhanced computed tomography (CT) for determining fatty liver and have shown that there is a strong linear correlation between CT attenuation and PDFF, using this imaging modality to investigate the prevalence of fatty liver in large screening cohorts [7–9]. While strong correlation is shown in specific settings, the accuracy of CT hepatic fat quantification can be affected by several factors including iron levels in the liver and medication [10–12]. Additionally, quantification accuracy through CT attenuation decreases at low levels of fattiness and may not be suitable for detection of mild steatosis [4].

Radiomics is a quantitative approach that extracts features from clinical imaging data and aims to assist in diagnosis, prognostication, and decision support in many clinical areas including oncology, immunology, neurodegenerative disorders [13–16]. Conventional radiomics workflows typically extract predetermined features, known as hand-crafted features, from segmented regions of interest. With recent advances in artificial intelligence, data-driven approaches to radiomics have been studied in cancer, cardiovascular disease, and infectious disease detection using deep learning [17–23].

In this study, we developed a data-driven classification framework for fatty liver disease from unenhanced abdominal CT scans in a cohort of over 2,000 individuals. We proposed a two-stage pipeline that included a 3D liver segmentation model and feature extractor, followed by a feedforward network to classify fatty liver disease. We compared the classification performance of different deep learning feature representations to volumetric liver attenuation. Additionally, we performed a follow-up analysis to investigate the predictive capability of our classifier for the future occurrence of fatty liver disease.

2 Methods

2.1 Study Population and Data Preprocessing

The data in this study was obtained from a cohort of individuals who underwent comprehensive health check-ups in the Korean population (Table 1). Detailed information regarding the dataset can be found elsewhere [24]. The data was obtained from the Health and Prevention Enhancement (H-PEACE) study, a retrospective, population-based cohort study conducted at the Seoul National University Hospital Gangnam Center in South Korea. The comprehensive health screening encompassed various clinical tests, including abdominal computed tomography (CT) scans, to evaluate anthropometric, cardiovascular, digestive, endocrinological, metabolic, hematologic, lung, and renal conditions. Additionally, we collected the 5-year follow-up data for over 1,300 individuals. In this study, we used the non-contrast (unenhanced) abdominal CT scan stored in the Digital Imaging and Communications in Medicine (DICOM) file format.

Fatty liver diagnosis was established through the utilization of abdominal ultrasound. An experienced radiologist conducted the procedure, assessing vascular blurring, attenuation, hepatorenal echo contrast, and liver brightness in ultrasonography images to define the presence of fatty liver [25]. Radiologists provided a grading from 0 to 3, where 0 represents normal liver and 1, 2, 3 represent mild, moderate, and severe fattiness, respectively. For the fatty liver label, we classified the liver as either "normal" or indicating the presence of "at least a mild degree of fatty liver."

Table 1. Cohort characteristics of the data collected at baseline and follow-up. (Note the follow-up cohort only consists of individuals who were not diagnosed with fatty liver at baseline. The grading data was not available for the follow-up cohort.)

Cohort (N)	Fatty liver	Non-fatty liver
Baseline (2,264)	Mild: 575 Moderate: 354 Severe: 55	1,280
Follow-up (725)	226	499

All DICOM series were converted to 3D Neuroimaging Informatics Technology Initiative (NIfTI) file format using the *dicom2nifti* python package (https://icometrix.git hub.io/dicom2nifti/). 3D volumes were preprocessed using the nnU-Net software [26]. Briefly, the intensity values were clipped at the 0.5 and 99.5 percentiles, and each volume was resampled to the median values of the training data spacing in each axis.

2.2 Liver Segmentation

We followed the standard 3D U-Net architecture for liver segmentation (Fig. 1) [27]. This model adopts an encoder-decoder structure with skip connections aiming to learn a dense embedding of the input data (encoder) and effectively map relevant information back to the original input dimensions (decoder). The skip connections connect the corresponding encoder and decoder layers to allow the direct flow of information across different resolutions. We used the nnU-Net software to optimize the network topology and the input patch size [26]. The final patch size was set to 28 x 256 x 256.

The segmentation model was trained and validated on 358 abdominal CT scans with the liver regions annotated. nnU-Net performed 5-fold cross validation and used the Dice coefficient as the performance metric. The top-performing segmentation model was used to create liver segmentation masks for all samples in the dataset.

2.3 Liver Attenuation

We compared our deep learning approach to volumetric liver attenuation (expressed as Hounsfield units). Liver attenuation is a hand-crafted feature that is linearly correlated with hepatic fat determined by MRI-derived proton density fat fraction [7, 8]. Liver attenuation values were extracted by applying the segmentation mask to the original CT volume and calculating the median attenuation of the voxels predicted as the liver. We used the median liver attenuation in a logistic regression to classify fatty liver.

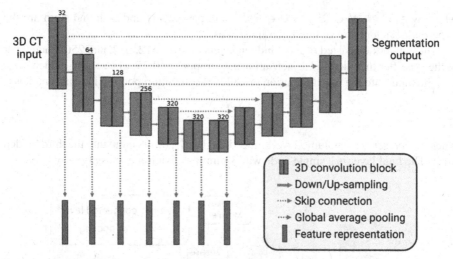

Fig. 1. Overview of the segmentation model and feature extraction. The segmentation model follows a 3D UNet architecture, with the entire CT volume as input. After the segmentation model is trained, the segmentation outputs are used to crop and mask the CT scans and features are extracted through the pretrained encoder at different depths.

2.4 3D Feature Extraction

We used the outputs of the segmentation model to crop and mask the CT scans, selecting only the predicted liver region. In order to extract informative features related to the liver, we utilized the pretrained encoder from the segmentation model to encode the cropped and masked 3D CT scan into latent vector representations. Specifically, we performed global average pooling over the outputs of the encoder blocks to aggregate spatial information. We experimented with different depths of the encoder, from using the first encoder block only to including the outputs from all encoder blocks (Fig. 1). We concatenated the pooled encoder blocks for the final feature representations. In total, we generated seven feature representations of sizes 32, 96, 224, 480, 800, 1120, and 1440.

2.5 Deep Learning Classifier

The segmentation model was optimized for patch-based input, so each CT scan was divided into N 3D patches and a feature representation was created for each patch. To aggregate the patch-level representations into a volume-level representation to classify fatty liver, we used an attention-based multiple instance learning framework (Fig. 2) [28]. This consisted of a feedforward network containing a gated-attention module. For a scan consisting of $H = \{\mathbf{h}_1,..., \mathbf{h}_N\}$ embeddings, the gated-attention can be represented as:

$$a_n = \frac{\exp\left\{\mathbf{w}^{\mathrm{T}}\left(\tanh\left(\mathbf{V}\mathbf{h}_n^{\mathrm{T}}\right) \odot \mathrm{sigm}\left(\mathbf{U}\mathbf{h}_n^{\mathrm{T}}\right)\right)\right\}}{\sum_{j=1}^{N} \exp\left\{\mathbf{w}^{\mathrm{T}}\left(\tanh\left(\mathbf{V}\mathbf{h}_j^{\mathrm{T}}\right) \odot \mathrm{sigm}\left(\mathbf{U}\mathbf{h}_j^{\mathrm{T}}\right)\right)\right\}} \tag{1}$$

where $\mathbf{w} \in \mathbb{R}^{L \times 1}$, $\mathbf{V} \in \mathbb{R}^{L \times M}$, $\mathbf{U} \in \mathbb{R}^{L \times M}$ are parameters and tanh and sigm are the hyperbolic tangent and sigmoid non-linearities.

The network consisted of three hidden layers of sizes 512, 512, and 256. Each layer in the classifier included ReLU activation and dropout of 0.25. The second hidden layer was the input into the gated-attention module. The attention pooling is represented as:

$$z = \sum_{n=1}^{N} a_n \mathbf{h}_n \qquad (2)$$

where $\mathbf{z}$ represents the volume-level embedding. Lastly, this is input into the third hidden layer. The final layer is a single node with sigmoid activation to classify fatty liver.

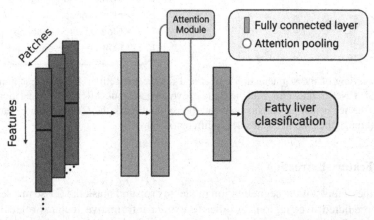

Fig. 2. Deep learning classifier for fatty liver disease. The input to the model is the feature representations for all patches in a CT volume. They are passed through two fully-connected layers and aggregated into a volume-level embedding through attention pooling to classify fatty liver.

3   Experiment

3.1   Experimental Setup

The pretrained encoder from the segmentation model was used to generate feature representations of the masked CT scans. The features were used as the input into the classifier. All models were implemented by PyTorch. The classifier was trained for 100 epochs using the AdamW optimizer in PyTorch version 1.12.1 with default parameters, and a batch size of 10 was used. Early stopping was implemented using a validation subset of the training data. The experiments were performed on a machine equipped with two Intel Xeon Gold 6226 Processors and 10 NVIDIA GeFORCE RTX 2080Ti graphics cards.

We used 10-times repeated 5-fold cross validation to evaluate the models and reported the area under the receiver operating curve (AUC) and the area under the precision-recall curve (AUPRC). For each split we reserved 20% of the total dataset for testing and evaluation, and with the remaining data, we trained on 80% and used 20% as a validation set to tune the hyperparameters. We compared the classification performance of the deep learning models to the baseline volumetric liver attenuation model. Additionally, we evaluated the performance within each subgroup of degree of fatty liver.

In addition to baseline fatty liver prediction, we performed a follow-up analysis to evaluate how the features extracted from baseline CT scans can predict the future occurrence of fatty liver. Specifically, we removed all individuals with fatty liver at baseline, extracted the baseline attenuation and output of the classifier from the top performing 3D feature representation from all remaining individuals, and used these features in a logistic regression to predict future occurrence of fatty liver.

3.2 Results

The top performing liver segmentation model achieved a Dice score of 0.98. Table 2 shows the results for baseline fatty liver prediction of all models. We compared each of the deep learning feature representations to liver attenuation. The liver attenuation model achieved an AUC of 0.860 and AUPRC of 0.636. All of the deep learning models outperformed the attenuation model, with the features generated from encoder depth 6 showing the top performance with an AUC of 0.900 and AUPRC of 0.710. There was a modest performance improvement as the encoder depth increased until a certain point, suggesting that the model was learning the majority of informative features in the early convolutional stages.

Table 2. Performance results for baseline fatty liver prediction. The encoder depths are the different 3D feature representations described in Sect. 2.4. Results are shown as mean (standard deviation) over 50 models.

Feature	AUC	AUPRC
Liver attenuation	0.860 (0.027)	0.636 (0.052)
Encoder depth 1	0.884 (0.024)	0.679 (0.051)
Encoder depth 2	0.886 (0.022)	0.686 (0.046)
Encoder depth 3	0.890 (0.021)	0.693 (0.045)
Encoder depth 4	0.898 (0.020)	0.701 (0.045)
Encoder depth 5	0.899 (0.019)	0.707 (0.043)
Encoder depth 6	**0.900 (0.019)**	**0.710 (0.043)**
Encoder depth 7	0.898 (0.020)	0.707 (0.044)

Table 3 shows the evaluation of the model within the different subgroups of fatty liver degree. For our proposed approach, we used the feature representations from the encoder depth 6. For each of the fatty liver subgroups, the proposed model outperforms attenuation. We see the strongest performance improvement for the mild fatty liver group, from an AUC of 0.679 for liver attenuation to 0.761 for our proposed model.

Table 3. Performance results within the different degrees of fatty liver diagnosis. The proposed approach is the feature representation from the encoder depth 6 (Sect. 2.4). Results are shown as mean (standard deviation) over 50 models.

Fatty liver degree	AUC		AUPRC	
	Attenuation	Proposed	Attenuation	Proposed
Mild	0.679 (0.027)	**0.761 (0.021)**	0.539 (0.040)	**0.610 (0.042)**
Moderate	0.889 (0.026)	**0.937 (0.018)**	0.774 (0.041)	**0.846 (0.028)**
Severe	0.953 (0.032)	**0.963 (0.035)**	0.617 (0.141)	**0.697 (0.128)**

Table 4 shows the results for the follow-up fatty liver prediction analysis. The proposed model outperformed the liver attenuation model, achieving an AUC of 0.676 versus 0.581 and AUPRC of 0.488 compared to 0.402.

Table 4. Performance results for the follow-up analysis. The proposed approach is the feature representation from the encoder depth 6 (Sect. 2.4). Results are shown as mean (standard deviation) over 50 models.

Feature	AUC	AUPRC
Liver attenuation	0.581 (0.046)	0.402 (0.070)
Proposed	**0.676 (0.052)**	**(0.068)**

4 Conclusion

This study demonstrated the potential of a data-driven classification framework for the identification of fatty liver disease from unenhanced abdominal CT scans. We developed a pipeline that incorporated a 3D liver segmentation model and feature extractor followed by a deep learning classifier. By comparing different deep learning feature representations to the traditional volumetric liver attenuation approach, we observed improved classification performance. Notably, we see a strong performance improvement for individuals with a mild degree of fatty liver. Furthermore, our classifier exhibited promising predictive capability for the future occurrence of fatty liver disease. These findings suggest that our approach has the potential to enable early and accurate detection of fatty liver, allowing for timely clinical intervention and prevention of adverse outcomes. The integration of advanced deep learning techniques with non-invasive imaging provides a promising pathway for enhancing liver disease assessment and management. As the availability of large-scale imaging datasets continues to grow, leveraging data-driven methods will likely open new avenues for comprehensive disease characterization and individualized treatment strategies. Our work adds to the growing body of evidence supporting the value of artificial intelligence in medical imaging and highlights its potential to revolutionize the field of hepatology.

References

1. Younossi, Z., et al.: Global burden of NAFLD and NASH: trends, predictions, risk factors and prevention. Nat. Rev. Gastroenterol. Hepatol. **15**, 11–20 (2017). https://doi.org/10.1038/nrgastro.2017.109

2. Farrell, G.C., Larter, C.Z.: Nonalcoholic fatty liver disease: from steatosis to cirrhosis. Hepatol. **43**, S99–S112 (2006)

3. Kumar, R., Priyadarshi, R.N., Anand, U.: Non-alcoholic fatty liver disease: growing burden, adverse outcomes and associations. J. Clin.Transl. Hepatol. **8**, 1–11 (2019). https://doi.org/10.14218/JCTH.2019.00051

4. Lee, S.S., Park, S.H.: Radiologic evaluation of nonalcoholic fatty liver disease. World J. Gastroenterol. **20**, 7392 (2014)

5. Cotler, S.J., Guzman, G., Layden-Almer, J., Mazzone, T., Layden, T.J., Zhou, X.J.: Measurement of liver fat content using selective saturation at 3.0 T. J. Magn. Reson. Imaging **25**, 743–748 (2007)

6. Reeder, S.B., Cruite, I., Hamilton, G., Sirlin, C.B.: Quantitative assessment of liver fat with magnetic resonance imaging and spectroscopy. J. Magn. Reson. Imaging **34**, 729–749 (2011)

7. Pickhardt, P.J., Graffy, P.M., Reeder, S.B., Hernando, D., Li, K.: Quantification of liver fat content with unenhanced MDCT: phantom and clinical correlation with MRI proton density fat fraction. Am. J. Roentgenol. Am. Roentgen Ray Soc. **211**, W151–W157 (2018)

8. Guo, Z., et al.: Liver fat content measurement with quantitative CT validated against MRI proton density fat fraction: a prospective study of 400 healthy volunteers. Radiol. Radiol. Soc. North Am. (RSNA) **294**, 89–97 (2020)

9. Boyce, C.J., et al.: Hepatic steatosis (fatty liver disease) in asymptomatic adults identified by unenhanced Low-Dose CT. Am. J. Roentgenol. Am. Roentgen Ray Soc. **194**, 623–628 (2010)

10. Pickhardt, P.J., Park, S.H., Hahn, L., Lee, S.-G., Bae, K.T., Yu, E.S.: Specificity of unenhanced CT for non-invasive diagnosis of hepatic steatosis: implications for the investigation of the natural history of incidental steatosis. Eur. Radiol. **22**, 1075–1082 (2011). https://doi.org/10.1007/s00330-011-2349-2

11. Park, Y.: Biopsy-proven nonsteatotic liver in adults: estimation of reference range for difference in attenuation between the liver and the spleen at nonenhanced CT. Radiol. Radiol. Soc. North Am. (RSNA) **258**, 760–766 (2011)

12. Patrick, D., White, F.E., Adams, P.C.: Long-term amiodarone therapy: a cause of increased hepatic attenuation on CT. Br. J. Radiol. **57**, 573–576 (1984)

13. Limkin, E., et al.: Promises and challenges for the implementation of computational medical imaging (radiomics) in oncology. Ann. Oncol. **28**, 1191–1206 (2017)

14. Kang, C.Y., et al.: Artificial intelligence-based radiomics in the era of immune-oncology. Oncologist **27**(6), e471–e483 (2022). https://doi.org/10.1093/oncolo/oyac036

15. Ranjbar, S., Velgos, S.N., Dueck, A.C., Geda, Y.E., Mitchell, J.R.: Brain MR radiomics to differentiate cognitive disorders. J. Neuropsychiatry Clin. Neurosci. **31**, 210–219 (2019)

16. Tupe-Waghmare, P., Rajan, A., Prasad, S., Saini, J., Pal, P.K., Ingalhalikar, M.: Radiomics on routine T1-weighted MRI can delineate Parkinson's disease from multiple system atrophy and progressive supranuclear palsy. Eur. Radiol. **31**, 8218–8227 (2021). https://doi.org/10.1007/s00330-021-07979-7

17. Afshar, P., Mohammadi, A., Plataniotis, K.N., Oikonomou, A., Benali, H.: From handcrafted to deep-learning-based cancer radiomics: challenges and opportunities. IEEE Signal Process. Mag. **36**, 132–160 (2019)

18. Zheng, X., et al.: Deep learning radiomics can predict axillary lymph node status in early-stage breast cancer. Nature Communications, Springer Science and Business Media LLC, **11**, 1236 (2020). https://doi.org/10.1038/s41467-020-15027-z

19. Lee, S., Choe, E.K., Kim, S.Y., Kim, H.S., Park, K.J., Kim, D.: Liver imaging features by convolutional neural network to predict the metachronous liver metastasis in stage I-III colorectal cancer patients based on preoperative abdominal CT scan. BMC Bioinformatics, Springer Science and Business Media LLC, **21**, 382 (2020). https://doi.org/10.1186/s12859-020-03686-0

20. Andrearczyk, V., et al.: Multi-task deep segmentation and radiomics for automatic prognosis in head and neck cancer. In: Rekik, I., Adeli, E., Park, S.H., Schnabel, J. (eds.) Predictive Intelligence in Medicine: 4th International Workshop, PRIME 2021, Held in Conjunction with MICCAI 2021, Strasbourg, France, October 1, 2021, Proceedings, pp. 147–156. Springer International Publishing, Cham (2021). https://doi.org/10.1007/978-3-030-87602-9_14

21. Poplin, R., et al.: Prediction of cardiovascular risk factors from retinal fundus photographs via deep learning. Nat. Biomed. Eng. **2**, 158–164 (2018). https://doi.org/10.1038/s41551-018-0195-0

22. Shorten, C., Khoshgoftaar, T.M., Furht, B.: Deep Learning applications for COVID-19. J. Big Data, Springer Science and Business Media LLC **8**, 18 (2021). https://doi.org/10.1186/s40537-020-00392-9

23. Zhao, W., Jiang, W., Qiu, X.: Deep learning for COVID-19 detection based on CT images. Scientific Reports, Springer Science and Business Media LLC **11**, 14353 (2021). https://doi.org/10.1038/s41598-021-93832-2

24. Lee, C., et al.: Health and prevention enhancement (H-PEACE): a retrospective, population-based cohort study conducted at the Seoul national university hospital Gangnam center, Korea. BMJ Open, BMJ **8**, e019327 (2018)

25. Hamaguchi, M., et al.: The severity of ultrasonographic findings in nonalcoholic fatty liver disease reflects the metabolic syndrome and visceral fat accumulation. Am. J. Gastroenterol. **102**, 2708–2715 (2007)

26. Isensee, F., Jaeger, P.F., Kohl, S.A.A., Petersen, J., Maier-Hein, K.H.: nnU-Net: a self-configuring method for deep learning-based biomedical image segmentation. Nat. Methods **18**, 203–211 (2020). https://doi.org/10.1038/s41592-020-01008-z

27. Ronneberger, O., Fischer, P., Brox, T.: U-Net: convolutional networks for biomedical image segmentation. arXiv:1505.04597 (2015)

28. Ilse, M., Tomczak, J.M., Welling, M.: Attention-based deep multiple instance learning. arXiv:1802.04712 (2018)

Replica-Based Federated Learning with Heterogeneous Architectures for Graph Super-Resolution

Ramona Ghilea and Islem Rekik[(✉)][iD]

BASIRA Lab, Imperial-X and Department of Computing, Imperial College London, London, UK
i.rekik@imperial.ac.uk
https://basira-lab.com/

Abstract. Having access to brain connectomes at various resolutions is important for clinicians, as they can reveal vital information about brain anatomy and function. However, the process of deriving the graphs from magnetic resonance imaging (MRI) is computationally expensive and error-prone. Furthermore, an existing challenge in the medical domain is the small amount of data that is available, as well as privacy concerns. In this work, we propose a new federated learning framework, named *RepFL*. At its core, *RepFL* is a replica-based federated learning approach for heterogeneous models, which creates replicas of each participating client by copying its model architecture and perturbing its local training dataset. This solution enables learning from limited data with a small number of participating clients by aggregating multiple local models and *diversifying* the data distributions of the clients. Specifically, we apply the framework for *graph super-resolution using heterogeneous model architectures*. In addition, to the best of our knowledge, this is the first federated multi-resolution graph generation approach. Our experiments prove that the method outperforms other federated learning methods on the task of brain graph super-resolution. Our RepFL code is available at https://github.com/basiralab/RepFL.

Keywords: Graph Neural Networks · graph super-resolution · multi-resolution · federated learning · heterogeneous architectures · brain connectome

1 Introduction

The brain can be naturally modelled as a graph, where the nodes represent anatomical regions of interest (ROIs) and the edges denote biological connections between ROIs [1]. The goal of representing the brain as a network is to better understand its complex structure. Brain connectomes at different *resolutions* (where the resolution denotes the number of nodes in the brain graph) can provide important insights regarding brain

Supplementary Information The online version contains supplementary material available at https://doi.org/10.1007/978-3-031-45676-3_28.

structure and function [2]. Graphs having different resolutions are derived from different brain magnetic resonance imaging (MRI) atlases and the process of creating the connectome involves a range of image pre-processing steps, such as registering the MRI scan and labelling the ROIs in the graph. Additionally, connectomes from different domains offer a holistic view of the brain, which has the potential to improve the management of neurological disorders. The *domain* is given by the imaging modality that was used to derive the graph, and can be morphological, structural or functional. However, generating brain graphs at multiple resolutions involves a large amount of computation and it is error-prone, while acquiring images from different modalities involves high costs and long processing time. Graph Neural Networks (GNNs) are powerful algorithms for graph-related tasks [3] and they have been applied on data such as molecules, social networks or transportation networks [4]. Recently, GNNs have also been used on brain graphs for tasks such as classification, prediction, generation or integration [2]. These models are very promising for the brain connectomics domain, as they could automate the process of creating brain graphs at different resolutions, as well as generating networks from different domains. A principal challenge in the medical field is the small amount of data. Federated learning aims to train a global model using private datasets of multiple distributed clients without transferring the local data to a centralized machine [5]. It is an approach that can be used to address this issue, as data from multiple hospitals could be leveraged for training a global model without compromising the data privacy [6].

In this study, we propose a new federated learning framework, called *RepFL*. In this setting, clients have different GNN architectures and each client is replicated for a number of times in order to create diverse models and enable learning from small local datasets. On each replica, which is a virtual copy of the original client, the data distribution is perturbed by removing a small percentage of the local samples to obtain better generalisability. The intuition is that a larger amount of clients with diverse data distributions could improve the generalisability of the global model. However, in a realistic scenario, there is not a very large number of participating hospitals (clients) and data is limited. Specifically, we apply the method for brain graph super-resolution with heterogeneous model architectures, where each client generates a target brain connectome of a different resolution and different domain from the source brain graph. The main contributions of this work are:

1. From a methodological perspective, we propose a novel federated learning approach, *RepFL*, where each client is replicated for a specific number of times and the data distribution on each replica is perturbed.
2. On a conceptual level, this is the first work that focuses on multi-resolution graph generation while federating diverse GNN architectures.
3. Experimentally, we demonstrate the ability of the proposed method to improve model performance in a limited data setting with a small number of participating clients. Furthermore, we provide an analysis of the rate of perturbation of the data distribution on the replicas, as well as an investigation of the number of replicas for each client.

Related Work. Graph super-resolution was first proposed in GSR-Net [7], a framework which predicts a high-resolution brain network from a low-resolution graph. Later, Mhiri et al. [8,9] primed inter-modality brain graph synthesis frameworks. They proposed ImanGraphNet [8] super-resolves brain graphs including a non-isomorphic domain alignment network to account for shifts in resolutions and domains. StairwayGraphNet (SG-Net) [9] super-resolves brain graphs at cascaded resolutions and across different domains. In another work, [10] proposed a template-guided brain graph super-resolution framework using a GNN architecture that is trained using one-shot connectional brain template [11]. [12] leveraged the teacher-student learning paradigm for super-resolving brain graphs across varying connectivity domains. However, to the best of our knowledge, none of these works were designed within a federated learning context [2].

Federated learning is an approach that trains a global model using data stored on multiple devices without sharing the local datasets [5]. The framework has significant potential in the medical field, as hospitals do not have very large amounts of medical data and the privacy is extremely important [6]. The most popular federated learning method is FedAvg [13], which trains a global model by averaging the parameters of the clients. Motivated by the differences in data distribution and in system capacity among clients, methods that introduce an additional term in the local objective were proposed, such as FedProx [14] and FedDyn [15]. Focusing on the same statistical heterogeneity problem, recent studies developed methods for personalized federated learning [16], which aims to train a global model while personalizing it for each client. Popular approaches include model mixing, where each client is a combination of local and global models [17,18], clustered federated learning, where each client performs federation only with a cluster of relevant clients [19,20], and dividing the model into base and personalized layers [21].

Even though the data heterogeneity challenge has been extensively analysed in the federated learning context, there are few studies that focused on federating clients with *different model architectures*. One of the first works that addressed the model heterogeneity limitation is HeteroFL [22]. However, the main limitation of this framework is the requirement that the clients belong to the same model class. Other studies proposed hypernetworks for generating the personalized weights for the clients. The authors of pFedHN [23] used such a hypernetwork, while Litany et al. [24] suggested a graph hypernetwork for federating heterogeneous architectures. However, while pFedHN employs a large number of clients, the solution implementing graph hypernetworks uses models having 10 layers or more, which is not applicable to our setting. Another category of methods propose knowledge distillation as a solution for federating heterogeneous models [25,26]. However, these approaches require a public dataset, which is not realistic when working with medical data.

While learning from small datasets has been an important area of research in machine learning, federated learning with limited local data is an underexplored topic. The method which is most related to the current work is FedDC [27], which combines classic aggregation with permutations of local models, such that models are moved from client to client and trained on multiple local datasets. Although the solution is developed for federated learning in a limited data setting, we are also interested in learning models with a small number of clients.

2   Method

Figure 1 provides an overview of the proposed method. There are three main components, which we detail below: the proposed *replica-based* federated learning method, *RepFL*, and the two types of local model architecture, G_1 and G_2.

Problem Definition. In classic federated learning, the aim is to train and federate m clients, where each client has its local dataset D_i. In *RepFL*, there are m original clients having heterogeneous model architectures and each client is replicated r_i times, where i is the index of the client. We define the original client as the *anchor*, and the copy as the *replica*. A replica has an identical model architecture to the anchor architecture, while the dataset of the replica is modified by leveraging a perturbation technique of the anchor data distribution. The perturbation consists in removing a percentage p_i of the samples from the anchor dataset, where p_i is the rate of perturbation of anchor client i. Therefore, we federate $m + \sum_{i=1}^{m} r_i$ models. The local dataset D_i of hospital i contains n_i samples, where each sample j is represented by a low-resolution (LR) source graph $X_{i,j}^s$ and a high-resolution (HR) target graph $X_{i,j}^t$.

Model Architecture. A client in the federated learning setting can have one of two models: G_1, which takes as input a morphological brain connectome X^s of resolution n_r and outputs a functional brain graph X^{t_1} of resolution n_r', or G_2, which takes as input a morphological brain connectome X^s of resolution n_r and outputs a functional brain graph X^{t_2} of resolution n_r'', where $n_r < n_r' < n_r''$. Two clients having the same model differ in their local datasets.

The architecture is an encoder-decoder graph neural network (GNN) model containing multiple groups of three types of layers: an edge-based graph convolutional network (GCN) layer [28], followed by batch normalization and dropout. While batch normalization accelerates the convergence of the model, dropout overcomes overfitting. The models G_1 and G_2 take as input the LR morphological brain graph (represented by the adjacency matrix) of dimension $n_r \times n_r$ and output a HR functional brain network of dimension $n_r' \times n_r'$ and $n_r'' \times n_r''$, respectively. Both models implement the mapping function defined in SG-Net [9][1], which generates a symmetric brain graph after the last layer in the network. More specifically, the mapping function $T_r = (Z_h)^T Z_h$ takes as input the embedding of the last layer Z_h of the model, having the size $d_h \times d_h'$, and outputs a symmetric graph of size $d_h' \times d_h'$. In order to get the desired target resolution, d_h' has to be set to this specific resolution. The model architectures need to be adapted to the specific resolutions used, as the dimensions are hard-coded. Considering that G_1 and G_2 take as input LR graphs of the same dimension and output connectomes of different resolutions, a part of the layers have the same dimensions, while the rest of the layers differ. We denote the weight matrices of the common layers as $W^C = \{W_1^C, W_2^C, \ldots, W_{n_c}^C\}$ (where n_c is the number of common layers), the layers that are present only in G_1 as $W^{D_1} = \{W_1^{D_1}, W_2^{D_1}, \ldots, W_{n_{d_1}}^{D_1}\}$ (where n_{d_1} represents the number of layers specific for G_1), and the layers that exist only in G_2 as $W^{D_2} = \{W_1^{D_2}, W_2^{D_2}, \ldots, W_{n_{d_2}}^{D_2}\}$ (where n_{d_2} denotes the number of layers specific for G_2). When aggregating the parameters, only the common layers $\{W_1^C, W_2^C, \ldots, W_{n_c}^C\}$ across local heterogeneous GNN architectures are considered, while the different, unmatched layers are trained only

---

[1] https://github.com/basiralab/SG-Net.

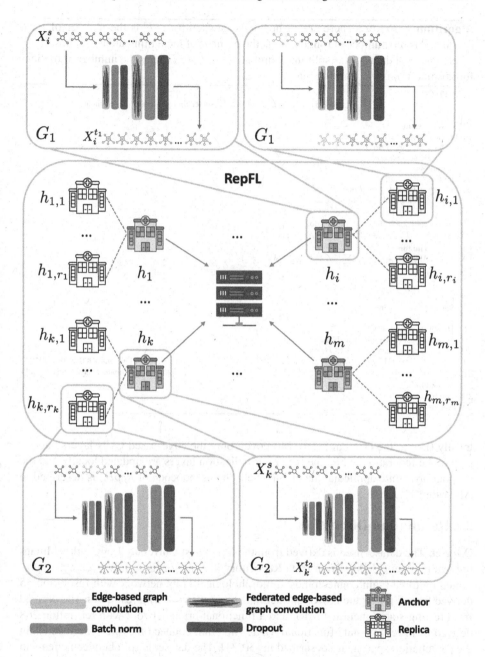

Fig. 1. *Proposed framework.* **RepFL.** Each client (hospital) is replicated r_i times, where i is the index of the client. h_i is the ith client, while $h_{i,l}$ is the lth replica of client i. In the figure, h_i and h_k are arbitrary clients. The replica copies the anchor architecture and it perturbs its local dataset. The anchors can either have architecture G_1 or G_2. **G_1.** G_1 takes as input the LR morphological brain graph X^s of dimension $n_r \times n_r$ and outputs a HR functional brain network X^{t_1} of dimension $n_r' \times n_r'$. **G_2.** G_2 takes as input the LR morphological brain graph X^s of dimension $n_r \times n_r$ and outputs a HR functional brain network X^{t_2} of dimension $n_r'' \times n_r''$.

Algorithm 1. *RepFL*. There are m hospitals indexed by i; E is the number of local epochs; R is the number of rounds; B_i is the number of local minibatches on hospital i, where the local dataset was split into minibatches of size B; r_i is the number of replicas for hospital i; η is the learning rate;

1: INPUTS:
    $X_{i,j} = \{X_{i,j}^s, X_{i,j}^t\}$: source and target graphs of the jth subject in local dataset X_i of client i

2: **Server executes:**
3: **for** each hospital $i = 1$ **to** m **do**
4:     **for** $r = 1$ **to** r_i **do**
5:        create replica r
6:     **end for**
7: **end for**
8: **for** each round $t = 1$ **to** R **do**
9:     **for** each hospital $i = 1$ **to** m **do**
10:        $W_{t+1,i}^C \leftarrow HospitalUpdate(i, W_{t,i}^C)$
11:        **for** each replica $r = 1$ **to** r_i **do**
12:           $W_{t+1,r}^C \leftarrow HospitalUpdate(r, W_{t,r}^C)$
13:        **end for**
14:        $W_{t+1,i}^C \leftarrow \frac{1}{r_i+1}(\sum_{r=1}^{r_i} W_{t+1,r}^C + W_{t+1,i}^C)$
15:     **end for**
       $W_{t+1}^C \leftarrow \frac{1}{m}\sum_{i=1}^m W_{t+1,i}^C$
16: **end for**
17: Send W^C to hospitals

18: **HospitalUpdate**(i, W^C):
19: **for** each local epoch $e = 1$ **to** E **do**
20:     **for** each minibatch $b = 1$ **to** B_i **do**
21:        $W^C \leftarrow W^C - \eta \bigtriangledown l(W^C, b)$ ▷ update the common (federated) layers
22:        $W_i^D \leftarrow W_i^D - \eta \bigtriangledown l(W_i^D, b)$ ▷ update the personalized layers on hospital i
23:     **end for**
24: **end for**
25: Return W^C to server

---

locally (i.e., at the hospital level). Moreover, only the edge-based GCN layers participate in the federation, while the batch normalization layers are updated locally and the dropout layers are applied at the client level. The pseudocode of *RepFL* is described in Algorithm 1.

3 Results and Discussion

Dataset. The dataset used is derived from the Southwest University Longitudinal Imaging Multimodal (SLIM) Brain Data Repository [29]. It consists of 279 subjects represented by three brain connectomes: a morphological brain network with resolution 35 derived from T1-weighted MRI, a functional brain network with resolution 160 derived from resting-state functional MRI, and a functional brain network with resolution 268 derived from resting-state functional MRI. The brain connectomes were derived using the parcellation approaches described in [30,31]. The dataset is split between clients in order to simulate the federated learning setting using 5-fold cross-validation.

Parameter Setting. Four clients are used in the experiments: two clients which have the task of predicting a brain graph with resolution $n_r' = 160$ from a graph with resolution $n_r = 35$ and two clients which have the task of predicting a brain graph with resolution $n_r'' = 268$ from a graph with resolution $n_r = 35$. The number of local epochs is $E = 10$ and the number of rounds is $R = 10$. In the standalone version (i.e., no federation), the models are trained locally for 100 epochs. A batch size of 5 is used. The learning rate is set to 0.1. For FedDyn, alpha is set to 0.01, and for FedDC, the daisy-chaining period is 1. The model is evaluated using five fold cross-validation: for each fold configuration, each client receives one fold as its local dataset and the remaining fold is used as a global test set. SGD is used as optimizer and as loss function, the L1 loss is applied. The L1 loss for a sample j is computed as $l_j = |X_j^t - X_j^p|$, where X_j^t is the target graphs and X_j^p is the predicted graph.

Evaluation and Comparison Methods. We compare the proposed method against three other federated learning algorithms: FedAvg [13], FedDyn [15] and FedDC [27]. In the standalone setting, each client is trained using only its local dataset. As evaluation metric, the Mean Absolute Error (MAE) is applied. The comparison to the baselines is shown in Fig. 2 (see Supplementary Table 1 for detailed values). It can be observed that the federated learning algorithms, FedAvg [13] and FedDyn [15], improve the performance of the models compared to the standalone version. Therefore, by using this approach, clients can benefit from training using a larger amount of data without having to transfer their local datasets to a centralized location. FedDyn achieves almost the same performance as FedAvg. This can be explained by the fact that the data is independent and identically distributed (IID) in our setting, therefore FedDyn, which was developed to overcome the issue of non-IID data, cannot enhance the method further. The proposed method, $RepFL^2$, outperforms all baselines in terms of MAE and achieves the smallest standard deviation across folds as well.

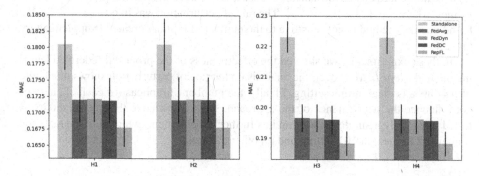

Fig. 2. Comparison against baselines - MAE mean and standard deviation across five folds. The rate of perturbation in RepFL is $p = 30\%$. Clients H1 and H2 have as super-resolution GNN model G_1, while H3 and H4 have architecture G_2.

Analysis of the Rate of Perturbation. Furthermore, we studied how sensitive the method is with respect to the number of perturbed samples on the replicas. In this experiment, all replicas had the same rate of perturbation $p \in \{5\%, 10\%, 15\%, 20\%, 25\%, 30\%, 35\%, 40\%\}$. The number of replicas was fixed to $r = 5$ across all anchors. The results can be seen in Fig. 3. For both architectures, as the number of perturbed samples increases, the MAE decreases up to one point. While for the first architecture (G_1), the

---

2 https://github.com/basiralab/RepFL.

MAE starts to become stable after 35%, when analysing the second architecture (G_2), the MAE increases starting at 35%. Therefore, the best perturbation rate across both architectures is $p = 30\%$.

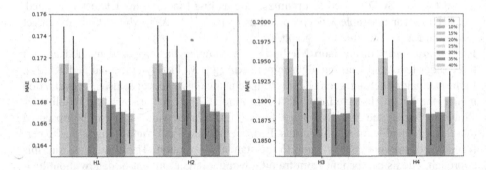

Fig. 3. Analysis of the rate of anchor training data perturbation - MAE mean and standard deviation across five folds. Clients H1 and H2 have as model G_1, while H3 and H4 have architecture G_2. The number of replicas is set to $r = 5$.

Analysis of the Number of Replicas. Additionally, we studied how the number of replicas influences the performance. In this experiment, all anchors have been replicated r times, where $r \in \{2, 5, 10, 15, 20\}$. The rate of perturbation was fixed to $p = 30\%$. We found that the method is not sensitive to the number of replicas created (Supplementary Fig. 2).

In these experiments, we showed the effectiveness of the proposed federated learning method, *RepFL*. Moreover, the analysis on the rate of perturbation shows that a rate of $p = 30\%$ is ideal in this setting. Finally, the number of replicas for each client does not influence the performance of the approach. We acknowledge that the method was tested using only a single dataset, and as further development, we plan to evaluate the framework in the non-IID setting and on other datasets.

4 Conclusion

We proposed *RepFL*, a replica-based federated learning framework for heterogeneous generative model architectures. *RepFL* replicates each client by copying its model architecture and perturbing its local dataset in order to create a larger amount of clients having diverse data distributions, and therefore, enhance the generalisability of the global model. This solution is particularly suitable in the context where there is a small amount of clients having limited data. In our experimental results, we applied the method for brain graph super-resolution using two different model architectures. To the best of our knowledge, this is the first study which focuses on graph super-resolution in the federated learning setting. The empirical evaluation demonstrated that the framework outperforms the comparison methods. While replicating the client models and perturbing their data distributions enables learning from small datasets and a limited number of clients, it leads to additional computational costs for hospitals, which remains a topic of further investigation. A future direction lies in leveraging dynamic weights when aggregating anchors and replicas, which could enhance the proposed method.

References

1. Fornito, A., Zalesky, A., Bullmore, E.: Fundamentals of Brain Network Analysis. Academic Press (2016)
2. Bessadok, A., Mahjoub, M.A., Rekik, I.: Graph neural networks in network neuroscience. IEEE Trans. Pattern Anal. Mach. Intell. **45**, 5833–5848 (2022)
3. Zhou, J., et al.: Graph neural networks: a review of methods and applications. AI Open **1**, 57–81 (2020)
4. Veličković, P.: Everything is connected: graph neural networks. Curr. Opin. Struct. Biol. **79**, 102538 (2023)
5. Kairouz, P., et al.: Advances and open problems in federated learning. Found. Trends® Mach. Learn. **14**, 1–210 (2021)
6. Pfitzner, B., Steckhan, N., Arnrich, B.: Federated learning in a medical context: a systematic literature review. ACM Trans. Internet Technol. (TOIT) **21**, 1–31 (2021)
7. Isallari, M., Rekik, I.: GSR-net: Graph super-resolution network for predicting high-resolution from low-resolution functional brain connectomes. In: Liu, M., Yan, P., Lian, C., Cao, X. (eds.) MLMI 2020. LNCS, vol. 12436, pp. 139–149. Springer, Cham (2020). https://doi.org/10.1007/978-3-030-59861-7_15
8. Mhiri, I., Nebli, A., Mahjoub, M.A., Rekik, I.: Non-isomorphic inter-modality graph alignment and synthesis for holistic brain mapping. In: Feragen, A., Sommer, S., Schnabel, J., Nielsen, M. (eds.) IPMI 2021. LNCS, vol. 12729, pp. 203–215. Springer, Cham (2021). https://doi.org/10.1007/978-3-030-78191-0_16
9. Mhiri, I., Mahjoub, M.A., Rekik, I.: Stairwaygraphnet for inter-and intra-modality multi-resolution brain graph alignment and synthesis. In: Machine Learning in Medical Imaging: 12th International Workshop, MLMI 2021, Held in Conjunction with MICCAI 2021, Strasbourg, 27 September 2021, pp. 140–150. Springer, Cham (2021)
10. Pala, F., Mhiri, I., Rekik, I.: Template-based inter-modality super-resolution of brain connectivity. In: Rekik, I., Adeli, E., Park, S.H., Schnabel, J. (eds.) PRIME 2021. LNCS, vol. 12928, pp. 70–82. Springer, Cham (2021). https://doi.org/10.1007/978-3-030-87602-9_7
11. Chaari, N., Akdağ, H.C., Rekik, I.: Comparative survey of multigraph integration methods for holistic brain connectivity mapping. Med. Image Anal. 102741 (2023)
12. Demir, B., Bessadok, A., Rekik, I.: Inter-domain alignment for predicting high-resolution brain networks using teacher-student learning. In: Albarqouni, S., et al. (eds.) DART/FAIR - 2021. LNCS, vol. 12968, pp. 203–215. Springer, Cham (2021). https://doi.org/10.1007/978-3-030-87722-4_19
13. McMahan, B., Moore, E., Ramage, D., Hampson, S., Arcas, B.A.: Communication-efficient learning of deep networks from decentralized data. In: Artificial Intelligence and Statistics, PMLR, pp. 1273–1282 (2017)
14. Li, T., Sahu, A.K., Zaheer, M., Sanjabi, M., Talwalkar, A., Smith, V.: Federated optimization in heterogeneous networks. Proc. Mach. Learn. Syst. **2**, 429–450 (2020)
15. Acar, D.A.E., Zhao, Y., Matas, R., Mattina, M., Whatmough, P., Saligrama, V.: Federated learning based on dynamic regularization. In: International Conference on Learning Representations (2020)
16. Kulkarni, V., Kulkarni, M., Pant, A.: Survey of personalization techniques for federated learning. In: 2020 Fourth World Conference on Smart Trends in Systems, Security and Sustainability (WorldS4), pp. 794–797. IEEE (2020)

17. Hanzely, F., Richtárik, P.: Federated learning of a mixture of global and local models. arXiv preprint arXiv:2002.05516 (2020)
18. Deng, Y., Kamani, M.M., Mahdavi, M.: Adaptive personalized federated learning. arXiv preprint arXiv:2003.13461 (2020)
19. Sattler, F., Müller, K.R., Samek, W.: Clustered federated learning: model-agnostic distributed multitask optimization under privacy constraints. IEEE Trans. Neural Netw. Learn. Syst. **32**, 3710–3722 (2020)
20. Ghosh, A., Chung, J., Yin, D., Ramchandran, K.: An efficient framework for clustered federated learning. Adv. Neural Inf. Process. Syst. **33**, 19586–19597 (2020)
21. Arivazhagan, M.G., Aggarwal, V., Singh, A.K., Choudhary, S.: Federated learning with personalization layers. arXiv preprint arXiv:1912.00818 (2019)
22. Diao, E., Ding, J., Tarokh, V.: Heterofl: Computation and communication efficient federated learning for heterogeneous clients. In: International Conference on Learning Representations (2020)
23. Shamsian, A., Navon, A., Fetaya, E., Chechik, G.: Personalized federated learning using hypernetworks. In: International Conference on Machine Learning, PMLR, pp. 9489–9502 (2021)
24. Litany, O., Maron, H., Acuna, D., Kautz, J., Chechik, G., Fidler, S.: Federated learning with heterogeneous architectures using graph hypernetworks. arXiv preprint arXiv:2201.08459 (2022)
25. Li, D., Wang, J.: Fedmd: heterogenous federated learning via model distillation. arXiv preprint arXiv:1910.03581 (2019)
26. Wang, S., Xie, J., Lu, M., Xiong, N.N.: Fedgraph-kd: an effective federated graph learning scheme based on knowledge distillation. In: 2023 IEEE 9th International Conference on Big Data Security on Cloud (BigDataSecurity), IEEE International Conference on High Performance and Smart Computing, (HPSC) and IEEE International Conference on Intelligent Data and Security (IDS), pp. 130–134. IEEE (2023)
27. Kamp, M., Fischer, J., Vreeken, J.: Federated learning from small datasets. In: The Eleventh International Conference on Learning Representations (2022)
28. Simonovsky, M., Komodakis, N.: Dynamic edge-conditioned filters in convolutional neural networks on graphs. In: Proceedings of the IEEE Conference on Computer Vision and Pattern Recognition, pp. 3693–3702 (2017)
29. Liu, W., et al.: Longitudinal test-retest neuroimaging data from healthy young adults in Southwest China. Sci. Data **4**, 1–9 (2017)
30. Dosenbach, N.U., et al.: Prediction of individual brain maturity using FMRI. Science **329**, 1358–1361 (2010)
31. Shen, X., Tokoglu, F., Papademetris, X., Constable, R.: Groupwise whole-brain parcellation from resting-state FMRI data for network node identification. Neuroimage **82**, 403–415 (2013)

A Multitask Deep Learning Model for Voxel-Level Brain Age Estimation

Neha Gianchandani[1,3]($\boxtimes$) (iD), Johanna Ospel[4,5] (iD), Ethan MacDonald[1,2,3,4] (iD),
and Roberto Souza[2,3] (iD)

[1] Department of Biomedical Engineering, University of Calgary,
Calgary, AB, Canada
neha.gianchandani@ucalgary.ca
[2] Department of Electrical and Software Engineering, University of Calgary,
Calgary, AB, Canada
[3] Hotchkiss Brain Institute, University of Calgary, Calgary, AB, Canada
[4] Department of Radiology, University of Calgary, Calgary, AB, Canada
[5] Department of Clinical Neurosciences, University of Calgary, Calgary, AB, Canada

Abstract. Global brain age estimation has been used as an effective biomarker to study the correlation between brain aging and neurological disorders. However, it fails to provide spatial information on the brain aging process. Voxel-level brain age estimation can give insights into how different regions of the brain age in a diseased versus healthy brain. We propose a multitask deep-learning-based model that predicts voxel-level brain age with a Mean Absolute Error (MAE) of 5.30 years on our test set (n=50) and 6.92 years on an independent test set (n = 359). The results of our model outperformed a recently proposed voxel-level age prediction model. The source code and pre-trained models will be made publicly available to make our research reproducible.

Keywords: Brain Age · Deep Learning · Voxel-level predictions

1 Introduction

Brain age can be thought of as a hypothetical index that reflects the maturity level of the brain. The brain age gap is computed as the predicted and chronological brain age difference. For brain age prediction studies, it is assumed that for presumed healthy subjects the brain age is equal to the chronological age. Previous works suggest that the brain age gap can be correlated with the possible onset and presence of neurological disorders in individuals [8,19,29]. Often a positive brain age gap is seen in subjects with neurological disorders, meaning the brain tends to follow an accelerated aging pattern for subjects with underlying neurological disorders [8,10]. Neurological disorders are often correlated with atrophy in specific regions of the brain. For example, Alzheimer's is associated

Supplementary Information The online version contains supplementary material available at https://doi.org/10.1007/978-3-031-45676-3_29.

with atrophy in the hippocampus and temporal lobe [23], and Parkinson's with atrophy in the basal ganglia [6]. Regional estimation of brain age can prove to be useful in predicting the early onset of specific neurological disorders.

Early works on brain age prediction were global in nature. They used traditional machine learning techniques combined with engineered features to obtain an age prediction. Such techniques, despite achieving good prediction results (Mean Absolute Error (MAE) ~3–5 years [2,13,20]), have two limitations: 1) Extraction of manual features is time-consuming. 2) Manual selection of relevant features for the task can often lead to missing out on crucial information.

To overcome these limitations, the focus of brain age prediction studies shifted to deep learning (DL) techniques that enable automatic feature extraction. DL has proven powerful, learning rich and relevant features producing state-of-the-art results with MAE as low as 2–4 years [15,18]. Global brain age is an effective biomarker to study the maturation process of the brain, however, it lacks spatial information on regional aging processes [9].

To achieve more fine-grained information about the mechanisms of aging, studies have attempted to use patches or blocks to predict brain age [4,5], where individual predictions are assigned to each block of the brain. They achieve results in the MAE ~ 1.5–2 years range while acknowledging the future scope of improvement in obtaining more fine-grained results. Voxel-level brain age can be predicted, leading to a localized analysis of how various brain regions are aging in healthy and diseased subjects. There has been limited research effort for voxel-level brain age predictions. The model proposed in [9] predicts voxel-level brain age using a U-Net-type neural network [25]. They achieved an MAE of 9.94 ± 1.73 years, which is higher than the state-of-the-art global predictions but provides additional insights into the regional aging process as compared to other studies with lower MAEs. This study is the first and the only study (as per our knowledge at the time of writing) that attempts voxel-level brain age prediction and hence will be the baseline for comparison in this article.

Our contribution in this work is the proposal and initial validation of a DL model for voxel-level brain age prediction that outperforms the other existing method [9]. The proposed model is a multitask version of the U-net and we perform an ablation study to observe the impact of adding additional tasks to the architecture. The model receives as input T1-weighted magnetic resonance (MR) images and produces three outputs: Brain tissue segmentation masks (Gray Matter (GM), White Matter (WM), and Cerebrospinal Fluid (CSF)), voxel-level brain age, and global level brain age.

2   Materials and Methods

2.1   Data

We used 3D T1-weighted MR scans from the publicly available Cambridge Centre for Ageing Neuroscience (Cam-CAN) study [28] (https://camcan-archive.mrc-cbu.cam.ac.uk/dataaccess/) acquired on a Siemens 3T scanner. We will refer to this dataset as D_{cm} hereafter in this article. All data corresponds to presumed healthy subjects. D_{cm} has 651 samples aged 18–88 years (mean age of 53.47 ± 7.84) with a male:female sex ratio of 55%:45%. The data is uniformly distributed to ensure that the model is exposed to a balanced number of samples across all age ranges to avoid any biases.

We also sourced data (n = 359 samples) corresponding to presumed healthy subjects from the publicly available Calgary-Campinas study [27] (D_{cc}) with an age range of 36–69 years with a mean age of 53.46 ± 9.72 years and male:female sex ratio of 49%:51% to use as an external test set for the multitask model. D_{cc} includes MR scans acquired on three different scanner vendors (Philips, Siemens, and GE) at two magnetic field strengths (1.5T and 3T). We also utilize this dataset to train two 3D Residual U-Net [17] models for the skull stripping and brain tissue segmentation (into GM, WM, and CSF) tasks respectively. The trained models are used to obtain skull stripping and tissue segmentation masks (pseudo-labels) for the D_{cm} dataset for the proposed voxel-level age prediction model.

2.2 Baseline

We reproduced baseline results [9] on the D_{cm} dataset for fair comparison. Input GM and WM segmentation masks to the model were obtained using the Statistical Parametric Mapping (SPM12) software [22] and then registered to the MNI152 template using the Dartel algorithm [1]. The authors provide open-source code which enabled the replication of their pre-processing and training methodology.

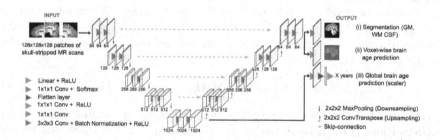

Fig. 1. The proposed multitask model used for voxel-level brain age prediction.

2.3 Proposed Model

Neural Network Architecture. The proposed model follows the 3D U-Net architecture depicted in Fig. 1. The model has three different outputs. First, a segmentation output to segment GM, WM, CSF, and background in the brain structure. The segmentation output head consists of four $1 \times 1 \times 1$ convolutions followed by a softmax activation that outputs class probabilities.

The second output is the voxel-level brain age prediction. Resolution is maintained at the output to obtain predictions at the level of each voxel in the input image. This is achieved using one $1 \times 1 \times 1$ convolution that reduces the channel dimensionality to a single channel at the output. A Rectified Linear Unit (ReLU) activation is then applied to obtain positive age predictions.

The third output is the global-level brain age prediction, computed from the bottleneck of the network. This is done using a $1 \times 1 \times 1$ convolution, followed by a flatten layer to convert the multi-dimensional feature map to a 1D map, followed by a ReLU activation to compute the global age prediction.

The global age and segmentation outputs are primarily added as additional tasks to assist with the voxel-level age prediction task. Global-level brain age and segmentation tasks can be thought of as the preliminary steps that help the model learn appropriate features from the input to predict voxel-level brain age.

Loss Function. To accommodate multitask learning, our loss function consists of three terms. The segmentation task utilizes the Dice loss ($Dice_{loss}$), which is inversely related to the Dice coefficient. The Dice Coefficient measures the overlap between the ground truth and predicted segmentation. Hence, as the model trains to minimize the loss, the Dice Coefficient consequently improves, making the predictions more accurate.

We use MAE as the loss function at two levels to accommodate voxel level and global level predictions (as done in previous works [3,21]). Voxel-level MAE is first averaged across all brain voxels in the input, followed by batch average, whereas global-level MAE is averaged over the batch. The ground truth for the voxel-level brain age prediction task is based on the underlying assumption that for presumed healthy subjects, brain age is equal to chronological age. To ensure that the model does not simply learn to predict global age as the voxel-level age for all brain voxels, we add noise to the ground truth. A random value in the range $[-2, 2]$ is added to all brain voxels before voxel-level loss computation to ensure the model learns voxel-level brain age robustly without significantly impacting the MAE.

$$L_{overall} = \alpha \cdot Dice_{loss} + \beta \cdot MAE_{global} + \gamma \cdot MAE_{voxel} \tag{1}$$

The weighted sum of the three loss function terms is calculated to obtain the overall loss (Eq. 1). α, β, and γ are the weights for the three loss terms. The three loss terms have different ranges of values, with the Dice Loss ranging between 0 and 1 and MAE ranging from 0 to ∞. To accommodate the different ranges, α decreases and β, and γ increase as the training progresses. The weights are empirically altered during the training process for optimized results. Initial values, along with changes made to the values of α, β, and γ during training, can be found in our source code.

2.4 Ablation Study

We chose to utilize the multitask learning paradigm to train the voxel-level age prediction model, with two additional tasks, global-age prediction and brain tissue (GM, WM, and CSF) segmentation. We perform an ablation study to verify the impact of using a multitask architecture instead of a traditional DL model trained for a single task. Multiple experiments are created following the proposed model architecture, the differentiating factor being the number of output tasks. We start with a single output voxel-level brain age prediction model, iteratively adding the two additional tasks, eventually matching the proposed model architecture with three tasks that the model learns features for simultaneously.

2.5 Experimental Setup

All models (baseline, ablation study models, and proposed model) are trained on 553 samples (85% of D_{cm} dataset) and tested on 50 samples from D_{cm}, and 359 samples from D_{cc} dataset for fair comparison of results. For the baseline, we use the hyperparameters suggested in [9]. No additional hyperparameter tuning is performed for model training.

All MR scans are reoriented to the MNI152 template [12] using the FSL [16] utility 'reorient2std' to ensure consistency across the dataset. We use the Adam optimizer with beta values = (0.5, 0.999) for the proposed model and default betas for the ablation experiment models. The loss function (Eq. 1) is used to train the proposed model, and the weights (α, β and γ) are altered to train the ablation experiment models such that only the loss terms relevant to the output tasks of the model are considered. The initialization of the loss function weights is described in the Supplementary Material. The models are trained for 300 epochs with an initial learning rate of 1e−3 and a 'step' scheduler which decreases the learning rate by a multiplicative factor of 0.6 every 70 epochs.

Additionally, we apply random rotation as an augmentation step on the MR samples and use cropped blocks of size $128 \times 128 \times 128$ (1 crop per input image) as the model input. Random crops are done such that the majority of samples have significant brain regions ensuring the model learns relevant features. The data ingestion and pre-processing pipeline are implemented using MONAI [7], and the proposed model is implemented using PyTorch.

3 Results

The test results are described in Table 1. We do not consider the model's segmentation and global age prediction output for quantitative comparison with the baseline as they are helper tasks to assist the model in learning rich features for the main task (i.e. voxel-level age predictions).

Model evaluation is done on n = 50 samples from D_{cm} dataset (internal test set). We also validate all models on the D_{cc} dataset (which is multi-vendor and acquired at two different magnetic field strengths) and obtain consistent results as D_{cm} dataset using the proposed multitask model confirming the robustness of our model to handle previously unseen and diverse data. The baseline failed to process 12 out of 359 samples, so we reported results for 347 successful samples. It is evident from Table 1, that the 3-task proposed model outperforms the 1-task and 2-task model with a 47% and 9% improvement in MAE observed on

Table 1. Model performance (MAE±S.D.) on an internal and external test set.

Model (output tasks)	D_{cm} (n = 50)	D_{cc} (n = 359)
Baseline (G+V) [9]	8.84 ± 4.82	16.74 ± 3.71
1 output model (V)	10.11 ± 5.68	7.63 ± 4.53
2 output model (G+V)	7.90 ± 4.3	7.93 ± 4.73
2 output model (S+V)	6.75 ± 3.94	7.84 ± 4.73
Proposed model (S+G+V)	$\mathbf{5.30 \pm 3.29^{*}}$	$\mathbf{6.92 \pm 4.28^{*}}$

Abbreviations: V - voxel-level brain age prediction task, S - segmentation task (GM, WM, CSF), G - global-level brain age prediction task, *- p<0.05

the D_{cm} and D_{cc} test results respectively on iteratively adding additional tasks with the voxel-level age prediction task. This shows that a multitask approach with added segmentation and global brain age prediction task is beneficial for efficient feature extraction for the voxel-level brain age prediction task. The Wilcoxon-Signed Rank test [30] was done to compare the pair-wise performance of the proposed model against the 1-output and 2-output models as well as the baseline with a Holm-Bonferroni correction [14] with an initial $\alpha = 0.05$ to account for multiple comparisons. All p-values obtained were <0.05 indicating statistical significance along with the proposed model outperforming the ablation experiment models as well as the baseline.

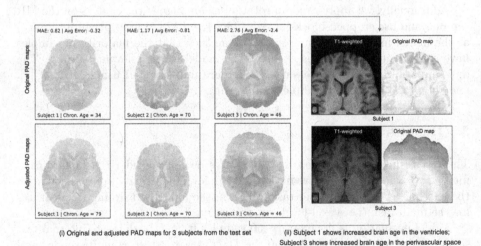

(i) Original and adjusted PAD maps for 3 subjects from the test set

(ii) Subject 1 shows increased brain age in the ventricles; Subject 3 shows increased brain age in the perivascular space

Fig. 2. (i) Row 1 - PAD maps based on the voxel-level difference between chronological and predicted age, Row 2 - adjusted PAD maps by subtracting the overall MAE of the brain volume from each voxel PAD value. (ii) T1-weighted MR images and corresponding original PAD maps showing increased brain age in the ventricles in Subject 1 (top) and in the basal ganglia in Subject 3 (bottom).

4   Discussion

The proposed framework offers improved MAE over the baseline and simplified reproducibility without extensive preprocessing. The performance of the baseline using GM and WM masks as input (obtained using the SPM12 software) is partially dependent on the quality of segmentation done by the software. To avoid the loss of any useful structural information relevant to the age prediction task, we use skull-stripped T1-weighted MR images as input. This makes the proposed framework easier to implement and reproduce. We also test our proposed model on the D_{cc} dataset which includes samples from 3 different scanners (Philips, General Electric and Siemens), each at two different magnetic field strengths (1.5T and 3T). Our model has comparable performance to the internal D_{cm} test set and improved performance over the baseline which reflects on the model's ability to effectively work on a diverse dataset confirming robustness.

Figure 2 (i) shows D_{cm} test set results using PAD maps. The first row shows the raw PAD maps where each pixel's intensity is the difference between the chronological age and the predicted brain age for that voxel. The second-row shows adjusted PAD maps where each voxel intensity is the difference between the prediction error at the level of each voxel and the MAE of the brain volume. Subtracting the overall MAE of the brain volume helps in normalizing the values by highlighting the relative variations in the predicted age difference across different voxels (or regions of the brain). Additionally, the adjustment can also help reduce the impact of the model error (MAE) while visualizing the brain PAD maps, allowing us to concentrate on the spatial variations in the predictions made by the model. A variation in PAD values is seen across regions of the brain, with most values centered around 0 (Subject 1–3). We also had outliers in our test set, but despite the differences in PAD values across samples, adjusted PAD maps in the second row show similar faint contrast patterns across different regions of the brain for all subjects highlighting small variations in the healthy brain aging process.

We add noise to the ground truths to ensure the model learns spatial variations in brain age (discussed in Sect. 2.3). Despite the addition of a random noise to each voxel, we observe gradual variations in PAD values, rather than sharp differences at the voxel-level, which indicates the model's ability to learn age specific features at a region level. This figure supports our hypothesis and findings in previous works [24, 26] that regions of the brain age differently and follow different aging patterns. This is reflected by variations in PAD values across different regions of the brain.

The PAD maps obtained were reviewed by a radiologist to evaluate the clinical significance of our findings. Although a pattern was hard to decipher across the test set in terms of regional aging processes in PAD masks of healthy subjects, a notable observation was the contribution of CSF (within the ventricles) to the brain age predictions with the ventricles showing an increased brain age in Subject 1 (Fig. 2 (ii)). This indicates the contribution of brain atrophy in predicting brain age. In Subject 3 (Fig. 2 (ii)), a large perivascular space was observed with dark spots in the Basal ganglia in the T1-weighted image which is generally considered a risk factor for Alzheimer's [11]. The corresponding PAD map showed red voxels (increased brain age) in the region confirming the presence of abnormalities. This confirms the model's ability to identify regions with underlying structural changes that can correlate to the possible onset of neurological disorders in the future.

The voxel-level approach in brain age prediction studies is still new, and the proposed framework is an attempt to improve the results over the existing studies as well as to assess the clinical importance of voxel-level brain PAD maps. Despite achieving promising results, there is still room for improvement. First, a larger dataset can be utilized for training to capture variability and structural uniqueness in MR samples. We utilize non-registered MR scans as input to our model to capture the structural uniqueness of each brain during the feature learning process and show that variation is the brain aging process across different regions of the brain is observed in the PAD maps. Further, inference on samples from disease groups like Alzheimer's, Parkinsons', and other neurological disorders can be done to verify if the results resonate with other aging studies. Additionally, the PAD maps can be further used to obtain average brain age predictions for specific anatomical regions in the brain for easier correlation to existing aging studies since most aging studies are done at a regional level rather than a voxel-level.

5   Conclusion

We proposed a voxel-level brain age prediction technique using a DL model. We show an improved performance over the state-of-the-art with an MAE of 5.30 years and 6.92 years on two different test sets using a multitask approach where additional tasks help the model learn proper features. This work is a feasibility study where preliminary results outperform state-of-the-art baseline to show that voxel-level PAD maps can provide an enhanced understanding of regional brain aging processes. Future experiments are needed to verify the correlation to various neurological disorders.

Acknowledgements. We would like to extend our thanks to Alberta Innovates, Hotchkiss Brain Institute at the University of Calgary and the Natural Sciences and Engineering Research Council of Canada (NSERC) for funding this research. We are also grateful for the computational data resources provided by the Digital Research Alliance of Canada and DENVR Dataworks.

References

1. Ashburner, J.: A fast diffeomorphic image registration algorithm. Neuroimage **38**(1), 95–113 (2007)
2. Baecker, L., et al.: Brain age prediction: a comparison between machine learning models using region-and voxel-based morphometric data. Hum. Brain Mapp. **42**(8), 2332–2346 (2021)

3. Beheshti, I., Ganaie, M., Paliwal, V., Rastogi, A., Razzak, I., Tanveer, M.: Predicting brain age using machine learning algorithms: a comprehensive evaluation. IEEE J. Biomed. Health Inform. **26**(4), 1432–1440 (2021)
4. Beheshti, I., Gravel, P., Potvin, O., Dieumegarde, L., Duchesne, S.: A novel patch-based procedure for estimating brain age across adulthood. Neuroimage **197**, 618–624 (2019)
5. Bintsi, K.-M., Baltatzis, V., Kolbeinsson, A., Hammers, A., Rueckert, D.: Patch-based brain age estimation from MR images. In: Kia, S.M., et al. (eds.) MLCN/RNO-AI -2020. LNCS, vol. 12449, pp. 98–107. Springer, Cham (2020). https://doi.org/10.1007/978-3-030-66843-3_10
6. Caligiore, D., et al.: Parkinson's disease as a system-level disorder. NPJ Parkinson's Disease **2**(1), 1–9 (2016)
7. Cardoso, M.J., et al.: MONAI: an open-source framework for deep learning in healthcare. arXiv preprint arXiv:2211.02701 (2022)
8. Cole, J.H., Franke, K.: Predicting age using neuroimaging: innovative brain ageing biomarkers. Trends Neurosci. **40**(12), 681–690 (2017)
9. Cole, J.H., Popescu, S.G., Glocker, B., Sharp, D.J.: Local brain-age: a U-Net model. Front. Aging Neurosci. 838 (2021)
10. Davatzikos, C., Xu, F., An, Y., Fan, Y., Resnick, S.M.: Longitudinal progression of Alzheimer's-like patterns of atrophy in normal older adults: the SPARE-AD index. Brain **132**(8), 2026–2035 (2009)
11. Ding, J., et al.: Large perivascular spaces visible on magnetic resonance imaging, cerebral small vessel disease progression, and risk of dementia: the age, gene/environment susceptibility-reykjavik study. JAMA Neurol. **74**(9), 1105–1112 (2017)
12. Fonov, V.S., Evans, A.C., McKinstry, R.C., Almli, C., Collins, D.: Unbiased nonlinear average age-appropriate brain templates from birth to adulthood. Neuroimage **47**, S102 (2009)
13. Franke, K., Ziegler, G., Klöppel, S., Gaser, C., Initiative, A.D.N., et al.: Estimating the age of healthy subjects from T1-weighted MRI scans using kernel methods: exploring the influence of various parameters. Neuroimage **50**(3), 883–892 (2010)
14. Holm, S.: A simple sequentially rejective multiple test procedure. Scandinavian J. Stat. 65–70 (1979)
15. Ito, K., et al.: Performance evaluation of age estimation from T1-weighted images using brain local features and CNN. In: IEEE Engineering in Medicine and Biology Society (EMBC), pp. 694–697. IEEE (2018)
16. Jenkinson, M., Beckmann, C.F., Behrens, T.E., Woolrich, M.W., Smith, S.M.: FSL. Neuroimage **62**(2), 782–790 (2012)
17. Kerfoot, E., Clough, J., Oksuz, I., Lee, J., King, A.P., Schnabel, J.A.: Left-ventricle quantification using residual U-net. In: Pop, M., et al. (eds.) STACOM 2018. LNCS, vol. 11395, pp. 371–380. Springer, Cham (2019). https://doi.org/10.1007/978-3-030-12029-0_40
18. Kolbeinsson, A., et al.: Accelerated MRI-predicted brain ageing and its associations with cardiometabolic and brain disorders. Sci. Rep. **10**(1), 1–9 (2020)
19. Koutsouleris, N., et al.: Accelerated brain aging in Schizophrenia and beyond: a neuroanatomical marker of psychiatric disorders. Schizophr. Bull. **40**(5), 1140–1153 (2014)
20. de Lange, A.M.G., et al.: Mind the gap: performance metric evaluation in brain-age prediction. Hum. Brain Mapp. **43**(10), 3113–3129 (2022)
21. MacDonald, M.E., Pike, G.B.: MRI of healthy brain aging: a review. NMR Biomed. **34**(9), e4564 (2021)

22. Penny, W.D., Friston, K.J., Ashburner, J.T., Kiebel, S.J., Nichols, T.E.: Statistical Parametric Mapping: The Analysis of Functional Brain Images. Elsevier, Amsterdam (2011)

23. Rao, Y.L., Ganaraja, B., Murlimanju, B., Joy, T., Krishnamurthy, A., Agrawal, A.: Hippocampus and its involvement in Alzheimer's disease: a review. 3 Biotech **12**(2), 55 (2022)

24. Raz, N., et al.: Regional brain changes in aging healthy adults: general trends, individual differences and modifiers. Cereb. Cortex **15**(11), 1676–1689 (2005)

25. Ronneberger, O., Fischer, P., Brox, T.: U-net: convolutional networks for biomedical image segmentation. In: Navab, N., Hornegger, J., Wells, W.M., Frangi, A.F. (eds.) MICCAI 2015. LNCS, vol. 9351, pp. 234–241. Springer, Cham (2015). https://doi.org/10.1007/978-3-319-24574-4_28

26. Scahill, R.I., Frost, C., Jenkins, R., Whitwell, J.L., Rossor, M.N., Fox, N.C.: A longitudinal study of brain volume changes in normal aging using serial registered magnetic resonance imaging. Arch. Neurol. **60**(7), 989–994 (2003)

27. Souza, R., et al.: An open, multi-vendor, multi-field-strength brain MR dataset and analysis of publicly available skull stripping methods agreement. Neuroimage **170**, 482–494 (2018)

28. Taylor, J.R., et al.: The Cambridge centre for ageing and neuroscience (Cam-CAN) data repository: structural and functional MRI, MEG, and cognitive data from a cross-sectional adult lifespan sample. Neuroimage **144**, 262–269 (2017)

29. Wang, J., et al.: Gray matter age prediction as a biomarker for risk of dementia. Proc. Natl. Acad. Sci. **116**(42), 21213–21218 (2019)

30. Wilcoxon, F.: Individual comparisons by ranking methods. In: Kotz, S., Johnson, N.L. (eds.) Breakthroughs in Statistics. Springer Series in Statistics, pp. 196–202. Springer, New York (1992). https://doi.org/10.1007/978-1-4612-4380-9_16

Deep Nearest Neighbors for Anomaly Detection in Chest X-Rays

Xixi Liu$^{(\boxtimes)}$ⓘ, Jennifer Alvénⓘ, Ida Häggströmⓘ, and Christopher Zachⓘ

Chalmers University of Technology, Gothenburg, Sweden
{xixil,alven,idah,zach}@chalmers.se

Abstract. Identifying medically abnormal images is crucial to the diagnosis procedure in medical imaging. Due to the scarcity of annotated abnormal images, most reconstruction-based approaches for anomaly detection are trained only with normal images. At test time, images with large reconstruction errors are declared abnormal. In this work, we propose a novel feature-based method for anomaly detection in chest x-rays in a setting where only normal images are provided during training. The model consists of lightweight adaptor and predictor networks on top of a pre-trained feature extractor. The parameters of the pre-trained feature extractor are frozen, and training only involves fine-tuning the proposed adaptor and predictor layers using Siamese representation learning. During inference, multiple augmentations are applied to the test image, and our proposed anomaly score is simply the geometric mean of the k-nearest neighbor distances between the augmented test image features and the training image features. Our method achieves state-of-the-art results on two challenging benchmark datasets, the RSNA Pneumonia Detection Challenge dataset, and the VinBigData Chest X-ray Abnormalities Detection dataset. Furthermore, we empirically show that our method is robust to different amounts of anomalies among the normal images in the training dataset. The code is available at: https://github.com/XixiLiu95/deep-kNN-anomaly-detection.

Keywords: Anomaly detection · Siamese representation learning · k-Nearest neighbor

1 Introduction

Chest X-rays (CXRs) are commonly considered the main imaging study for the evaluation of many conditions because of their cost-effectiveness, low radiation dose, and versatility as a diagnostic tool [22]. Deep learning image analysis methods, with fast inference and high accuracy, can help improve the efficiency of image evaluation and the diagnostic accuracy as well as reduce the workload for radiologists [9]. However, for such methods to be safe and reliable, we need robust methods for detecting anomalies in the input data. Consequently,

Supplementary Information The online version contains supplementary material available at https://doi.org/10.1007/978-3-031-45676-3_30.

anomaly detetcion has been extensively studied and is an important sub-routine in many computer-aided diagnosis methods [9]. A recent paper on anomaly detection in medical images summarizes several scenarios requiring anomaly detection including but not limited to rejecting inputs that are incorrectly prepared (e.g. blurry images, poor contrast, and incorrect view) and rejecting inputs that are unseen in the training data (e.g. images with unseen diseases) [5]. In this work, we focus on the second case, where the goal is to identify medically abnormal images. However, the proposed method is general, and can with ease be applied to other scenarios (e.g. image artifacts, unusual anatomies, and artificial implants).

Anomaly Detection refers to the task of distinguishing abnormal data from normal data. In this study, our focus is on the scenario where only healthy images are accessible during the training phase. The existing methods can be roughly divided into two categories including reconstruction-based methods and self-supervised learning-based methods. Reconstruction-based methods assume that normal samples tend to produce lower reconstruction errors compared to abnormal samples. Several reconstruction-based methods are devised for anomaly detection, e.g. autoencoders (AEs) and their variants [11,12], and generative adversarial networks (GANs) such as f-AnoGAN [15]. Recently, diffusion models and their variants have gotten attention due to their powerful mode coverage over GANs [17,18] and due to the more realistic sample quality compared to variational autoencoders (VAEs). Most reconstruction-based methods rely on large amounts of normal training data, however, the authors in [4] argue that a large amount of unlabeled data containing outliers could be beneficial when learning anomaly detection. Their reconstruction-based dual distribution anomaly detection (DDAD) method utilizes the unlabeled data to learn the inter-discrepancy of two modules, where one is accessible to normal healthy images, and the other is accessible to the unlabeled images. While the DDAD method achieves impressive results, the performance is significantly correlated with the fraction of outliers in the unlabeled data. If the unlabeled data are all outliers (requiring known labels), DDAD boils down to a supervised method, which limits its wide application. All aforementioned reconstruction-based methods rely heavily on massive amounts of normal images, and in addition, require a high computational load due to the pixel-level comparison. Instead, our method is feature-based, which is free from the reconstruction of the whole image while achieves significant better performance with same amount of data.

Several works focus on devising self-supervised learning methods for anomaly detection [3,10,16]. The authors in [10] propose to train a multi-class model to discriminate between several geometric transformations applied on all the given images. The method in [16] additionally applies a set of pre-defined shifted transformations to images to create negative samples in the framework of contrastive learning. Meanwhile, a classification head is added to predict which shifting transformation is applied to the given image. However, classification-based methods require a sophisticated design of data transformations. The method in [3], inspired by [21], tries to learn the prototypical patterns of normal training

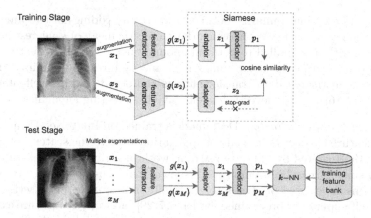

Fig. 1. The proposed pipeline. In the training stage (top), two random augmentations are applied to the input images, and feature banks are learned using the Siamese architecture. At test time (bottom) the geometric mean of the k-NN distances between augmented test embeddings and training feature banks yields the anomaly score.

samples and anomalous pattern via a memory bank and the anomaly score is calculated as a weighted combination of normal prototypical patterns. However, this method necessitates the availability of diverse augmented views of images as well as a limited number of labeled anomalous images for training. Our method replies solely on two random augmentations from the same augmentation distribution and does not require any labelled abnormal images.

Contributions. In this work, a *feature-based* method for anomaly detection is proposed, which employs the structure of a Siamese network and consists of a pre-trained backbone, an adaptor layer, and a predictor layer. A schematic overview of the method is shown in Fig. 1. Our method 1) is entirely feasible to use various pre-trained backbones, 2) achieves state-of-the-art results compared to other reconstruction-based methods [1,4,11,12,15] when only using normal images for (semi-supervised) training, 3) is robust to different amounts of abnormal images among the normal images in the training data.

2   Method

In this work, we use a similarity measure of feature embeddings to compute the anomaly score for differentiating abnormal samples from normal samples. Firstly, it utilizes a powerful pre-trained backbone. This allows us to leverage existing well-established pre-trained models (e.g. on ImageNet-1k [8]). Secondly, we only need to learn light-weight feature adaptors on the domain of interest (i.e. chest X-ray images in our case) by employing the Siamese [7] architecture in a self-supervised way, where input images are the two augmented views of the same image. Importantly, our method does not necessitate a large batch size or the use of a pair of strong and weak augmentations. The training loss is the negative cosine similarity between feature embeddings of two augmented views of the input image. At test time, the geometric mean of k-nearest neighbor (k-NN) algorithm is applied to perform anomaly detection in the feature space. The proposed method is easy to implement and requires only minimal training (e.g. less sophisticated data transformations compared to other self-supervised methods [10,16]), but is nevertheless effective in detecting anomalies.

Pre-trained Feature Extractor. Deep models trained on ImageNet-1K are widely used as feature extractors in medical applications due to the scarcity of labelled data [13,19]. Inspired by [2,6], we use a network pre-trained in a self-supervised manner, since it has been empirically shown that self-supervised pre-training outperforms supervised pre-training, especially in semi-supervised settings [6].

Any self-supervised pre-trained model [6,7,20] can work as a feature extractor in our framework. In this work, we use two of the most popular pre-trained models, ResNet-50 trained by SimCLRv2 [6] and Barlow [20] on ImageNet-1k.

Training. The pre-trained feature extractor on ImageNet-1k might preserve the general representation of natural images. To obtain a domain-specific representation of the target data (CXRs), an adaptor layer is added on top of the pre-trained feature extractor to distill the knowledge of CXRs. We use the Siamese [7] architecture to learn the representation of normal samples in a self-supervised way. The training architecture is shown in Fig. 1. An input image x is perturbed by sampling two different augmentations from the same augmentation distribution, denoted as $\mathcal{T}$, yielding x_1 and x_2. In particular, we use random crops and horizontal flips as our pool of augmentations.

In the first stage, x_1 and x_2 are processed by the pre-trained feature extractor g. The resulting features are subsequently transformed by an adaptor network f (consisting of one fully connected layer) and one set of features is additionally processed by a predictor network h (a fully connected layer). Specifically, the two resulting feature vectors are given by $p_1 = h(f(g(x_1)))$ and $z_2 = f(g(x_2))$. The training loss is the negative cosine similarity $\mathcal{D}$:

$$\mathcal{D}(p, z) = -\left\langle \frac{p}{\|p\|}, \frac{z}{\|z\|} \right\rangle, \tag{1}$$

where z and p denote features from the adaptor layer and predictor layer of the two augmentation branches, respectively. As suggested in [7], a symmetrical loss $\mathcal{L}$ is used, and the stop-gradient technique is adopted to tackle the issue of requiring a momentum encoder, negative samples, or larger batches, which are very common components in self-supervised training. The final training loss is given as follows,

$$\mathcal{L} = \tfrac{1}{2}\mathcal{D}(p_1, \text{stopgrad}(z_2)) + \tfrac{1}{2}\mathcal{D}(p_2, \text{stopgrad}(z_1)), \tag{2}$$

where the adaptor f receives only gradient information from the branch incorporating the predictor h. After the training process, the training feature bank $\mathcal{P}$ is constructed by applying multiple augmentations sourced from the augmentation distribution $\mathcal{T}$ to the training data. This is done with the aim of capturing a wider range of variations present in the training data.

Inference. At test time, an X-ray image is processed by the trained network and its feature representation from the predictor layer is used to compute an anomaly score. Specifically, we apply multiple augmentations drawn from the augmentation distribution T and obtain M augmented feature maps for the test image. To ensure the learned features are scale invariant, unit normalization is applied to the learned features before calculating the score. The score is calculated as the geometric mean of the k-NN using the Euclidean distance between the feature banks $\mathcal{P}$ and the feature extracted from the test image x^*. Specifically, let $(p_1^*, p_2^*, \cdots, p_M^*)$ be the features resulting from the multiple augmentations, then the anomaly score S is defined as

$$S(x^*) = \sqrt[M]{\|p_1^* - \mathcal{P}^{[k]}\| \cdot \|p_2^* - \mathcal{P}^{[k]}\| \cdots \|p_M^* - \mathcal{P}^{[k]}\|}, \tag{3}$$

where $\|p_i^* - \mathcal{P}^{[k]}\|$ is the Euclidean distance to the k-th NN in feature bank $\mathcal{P}$. The k-NN computation is performed independently for each augmentation.

3 Experiments

Datasets. We evaluate our method on the RSNA Pneumonia Detection Challenge dataset[1] and the VinBigData Chest X-ray Abnormalities Detection dataset[2], and we use the exact the same split between training and testing as in [4] to enable a fair comparison. The performance is evaluated by the area under the receiver operating characteristic curve (AUROC) and average precision (AP). The **RSNA dataset** includes 8,851 normal and 6,012 abnormal images. 1,000 normal and 1,000 abnormal images (i.e. lung opacity) are combined to create the RSNA test set. The **VinBigData dataset** is a much more challenging dataset, which consists of 10,601 normal and 4,394 abnormal images that cover 14 types of thoracic abnormalities (e.g. calcification and pleural thickening). The VinBigData test data includes 1,000 normal images as well as 1,000 abnormal images.

Data Augmentation. All input images are resized to $224 \times 224 \times 3$. Since the original feature extractor trained by simCLRv2 [6] and Barlow [20] uses RGB images, we create 3-channel images by duplicating the 1-channel grayscale images. In our approach, the success of training does not rely on the use of strong and weak augmentations. Instead, we utilize two augmentations from the same augmentation distribution T, i.e. random crops and random horizontal flips. During inference, the training feature bank $\mathcal{P}$ is M times size of the original training data, where M is the number of augmentations performed on each training image.

Implementation Details. The architecture of the pre-trained feature extractor is ResNet50 trained by SimCLRv2 [6] and Barlow [20] on ImageNet-1k in our

---

[1] https://www.kaggle.com/c/rsna-pneumonia-detection-challenge.
[2] https://www.kaggle.com/c/vinbigdata-chest-xray-abnormalities-detection.

Table 1. *Comparison with SOTA methods* (using $DR = 0.49$ for RSNA and $DR = 0.5$ for VinBigData). Values are AUROC and AP, where boldface indicates the best, underline indicates the second best. * represents adding the proposed adaptor layer and predictor layer.

Methods	RSNA		VinBigData	
	AUROC ↑	AP ↑	AUROC ↑	AP ↑
AE [1]	0.669	–	0.559	–
MemAE [11]	0.680	–	0.558	–
f-AnoGAN [15]	0.798	–	0.763	–
AE-U [12]	0.867	–	0.738	–
DDAD-AE-U [4]	0.873	–	0.743	–
SimCLRv2* (Ours)	<u>0.882</u>	<u>0.863</u>	**0.846**	**0.824**
Barlow* (Ours)	**0.905**	**0.908**	<u>0.809</u>	<u>0.802</u>

experiment. The adaptor of the proposed models consists of one fully connected layer with the size of ($2048 \rightarrow 1024$), and the predictor consists of one fully connected layer with the size of ($1024 \rightarrow 1024$). Our model is trained for 100 epochs using the Adam optimizer with a learning rate 10^{-5} for simCLRv2 [6], and $2 \cdot 10^{-7}$ for Barlow [20]. The number of augmentations applied is $M = 5$ and $k = 1$. All experiments were run on a single NVIDIA GeForce RTX 2080Ti, CUDA 11.2, using PyTorch 1.9.0+cu111 [14]. The inference time per image is approximately 50ms for both datasets.

3.1　Experimental Results

Comparison with SOTA Methods. We first compare our method trained on only normal images with a line of reconstruction-based methods including AE [1], MemAE [11], f-AnoGAN [15], AE-U [12], and DDAD-AE-U [4]. DDAD-AE-U refers to AE-U combined with the DDAD method. Due to the special setting of DDAD-AE-U [4], all methods are trained with partial training data, following the same dataset splits as in [4], i.e. implemented using their provided data list. Specifically, 3,851 normal images and 4,000 normal images are used as training data[3] for the RSNA dataset and the VinBigData dataset, respectively. The results in Table 1 show that our method consistently outperforms the other semi-supervised methods[4] on both datasets. In particular, by AUROC our method surpasses f-AnoGAN, which is SOTA for the more challenging VinBigData dataset, with a large margin of 8.3%. When using inter-discrepancy as a score, DDAD-AE-U is essentially a supervised method and these results are therefore not included in our comparison.

Different Training Data Amount. We explored the influence of different amounts of clean training data on our anomaly detection method's performance by

---

[3] 3951 images in RSNA corresponds to $DR = 0.49$, 4000 images in VinBigData corresponds to $DR = 0.5$.

[4] We use the results reported in [4].

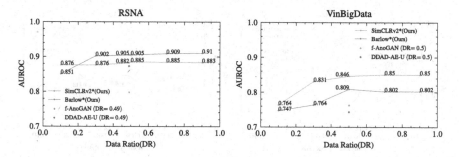

Fig. 2. The performance of our method varies with the amount of training data.

training the network with various training data ratios (DR) relative to the total amount of training data (n = X). Specifically, we selected DR values of 10%, 30%, 50%, 70%, and 90%. The results for the two datasets are shown in Fig. 2. For the RSNA dataset, even with only 30% of the training data, our method could obtain much better results (AUROC = 0.902) than DDAD-AE-U [4] with 49% of the training data (AUROC = 0.873). For the VinBigData dataset, the figure indicates that our method could achieve significantly better results even with 30% of the training data (AUROC = 0.831) compared with SOTA f-AnoGAN [15] with 50% of the training data (AUROC = 0.763). Notably, as the number of training images increases, the performance on both datasets with different backbones appears to reach a saturation point. One possible explanation for this observation is that the augmentations applied to the training/test data already encompass a wide range of variations, leaving little room for further improvement. Additional information regarding the impact of varying the number of augmentations can be found in the supplementary material.

Different Anomaly Amount. A more realistic setting is that the training data includes both normal and some amount of anomalous data, resembling unsupervised learning conditions. Hence, it is crucial that the proposed method is robust to different amounts of abnormal images in the training data, and therefore, we examine the influence of varying anomaly ratios (AR). The results can be found in Fig. 3. The results of the baseline models including f-AnoGAN and DDAD-AE-U corresponds to $AR = 0$, i.e. only known normal images in the training data. For the RSNA dataset, our method Barlow* with a 10% anomaly ratio achieves higher AUROC value than DDAD-AE-U [4] without any anomalies in the training data. Our method, regardless of the anomaly ratio, consistently obtain better results on the VinBigData compared to the DDAD-AE-U and f-AnoGAN methods trained without anomalies.

Importance of model components. We evaluate the relevance of the following three components of our method: using the pre-trained feature extractor (i.e. ResNet50 trained by simCLRv2), training with Siamese style, and the necessity of feature normalization in the feature banks $\mathcal{P}$. The impact of each component is shown in Table 2. No pre-trained means that the pre-trained backbone

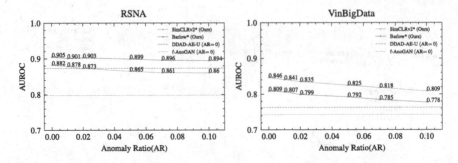

Fig. 3. The performance of our method varies with the amount of anomalies in the training data. (using $DR = 0.49$ for RSNA and $DR = 0.5$ for VinBigData.)

is replaced with a trainable encoder of the same size as the baseline AE [4], and no Siamese corresponds to directly using features extracted from the pre-trained backbone without fine-tuning on the target dataset, i.e. removing the adaptor and predictor. Finally, the impact of adding scale invariance by feature normalization to the stored features in $\mathcal{P}$ is investigated.

It is noteworthy that training the proposed layers significantly improve the performance on the VinBig dataset compared to the RSNA dataset. This could be attributed to the fact that the anomalies present in the RSNA dataset are limited to lung opacities, and the pre-trained models already possess sufficient capability to differentiate them from healthy images. Conversely, the diverse nature of anomalies in the VinBigData, making them more challenging to distinguish. Similar observations were made when using Barlow as the backbone. For additional results using Barlow [20] as the backbone can be found in supplementary material.

Table 2. The importance of different model components (using $DR = 0.5$ and $AR = 0$) for the two datasets. Values are AUROC and AP.

Backbone	Pre-trained	Siamese	Normalization	RSNA		VinBigData	
				AUROC ↑	AP ↑	AUROC ↑	AP ↑
simCLRv2	✗	✓	✓	0.689	0.651	0.678	0.666
	✓	✗	✓	0.754	0.688	0.623	0.619
	✓	✓	✗	0.798	0.776	0.813	0.800
	✓	✓	✓	**0.885**	**0.878**	**0.846**	**0.824**

4   Conclusion and Future Work

In this work, we propose a feature-based method for anomaly detection, comprising of a pre-trained backbone extended with an adaptor and a predictor layer. Our approach is simple, easy to train, and exhibits improved performance compared to reconstruction-based methods in a semi-supervised setting. The method is versatile by allowing the use of various backbones, and the suggested adaptor and predictor layers are shown to enhance anomaly detection performance. Additionally, the proposed method is able to cope with varying levels of outliers (abnormal images) in the training data, making it suitable for realistic unsupervised learning conditions. We train and test our model to detect medical anomalies, but the method can with ease be applied to more general and diverse outlier detection tasks as well, e.g. detecting non-pathological anomalies such as image artifacts or artificial implants. Furthermore, the current applied augmentation are the most basic type of augmentations, it is also interesting to investigate various augmentations tailored to specific target anomalies, aiming to enhance the performance.

Acknowledgement. This work is partially supported by the Wallenberg Artificial Intelligence, Autonomous Systems and Software Program (WASP) funded by Knut and Alice Wallenberg Foundation.

References

1. Akcay, S., Atapour-Abarghouei, A., Breckon, T.P.: GANomaly: semi-supervised anomaly detection via adversarial training. In: Jawahar, C.V., Li, H., Mori, G., Schindler, K. (eds.) ACCV 2018. LNCS, vol. 11363, pp. 622–637. Springer, Cham (2019). https://doi.org/10.1007/978-3-030-20893-6_39

2. Azizi, S., et al.: Big self-supervised models advance medical image classification. In: International Conference on Computer Vision (ICCV) (2021)

3. Bozorgtabar, B., Mahapatra, D., Vray, G., Thiran, J.-P.: SALAD: self-supervised aggregation learning for anomaly detection on x-rays. In: Martel, A.L., et al. (eds.) MICCAI 2020. LNCS, vol. 12261, pp. 468–478. Springer, Cham (2020). https://doi.org/10.1007/978-3-030-59710-8_46

4. Cai, Y., Chen, H., Yang, X., Zhou, Y., Cheng, K.T.: Dual-distribution discrepancy for anomaly detection in chest x-rays. In: Wang, L., Dou, Q., Fletcher, P.T., Speidel, S., Li, S. (eds.) MICCAI 2022. LNCS, vol. 13433, pp. 584–593. Springer, Cham (2022). https://doi.org/10.1007/978-3-031-16437-8_56

5. Cao, T., Huang, C., Hui, D.Y., Cohen, J.P.: A benchmark of medical out of distribution detection. J. Mach. Learn. Biomed. Imaging (2020)

6. Chen, T., Kornblith, S., Swersky, K., Norouzi, M., Hinton, G.: Big self-supervised models are strong semi-supervised learners. In: International Conference on Neural Information Processing Systems (NeurIPS) (2020)

7. Chen, X., He, K.: Exploring simple Siamese representation learning. In: Computer Vision and Pattern Recognition Conference (CVPR) (2020)

8. Deng, J., Dong, W., Socher, R., Li, L.J., Li, K., Fei-Fei, L.: Imagenet: a large-scale hierarchical image database. In: Computer Vision and Pattern Recognition Conference (CVPR) (2009)

9. Fernando, T., Gammulle, H., Denman, S., Sridharan, S., Fookes, C.: Deep learning for medical anomaly detection - a survey. arXiv (2020)

10. Golan, I., El-Yaniv, R.: Deep anomaly detection using geometric transformations. In: NeurIPS (2018)

11. Gong, D., et al.: Memorizing normality to detect anomaly: memory-augmented deep autoencoder for unsupervised anomaly detection. In: International Conference on Computer Vision (ICCV) (2019)

12. Mao, Y., Xue, F.-F., Wang, R., Zhang, J., Zheng, W.-S., Liu, H.: Abnormality detection in chest x-ray images using uncertainty prediction autoencoders. In: Martel, A.L., et al. (eds.) MICCAI 2020. LNCS, vol. 12266, pp. 529–538. Springer, Cham (2020). https://doi.org/10.1007/978-3-030-59725-2_51

13. McKinney, S.M., et al.: International evaluation of an AI system for breast cancer screening. Nature (2020)

14. Paszke, A., et al.: Pytorch: an imperative style, high-performance deep learning library. In: Advances in Neural Information Processing Systems (NeurIPS) (2019)

15. Schlegl, T., Seeböck, P., Waldstein, S.M., Langs, G., Schmidt-Erfurth, U.: f-anogan: Fast unsupervised anomaly detection with generative adversarial networks. Med. Image Anal. (2019)

16. Tack, J., Mo, S., Jeong, J., Shin, J.: CSI: novelty detection via contrastive learning on distributionally shifted instances. In: Advances in Neural Information Processing Systems (2020)

17. Wolleb, J., Bieder, F., Sandkühler, R., Cattin, P.C.: Diffusion models for medical anomaly detection. In: Wang, L., Dou, Q., Fletcher, P.T., Speidel, S., Li, S. (eds.) MICCAI 2022. LNCS, vol. 13438, pp. 35–45. Springer, Cham (2022). https://doi.org/10.1007/978-3-031-16452-1_4

18. Wyatt, J., Leach, A., Schmon, S.M., Willcocks, C.G.: Anoddpm: anomaly detection with denoising diffusion probabilistic models using simplex noise. In: CVPR Workshops (2022)

19. Xie, H., et al.: Dual network architecture for few-view CT - trained on imagenet data and transferred for medical imaging. In: International Society for Optics and Photonics (2019)

20. Zbontar, J., Jing, L., Misra, I., LeCun, Y., Deny, S.: Barlow twins: self-supervised learning via redundancy reduction (2021)

21. Zhuang, C., Zhai, A.L., Yamins, D.: Local aggregation for unsupervised learning of visual embeddings. In: International Conference on Computer Vision (ICCV) (2019)

22. Çallı, E., Sogancioglu, E., van Ginneken, B., van, K.G.: Deep learning for chest x-ray analysis: a survey. Med. Image Anal. (2021)

CCMix: Curriculum of Class-Wise Mixup for Long-Tailed Medical Image Classification

Sirui Li[1], Fuheng Zhang[1], Tianyunxi Wei[1], Li Lin[1,2], Yijin Huang[1,3], Pujin Cheng[1], and Xiaoying Tang[1,4(✉)]

[1] Department of Electrical and Electronic Engineering, Southern University of Science and Technology, Shenzhen, China
[2] Department of Electrical and Electronic Engineering, The University of Hong Kong, Hong Kong, Hong Kong SAR, China
[3] School of Biomedical Engineering, University of British Columbia, Vancouver, Canada
[4] Jiaxing Research Institute, Southern University of Science and Technology, Jiaxing, China
tangxy@sustech.edu.cn

Abstract. Deep learning-based methods have been widely used for medical image classification. However, in clinical practice, rare diseases are usually underrepresented with limited labeled data, which result in long-tailed medical datasets and significantly degrade the performance of deep classification networks. Previous strategies employ re-sampling or re-weighting techniques to alleviate this issue by increasing the influence of underrepresented classes and reducing the influence of overrepresented ones. Still, poor performance may occur due to overfitting of the tail classes. Further, Mixup is employed to introduce additional information into model training. Despite considerable improvements, the significant noise in medical images means that random batch mixing may introduce ambiguity into training, thereby impair the performance. This observation motivates us to develop a fine-grained mixing approach. In this paper we present Curriculum of Class-wise Mixup (CCMix), a novel method for addressing the challenge of long-tailed distributions. CCMix leverages a novel curriculum that takes into account both the degree of mixing and the class-wise performance to identify the ideal Mixup proportions of different classes. Our method's simplicity enables its effortless integration with existing long-tailed recognition techniques. Comprehensive experiments on two long-tailed medical image classification datasets demonstrate that our method, requiring no modifications to the framework structure or algorithmic details, achieves state-of-the-art results across diverse long-tailed classification benchmarks. The source code is available at https://github.com/sirileeee/CCMix.

Supplementary Information The online version contains supplementary material available at https://doi.org/10.1007/978-3-031-45676-3_31.

X. Cao et al. (Eds.): MLMI 2023, LNCS 14349, pp. 303–313, 2024.
https://doi.org/10.1007/978-3-031-45676-3_31

Keywords: Long-tailed Learning · Medical Image Classification · Mixup

1   Introduction

With the advent of powerful deep neural networks (DNNs), the medical image classification realm [15,20,21] has made significant progresses. Despite impressive breakthroughs, recent advances are mainly driven by balanced datasets. However, as shown in Fig. 1, real-world clinical datasets are characterized by long-tailed distributions, since collecting labeled samples of rare diseases or unusual instances is challenging. Imbalanced datasets may result in over-represented classes dominating the training process, inducing decreased performance on the under-represented classes. Skewed datasets often arise from rare diseases or insufficient expert annotations, leading to decision biases towards head classes and weakened performance on tail classes. Therefore, developing methods that well accommodate data imbalance is essential for the advancement of the medical image classification field.

To address this imbalance or even long-tailed issue, common solutions include data re-balancing [11,30], module improvement [16,23], and information augmentation [4,25]. Data re-balancing techniques, including re-sampling [2] and re-weighting [10] of classes, have been proposed to address the issue of under-represented classes and improve model performance. However, such methods may introduce bias, and thus, careful evaluation is necessary. Module improvement strategies, such as representation learning [14] and decoupled training [17], aim to enhance the recognition performance of a network on under-represented classes. Nevertheless, the implementation of these techniques may require significant computational and time resources.

Recently, more and more works start to focus on utilizing information augmentation to enhance long-tailed learning, such as performing regular or balanced sampling to mix-up samples [12] or utilizing class-wise augmentation to identify the proper degree of augmentation for each class [1]. However, due to the potentially high level of noise presented in medical images, randomly mixing-up samples within a same batch might introduce ambiguity into the training process, which poses negative influence on separating classes. Also, the variation in strength of class-specific augmentation techniques not only ignores the partitioning of features across different classes but also introduces irrelevant information to the model, ultimately impairing its performance. In order to combine the advantages of the aforementioned two categories of solutions for long-tailed learning, we consider using Mixup strategies [3,19,27] to adaptively learn the features of different classes for addressing the class imbalance and long-tail issues.

To this end, we propose **C**urriculum of **C**lass-wise **Mix**up (CCMix) to tackle the long-tailed medical image classification problem. With the aid of a class-wise Mixup module and a curriculum module that updates the class-wise degree of mixing, CCMix achieves adaptive control of the mix proportion to handle the bias induced by the long-tailed distribution, along with extraordinary performance observed on several frameworks and datasets. To the best of our knowledge, this is the first study to introduce a class-wise Mixup approach for determining an appropriate mix proportion to address class imbalance issues.

The main contributions of this paper are three-fold: (1) We expand the Mixup strategy from batch-level approach to class-level approach, which is straightforward yet refocuses the model's attention on the inter-class feature differences. (2) We develop a curriculum learning algorithm to determine the appropriate class-wise mix-strength, which is crucial to specify the mix ratios between different classes. (3) We perform comprehensive experiments on two medical classification datasets with long-tailed distributions. Our results indicate that CCMix demonstrates robust usability and generalizability, and can be seamlessly integrated with various Mixup strategies. Additionally, our approach outperforms existing methods in effectively mitigating data imbalance across all evaluated scenarios.

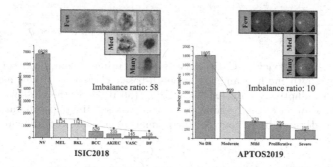

Fig. 1. Class distributions of our two medical datasets possessing long-tailed characteristics. To conduct more comprehensive evaluations of the effectiveness of our approach on both head and tail classes, we partition each dataset into three distinct subsets, categorized by the sample size of each class, namely many, med(ium), and few.

2   Methodology

The overall framework of CCMix is shown in Fig. 2, which comprises of a class-wise Mixup module and a curriculum evaluation module to adaptively adjust the mix-strength. Each component will be described in the following subsections. The algorithmic design of the entire framework is presented in Algorithm 1.

2.1   Class-Wise Mixup

Mixup Based on Class-Wise Mix-Strength. We propose to incorporate class-specific mix-strength, which represents the degree of mixing for each class, into the Mixup strategy to enable adaptive control of the mix proportion. As the proportion of a certain class decreases in mixed samples, the model faces

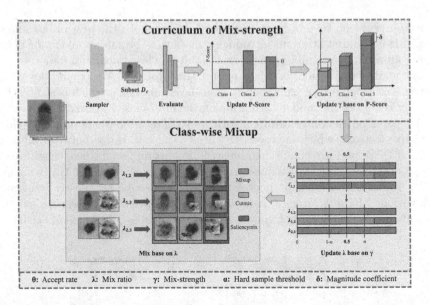

Fig. 2. Algorithm overview. CCMix consists of two primary components: (1) curriculum of mix-strength module and (2) strength-based class-wise Mixup module. The mix-strength is adaptively adjusted by utilizing the P-Score, the class-wise accuracy obtained from a subset of the training set. The class-wise Mixup incorporates a threshold parameter, denoted by α, to avoid the premature generation of difficult samples.

increased difficulty in accurately classifying it due to the reduced representation of this class. Therefore, we define the increase in mix-strength of a class as corresponding to the decrease of its proportion in mixed samples. For the training dataset with C classes, we formulate the mix-strength γ for each class as $\gamma = \{\gamma_1, \gamma_2, \ldots, \gamma_C\}$, where $\gamma_i \in (0, 1)$ presents the mix-strength of class c_i. With this notation, the mix proportion $\lambda'_{i,j}$ generated by mix-strength between class c_i and class c_j can be described as follows:

$$\lambda'_{i,j} = \frac{\gamma_j}{\gamma_i + \gamma_j}, \tag{1}$$

Compared to the original Mixup, our approach enhances model robustness by providing mixed samples with more variability for well-performing classes and greater representation for poorly performing classes, leading to improved training performance across all classes.

Control of Hard Sample Generation. To prevent the introduction of difficult samples, such as equally proportioned mixed images, during the early stages of training that can hinder accurate feature extraction for each class, we propose a novel approach that utilizes a linearly decreasing threshold parameter $\alpha \in (0.5, 1)$ to regulate the maximum allowable mix ratio per epoch. The final mix ratio $\lambda_{i,j}$ between class c_i and class c_j can be presented as:

$$\lambda_{i,j} = \begin{cases} \max(\lambda'_{i,j}, \alpha), & \lambda'_{i,j} \geqslant 0.5 \\ \min(\lambda'_{i,j}, 1 - \alpha), & \text{otherwise} \end{cases}, \tag{2}$$

and the class-wise Mixup between samples of the two classes can be found as:

$$\hat{x} = \lambda_{i,j} x_i + (1 - \lambda_{i,j}) x_j, \ \hat{y} = \lambda_{i,j} y_i + (1 - \lambda_{i,j}) y_j, \tag{3}$$

Combine with Multiple Mixup Techniques. As a class-wise mix-strength based curriculum learning method, CCMix can be conveniently and quickly combined with various Mixup techniques. In this paper, we integrate three classic Mixup techniques, including the original Mixup [29] technique that generates new training samples by mixing two different samples in a certain proportion, the Cutmix [28] technique that creates samples by randomly cutting a region from one sample and pasting it into another sample, and the Saliencymix [26] technique that produces samples by mixing the saliency regions of one image into another. CCMix enables a flexible and targeted mixing strategy for medical images, extracting additional features from underrepresented classes in a progressively challenging manner, systematically enhancing the model's robustness and generalization capabilities, ultimately resulting in improved performance.

2.2 Curriculum Evaluation of Mix-Strength

Precision Score (P-Score) Evaluation. To regulate the mix-strength value, we use a subset of the training dataset to verify the model's learning performance for each class. The P-Score P is introduced for each class c at epoch e to reflect the model's accuracy on the subset. This approach is adopted mainly because a high P-Score for class c in the previous epoch suggests that the model has learned the class well under the current mix-strength and requires further validation for model generalization using additional random training samples. The evaluation of P-Score is defined as follows:

$$P_c^e = \text{Accuracy}(P_c^{e-1}, D_c, \beta), \tag{4}$$

where we consider a training dataset D consisting of N samples and C classes, represented as $D = \{(x_i, y_i)_{i=1}^N\}$, where $x_i \in \mathbb{R}^N$ and $y_i \in \{1, 2, ..., C\}$. The subset of class c is denoted as $D_c \subset D$, where $D_c = \{(x_i, y_i) \mid y_i = c, (x_i, y_i) \in D\}$. The coefficient β determines the number of random samples, and the subset contains $\lceil \beta P_c^{e-1} \rceil$ samples from D_c. The mix-strength is then updated based on the P-Score value in the current epoch during subsequent procedures.

Curriculum Mixup. CCMix utilizes P-Score and mix-strength to achieve adaptive Mixup on a class-wise basis, facilitating better feature learning from both well-represented and underrepresented classes. Our core idea is that when the model learns sufficient information from class c, as indicated by a higher

---

Algorithm 1. Curriculum of Class-wise Mixup

Input: Training dataset $D = \{(x_i, y_i)_{i=1}^N\}$, epochs E, threshold α, θ, coefficient β
Output: mixed samples

   Initialization: $P_c^0 = 0.1, \gamma_c^0 = 0.5 \ \forall c \in \{1, 2, ..., C\}$
   for $e \leq E$ **do**
      $P_c^e = \text{Accuracy}(P_c^{e-1}, D_c, \beta) \ \forall c \in \{1, 2, ..., C\}$
      $\gamma_c^e = \begin{cases} \gamma_c^{e-1} + \delta, & P_c^e \geqslant \theta^e \\ \gamma_c^{e-1} - \delta, & \text{otherwise} \end{cases}$

      Class-wised Mixup with $\lambda_{i,j} = \begin{cases} \max(\lambda'_{i,j}, \alpha), & \lambda'_{i,j} \geqslant 0.5 \\ \min(\lambda'_{i,j}, 1 - \alpha), & \text{otherwise} \end{cases}$ where $\lambda'_{i,j} = \frac{\gamma_j^e}{\gamma_i^e + \gamma_j^e}$

      Update α, θ
   end for

P-Score, it can handle a higher mix-strength and learn more challenging mixed samples with reduced information from this class. As shown in Fig. 2, we utilize P-Score P_c to evaluate the training performance of class c, and introduce a linearly increasing accept threshold parameter $\theta \in (0, 1)$ to determine whether the performance is satisfactory. By combining P_c^e and θ^e, we can adaptively measure the mix-strength γ_c^e for each class c at epoch e as follows:

$$\gamma_c^e = \begin{cases} \gamma_c^{e-1} + \delta, & P_c^e \geqslant \theta^e \\ \gamma_c^{e-1} - \delta, & \text{otherwise} \end{cases}, \tag{5}$$

where δ is adjustable coefficient used to modify the magnitude of changes in γ_c.

3   Experiments and Results

Datasets and Evaluation. We conduct experiments on two publicly available datasets, namely the ISIC2018 dataset [5] and the APTOS2019 dataset [18]. The ISIC2018 dataset is a large-scale collection of dermoscopy images published by the International Skin Imaging Collaboration (ISIC) organizers, while the APTOS2019 dataset is a fundus dataset from the APTOS2019 blindness detection competition. To ensure a fair evaluation of the learning performance of each class, we divide each dataset into training and testing sets at a ratio of 9:1, with a balanced distribution of samples for all classes ensured in the testing set. This approach enables fair evaluation of all classes since imbalanced testing sets may bias evaluation metrics towards head classes, even if tail classes perform well.

   The distribution and partitioning details of the datasets are provided in Fig. 1 and Table A1, where we sort each class in descending order based on the number of samples. The imbalance ratio, represented by N_{max}/N_{min} where N is the number of samples in each class, indicates the difference between the largest and smallest classes based on their capacities. To conduct a thorough assessment of the efficacy of our method on both head and tail classes, we use four categories of accuracy for both datasets: all, many, med(ium), and few. These correspond to the average accuracy of samples from all classes, classes with over 45% representation, classes with representation between 10% and 45%, and classes with less than 10% representation in the training dataset, respectively.

Implementation Details. We conduct all experiments using Pytorch with NVIDIA RTX 2080Ti GPUs. We use RandAugment [6] as our data augmentation policy and ResNet50 [13] as our backbone. The initial learning rate is set to 0.01 with a batch size of 128. We utilize SGD as the default optimizer with a momentum of 0.9 and a weight decay of 0.0005. The hyperparameters α and θ vary in the range of $[0.6, 1]$ and $[0.1, 0.7]$, respectively. Coefficient hyperparameters β and δ are set to be 5 and 0.05, respectively. Training epochs are set as 500 for all experiments. Unless stated otherwise, we use the original Mixup technique to implement the class-wise Mixup module.

Table 1. Top-1 validation accuracy of two long-tailed medical datasets. The optimal accuracy across all comparisons is denoted in **bold**.

Algorithm	ISIC2018				APTOS2019			
	Many	Med	Few	Total	Many	Med	Few	Total
BALLAD [24]	60.5	75.0	54.8	61.4	96.0	22.0	60.0	59.6
PaCo [8]	35.0	37.5	80.0	61.4	48.1	31.8	58.7	51.2
GPaCo [9]	45.0	45.0	76.2	62.9	80.0	38.6	47.8	52.4
CE	**85.0**	**40.0**	37.5	45.0	**92.0**	**76.0**	39.3	57.2
CE + CCMix	80.0	37.5	**43.8**	**47.1**$_{+2.1}$	90.0	72.0	**45.3**	**59.6**$_{+2.4}$
Focal Loss [22]	70.0	**65.0**	45.0	54.3	**98.0**	66.0	43.3	58.8
Focal Loss [22] + CCMix	**90.0**	37.5	**55.0**	**55.0**$_{+0.7}$	94.0	**76.0**	**44.0**	**60.4**$_{+1.8}$
ResLT [7]	90.0	40.0	53.8	55.0	96.0	**88.1**	13.3	44.8
ResLT [7] + CCMix	**96.0**	**43.1**	**54.4**	**57.1**$_{+2.1}$	**98.0**	86.0	**16.7**	**46.8**$_{+2.0}$
BCL [31]	**85.0**	65.0	81.2	77.1	96.0	48.2	55.9	62.4
BCL [31] + CCMix	80.0	**72.5**	**82.5**	**79.3**$_{+2.2}$	96.0	**54.5**	**59.8**	**66.0**$_{+3.6}$

Comparison with State-of-the-Art. We apply CCMix to Cross-Entropy (CE) and various long-tail oriented algorithms: Focal Loss [22], ResLT [7] and BCL [31]. After applying CCMix to all algorithms in both datasets, we observe a nearly 2% improvement in overall accuracy. Despite oscillations in accuracy of many and medium classes, consistent improvement in few classes performance is observed in all experiments, indicating that CCMix enhances the model's ability to learn features of tail classes.

To assess the impact of CCMix, we compare its performance with state-of-the-art methods, including BALLAD [24], PaCo [8], and GPaCo [9]. As the results shown in Table 1, the application of CCMix improves the performance of all algorithms to, and even surpass, the level of the state-of-the-art methods. This serves as evidence of the generalizability and efficacy of our framework.

Table 2. The performance of applying CCMix to BCL with different components on ISIC2018. The best ones are **bolded**.

Class-wise Mixup	Threshold α	Curriculum	Many	Med	Few	Total
✓			**90.0**	52.5	75.0	70.7
✓	✓		80.0	65.0	76.2	73.6
✓		✓	**90.0**	70.0	76.3	76.4
✓	✓	✓	80.0	**72.5**	**82.5**	**79.3**

Contribution of Each Module. This work consists of two essential components: the class-wise Mixup module and the curriculum module. However, we argue that the threshold parameter α in the class-wise Mixup module, which controls the generation of hard mixed samples at an appropriate time, plays a crucial role in the training process. Thus, we consider it as a distinct component and conduct ablation studies to evaluate the effectiveness of all three components separately. We perform four additional experiments to integrate and synthesize the three constituent elements of our proposed framework.

Table 2 presents the performance obtained through BCL on the ISIC2018 dataset, while Table A2 provides the results obtained through other frameworks. The results indicate that integrating the threshold parameter α into the class-wise Mixup module leads to an effective delay in introducing challenging samples, resulting in improved accuracy. Furthermore, the observed increase in accuracy can be attributed to the curriculum module's ability to assist the model in acquiring more in-depth features of the tail classes. Finally, incorporating all three modules in the proposed framework achieves optimal performance.

Contribution of Each Mixup Strategy. To demonstrate CCMix's generalization, we combine it with Mixup [29], Cutmix [28], and Saliencymix [26] within the BCL framework. We then compare the results with those using only the three mixing strategies to prove CCMix's effectiveness. As the results presented in Table 3, CCMix improves the performance of all three initial mixing strategies. Moreover, combining Saliencymix with CCMix achieves remarkable performance, highlighting CCMix's potential to generate mixed samples with higher effective feature density for representation learning. These findings suggest opportunities for future research by demonstrating the potential of integrating CCMix with multiple Mixup strategies.

Table 3. Comparative analysis of three mixing strategies within the BCL framework. The better accuracy for each pair of comparisons is **bolded**. When applying CCMix, the currently used Mixup technique will be indicated in parentheses.

Algorithm	ISIC2018				APTOS2019			
	Many	Med	Few	Total	Many	Med	Few	Total
BCL	85.0	65.0	81.2	77.1	96.0	48.2	55.9	62.4
BCL + Mixup [29]	**80.0**	**80.0**	77.5	78.6	92.0	38.0	**63.3**	64.0
BCL + CCMix (Mixup)	**80.0**	72.5	**82.5**	**79.3**	**96.0**	**54.5**	59.8	**66.0**
BCL + Cutmix [28]	74.5	71.0	77.1	75.0	**98.0**	58.0	**50.0**	61.2
BCL + CCMix (Cutmix)	**77.5**	**76.7**	**80.0**	**78.7**	96.0	**74.0**	49.3	**63.6**
BCL + Saliencymix [26]	**90.0**	72.5	81.2	80.0	96.0	48.0	**60.0**	64.8
BCL + CCMix (Saliencymix)	80.0	**82.5**	**82.5**	**81.4**	**98.0**	**52.0**	**60.0**	**66.0**

4   Conclusion

In this study, we propose Curriculum of Class-wise Mixup (CCMix) to address the problem of long-tailed medical image classification. CCMix integrates a class-wise Mixup module and a curriculum module that updates the class-wise mix-strength to adaptively control the mix proportion and mitigate bias caused by the long-tailed distribution. To the best of our knowledge, this study represents the first effort to introduce a class-wise Mixup approach for addressing class imbalance issues. Our experimental results demonstrate the superiority of CCMix over existing methods on multiple frameworks and datasets. Furthermore, the potential applicability of CCMix to combine with more tasks suggests promising avenues for further research.

Acknowledgement. This study was supported by the Shenzhen Basic Research Program (JCYJ20190809120205578); the National Natural Science Foundation of China (62071210); the Shenzhen Science and Technology Program (RCYX20210609 103056042); the Shenzhen Basic Research Program (JCYJ20200925153847004); the Shenzhen Science and Technology Innovation Committee (KCXFZ2020122117340001).

References

1. Ahn, S., Ko, J., Yun, S.Y.: CUDA: curriculum of data augmentation for long-tailed recognition. arXiv preprint arXiv:2302.05499 (2023)
2. Chawla, N.V., Bowyer, K.W., Hall, L.O., Kegelmeyer, W.P.: SMOTE: synthetic minority over-sampling technique. J. Artif. Intell. Res. **16**, 321–357 (2002)
3. Choi, H.K., Choi, J., Kim, H.J.: TokenMixup: efficient attention-guided token-level data augmentation for transformers. Adv. Neural. Inf. Process. Syst. **35**, 14224–14235 (2022)
4. Chu, P., Bian, X., Liu, S., Ling, H.: Feature space augmentation for long-tailed data. In: Vedaldi, A., Bischof, H., Brox, T., Frahm, J.-M. (eds.) ECCV 2020. LNCS, vol. 12374, pp. 694–710. Springer, Cham (2020). https://doi.org/10.1007/978-3-030-58526-6_41

5. Codella, N., et al.: Skin lesion analysis toward melanoma detection 2018: a challenge hosted by the international skin imaging collaboration (ISIC). arXiv preprint arXiv:1902.03368 (2019)

6. Cubuk, E.D., Zoph, B., Shlens, J., Le, Q.V.: RandAugment: practical automated data augmentation with a reduced search space. In: Proceedings of the IEEE/CVF Conference on Computer Vision and Pattern Recognition Workshops, pp. 702–703 (2020)

7. Cui, J., Liu, S., Tian, Z., Zhong, Z., Jia, J.: ResLT: residual learning for long-tailed recognition. IEEE Trans. Pattern Anal. Mach. Intell. **45**(3), 3695–3706 (2022)

8. Cui, J., Zhong, Z., Liu, S., Yu, B., Jia, J.: Parametric contrastive learning. In: Proceedings of the IEEE/CVF International Conference on Computer Vision, pp. 715–724 (2021)

9. Cui, J., Zhong, Z., Tian, Z., Liu, S., Yu, B., Jia, J.: Generalized parametric contrastive learning. IEEE Trans. Pattern Anal. Mach. Intell. (2023)

10. Cui, Y., Jia, M., Lin, T.Y., Song, Y., Belongie, S.: Class-balanced loss based on effective number of samples. In: Proceedings of the IEEE/CVF Conference on Computer Vision and Pattern Recognition, pp. 9268–9277 (2019)

11. Estabrooks, A., Jo, T., Japkowicz, N.: A multiple resampling method for learning from imbalanced data sets. Comput. Intell. **20**(1), 18–36 (2004)

12. Galdran, A., Carneiro, G., González Ballester, M.A.: Balanced-MixUp for highly imbalanced medical image classification. In: de Bruijne, M., et al. (eds.) MICCAI 2021. LNCS, vol. 12905, pp. 323–333. Springer, Cham (2021). https://doi.org/10.1007/978-3-030-87240-3_31

13. He, K., Zhang, X., Ren, S., Sun, J.: Deep residual learning for image recognition. In: Proceedings of the IEEE Conference on Computer Vision and Pattern Recognition, pp. 770–778 (2016)

14. Huang, C., Li, Y., Loy, C.C., Tang, X.: Learning deep representation for imbalanced classification. In: Proceedings of the IEEE Conference on Computer Vision and Pattern Recognition, pp. 5375–5384 (2016)

15. Huang, Y., Lin, L., Cheng, P., Lyu, J., Tam, R., Tang, X.: Identifying the key components in ResNet-50 for diabetic retinopathy grading from fundus images: a systematic investigation. Diagnostics **13**(10) (2023). https://doi.org/10.3390/diagnostics13101664. https://www.mdpi.com/2075-4418/13/10/1664

16. Kang, B., Li, Y., Xie, S., Yuan, Z., Feng, J.: Exploring balanced feature spaces for representation learning. In: International Conference on Learning Representations (2020)

17. Kang, B., et al.: Decoupling representation and classifier for long-tailed recognition. arXiv preprint arXiv:1910.09217 (2019)

18. Karthick, M., Sohier, D.: APTOS 2019 blindness detection (2019). Kaggle https://kaggle.com/competitions/aptos2019-blindness-detection. Go to reference in chapter

19. Kim, J.H., Choo, W., Jeong, H., Song, H.O.: Co-Mixup: saliency guided joint mixup with supermodular diversity. arXiv preprint arXiv:2102.03065 (2021)

20. Li, Q., Cai, W., Wang, X., Zhou, Y., Feng, D.D., Chen, M.: Medical image classification with convolutional neural network. In: 2014 13th International Conference on Control Automation Robotics & Vision (ICARCV), pp. 844–848. IEEE (2014)

21. Lin, L., et al.: BSDA-net: a boundary shape and distance aware joint learning framework for segmenting and classifying OCTA images. In: de Bruijne, M., et al. (eds.) MICCAI 2021. LNCS, vol. 12908, pp. 65–75. Springer, Cham (2021). https://doi.org/10.1007/978-3-030-87237-3_7

22. Lin, T.Y., Goyal, P., Girshick, R., He, K., Dollár, P.: Focal loss for dense object detection. In: Proceedings of the IEEE International Conference on Computer Vision, pp. 2980–2988 (2017)
23. Liu, Z., Miao, Z., Zhan, X., Wang, J., Gong, B., Yu, S.X.: Large-scale long-tailed recognition in an open world. In: Proceedings of the IEEE/CVF Conference on Computer Vision and Pattern Recognition, pp. 2537–2546 (2019)
24. Ma, T., et al.: A simple long-tailed recognition baseline via vision-language model. arXiv preprint arXiv:2111.14745 (2021)
25. Tan, C., Sun, F., Kong, T., Zhang, W., Yang, C., Liu, C.: A survey on deep transfer learning. In: Kůrková, V., Manolopoulos, Y., Hammer, B., Iliadis, L., Maglogiannis, I. (eds.) ICANN 2018. LNCS, vol. 11141, pp. 270–279. Springer, Cham (2018). https://doi.org/10.1007/978-3-030-01424-7_27
26. Uddin, A., Monira, M., Shin, W., Chung, T., Bae, S.H., et al.: SaliencyMix: a saliency guided data augmentation strategy for better regularization. arXiv preprint arXiv:2006.01791 (2020)
27. Verma, V., et al.: Manifold mixup: better representations by interpolating hidden states. In: International Conference on Machine Learning, pp. 6438–6447. PMLR (2019)
28. Yun, S., Han, D., Oh, S.J., Chun, S., Choe, J., Yoo, Y.: CutMix: regularization strategy to train strong classifiers with localizable features. In: Proceedings of the IEEE/CVF International Conference on Computer Vision, pp. 6023–6032 (2019)
29. Zhang, H., Cisse, M., Dauphin, Y.N., Lopez-Paz, D.: mixup: beyond empirical risk minimization. arXiv preprint arXiv:1710.09412 (2017)
30. Zhang, Z., Pfister, T.: Learning fast sample re-weighting without reward data. In: Proceedings of the IEEE/CVF International Conference on Computer Vision, pp. 725–734 (2021)
31. Zhu, J., Wang, Z., Chen, J., Chen, Y.P.P., Jiang, Y.G.: Balanced contrastive learning for long-tailed visual recognition. In: Proceedings of the IEEE/CVF Conference on Computer Vision and Pattern Recognition, pp. 6908–6917 (2022)

MEDKD: Enhancing Medical Image Classification with Multiple Expert Decoupled Knowledge Distillation for Long-Tail Data

Fuheng Zhang[1], Sirui Li[1], Tianyunxi Wei[1], Li Lin[1,2], Yijin Huang[1,3], Pujin Cheng[1], and Xiaoying Tang[1,4(✉)]

[1] Department of Electrical and Electronic Engineering, Southern University of Science and Technology, Shenzhen, China
[2] Department of Electrical and Electronic Engineering, The University of Hong Kong, Hong Kong, Hong Kong SAR, China
[3] School of Biomedical Engineering, University of British Columbia, Vancouver, Canada
[4] Jiaxing Research Institute, Southern University of Science and Technology, Jiaxing, China
tangxy@sustech.edu.cn

Abstract. Medical image classification is a challenging task, particularly when dealing with long-tailed datasets where rare diseases are underrepresented. The imbalanced class distribution in such datasets poses significant challenges in accurately classifying minority classes. Existing methods for alleviating the long-tail problem in medical image classification suffer from limitations such as noise introduction, loss of crucial information, and the need for manual tuning and additional computational resources. In this study, we propose a novel framework called Multiple Expert Decoupled Knowledge Distillation (MEDKD) to tackle the imbalanced class distribution in medical image classification. The knowledge distillation of multiple teacher models can significantly alleviate the class imbalance by partitioning the dataset into several subsets. However, current frameworks of this kind have not yet explored the integration of more advanced distillation methods. Our framework incorporating TCKD and NCKD concepts to improve classification performance. Through comprehensive experiments on publicly available datasets, we evaluate the performance of MEDKD and compare it with state-of-the-art methods. Our results demonstrate remarkable accuracy improvements achieved by the proposed method, highlighting its effectiveness in alleviating the challenges of medical image classification with long-tailed datasets.

Keywords: Long-tailed Classification · Knowledge Distillation · Multiple Expert

---

F. Zhang and S. Li—Contributed equally to this work.

X. Cao et al. (Eds.): MLMI 2023, LNCS 14349, pp. 314–324, 2024.
https://doi.org/10.1007/978-3-031-45676-3_32

1   Introduction

Recent studies have demonstrated the successful application of deep learning models in medical image analysis tasks, such as disease diagnosis [2,10,15] and lesion detection [9,11], benefiting from large-scale, balanced, and high-quality labeled data. However, the presence of rare diseases leads to the existence of long-tailed datasets, where the majority of samples belong to a few common classes while the minority classes have limited samples [17]. This poses a significant challenge in accurately classifying the minority classes.

Existing approaches to alleviate the long-tail problem in medical image classification often rely on re-sampling [21] or re-weighting [8] strategies. These methods include oversampling [6] or undersampling [19] techniques, which have their limitations. Over-sampling may introduce noise or duplicate information, while under-sampling may lead to the loss of crucial information. Re-weighting methods aim to assign appropriate weights to different samples based on their class frequencies to mitigate the class imbalance issue. However, determining the optimal weight allocation for each class necessitates prior knowledge or parameter tuning, which can be challenging in real-world scenarios. Additionally, other common methods include class-balanced loss functions and model improvement [13]. However, these methods may require manual tuning, domain expertise, or additional computational resources.

In recent research, knowledge distillation (KD) techniques have gained attention as a means to alleviate the long-tail problem [12]. KD aims to transfer knowledge from a teacher model to a student model, optimizing computational efficiency while enhancing performance. Notably, the Learning From Multiple Experts (LFME) framework [22] and the concept of relational subsets have demonstrated promising outcomes in multi-teacher knowledge distillation. However, the traditional logits KD methods commonly employed in these frameworks [1] may not fully leverage the potential benefits of novel distillation techniques, as the coupling nature of the basic KD loss function tends to suppress the effectiveness of logit distillation.

Motivated by recent advancements in knowledge distillation (KD) techniques, particularly the Decoupled Knowledge Distillation (DKD) method, which effectively resolves the coupling issue in the KD loss function, we propose a novel framework called Multiple Expert Decoupled Knowledge Distillation (MEDKD) to alleviate the long-tail problem in medical image classification. Our framework introduces the concepts of target class knowledge distillation (TCKD) and non-target class knowledge distillation (NCKD). By explicitly treating the target and non-target classes differently, MEDKD mitigates potential confusion introduced by multiple experts. Additionally, in line with the experimental findings in LFME [22], which demonstrated the performance enhancement achieved by utilizing superior loss functions during the training process, we explore the utilization of a novel loss function called Large Margin aware Focal loss (LMF loss) [20]. By incorporating the LMF loss into our framework, we aim to further improve the overall classification performance.

Our study makes the following key contributions: (1) We introduce a novel framework for multiple expert knowledge distillation, which incorporates the concepts of TCKD and NCKD. This framework effectively alleviates the long-tail problem in medical image classification. (2) We further enhance the framework's performance by integrating the LMF loss function, which improves the classification performance through Multiple Expert Decoupled Knowledge Distillation. (3) We demonstrate the flexibility and adaptability of our approach through extensive experiments on diverse datasets, showcasing its capability to adjust the number of teachers and adapt to varying dataset characteristics.

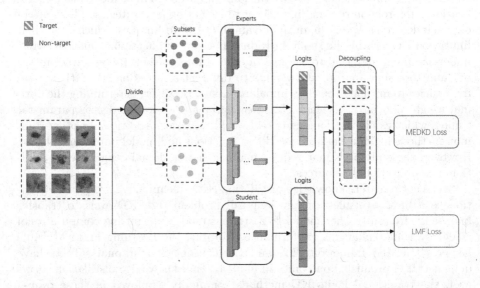

Fig. 1. The overall framework of our work. We partitioning the dataset into subsets and training experts within each subset, followed by utilizing these experts to guide the training of the student model. The loss function comprises MEDKD and LMF loss.

2   Methodology

2.1   Overview

In Fig. 1, we present an illustrative diagram of the model framework. Our methodology involves initially partitioning the dataset into subsets based on sample quantities and training experts within each subset. After the completion of training, these experts are utilized to guide the training of the student model. During the training process of the student model, the loss function comprises two components: Multiple Expert Decoupled Knowledge Distillation (MEDKD) and LMF loss. MEDKD incorporates the concepts of TCKD and NCKD into the multiple-expert framework, effectively resolves the coupling issue in the traditional KD loss function, achieving a more effective knowledge transfer from the experts to the student model and leading to improved distillation performance. The incorporation of LMF loss facilitates the comparison of the student model's output with the ground truth, aiding in the training process. Subsequently, we provide a comprehensive description of the constituents of our framework.

2.2 Multiple Expert Decoupled Knowledge Distillation

We decouple the distillation of the experts into TCKD and NCKD. Building upon this distillation technique, we construct a new multi-expert knowledge distillation framework, MEDKD, which achieves improved performance by effectively decoupling the logits and avoiding potential confusion introduced by multiple experts. The TCKD and NCKD term can be expressed as follows:

$$L_{TCKD} = -\sum_{c=1}^{C} p^{M_E}(c) \log(p^{M}(c)), \tag{1}$$

$$L_{NCKD} = -\sum_{c=1}^{C} p^{M_E}(c) \log(p^{M}(c)) - \sum_{c=1}^{C} (1 - p^{M_E}(c)) \log(1 - p^{M}(c)). \tag{2}$$

Here, $p^{M_E}(c)$ represents the predicted probability of the class c by the teacher model, and $p^{M}(c)$ represents the corresponding predicted probability by the student model. Their terms can be expressed as follows:

$$p^{M_E}(c) = \frac{\exp(z^{M_E}(c)/T)}{\sum_j \exp(z^{M_E}(j)/T)}, p^{M}(c) = \frac{\exp(z^{M}(c)/T)}{\sum_j \exp(z^{M}(j)/T)}. \tag{3}$$

Then we define the loss function for our proposed framework, denoted as L_{MEDKD}, which combines the TCKD and NCKD terms with weighting parameter α :

$$L_{MEDKD} = L_{TCKD}(target(p^{M_E}, p^{M})) + \alpha L_{NCKD}(non - target(p^{M_E}, p^{M})). \tag{4}$$

In this equation, $target(p^{M_E}, p^{M})$ represents the logits and probabilities for the target class, and $non\_target(p^{M_E}, p^{M})$ represents the logits and probabilities for the non-target classes.

2.3 Large Margin Aware Focal Loss

In the exploration of previous multi-expert knowledge distillation frameworks [12,22], the use of different loss functions as auxiliary components, such as Focal loss [16] and LDAM loss [3], has been considered. Since loss functions with outstanding performance in addressing the long-tail problem can provide better assistance, we adopt the Large Margin aware Focal (LMF) loss [20] as a powerful and recent addition to our framework to achieve improved distillation results. The loss function, formulated as:

$$L_{LMF} = L_{LDAM} + L_{FL}, \tag{5}$$

This choice stems from the necessity to alleviate the class imbalance problem prevalent in medical image classification tasks. It combines two powerful loss functions, LDAM loss and Focal loss.

The inclusion of the LMF loss in our framework is motivated by its ability to effectively handle class imbalance. The LDAM loss focuses on enlarging the margins between different classes, enabling the model to better distinguish minority and majority classes. This aligns well with our objective of improving the classification performance of under-represented classes. On the other hand, the Focal loss reduces the influence of well-classified samples during training, allowing the model to prioritize challenging and misclassified samples.

By incorporating both LDAM and Focal loss components, the LMF loss leverages the strengths of these methods. It alleviate the class imbalance problem while emphasizing the importance of challenging samples, ultimately leading to improved classification performance, particularly for minority classes. The LMF loss strikes a balance between prioritizing difficult samples and maintaining discriminative margins between classes.

2.4 Self-paced Expert Selection

We incorporate the Self-paced Expert Selection technique to dynamically adjust the weights assigned to each expert model during the computation of the loss function based on their performance on the training set. The objective is to allocate greater attention to the classes in which the student model exhibits relatively lower performance, aiming to improve the overall classification accuracy. The weight calculation for each expert model takes into account the accuracies of both the teacher models and the student model on the training set. Specifically, we utilize the following weight assignment formulation:

$$w_k = \begin{cases} 1.0, & Acc_M \leq \gamma Acc_{M_{E_k}} \\ \frac{Acc_{M_{E_k}} - Acc_M}{(1-\gamma)Acc_{M_{E_k}}}, & \text{otherwise} \end{cases}, \tag{6}$$

where w_k represents the weight assigned to the k-th expert model, Acc_M is the accuracy of the student model, $Acc_{M_{E_k}}$ is the accuracy of the k-th expert model, and γ is a hyperparameter. Through the Self-paced Expert Selection approach, we alleviate the imbalance in the student model's performance across different classes. This adaptive weighting mechanism enables a more focused learning process for the student model, emphasizing the improvement of its performance on classes where it lags behind. As a result, the student model effectively leverages the expertise of the teacher models to overcome the challenges posed by the long-tail problem. Furthermore, to integrate the self-paced expert selection into the distillation process, we modify the loss function as follows:

$$L = \sum_{i=1}^{N} (\varphi L_{LMF}(xi, yi) + (1 - \varphi) \sum_{k=1}^{K} w_k L_{MEDKD}(M, M_{E_k}; x_i)), \tag{7}$$

where L_{LMF} denotes the LMF loss function, L_{MEDKD} represents the MEDKD loss function, and w_k is the weight assigned to the k-th expert model. The parameter φ and $(1 - \varphi)$ reflect the proportion of the two loss components.

3   Experiments and Results

3.1   Dataset

We conducted experiments on two publicly available datasets: the ISIC2018 dataset [4] provided by the International Skin Imaging Collaboration (ISIC) and the APTOS2019 dataset [14] from the APTOS 2019 blindness detection competition. The ISIC2018 dataset comprises a large-scale collection of dermoscopy images, while the APTOS2019 dataset consists of fundus images. To alleviate the long-tail problem, we split the original datasets into train and test sets in a 9:1 ratio, ensuring balanced sample sizes for all classes in the test sets. The evaluation metric used for experiments is classification accuracy, which measures the proportion of correctly classified samples in the test set.

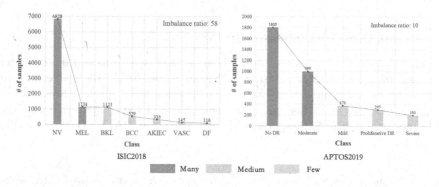

Fig. 2. Long-tailed distributions of the ISIC2018 and APTOS2019 datasets. The imbalance ratio is computed as Nmax/Nmin, where N is the number of samples in each class.

Regarding the subset division, we categorize the ISIC2018 dataset into three subsets: "Many", "Medium", and "Few". Similarly, we divide the APTOS2019 dataset into two subsets: "Many" and "Few". The specific division details are illustrated in Fig. 2. It is worth noting that we have the flexibility to categorize the datasets into an appropriate number of relational subsets based on their specific characteristics. A careful division allows us to mitigate the imbalance ratio and minimize confusion induced by training the student model with multiple teacher models. By tailoring the subset division to each dataset, we can effectively alleviate the challenges of class imbalance. This approach ensures that the student model learns from a diverse range of teacher models without being overwhelmed by conflicting information. Consequently, it enhances the learning process and improves the overall performance of the classification system.

The precise determination of the number and composition of the relational subsets depends on the characteristics of each dataset, such as the distribution of class frequencies and the specific goals of the study. This adaptive subset division strategy enables us to create a balanced training environment that better captures the underlying patterns and complexities within the dataset.

3.2 Implementation Details

Our experiments were conducted on a workstation equipped with an NVIDIA RTX 2080 Ti GPU. To augment the training data, we applied various data augmentation techniques, including random crop, horizontal flip, vertical flip, color distortion, and rotation. The backbone network used in our experiments is ResNet50 [7], and we set the batch size to 32 with an initial learning rate of 0.001. The ADAM optimizer with default settings was employed for optimization. The input images were resized to 224×224 pixels. Regarding the hyper-parameter configuration, we set the value of α to 8.0 to emphasize the importance of the NCKD component. Additionally, we set γ to 0.6 and φ to 0.83. The distillation temperature was set to 7. The training process consisted of 1,000 epochs for all datasets.

3.3 Comparison with State-of-the-Art

We conducted a comprehensive comparison between our proposed method and nine state-of-the-art approaches on the ISIC2018 and APTOS2019 datasets. In addition to some common long-tail issue mitigation techniques such as resampling and various types of loss functions, we also conducted comparative analyses with three emerging and high-performing methodologies: BALLAD [18], Paco [5], and BCL [23].

Table 1. Comparisons with state-of-the-art methods. The best ones are bolded while the second best ones are underlined.

Method	Publication	ISIC				APTOS		
		All	Many	Med	Few	All	Many	Few
CE (baseline)	–	0.621	0.817	0.401	0.382	0.552	**0.925**	0.405
RS [21]	MICCAI'2022	0.742	0.775	0.595	0.420	0.536	0.850	0.450
RW [8]	CVPR2016	0.721	0.760	0.590	0.430	0.576	0.880	0.370
Focal Loss [16]	ICCV'2017	0.707	0.725	0.600	0.475	0.592	0.900	0.450
LDAM [3]	NeurIPS'2019	0.750	0.770	0.600	0.460	0.600	0.850	0.400
BALLAD [18]	–	0.614	0.605	**0.750**	0.500	0.536	0.820	0.548
Paco [5]	ICCV'2021	0.764	0.839	0.646	0.726	0.596	0.683	0.538
BCL [23]	CVPR'2022	0.786	0.825	0.730	0.820	0.660	0.724	0.617
LMF [20]	–	0.779	0.840	0.710	0.815	0.624	0.780	0.520
MEDKD (ours)	–	**0.800**	**0.875**	0.730	**0.825**	**0.668**	0.736	**0.623**

The experimental results, presented in Table 1, showcase the performance of each method in terms of accuracy. Notably, our method achieved significant accuracy improvements compared to the second-best approach. Specifically, we observed a substantial 2.1% accuracy improvement on the ISIC2018 dataset and a notable 0.8% improvement on the APTOS2019 dataset.

Moreover, our framework demonstrates remarkable improvements in the performance of minority classes, while maintaining high accuracy for the majority classes. This is particularly evident in the "Few" subsets of both datasets, where we observed substantial accuracy improvements. Our method effectively enhances the classification performance of under-represented classes in the "Few" subsets without sacrificing the performance of majority classes. These results confirm the effectiveness and superiority of our proposed method in alleviating the challenges of medical image classification.

Table 2. Comparison of different loss function combine with KD and DKD.

	ISIC		APTOS	
	KD	DKD	KD	DKD
CE	0.628	0.646	0.580	0.592
Focal	0.692	0.714	0.628	0.640
LDAM	0.786	0.779	0.656	0.660
LMF	0.750	**0.800**	0.648	**0.668**

3.4 Ablation Study

In the ablation study, we compared the performance of several different loss functions, namely Cross-Entropy(CE), LDAM loss, Focal loss, and LMF loss, in conjunction with two different distillation methods, KD and DKD. The goal was to assess their impact on the classification performance in the context of the long-tail problem. The experimental results are presented in Table 2.

Our experimental results demonstrated that the combination of lmf loss function and DKD distillation method achieved the best performance among all the evaluated configurations. This finding suggests that lmf loss, which is specifically designed to alleviate the challenges posed by long-tail datasets, effectively improves the classification accuracy for minority classes. Moreover, the DKD method, which leverages the logits of the teacher models to guide the training of the student model, facilitates knowledge transfer and further enhances the overall performance.

During the hyperparameter tuning process, we thoroughly investigate two critical hyperparameters in our framework: the distillation temperature (T) and the ratio $\varepsilon/(1 - \varepsilon)$ (referred to as "ratio"). We conduct extensive experiments on the ISIC2018 dataset to evaluate the impact of these hyperparameters on the top-1 accuracy (Acc). For different values of T, specifically T values of 3, 5, 7, and 10, we observe the corresponding top-1 accuracies as 0.729, 0.757, 0.800, and 0.743, respectively. Remarkably, the optimal performance is achieved when T is set to 7. Likewise, when exploring various values of the ratio parameter, specifically ratio values of 0.5, 2, 5, and 10, we observe the corresponding top-1 accuracies as 0.750, 0.786, 0.800, and 0.786, respectively. Intriguingly, the best result is obtained when the ratio is set to 5. These compelling findings underscore the paramount importance of meticulous hyperparameter tuning to optimize the performance of our framework. By selecting T as 7 and the ratio as 5, we achieve the finest performance on the ISIC2018 dataset, solidifying the significance of hyperparameter optimization in our proposed framework.

4 Discussion and Conclusion

In summary, we have introduced a novel framework called MEDKD to alleviate the challenges of medical image classification, particularly in the presence of long-tailed datasets. We incorporate the concepts of TCKD and NCKD to construct a new framework for multiple expert knowledge distillation. This framework effectively transfers knowledge from multiple teacher models to the student model. Additionally, we use a powerful loss function, LMF loss, to further enhance the classification performance. Our experimental results on publicly available datasets have demonstrated the superiority of MEDKD, achieving significant improvements in accuracy compared to state-of-the-art methods.

Acknowledgement. This study was supported by the Shenzhen Basic Research Program (JCYJ20190809120205578); the National Natural Science Foundation of China (62071210); the Shenzhen Science and Technology Program (RCYX202106091030 56042); the Shenzhen Basic Research Program (JCYJ20200925153847004); the Shenzhen Science and Technology Innovation Committee Program (KCXFZ20201221 17340001).

References

1. Ba, J., Caruana, R.: Do deep nets really need to be deep? In: Advances in Neural Information Processing Systems, vol. 27 (2014)
2. Cai, Z., Lin, L., He, H., Tang, X.: Corolla: an efficient multi-modality fusion framework with supervised contrastive learning for glaucoma grading. In: 2022 IEEE 19th International Symposium on Biomedical Imaging (ISBI), pp. 1–4. IEEE (2022)
3. Cao, K., Wei, C., Gaidon, A., Arechiga, N., Ma, T.: Learning imbalanced datasets with label-distribution-aware margin loss. In: Advances in Neural Information Processing Systems, vol. 32 (2019)
4. Codella, N., et al.: Skin lesion analysis toward melanoma detection 2018: a challenge hosted by the international skin imaging collaboration (ISIC). arXiv preprint arXiv:1902.03368 (2019)
5. Cui, J., Zhong, Z., Liu, S., Yu, B., Jia, J.: Parametric contrastive learning. In: Proceedings of the IEEE/CVF International Conference on Computer Vision, pp. 715–724 (2021)

6. Han, H., Wang, W.-Y., Mao, B.-H.: Borderline-SMOTE: a new over-sampling method in imbalanced data sets learning. In: Huang, D.-S., Zhang, X.-P., Huang, G.-B. (eds.) ICIC 2005. LNCS, vol. 3644, pp. 878–887. Springer, Heidelberg (2005). https://doi.org/10.1007/11538059_91

7. He, K., Zhang, X., Ren, S., Sun, J.: Deep residual learning for image recognition. In: Proceedings of the IEEE Conference on Computer Vision and Pattern Recognition, pp. 770–778 (2016)

8. Huang, C., Li, Y., Loy, C.C., Tang, X.: Learning deep representation for imbalanced classification. In: Proceedings of the IEEE Conference on Computer Vision and Pattern Recognition, pp. 5375–5384 (2016)

9. Huang, Y., Huang, W., Luo, W., Tang, X.: Lesion2void: unsupervised anomaly detection in fundus images. In: 2022 IEEE 19th International Symposium on Biomedical Imaging (ISBI), pp. 1–5. IEEE (2022)

10. Huang, Y., Lin, L., Cheng, P., Lyu, J., Tam, R., Tang, X.: Identifying the key components in resnet-50 for diabetic retinopathy grading from fundus images: a systematic investigation. Diagnostics 13(10), 1664 (2023)

11. Huang, Y., et al.: Automated hemorrhage detection from coarsely annotated fundus images in diabetic retinopathy. In: 2020 IEEE 17th International Symposium on Biomedical Imaging (ISBI), pp. 1369–1372. IEEE (2020)

12. Ju, L., et al.: Relational subsets knowledge distillation for long-tailed retinal diseases recognition. In: de Bruijne, M., et al. (eds.) MICCAI 2021. LNCS, vol. 12908, pp. 3–12. Springer, Cham (2021). https://doi.org/10.1007/978-3-030-87237-3_1

13. Kang, B., Li, Y., Xie, S., Yuan, Z., Feng, J.: Exploring balanced feature spaces for representation learning. In: International Conference on Learning Representations (2020)

14. Karthick, M., Sohier, D.: Aptos 2019 blindness detection. Kaggle (2019). https://kaggle.com/competitions/aptos2019-blindness-detection Go to reference in chapter

15. Lin, L., et al.: BSDA-Net: a boundary shape and distance aware joint learning framework for segmenting and classifying OCTA images. In: de Bruijne, M., et al. (eds.) MICCAI 2021. LNCS, vol. 12908, pp. 65–75. Springer, Cham (2021). https://doi.org/10.1007/978-3-030-87237-3_7

16. Lin, T.Y., Goyal, P., Girshick, R., He, K., Dollár, P.: Focal loss for dense object detection. In: Proceedings of the IEEE International Conference on Computer Vision, pp. 2980–2988 (2017)

17. Liu, Z., Miao, Z., Zhan, X., Wang, J., Gong, B., Yu, S.X.: Large-scale long-tailed recognition in an open world. In: Proceedings of the IEEE/CVF Conference on Computer Vision and Pattern Recognition, pp. 2537–2546 (2019)

18. Ma, T., et al.: A simple long-tailed recognition baseline via vision-language model. arXiv preprint arXiv:2111.14745 (2021)

19. Mohammed, R., Rawashdeh, J., Abdullah, M.: Machine learning with oversampling and undersampling techniques: overview study and experimental results. In: 2020 11th International Conference on Information and Communication Systems (ICICS), pp. 243–248. IEEE (2020)

20. Sadi, A.A., et al.: Lmfloss: a hybrid loss for imbalanced medical image classification. arXiv preprint arXiv:2212.12741 (2022)

21. Shen, L., Lin, Z., Huang, Q.: Relay backpropagation for effective learning of deep convolutional neural networks. In: Leibe, B., Matas, J., Sebe, N., Welling, M. (eds.) ECCV 2016. LNCS, vol. 9911, pp. 467–482. Springer, Cham (2016). https://doi.org/10.1007/978-3-319-46478-7_29

22. Xiang, L., Ding, G., Han, J.: Learning from multiple experts: self-paced knowledge distillation for long-tailed classification. In: Vedaldi, A., Bischof, H., Brox, T., Frahm, J.-M. (eds.) ECCV 2020. LNCS, vol. 12350, pp. 247–263. Springer, Cham (2020). https://doi.org/10.1007/978-3-030-58558-7_15
23. Zhu, J., Wang, Z., Chen, J., Chen, Y.P.P., Jiang, Y.G.: Balanced contrastive learning for long-tailed visual recognition. In: Proceedings of the IEEE/CVF Conference on Computer Vision and Pattern Recognition, pp. 6908–6917 (2022)

Leveraging Ellipsoid Bounding Shapes and Fast R-CNN for Enlarged Perivascular Spaces Detection and Segmentation

Mariam Zabihi[1,2]([✉])(iD), Chayanin Tangwiriyasakul[3], Silvia Ingala[4,5], Luigi Lorenzini[4], Robin Camarasa[7], Frederik Barkhof[4,6](iD), Marleen de Bruijne[7,8], M. Jorge Cardoso[3], and Carole H. Sudre[1,9](iD)

[1] Department of Population Science and Experimental Medicine, University College London, London, UK
{m.zabihi,c.sudre}@ucl.ac.uk
[2] Donders Institue, Radboud University, Nijmegen, The Netherlands
m.zabihi@donders.ru.nl
[3] School of Biomedical Engineering, King's College London, London, UK
{chayanin.tangwiriyasakul,m.jorge.cardoso}@kcl.ac.uk
[4] Department of Radiology and Nuclear Medicine, VU University Medical Center, Amsterdam, The Netherlands
{s.ingala,l.lorenzini}@amsterdamumc.nl
[5] Department of Radiology, Copenhagen University Hospital Rigshopitalet, Copenhagen, Denmark
[6] Department of Medical Physics and Biomedical Engineering, University College London, London, UK
f.barkhof@ucl.ac.uk
[7] Department of Radiology and Nuclear Medicine, Erasmus Medical Center, Rotterdam, The Netherlands
{r.camarasa,marleen.debruijne}@erasmusmc.nl
[8] Department of Computer Science, University of Copenhagen, Copenhagen, Denmark
[9] Department Computer Engineering, University College London, London, UK

Abstract. Enlarged perivascular spaces (EPVS) are small fluid-filled spaces surrounding blood vessels in the brain. They have been found to be important in the development and progression of cerebrovascular disease, including stroke, dementia, and cerebral small vessel disease. Their accurate detection and quantification are crucial for early diagnosis and better management of these diseases.

In recent years, object detection techniques such as Mask R-CNN approach have been widely used to automate the detection and segmentation of small objects. To account for the tubular shape of these markers we use ellipsoid shapes instead of bounding boxes to express the location of individual elements in the implementation of the Fast R-CNN. We investigate the performance of this model under different modality combinations and find that the T2 modality alone, as well as the combination of T1+T2, deliver better performance.

© The Author(s), under exclusive license to Springer Nature Switzerland AG 2024
X. Cao et al. (Eds.): MLMI 2023, LNCS 14349, pp. 325–334, 2024.
https://doi.org/10.1007/978-3-031-45676-3_33

Keywords: Ellipsoid bounding shapes · Ellipsoid bounding shapes · Fast R-CNN · Cerebrovascular diseases · enlarged perivascular spaces

1   Introduction

Enlarged perivascular spaces (EPVS) are minute, fluid-filled cavities that cradle the blood vessels coursing through the brain. Despite their minuscule size, their presence is associated with cognitive decline and dementia onset. They are also associated with the presence and progression of other markers of cerebral small vessel disease such as microbleeds and white matter hyperintensities (WMH) [1–3]. Accurate detection and quantification of these diminutive markers are crucial, as they pave the way for early diagnosis and improved management strategies for such ailments. Although mainly tubular, EPVS are highly diverse in spatial distribution. This variation in location adds a degree of difficulty in EPVS detection, that, combined with their small size renders their process arduous. In addition, while they are often visually graded on T2-weighed images, EPVS may be mimicked by that of white matter hyperintensities leading to possible misdetection of punctuate WMH as EPVS due to similarities in intensity signatures.

3D object detection models are revolutionizing medical imaging by precisely identifying and localizing anomalies in volumetric scans such as CT, MRI, and mammograms. These models excel in detecting larger structures like breast cancer lesions, brain tumors, or lung nodules, assisting in early diagnosis [4–6]. However, for smaller features like Enlarged Perivascular Spaces (EPVS), traditional 3D detection methods may not provide the same level of accuracy due to their small size and complex morphology. Despite these challenges, recent advancements in 3D object detection methods have significantly propelled the field of automated EPVS detection [7]. A larger scale study [8] offered an automated method for detecting and quantifying enlarged PVS across a broad population base and studying the relationship between EPVS and cognitive decline. In parallel, Rashid et al. [9] used a multi-scale 3D convolutional neural network incorporated deep learning to detect enlarged PVS in brain MRI scans accurately. [10] used distance transforms to detect EPVS, however, this is based on point annotations only and hence, it does not capture the shape. Despite these advances, common to many of these methods is the use of axis-oriented bounding boxes for the identification of EPVS. The tubular nature of PVS structures may lead to inaccuracies in this approach.

To bridge this gap, we developed a model in this paper, bringing together the strengths of advanced object detection methods and more representative geometric shapes. We delve into a thorough discussion of the implications and benefits of our approach, highlighting how it stands as a valuable contribution to the evolving field of automated EPVS detection. In light of its potential to provide a more accurate and efficient way to quantify and analyze EPVS, our approach opens new pathways for robust and reliable neurological research, contributing significantly to the ongoing efforts in understanding cerebrovascular disease.

2   Methods

2.1   Framework

In our study, we employed a 3D Region Convolutional Neural Network (RCNN) composed of four key stages for the detection and segmentation of Enlarged Perivascular Spaces (EPVS).

Feature Learning: The first step is to train a 3D convolutional U-Net architecture [11] using our input images. This U-Net serves as the backbone network and is tasked with learning the key features from these images. A distance map of the objects of interest forms the target for this network, enhancing its ability to extract relevant characteristics.

Region Proposal Network (RPN): The feature maps derived from the U-Net are then supplied to a Region Proposal Network (RPN). The RPN, essentially a 3D convolutional network, uses these feature maps to generate score maps. These score maps depict the probability of an object's presence at each voxel location. Hence, the RPN primarily serves as a classifier, identifying potential regions in the 3D space that are likely to contain target objects.

Non-maximum Suppression: A non-maximum suppression (NMS) process is then applied to these score maps. This process includes score thresholding to filter out regions of interest (ROIs) with low confidence. The ROIs that survive the thresholding are extracted as ellipsoids and subjected to pairwise comparison. If the overlap between any two ellipsoids exceeds a certain threshold, the ellipsoid with the lower score is suppressed. This ensures that the remaining, or surviving, ellipsoids are those with high confidence and minimal overlap, optimizing the selection process for the ROIs. We used the Hellinger distance [12] to measure the overlap between ROIs (when overall > 0.2), and we used the Euclidean distance from the center of elements as an additional measure (when the distance > 2 voxels).

Object Detection and Segmentation: The selected ellipsoids are then provided as input to another convolutional network for detailed object detection. This network includes four layers that help to fine-tune the mean object classification and object shape.

Finally, to enhance segmentation results, the distance map generated from the U-Net is integrated with the convolutional network just before its final layer. This step aids a subsequent segmentation network in generating a segmented representation of the original 3D image.

2.2   Object Shape Encoding

To represent the shape of each candidate object, we adopted a seven-parameter simplified encompassing ellipsoid. Given the expected tubular shape of EPVS, we only considered the first eigenvalue to provide us with the scale of the investigated element. Specifically, the largest eigenvalue was used for the indication of scale along the main axis, the first two components of the associated eigenvector to indicate direction, and the fractional anisotropy value of the associated tensor to represent the spread in the plane perpendicular to the main axis. The center of mass parameters (x,y,z) encoded the location of the elements of interest. From these ellipsoid characteristics, the smallest encompassing patches are iteratively fed into the RCNN component to confirm the presence of an element of interest and to perform the final segmentation. The method's pipeline is shown in Fig. 1).

2.3   Data

The Where is VALDO? challenge was run in 2021 as satellite event of MICCAI [13]. It featured 3 tasks focusing on the detection and segmentation of small markers of cerebral small vessel disease namely EPVS, cerebral microbleeds, and lacunes. For this study, we used the six subjects from the SABRE dataset available for training. The elements were segmented manually using structural MR sequences coregistered to the $1\,\text{mm}^3$ isotropic T1-weighted sequence (T1-weighted, T2-weighted, T2 FLAIR) by two raters. The final label was generated by taking the union of the objects that have been annotated by both raters. Furthermore, only objects larger than 2 voxels were considered, resulting in a database comprising 1864 EPVS elements.

3   Experiments and Results

For performance assessment, the code made available for the VALDO challenge was applied providing 2 outputs related to detection (Absolute element difference and F1 score) and 2 outputs related to segmentation quality (Mean dice over true positive elements and absolute volume difference). An element is considered a true positive if the Intersection over Union with the ground truth is more than 0.10. We addressed the following aspects in our experiments: 1) Segmentation performance according to modality combination 2) Segmentation and detection of performance variability according to location. 3) Segmentation performance on the VALDO test set (SABRE component).

To ascertain which imaging modality is more informative for EPVS detection, we trained the model using different modalities, including T1, T2, and FLAIR, as well as all possible combinations of these modalities. This approach resulted in a total of seven distinct models for evaluation. Given our limited sample size, we employed leave-one-out cross-validation. Therefore, for each modality combination, we generated six different models.

During training, each model was fed with random 3D patches (size = 64) of the training images. Figure 2 shows an example of the gold standard and the model's output in T2 modality.

Figure 3 displays the final segmented image.

The out-of-sample results presented in Table 1 show the average performance across these cross-validated models for the different modality combinations. We applied the T1+T2 model to data from the testset of the VALDO challenge including only 3D segmentations (SABRE), which resulted in an F1 score of .14 ± .5 and a Dice coefficient of .25 ± .3.

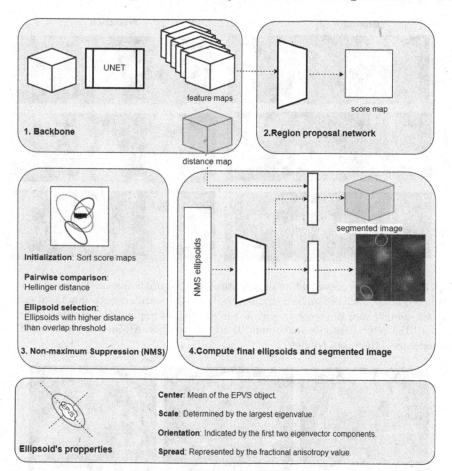

Fig. 1. Method's overview: (1) The 3D image undergoes processing via U-Net, yielding feature and distance maps. (2) These maps feed into the Region Proposal Network (RPN) to produce score maps. (3) Non-Maximum Suppression (NMS) sorts these score maps, filtering out low-confidence ROIs, leaving behind high-confidence, minimally overlapping ellipsoids. (4) These ellipsoids are used by a convolutional network, guided by the distance map from U-Net, for precise object detection.

3.1 EPVS Detection According to Location: Centrum Semi-ovale vs Basal Ganglia

EPVS can occur anywhere in the brain but regional considerations are often adopted separating in particular basal ganglia (BG) from the centrum semi-ovale (CSO) but also hippocampus and midbrain. While the EPVS load across regions is highly correlated and generally increases with age, differences in the association with risk factors are observed across regions. For instance, the relationship with hypertension is particularly notable in the basal ganglia [14]. Figure 4 shows the comparative ratios of the voxel-wise appearance of EPVS and the corresponding

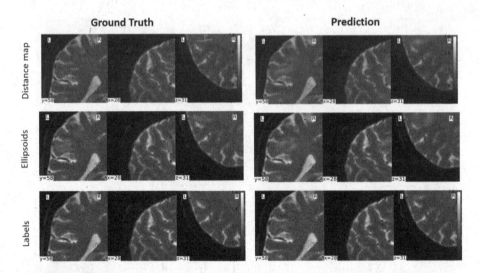

Fig. 2. Comparative visualization of gold standard and predictive model outputs on an example patch. The first column exhibits the 'gold standard', expert-annotated examples for comparison, the second column displays the corresponding outputs generated by our 3D-UNET (distance map) and 3D RCNN models. All images are overlaid on T2-weighted MR image patches.

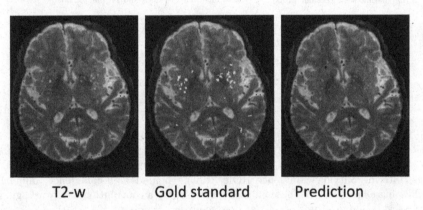

Fig. 3. Segmentation visualization Overlaid on a Full 3D T2-weighted MRI Scan.

total number of EPVS in the BG and CSO regions for the gold standard and prediction, true positive prediction (of modality: T1+T2).

Table 2 provides a comprehensive summary of the performance of the model for different modality combinations, when tested on the left-out samples in a cross-validation setting.

Table 1. Model Performance Summary (mean and standard deviation across six subjects for each measure) for the different modality combinations on the left-out sample in the cross-validation setting.

Modality	Absolute difference	F1	Dice	Absolute volume difference
T1	167 ± 78	0.16 ± 0.04	0.39 ± 0.03	2014 ± 1815
T2	132 ± 80	0.21 ± 0.04	$\mathbf{0.43 \pm 0.04}$	$\mathbf{2107 \pm 1978}$
FLAIR	192 ± 105	0.02 ± 0.01	0.21 ± 0.1	2759 ± 225
T1+T2	$\mathbf{114 \pm 82}$	$\mathbf{0.22 \pm 0.03}$	0.43 ± 0.04	2180 ± 1810
T1+FLAIR	188 ± 91	0.09 ± 0.09	0.35 ± 0.05	2529 ± 2409
T2+FLAIR	151 ± 104	0.13 ± 0.06	0.37 ± 0.04	2366 ± 2109
T1+T2+FLAIR	140 ± 78	0.20 ± 0.05	$\mathbf{0.43 \pm 0.04}$	2642 ± 1435

4 Discussion

Our findings indicate that among the examined modalities, T2, and specifically the combination of T1 and T2, combination of modalities used by the annotators, showed superior performance. This suggests that T2-weighted imaging provides more relevant information for the identification of EPVS within the framework of our study and aligns with successful strategies used elsewhere (e.g. winner of VALDO challenge).

Our results demonstrate that the model's performance in detecting EPVS elements was slightly superior in ganglionic regions compared to others. This may be attributed to the higher prevalence of EPVS voxels in these regions, thereby affecting the balance of examples during training. The difference in performance may be related to 1) the smaller amount of training examples in the CSO compared to the BG region and 2) the higher variability of the background tissue when comparing CSO to BG. This could be addressed in the future by either optimizing the patch sampling or separating the training for the two regions.

Given our multi-stage approach, tracking down the model's performance across different stages provides insights into where improvements can be made. For instance, engineering the sampling during training to ensure that regions like the CSO are adequately represented could be a beneficial adaptation. While our method currently achieves a lower performance than the contenders of the VALDO challenge overall for the SABRE dataset (winner with F1 around 0.35), it was similar to inter-rater F1 and performance in the BG appears higher. In addition, the drop in performance we observed between cross-validation testing and hold-out data testing may be due to our choice to only train on elements larger than 3 voxels and on examples where raters agreed.

Furthermore, this model, which takes into account the unique geometric morphology of EPVS, uses four parameters to encapsulate the shape of each candidate object. This offers a good geometrical fit to the general presentation of EPVS, a factor that could be clinically relevant when capturing the shape of EPVS.

Future Direction: Despite the promising results, there are still opportunities for further optimization. Refinements could be made through fine-tuning model

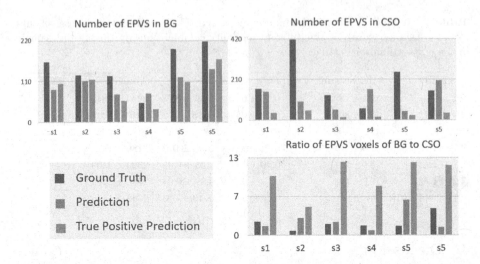

Fig. 4. Top-left: the number of EPVS in the BG, Top-right: the number of EPVS in CSO, Bottom: voxel-wise ratios of EPVS in BG to CSO for the gold standard, prediction, and true positive prediction (T1+T2 model).

Table 2. Model Performance Summary (mean and standard deviation across six subjects for each measure) for the different modality combinations on the left-out sample in the cross-validation setting for Basel Ganglia and CSO regions.

Basal Ganglia				
Modality	Absolute difference	F1	Dice	Absolute volume difference
T1	53 ± 21	0.29 ± 0.08	0.39 ± 0.04	872 ± 777
T2	37 ± 15	0.35 ± 0.05	$\mathbf{0.46 \pm 0.05}$	923 ± 737
FLAIR	79 ± 34	0.05 ± 0.04	0.18 ± 0.14	1453 ± 1411
T1+T2	$\mathbf{28 \pm 16}$	$\mathbf{0.37 \pm 0.05}$	$\mathbf{0.47 \pm 0.03}$	1003 ± 353
T1+FLAIR	69 ± 28	0.14 ± 0.14	0.31 ± 0.09	1070 ± 476
T2+FLAIR	50 ± 22	0.22 ± 0.08	0.40 ± 0.04	1284 ± 750
T1+T2+FLAIR	39 ± 18	0.32 ± 0.09	0.43 ± 0.05	1294 ± 296
CSO				
Modality	Absolute difference	F1	Dice	Absolute volume difference
T1	118 ± 82	0.07 ± 0.01	$\mathbf{0.43 \pm 0.04}$	1069 ± 1120
T2	101 ± 88	0.09 ± 0.04	0.34 ± 0.00	992 ± 1156
FLAIR	123 ± 96	0.01 ± 0.01	0.14 ± 0.18	1096 ± 1173
T1+T2	$\mathbf{98 \pm 86}$	$\mathbf{0.1 \pm 0.03}$	0.36 ± 0.06	1186 ± 1053
T1+FLAIR	123 ± 88	0.03 ± 0.04	0.39 ± 0.33	1057 ± 958
T2+FLAIR	113 ± 98	0.05 ± 0.02	0.41 ± 0.15	1092 ± 1171
T1+T2+FLAIR	110 ± 82	0.09 ± 0.03	0.41 ± 0.04	1100 ± 1045

parameters, such as the number and types of features extracted, to enhance the precision of EPVS detection. Another potential improvement could involve implementing smart sampling based on the frequency of EPVS appearance in specific regions. By incorporating a prior probability map of EPVS occurrences, the overall process could be significantly improved. Lastly, we noted a tendency for oversegmentation in most of our models which could be alleviated by some restriction on the expected individual element volume.

Moreover, our approach could be extended beyond the detection of EPVS. Given the prevalence of elongated or irregularly shaped structures in various biological contexts, our model could be useful for the detection and analysis of other biomarkers. This positions our model as a potentially powerful tool in the broader field of medical imaging analysis.

Acknowledgements. Wellcome Trust (082464/Z/07/Z), British Heart Foundation (SP/07/001/23603, PG/08/103, PG/12/29/29497 and CS/13/1/30327), Erasmus MC University Medical Center, the Erasmus University Rotterdam, the Netherlands Organization for Scientific Research (NWO) Grant 918-46-615, the Netherlands Organization for Health Research and Development (ZonMW), the Research Institute for Disease in the Elderly (RIDE), and the European Union Seventh Framework Programme (FP7/2007-2013) under grant agreement No. 601055, VPHDARE@IT, the Dutch Technology Foundation STW. This study was also supported by (WT203148/Z/16/Z) and Wellcome Flagship Programme (WT213038/Z/18/Z).

MZ and CHS are supported by an Alzheimer's Society Junior Fellowship (AS-JF-17-011). SI and LL have received funding from the Innovative Medicines Initiative 2 Joint Undertaking under Amyloid Imaging to Prevent Alzheimer's Disease (AMYPAD) grant agreement No. 115952 and European Prevention of Alzheimer's Dementia (EPAD) grant No. 115736. This Joint Undertaking receives the support from the European Union's Horizon 2020 Research and Innovation Programme and EFPIA. Rc and MdB is supported by Netherlands Organisation for NWO project VI.C.182.042.

References

1. Bown C.W: Physiology and clinical relevance of enlarged perivascular spaces in the aging brain. Neurology **98**(3), 107–117 (2022)
2. Paradise, M.: Association of dilated perivascular spaces with cognitive decline and incident dementia. Neurology **96**(11), 1501–1511 (2021)
3. Ding, J.: Large perivascular spaces visible on magnetic resonance imaging, cerebral small vessel disease progression, and risk of dementia: the age, gene/environment susceptibility-Reykjavik study. JAMA Neurol. **74**(9), 1105–1112 (2017)
4. Asgari, T.S.: Deep semantic segmentation of natural and medical images: a review. Artif. Intell. Rev. **54**, 137–178 (2021)
5. Ranjbarzadeh R. : Brain tumor segmentation of MRI images: a comprehensive review on the application of artificial intelligence tools. Comput. Biol. Med. **152** (2023)
6. Ribli D.: Detecting and classifying lesions in mammograms with Deep Learning. Sci. Rep. **8**(1), 4165 (2018)
7. Williamson, B.: Automated grading of enlarged perivascular spaces in clinical imaging data of an acute stroke cohort using an interpretable, 3D deep learning framework. Sci. Rep. **12**(1), 1–7 (2023)
8. Dubost, F.: Enlarged perivascular spaces in brain MRI: automated quantification in four regions. NeuroImage **185**, 534–544 (2019)

9. Rashid, T.: Deep learning based detection of enlarged perivascular spaces on brain MRI. Neuroimage Rep. **3**(1), 100162 (2023)

10. van Wijnen, K.M.H., et al.: Automated lesion detection by regressing intensity-based distance with a neural network. In: Shen, D., et al. (eds.) MICCAI 2019. LNCS, vol. 11767, pp. 234–242. Springer, Cham (2019). https://doi.org/10.1007/978-3-030-32251-9_26

11. Çiçek, Ö., Abdulkadir, A., Lienkamp, S.S., Brox, T., Ronneberger, O.: 3D U-net: learning dense volumetric segmentation from sparse annotation. In: Ourselin, S., Joskowicz, L., Sabuncu, M.R., Unal, G., Wells, W. (eds.) MICCAI 2016. LNCS, vol. 9901, pp. 424–432. Springer, Cham (2016). https://doi.org/10.1007/978-3-319-46723-8_49

12. Fu, GH.: Hellinger distance-based stable sparse feature selection for high-dimensional class-imbalanced data. BMC Bioinform. **21**(121) (2020)

13. Sudre, Carole H.: Where is VALDO? VAscular lesions detection and segmentation challenge at MICCAI 2021. arXiv preprint arXiv:2208.07167 (2022)

14. Wardlaw, J.M.: Perivascular spaces in the brain: anatomy, physiology and pathology. Nat. Rev. Neurol. **16**, 137–153 (2020)

Non-uniform Sampling-Based Breast Cancer Classification

Santiago Posso Murillo(✉) ⓘ, Oscar Skean ⓘ, and Luis G. Sanchez Giraldo ⓘ

University of Kentucky, Lexington, USA
{spo,oscar.skean,luis.sanchez}@uky.edu

Abstract. The emergence of deep learning models and their remarkable success in visual object recognition and detection have fueled the medical imaging community's interest in integrating these algorithms to improve medical screening and diagnosis. However, natural images, which have been the main focus of deep learning models, and medical images, such as mammograms, have fundamental differences. First, breast tissue abnormalities are often smaller than salient objects in natural images. Second, breast images have significantly higher resolutions. To fit these images to deep learning approaches, they must be heavily downsampled. Otherwise, models that address high-resolution mammograms require many exams and complex architectures. Spatially resizing mammograms leads to losing discriminative details that are essential for accurate diagnosis. To address this limitation, we develop an approach to exploit the relative importance of pixels in mammograms by conducting non-uniform sampling based on task-salient regions generated by a convolutional network. Classification results demonstrate that non-uniformly sampled images preserve discriminant features requiring lower resolutions to outperform their uniformly sampled counterparts.

Keywords: Mammogram Classification · Non-Uniform Sampling · Saliency

1 Introduction

Breast cancer is one of the leading cancer-related causes of death among women. American Cancer Society (ACS) projected almost three hundred thousand new cases of breast cancer diagnosed in the United States in 2022 [26]. Although screening mammography can reveal suspicious lesions that may lead to the presence of cancer, the predictive accuracy of radiologists is low because of the variation of abnormalities in terms of texture, density, size, distribution, and shape [2]. Deep Learning models such as Convolutional Neural Networks have achieved outstanding results in computer vision [12,13,22]. Therefore, the breast imaging

University of Kentucky.

Supplementary Information The online version contains supplementary material available at https://doi.org/10.1007/978-3-031-45676-3_34.

X. Cao et al. (Eds.): MLMI 2023, LNCS 14349, pp. 335–345, 2024.
https://doi.org/10.1007/978-3-031-45676-3_34

community has shown interest in applying those models to solving the problem of poor diagnosis performance [32].

Approaches introduced in [11, 23] have achieved excellent results in breast cancer classification. However, deep learning architectures such as Resnet, VGG, and DenseNet, used in these works, were initially conceived for "natural images," where objects of interest occupy big areas relative to the entire image. This allows the networks to work well without requiring high-resolution images. On the other hand, salient objects in mammograms are usually tiny and sparsely distributed, requiring high-resolution images to preserve important details. Uniformly downsampling mammogram images to fit models trained for object recognition can be detrimental to classification performance [5]. Downsampled mammograms will likely make lesions hard to detect due to loss in resolution.

Several works address high-resolution restrictions using pixel-level annotations to crop patches that enclose benign and malignant findings [9, 16, 34]. Then, the patch datasets are utilized for training patch classifiers. However, these approaches neglect global information from the mammograms and lack interpretable results; extracted features might not be helpful for radiologists to understand associations between mammograms and the model's predictions. Other works [5, 24] address high-resolution images, but their results are based on the NYU breast cancer screening dataset [30], a proprietary large-scale dataset.

We propose a non-uniform sampling strategy for extracting discriminative features from high-resolution mammography to address these challenges. The rationale is that relevant information for discrimination is contained in local features from the regions where lesions are present and global features from the image as a whole. The non-uniform sampling reduces the image input size to the CNN but keeps important details by maintaining higher resolutions on areas containing lesions. Inspired by the work introduced in [20, 23], our system is composed of a patch-level network that highlights salient regions to be sampled at high resolution, a mapping that performs the non-uniform sampling and a deep CNN that classifies the non-uniformly sampled image. This model is evaluated on the CBIS-DDSM dataset [10]. Although [4] introduces a Graph Attention Network where a zooming operation is modeled as connections between neighboring scales for breast cancer classification, to the best of our knowledge, our work makes the first attempt to implement a non-uniform sampler directly on the whole image for mammogram screening.

The contributions of the proposed approach are: 1. It addresses mammograms at high resolution as input by reducing their size without downsampling the relevant details of the image. 2. Exploits the relative importance of pixels in mammograms by conducting non-uniform sampling based on the task-salient regions generated by a patch-level classifier. 3. Produces human-readable outputs in the form of warped images, which allow the model to justify its results.

2   Related Work

2.1   Breast Cancer Classification

Most of the Breast cancer classification tasks have followed two main paradigms: fully supervised, where models depend on pixel-level annotations; and weakly supervised models that rely on image-level labels. In the fully supervised paradigm, models extract smaller patches from the mammograms that enclose the breast lesions and train patch classifiers. Works such as [14–16] employ different transformations as feature extractors and then use traditional classifiers like Support Vector Machines and K-NNs to classify breast abnormalities. Several works also have used deep networks to form patch-level predictions [11,19]. A drawback of such architectures is the lack of interpretability and reliance on local information.

Some models utilize pixel-level labels in their pipeline's early or intermediate phases. [29] developed an interactive lesion locator that can localize and classify lesions on the mammogram through patch features. This locator is trained based on annotations and is evaluated on the MIAS dataset [27]. Alternatively, [23] proposed a ResNet-based model, which employs annotations in the first training stage to train a patch-level classifier that is subsequently used in a whole-image classifier. Their model is trained on the CBIS-DDSM dataset and achieves an AUC score of 0.86. [18] uses almost the same pipeline proposed by Shen et al. [23]. The main differences are twofold. Firstly, ResNet is replaced by an EfficientNet [28]. Secondly, rather than one whole-image classifier, they utilize two classifiers: A single-view classifier, whose weights are used to initialize a two-view classifier.

In the weakly supervised paradigm, models receive full images and are trained with only image-level labels. A deep network is generally employed to generate coarse feature maps. Then, different attention strategies are implemented to select the most important areas from the feature maps. [34] proposed a sparse deep Multi-instance Network that constrains the number of areas for categorization. Likewise, [25] proposed two pooling structures that can be aggregated to a CNN. These structures address the mammographic characteristic of large images with tiny lesions in a more suitable way than typical max and global average pooling. [24] presented an end-to-end model that learns global and local details for breast cancer screening. The global features are generated through a low-capacity CNN that works on high-resolution images. The network generates saliency maps to retrieve a fixed number of local regions via an attention mechanism. [17] follows a Multiple Instance Learning using a Graph Neural Network to learn local and global features from a mammogram in a weakly supervised manner.

2.2   Visual Attention Through Non-uniform Sampling

Instead of processing the entire scene instantly, humans selectively focus on parts of the visual space to acquire information [21]. Several models with similar sampling behaviors have been developed. [3,20,33] proposed attention-based

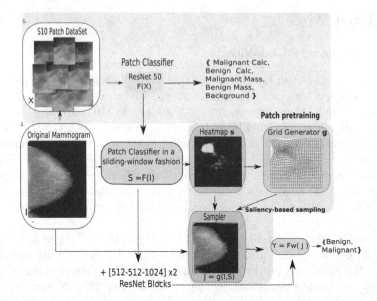

Fig. 1. Outline of the proposed Non-uniform sampling approach. The patch classifier F is first pre-trained on regions of interest and serves two functions: 1. Generating the heatmaps from the original images to guide the non-uniform sampling; and 2. As the backbone of the model since it is converted into a whole-image classifier F_w by adding two ResNet blocks as top layers [6].

samplers that highlight attended parts with high resolution guided by different attention maps. Unlike our study, these works are tested on fined-grained datasets like CUB-bird and iNaturalist. Related to medical imaging, [1] proposed an attention sampling network to localize high-resolution skin-disease regions.

Our work is similar to the models that utilize pixel-level labels in early phases. However, we only use the patch-level classifier to produce the heatmaps that guide the formation of non-uniformly sampled images. Salient regions are kept at high resolutions to preserve the fine details, while all the surrounding context is heavily downsampled. The model in [23] is used as a baseline.

3   Method

As shown in Fig. 1, a patch-level classifier is first trained using pixel-level annotations. Then, this classifier is used on the original mammograms to generate a heatmap to specify the location of saliency areas. The saliency sampler block uses the heatmap along with the original mammogram to sample the areas proportionally to their importance. Finally, the non-uniformly sampled images are passed into a whole-image classifier to determine the nature of the lesions in the breast. This framework can be divided into three stages: generation of heatmaps, image sampling, and whole-image classification.

3.1 Generation of Heatmaps

We consider the patch-level classifier as a function f, which has an input patch $X \in R^{p \times q}$ so that $f(X) \in R^c$, where c is the number of classes that f recognizes. The output of f satisfies $f(X)_i \in [0, 1]$ and $\sum_{i=1}^c f(X)_i = 1$. In our case, the number of classes $c = 5$: benign calcification, malignant calcification, benign mass, malignant mass, and background (void space and normal tissue). When f is applied in a sliding-window fashion to a whole image $M \in R^{h,w}$ where $h \gg p$ and $w \gg q$, we obtain $f(M) \in R^{a \times b \times c}$, where a and b is the height and width of the heatmap. This map is task-specific since the salient area focuses on lesions. The heatmap size (a, b) depends on the input size (h, w), the patch size (p, q), the stride s at which f is swept across the image, and padding m. Namely $a = (h + m - p)/s + 1$ and $b = (w + m - q)/s + 1$. As mentioned above, the patch-level classifier f generates a heatmap $H \in R^{a \times b \times c}$. Since we want the heatmap to indicate potential cues of cancerous lesions, we combine the outputs of the third dimension of H to obtain a single channel representing the probability of suspicious lesions in each image given by $S(x, y) = 1 - H(x, y, 0)$ where $H(x, y, 0)$ denotes the heatmap for class c_0, background at the x, y location.

3.2 Image Sampling and Classification

For the non-uniform sampling, we use the transformation introduced [20], which distorts the space based on the heatmap, described below:

$$u(x, y) = \frac{\sum_{x', y'} S(x', y') k((x, y), (x', y')) x'}{\sum_{x', y'} S(x', y') k((x, y), (x', y'))} \qquad (1)$$

$$v(x, y) = \frac{\sum_{x', y'} S(x', y') k((x, y), (x', y')) y'}{\sum_{x', y'} S(x', y') k((x, y), (x', y'))}, \qquad (2)$$

where k is a Gaussian kernel that measures the distance between a pixel (x, y) from a regular grid and its closer neighbors (x', y'), and $u(x, y)$ and $v(x, y)$ correspond to the new coordinates of (x, y) in the distorted grid D.

The map from the original image to the warped image is performed by a sampler introduced in [7]. The sampler g takes as input the distortion grid D along with the original image M and computes the warped image $J = g(M, D)$. Each (u_i^s, v_i^s) coordinate in D defines the spatial location in the input. The size of the output image depends on the size of D. Highly weighted areas in S will be represented by a larger extent in the output, and the surrounding context is held at a lower resolution. The sampled output is passed through a whole-image classifier, which uses the patch-level classifier as the backbone of a fully convolutional network to extract and combine both global and local information.

4 Experiments

To demonstrate the effectiveness of non-uniform sampling, we evaluate the performance of the whole-image classifier using different deformation degrees and resolutions. These results are compared with the performance of the whole-image classifier trained on the uniformly sub-sampled model to identify potential accuracy gains using low-resolution warped images.

4.1 Experimental Setup

Dataset. Our method is tested in the CBIS-DDSM dataset. CBIS-DDSN is a public mammography dataset that contains 3103 16-bit mammograms. 2458 mammograms (79.21%) belong to the training set, and 645 (20.79%) belong to the test set. Furthermore, 3568 crops of abnormalities and 3568 masks are included (Since some mammograms have multiple lesions, the number of masks is greater than the number of mammograms).

Processing Details. We convert mammograms from DICOM files into 16-bit PNG files. Then, we resize the mammograms to 1152×896 pixels. We split the dataset into training and test sets using an 85/15% split. We further split the training set to generate a validation set using a 90/10% split. The partitions are stratified to maintain the same proportion of cancer cases across all sets. We remove the watermarks of the mammograms and re-scale the pixel values to $[0, 1]$.

4.2 Patch Classifier Training

We generate a patch dataset (S10), which consists of 20 patches per mammogram: 10 patches randomly selected from each region of interest (ROI) with a minimum overlapping ratio of 0.9, plus 10 patches randomly selected from anywhere in the image other than the ROI. All patches have a size of 224×224. Patches are divided into one of the five classes: 0: Background, 1: Malignant Calcification, 2: Benign Calcification, 3: Malignant Mass, and 4: Benign Mass.

The patch-level classifier is based on Resnet50 [6]. A 3-stage training strategy, based on [23], is used to train the classifier on the S10 dataset. This strategy is described in the Supplementary Material. The model's parameters are initialized with the pre-trained weights in Imagenet.

To promote model generalization, during training, we apply the following augmentations: Horizontal and vertical flips, rotations in $[-25, 25]$ degrees, Zoom in $[0.8, 1.2]$ ratio, and intensity shift in $[-20, 20]\%$ of pixel values. We use ADAM optimizer [8] with $\beta_1 = 0.9$ and $\beta_2 = 0.999$ and batch size 256. The validation and test accuracy of the patch classifier is 0.97 and 0.967, respectively. Note these are not the accuracy on whole images, but only on patches.

4.3 Whole-Image Classifier

We add two identical Resnet blocks of $[512-512-1024]$ to the Patch Classifier to convert it into a whole-image classifier. Resnet blocks consist of repeated units of three convolutional layers with filter sizes 1×1, 3×3, and 1×1. The numbers in the brackets indicate the widths of the three convolutional layers in each block.

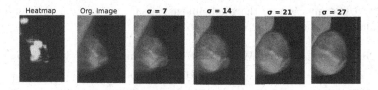

Fig. 2. Deformation levels at different σ values

Table 1. Best model performances at different deformation degrees and resolutions

Resolution	σ	Val_acc	Test_acc	AUC
(288, 224)	21	0.7197	0.7326	0.7169
(576, 448)	14	0.7727	0.7761	0.7627
(864, 672)	14	0.8106	0.8086	0.8045
(1152, 896)	14	0.8031	0.8217	0.8144

Before attaching the Resnet blocks to the patch classifier, the patch classification head is replaced with a Global Average Pooling. We connect the ResNet blocks to a fully connected layer that predicts benign or malignant classes.

Similar to the patch classifier, we employ a 2-stage training strategy described in the Supplementary Material. We use a batch size of 10. We apply the same augmentations as in the patch classifier. Performance is based on the ROC-AUC on the image-level labels (benign or malignant). We also report the validation and test accuracy. However, we evaluate models based on the ROC-AUC since it summarizes the sensitivity and specificity trade-off.

5   Results

5.1   Different Deformation Degrees in the Saliency Sampler

The size of the Gaussian kernel and the σ of the Gaussian distribution are the parameters we must set before warping the images. We set the size of the square Gaussian kernel to one-fourth of the width of the heatmap and test different sigmas that are proportional to the size of the Gaussian kernel. The sigmas tested are 7, 14, 21, and 27 (see Fig. 2). The best results are reported in Table 1. Since the sampling formulation in Eq. 1 and Eq. 2 has a bias to sample towards the image center, we address this by padding 28 pixels around the heatmap.

As shown in Table 1, we found the non-uniform sampled images at the lowest resolution are more accurate at higher deformations. It means the discriminative information is properly retained despite the heavy downsampling. At higher resolutions, a large degree of deformation ($\sigma = 21, 27$) affects the model's performance. See the supplementary material for trends and performance drop-offs through the different combinations of parameters. We also see that the resolution directly affects the model's classification performance, with high accuracy

Table 2. AUC comparison of the whole-image classifier on uniform and non-uniform sampling images guided by different saliency maps.

Resolution	Heatmap-Warp	Mask-Warp	Uniform	Random-Warp
(288,224)	0.7169	**0.7612**	0.6258	0.6661
(576,448)	0.7627	**0.7758**	0.7156	0.6879
(864,672)	0.8045	**0.8507**	0.7927	0.6887
(1152,896)	0.8144	**0.8524**	0.8456	0.7091

even at (864×672). To compare the model performance on non-uniform sampled images, we conduct the same experiment using uniform sampled images.

Additionally, to evaluate the consistency of the benefits of the non-uniform sampling approach, we directly utilize each mammogram's pixel-level annotations (masks) to guide the deformation. See column "Mask-Warp" in Table 2. Fine-tuning of σ is shown in the supplementary material. As a sanity check, we sample based on heatmaps obtained by randomly translating the mask in a range of 10–20% of its width and height.

As shown in Table 2, the non-uniform approach outperforms the uniform sampling for image classification at all resolutions. The gain in accuracy is considerable, especially for lower resolutions. It confirms our initial hypothesis: we can effectively exploit the relative importance of the saliency area, attaining discriminative information from the original resolution at lower resolutions. Since the non-uniform sampling guided by the masks outperformed the non-uniform sampling guided by the heatmaps, we conjecture that the maps generated by the patch classifier do not focus on the salient areas robustly (coarser maps). Initially, we hypothesized that the heatmaps could identify subtle areas where is difficult to determine the existence of hidden lesions. However, the resulting heatmaps were widespread, causing the compression of large areas of normal tissue, which also is used by the model to determine the nature of the lesions. A possible solution, although not explored here, would consider using attention modules to improve the quality of the heatmaps such as Deformation based Attention Consistency Loss introduced in [31] and the Gated Attention Mechanism utilized in [24]. Deformation at random areas is worse than localized warping, corroborating the utility of keeping discriminant details at high resolution.

6   Conclusion

We proposed a supervised non-uniform sampling approach to improve the classification performance of a CNN-based model on mammograms at lower resolutions. The experimental results demonstrate that preserving discriminant details from original images through non-uniform sampling enhances breast cancer classification performance. In future studies, we plan to develop a classification model that only relies on image-level labels since mammography datasets usually do not have pixel-level annotations. CNNs have been shown to naturally direct their attention to task-salient regions of the input data. These task-salient regions, in tandem with attention modules, can be used as heatmaps to apply the non-uniform sampling. By doing so, the model can be trained end-to-end without depending on pixel-level annotations to generate the heatmaps interactively.

References

1. Chen, X., Li, D., Zhang, Y., Jian, M.: Interactive attention sampling network for clinical skin disease image classification. In: Ma, H., et al. (eds.) PRCV 2021. LNCS, vol. 13021, pp. 398–410. Springer, Cham (2021). https://doi.org/10.1007/978-3-030-88010-1_33

2. Couture, H.D., et al.: Image analysis with deep learning to predict breast cancer grade, ER status, histologic subtype, and intrinsic subtype. NPJ breast cancer 4(1), 1–8 (2018)

3. Ding, Y., Zhou, Y., Zhu, Y., Ye, Q., Jiao, J.: Selective sparse sampling for fine-grained image recognition. In: Proceedings of the IEEE/CVF International Conference on Computer Vision, pp. 6599–6608 (2019)

4. Du, H., Feng, J., Feng, M.: Zoom in to where it matters: a hierarchical graph based model for mammogram analysis. arXiv preprint arXiv:1912.07517 (2019)

5. Geras, K.J., et al.: High-resolution breast cancer screening with multi-view deep convolutional neural networks. arXiv preprint arXiv:1703.07047 (2017)

6. He, K., Zhang, X., Ren, S., Sun, J.: Deep residual learning for image recognition. In: Proceedings of the IEEE Conference on Computer Vision and Pattern Recognition, pp. 770–778 (2016)

7. Jaderberg, M., Simonyan, K., Zisserman, A., et al.: Spatial transformer networks. In: Advances in Neural Information Processing Systems, vol. 28 (2015)

8. Kingma, D.P., Ba, J.: Adam: a method for stochastic optimization. arXiv preprint arXiv:1412.6980 (2014)

9. Kooi, T., Karssemeijer, N.: Classifying symmetrical differences and temporal change for the detection of malignant masses in mammography using deep neural networks. J. Med. Imaging 4(4), 044501–044501 (2017)

10. Lee, R.S., Gimenez, F., Hoogi, A., Miyake, K.K., Gorovoy, M., Rubin, D.L.: A curated mammography data set for use in computer-aided detection and diagnosis research. Scientific data 4(1), 1–9 (2017)

11. Lévy, D., Jain, A.: Breast mass classification from mammograms using deep convolutional neural networks. arXiv preprint arXiv:1612.00542 (2016)

12. Li, Y., Wang, L., Mi, W., Xu, H., Hu, J., Li, H.: Distracted driving detection by combining VIT and CNN. In: 2022 IEEE 25th International Conference on Computer Supported Cooperative Work in Design (CSCWD), pp. 908–913. IEEE (2022)

13. Liu, S., Cai, T., Tang, X., Zhang, Y., Wang, C.: Visual recognition of traffic signs in natural scenes based on improved retinanet. Entropy 24(1), 112 (2022)

14. Maqsood, S., Damaševičius, R., Maskeliūnas, R.: TTCNN: a breast cancer detection and classification towards computer-aided diagnosis using digital mammography in early stages. Appl. Sci. 12(7), 3273 (2022)

15. Mohanty, F., Rup, S., Dash, B., Majhi, B., Swamy, M.: Mammogram classification using contourlet features with forest optimization-based feature selection approach. Multimedia Tools Appl. **78**(10), 12805–12834 (2019)
16. Muduli, D., Dash, R., Majhi, B.: Automated breast cancer detection in digital mammograms: a moth flame optimization based elm approach. Biomed. Sig. Process. Control **59**, 101912 (2020)
17. Pelluet, G., Rizkallah, M., Tardy, M., Acosta, O., Mateus, D.: Multi-scale graph neural networks for mammography classification and abnormality detection. In: Yang, G., Aviles-Rivero, A., Roberts, M., Schönlieb, C.B. (eds.) Medical Image Understanding and Analysis - MIUA 2022. Lecture Notes in Computer Science, vol. 13413, pp. 636–650. Springer, Cham (2022)
18. Petrini, D.G., Shimizu, C., Roela, R.A., Valente, G.V., Folgueira, M.A.A.K., Kim, H.Y.: Breast cancer diagnosis in two-view mammography using end-to-end trained efficientnet-based convolutional network. IEEE Access **10**, 77723–77731 (2022)
19. Rahman, A.S.A., Belhaouari, S.B., Bouzerdoum, A., Baali, H., Alam, T., Eldaraa, A.M.: Breast mass tumor classification using deep learning. In: 2020 IEEE International Conference on Informatics, IoT, and Enabling Technologies (ICIoT), pp. 271–276. IEEE (2020)
20. Recasens, A., Kellnhofer, P., Stent, S., Matusik, W., Torralba, A.: Learning to zoom: a saliency-based sampling layer for neural networks. In: Ferrari, V., Hebert, M., Sminchisescu, C., Weiss, Y. (eds.) ECCV 2018. LNCS, vol. 11213, pp. 52–67. Springer, Cham (2018). https://doi.org/10.1007/978-3-030-01240-3_4
21. Rensink, R.A.: The dynamic representation of scenes. Vis. Cogn. **7**(1–3), 17–42 (2000)
22. Sha, M., Boukerche, A.: Performance evaluation of CNN-based pedestrian detectors for autonomous vehicles. Ad Hoc Netw. **128**, 102784 (2022)
23. Shen, L., Margolies, L.R., Rothstein, J.H., Fluder, E., McBride, R., Sieh, W.: Deep learning to improve breast cancer detection on screening mammography. Sci. Rep. **9**(1), 1–12 (2019)
24. Shen, Y., et al.: An interpretable classifier for high-resolution breast cancer screening images utilizing weakly supervised localization. Med. Image Anal. **68**, 101908 (2021)
25. Shu, X., Zhang, L., Wang, Z., Lv, Q., Yi, Z.: Deep neural networks with region-based pooling structures for mammographic image classification. IEEE Trans. Med. Imaging **39**(6), 2246–2255 (2020)
26. Siegel, R.L., Miller, K.D., Fuchs, H.E., Jemal, A.: Cancer statistics, 2022. CA Cancer J. Clin. **72**(1), 7–33 (2022)
27. Suckling, J., et al.: Mammographic image analysis society (MIAS) database v1. 21 (2015)
28. Tan, M., Le, Q.: Efficientnet: rethinking model scaling for convolutional neural networks. In: International Conference on Machine Learning, pp. 6105–6114. PMLR (2019)
29. Ting, F.F., Tan, Y.J., Sim, K.S.: Convolutional neural network improvement for breast cancer classification. Expert Syst. Appl. **120**, 103–115 (2019)
30. Wu, N., et al.: Deep neural networks improve radiologists' performance in breast cancer screening. IEEE Trans. Med. Imaging **39**(4), 1184–1194 (2019)
31. Xing, X., Yuan, Y., Meng, M.Q.H.: Zoom in lesions for better diagnosis: attention guided deformation network for WCE image classification. IEEE Trans. Med. Imaging **39**(12), 4047–4059 (2020)
32. Yu, X., Zhou, Q., Wang, S., Zhang, Y.D.: A systematic survey of deep learning in breast cancer. Int. J. Intell. Syst. **37**(1), 152–216 (2022)

33. Zheng, H., Fu, J., Zha, Z.J., Luo, J.: Looking for the devil in the details: learning trilinear attention sampling network for fine-grained image recognition. In: Proceedings of the IEEE/CVF Conference on Computer Vision and Pattern Recognition, pp. 5012–5021 (2019)
34. Zhu, W., Lou, Q., Vang, Y.S., Xie, X.: Deep multi-instance networks with sparse label assignment for whole mammogram classification. In: Descoteaux, M., Maier-Hein, L., Franz, A., Jannin, P., Collins, D.L., Duchesne, S. (eds.) MICCAI 2017. LNCS, vol. 10435, pp. 603–611. Springer, Cham (2017). https://doi.org/10.1007/978-3-319-66179-7_69

A Scaled Denoising Attention-Based Transformer for Breast Cancer Detection and Classification

Masum Shah Junayed$^{(\boxtimes)}$ (iD) and Sheida Nabavi (iD)

Department of CSE, University of Connecticut, Storrs, CT 06269, USA
{masumshah.junayed,sheida.nabavi}@uconn.edu

Abstract. Breast cancer significantly threatens women's health, and early, accurate diagnosis via mammogram screening has considerably reduced overall disease burden and mortality. Computer-Aided Diagnosis (CAD) systems have been used to assist radiologists by automatically detecting, segmenting, and classifying medical images. However, precise breast lesion diagnosis has remained challenging. In this paper, we propose a novel approach for breast cancer detection and classification in screening mammograms. Our model is a hybrid of CNN and Transformers, specifically designed to detect and classify breast cancer. The model first utilizes a depthwise convolution-based hierarchical backbone for deep feature extraction, coupled with an Enhancement Feature Block (EFB) to capture and aggregate multi-level features to the same scale. Subsequently, it introduces a transformer with Scale-Denoising Attention (SDA) to simultaneously capture global features. Finally, the model employs regression and classification heads for detecting and localizing lesions and classifying mammogram images. We evaluate the proposed model using the CBIS-DDSM dataset and compare its performance with those of state-of-the-art models. Our experimental results and extensive ablation studies demonstrate that our method outperforms others in both detection and classification tasks.

Keywords: Depthwise Convolutions · Enhancement Feature Block · Scaled Denoising Attention · Transformer · Mammograms

1 Introduction

Breast cancer is one of the most common diseases in the world for women, accounting for a significant number of deaths. According to the National Breast Cancer Foundation, the 5-year survival rate after early detection [12] is 99%. For an effective diagnosis and analysis of breast cancer, early detection is essential. Breast cancer can be detected using a variety of techniques, including MRI, mammography, and ultrasound [17]. Mammography is the most common breast imaging modality with annual screening recommendation by the American Cancer Society for all women starting at 40 years old. Radiologists have difficulties in recognizing probable malignant tumors due to the vast number of images and low prevalence of malignancy. Radiologists are using computer-aided diagnostic (CAD) systems to help them detect malignant breast tumors as a result of the necessity for an effective automated process [6]. However, current CAD systems cannot improve breast cancer detection significantly [26].

X. Cao et al. (Eds.): MLMI 2023, LNCS 14349, pp. 346–356, 2024.
https://doi.org/10.1007/978-3-031-45676-3_35

In recent years, AI-based CAD systems have become more popular due to their improved efficiency and ability to spot problems earlier. Because of this, researchers and developers have access to investigate machine learning (ML) algorithms for mammograms. Early ML methods were employed to extract features from mammograms which were then fed into a standard machine learning classifier [6]. However, with advancements in ML, particularly in deep learning (DL), these features can now be extracted automatically. This automated feature engineering, especially in the domain of medical image analysis, can lead to more accurate results [3,23]. The advanced DL methods especially Convolution Neural Networks (CNNs) have been employed in particular for tasks involving the localization, recognition, and classification of lesions in mammograms. DL methods are increasingly used to create automated CAD systems capable of analyzing various types of medical images [1,2,5]. These methods directly extract features by the hierarchical levels, making DL the most accurate medical imaging approach [2,18]. Numerous DL-based CAD systems have been developed for breast lesion identification, and they significantly outperform traditional methods. However, their inability to perform as fully automated detectors and their susceptibility to bias due to uniformly distributed noise in training samples remain unaddressed [2,24].

This research aims to develop a method for more precise breast cancer detection as well as classification using mammogram images. We present a hybrid (CNN+Transformer) model that can learn how to identify instances of breast lesions in mammograms by recognizing the existence of tumors, localizing them, and determining whether or not they are benign or malignant. To do the above objectives, the proposed model consists of a hierarchical backbone for feature extraction, an enhancement feature block (EFB) for capturing the details features from different levels, a transformer for correlating the local and global features, and finally, two different heads for detection and classification. The key contributions are as follows:

1. A novel CNN and transformer-based hybrid model is introduced for lesion detection and mammogram classification.
2. A depthwise-convolution-based hierarchical backbone is presented to extract both low-level and high-level features in depth.
3. An Enhancement Feature Block (EFB) is utilized, which allows for the aggregation of features from various scales to a single scale, thereby enhancing the key and pattern features of the image.
4. A scale denoising attention (SDA) is proposed to incorporate feature denoising and adaptive attention weighting that allows recording local and global feature dependency.

2 Proposed Method

Figure 1 depicts an overview of the proposed model. It primarily comprises four parts: a backbone network, an enhancement feature block (EFB), a transformer, and two heads. The CNN-based backbone is utilized for deep feature extraction. The EFB is applied to highlight the key differentiating characteristics that strengthen the regional features generated by the hierarchical backbone. Consequently, the model can concentrate on relevant data and patterns associated with breast cancer. Following that, a transformer is introduced to provide feature denoising and adaptive attention weighting, which enables the model to more effectively manage noise and capture local and global information. After that, two heads are applied for the classification and detection of breast cancer. In the following, we describe each part in more depth.

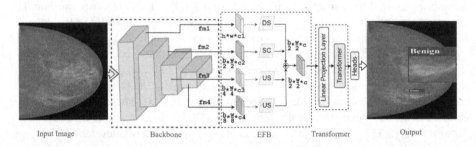

Fig. 1. Overview of the proposed architecture. It contains a CNN-based hierarchical backbone, an enhancement feature block (EFB), a transformer block, and a head. Here DS, SC, and US represent downsampling, standard convolution, and downsampling, respectively.

2.1 Depthwise-Convolution Backbone

We propose a hierarchical backbone made up of four depthwise convolutions (DC) [9] blocks to capture low-level and high-level features, more complex tumor structures, and abnormal mass patterns in depth. Figure 2 (a) represents each block, which consists of a stacked DC paired with an identity skip connection to simultaneously extract local details and the global context by capturing features at multiple scales, utilizing layers with progressively larger receptive fields. The block forms a DC with a 1×1 kernel, a stride of 2, and a filter of 32. In order to decrease the size of the second- to fourth-level feature maps, the DC layers er subsequently followed by batch normalization (BN) with a stride of 1 and the Gaussian Error Linear Unit (GELU) [14], resulting in the downsampling of the feature resolution by a factor of two. In Fig. 1, the $fm1 = h \times w \times c_1$, $fm2 = \frac{h}{2} \times \frac{w}{2} \times c_2$, $fm3 = \frac{h}{4} \times \frac{w}{4} \times c_3$, and $fm4 = \frac{h}{8} \times \frac{w}{8} \times c_4$ represents the output of 1st, 2nd, 3rd and 4th blocks, respectively. After each block, the model generates multi-level feature maps, which are then passed to the enhancement feature block (EFB).

2.2 Enhancement Feature Block (EFB)

Motivated by Scale Enhancement Module (SEM) [10], we introduced EFB to better comprehend the context of various features by correlating features of different scales. The four dilated convolution layers with different dilated rates are used to capture information at various scales from the levels 1 to 4 feature map inputs ($fm1$, $fm2$, $fm3$, and $fm4$). The features are downsampled (DS), and fine-grained characteristics are captured in the first layer, which has a dilation rate of 1. The second layer maintains the scale of the features and provides an adequate spatial context by using standard convolution (SC). A larger range of location information can be learned and upsampled (US) by using dilation rates of 2 and 3 in the third and fourth layers, respectively. The fourth layer has the largest receptive field, allowing the model to extract the most global properties and include both local and global contexts from the input information. After that, the output features of the four layers (i.e., DS, SC, US, and US) of EFB are concatenated which can be represented as $\frac{h}{2} \times \frac{w}{2} \times c$ and then fed into the transformer to detect and classify breast cancer.

2.3 Transformer

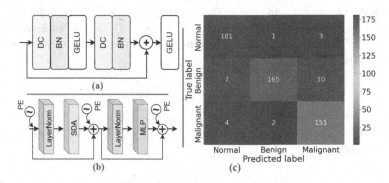

Fig. 2. The diagram depicts the backbone block (a), the transformer (b), and the confusion matrix (c). (a) The backbone block consists of DC, BN, and GELU, all of which are interconnected. (b) The transformer consists of positional embeddings (PE), LayerNorm, SDA, MLP, and skip connections.

Figure 2 (b) represents the transformer module consisting of several key components, including LayerNorm, scaled denoising attention (SDA), multi-layer perception (MLP), and skip connections. To capture contextual dependencies within the feature, the feature map $m \in \mathbb{R}^{\frac{h}{2} \times \frac{w}{2} \times c}$ is divided into fixed size query windows of size $g_h \times g_w$. For the input of the transformer, each of these query windows is subjected to a linear projection layer, generating the query window value denoted as $Q^i_s = P_q(m^s)$. Here, Q^i_s refers to the input query window value for the s-th query window, while P_q depicts the linear projection applied to the input features m^s. To construct the s-th query window for the subsequent SDA computation of query windows, denoted as $Q_s \in \mathbb{R}^{g_w \times g_w \times d}$, where d is the dimension of the query window, the input feature m is divided into several sub-windows. Each of these sub-windows has a dimension of $g^l \times g^l$. To create fine-grained and coarse-grained information for every single feature level $l \in \{1,\dots,L\}$, pooling operations are applied to aggregate the information from the sub-windows at that level. Following that, the denoising kernels are employed during the pooling operation to reduce the impact of noise and enhance the quality of the pooled features [4,27]. To spatially pool the sub-windows, we use the straightforward linear layer P^l_p, followed by:

$$m^l = P^l_p(\hat{m}) \in \mathbb{R}^{\frac{H}{g^l} \times \frac{W}{g^l} \times c}, \tag{1}$$

$$\hat{m} = Reshape(m) \in \mathbb{R}^{\left(\frac{H}{g^l} \times \frac{W}{g^l} \times c\right) \times (g^l \times g^l)}, \tag{2}$$

where $g^l \in \{g^1, \ldots, g^L\}$ represents the size of the sub-windows, P_p^l denotes the linear layer for spatial pooling, m^l represents the pooled feature at level l, H and W are the height and width of the input (m) feature patch, and c is the number of channels. To perform SDA, each query, key, and value token are extracted. We calculate queries within a query window $Q_s \in \mathbb{R}^{g_w \times g_w \times d}$. Next, the key ($K_s$) and value ($V_s$) are calculated, $K_s \in \mathbb{R}^{g_w \times g_w \times c}$ and $V_s \in \mathbb{R}^{g_w \times g_w \times c}$. Following the approach introduced in [20], a relative position bias $O^l \in \{O^1, \ldots, O^L\}$ is applied to this model for the learnable relative position in the attention computation. The relative positions along both the horizontal and vertical axes fall within the range of $[-g_w + 1, g_w - 1]$. In the first level, the bias $O^1 \in \mathbb{R}^{(2g_w-1) \times (2g_w-1)}$ is fine-tuned according to [20]. To obtain the SDA output, as expressed by the following equation [27].

$$SDA(Q_s, K_s, V_s) = Softmax\left(\frac{Q_s K_s^T}{\sqrt{c}} + O^l\right) V_s. \tag{3}$$

Let denote the updated queries, keys and values as Q_c, K_c, and V_c which can be expressed as follows:

$$Q_c = C_c + S_c, \quad K_c = C_c + S_k, \quad V_c = C_c, \tag{4}$$

where C_c refers to the learnable content query. The S_c, and S_k denote spatial query and spatial keys, which are the vectors of size dimension. The concatenated positional encodings $PE(x_c), PE(y_c), PE(w_c)$, and $PE(h_c)$ are sent through an MLP to produce the spatial query S_c. Here, c-th anchor generated by the query selection module is represented by the coordinates (x_c, y_c, w_c, h_c) in the positional encodings. Following that, the final output, denoted as P_c, is generated by concatenating the output of the MLP with the positional encoding, which can be expressed as follows:

$$P_c = MLP[Cat\{PE(x_c), PE(y_c), PE(w_c), PE(h_c)\}], \tag{5}$$

where Cat stands for the concatenation function, and PE refers to a positional encoding function that generates sinusoidal embeddings. The output of P_c is processed by the classification and detection heads.

2.4 Head

After the transformer, two more heads-regression and classification heads are applied to conduct the detection and classification tasks concurrently. The regression head predicts the bounding box coordinates encompassing the malignant and benign masses in the mammogram. Here, the feature vector P_c is not a generic input but is computed as the output of an MLP that takes as input the PE of the coordinates (x, y) and dimensions (width, height) of the proposed box by Eq. 5. The bounding box coordinates $\overrightarrow{B}$ are then calculated by applying a linear transformation to this feature vector P_c, given by $\overrightarrow{B} = W_d \cdot P_c + \overrightarrow{b}_d$. Here, W_d is a weight matrix of dimensions $4 \times D$, and $\overrightarrow{b}_d$ is a bias vector of size 4. Simultaneously, the classification head, a fully connected (FC) layer, transforms the feature vector P_c into a scalar value P_{cls}, representing the class probability distribution. Thus, the probability of the mass being malignant, benign, or normal is given by $P_{cls} = \sigma(W_c \cdot P_c + b_c)$, where W_c is a weight vector of size D, b_c is a scalar bias, and σ is the softmax function.

3   Experiments

3.1   Dataset and Preprocessing

In this study, we used the Curated Breast Imaging Subset of the Digital Database for Screening Mammography (CBIS-DDSM) [13], a renowned dataset comprising 2620 cases of breast cancer mammogram images from both left and right breasts, each with a resolution of 3000×4800 pixels. The INbreast [22] dataset also utilized to compare the effectiveness. Each case accompanies labels, providing patient details and classifications indicating whether the images represent normal, benign, or malignant. To evaluate the robustness of our model, we resized the mammogram images to two scales: 224×224, and 512×512 pixels. Subsequently, augmentation techniques are utilized to prevent overfitting in our model. We made the training set more diverse and the model more reliable by using augmentation techniques for each image. These techniques included flipping, rotating, cropping, normalization to make sure data scales were the same, and contrast adjustment to bring out differences in intensity.

3.2   Experimental Setup

All training and testing were done on the same computer with two Nvidia GeForce RTX 3080 Ti GPUs with 8 GB of video RAM. We used PyTorch, a CUDA environment, and a vs-code editor to put it into practice. The model was trained over a number of 20 epochs, with each epoch indicating a full pass through the dataset with a batch size of 8. We employed a combination of smooth L1 loss and cross-entropy loss for localization and classification [8]. To optimize our model, we utilized the AdamW optimizer with a 0.001 learning rate, which provides adaptive learning rates and accelerates training. To assure the robustness of the model, we implemented a 5-fold cross-validation strategy and reported the mean performance. The performance of the model in detecting cancer was assessed using Intersection over Union (IoU), accuracy, and classification efficacy using metrics such as accuracy, precision, recall, F1 Score, and area under the curve (AUC).

4   Results and Discussions

We compared the performance of our model with those of two recent breast cancer detection (DT) and classification (CF) methods [23, 26] and four breast cancer CF methods [1, 5, 18, 24]. The results are depicted in Table 1 for the DDSM dataset, which compares existing methods from 2017 to 2022 under the same experimental setup. In the CF, the Chougrad et al. [5] approach (deep CNN-based) achieves the highest sensitivity of

96.77% and the highest AUC of 94.09%, even though it doesn't participate in detection tasks. Meanwhile, the Li et al. [18] approach (DL based) delivers a commendable overall CF accuracy of 95.01%. In combined detection and classification (DT+CF), the proposed method outperforms all others in terms of DT accuracy (95.16%), IoU (88.71%), CF accuracy (96.56%), and specificity (97.88%). These performances demonstrate the effective classification of the detected regions into malignant, benign, and normal categories. This method also scores the highest in specificity (97.88%), which means it correctly identifies a high percentage of actual negative (normal and benign) cases. However, it falls slightly in terms of sensitivity and AUC, where it's surpassed by the CNN-based approaches of Platania et al. [23] and Ueda et al. [26], respectively.

Table 1. Comparative analysis with recent existing methods using 224×224 input size under the same experimental setups on the DDSM dataset. Here, CF and DT depict the classification and detection.

Type	Approaches	Detection		Classification			
		Accuracy	IoU	Accuracy	Sensitivity	Specificity	AUC
CF	Chougrad et al. [5]	–	–	96.41%	**96.77%**	96.87%	**94.09%**
	Shen et al. [24]	–	–	86.47%	86.93%	87.24%	89.38%
	Rahman et al. [1]	–	–	86.73%	86.40%	87.23%	84.91%
	Li et al. [18]	–	–	95.01%	94.85%	95.86%	91.57%
DT+CF	Platania et al. [23]	91.59%	80.40	94.14%	95.28%	94.15%	92.60%
	Ueda et al. [26]	94.32%	86.38	96.28%	96.04%	96.46%	93.58%
	Proposed Method	**95.16%**	**88.71**	**96.56%**	95.55%	**97.88%**	93.13%

Table 2 presents an ablation study of the proposed model, showing the effect of different modules on evaluation metrics such as CF accuracy (ACC), AUC, DT accuracy, and IoU, when tested on the DDSM dataset. The purpose of this study is to investigate the impact of different components of our model: the backbone network (which could be VGG-16 [25], DenseNet [15], ResNet-50 [11], and the proposed network), the existence of the EFB module, and two types of transformer attention (SDA or multihead self-attention (MSA) [21]) mechanisms. For the VGG-16 backbone, the highest CF ACC (83.24%), AUC (81.46%) and DT ACC of (87.24%), and an IoU (83.55%) is obtained when both the EFB module and the SDA transformer are utilized. Conversely, when the EFB module is omitted, the performance drops to a CF ACC of 80.88%, AUC of 78.05% and DT ACC of 85.27%, IoU of 79.01%. Similar trends are observed for DenseNet and ResNet-50 backbones. For the proposed backbone, the peak performance is obtained when the EFB module is used alongside the SDA transformer, leading to a CF accuracy of 96.56%, an AUC of 93.13%, and a DT accuracy of 95.16%. However, if the EFB module is excluded, while the SDA transformer is retained, the model's efficiency takes a slight hit, declining to a CF accuracy of 88.76%, an AUC of 92.68%, and a DT accuracy of 92.15%. The table indicates that across all backbone configurations, including the proposed one, the combination of the EFB module with the SDA transformer consistently delivers superior outcomes. This supports the importance of both the EFB module and SDA transformer in enhancing the model's ability to detect and classify breast cancer.

Table 2. Ablation study of the proposed model in terms of accuracy, AUC, and IoU on the DDSM dataset. Different combinations are used with/without considering the modules to evaluate this model.

Backbone	EFB	Transformer		Classification		Detection	
		SDA	MSA	Accuracy(%)	AUC	Accuracy(%)	IoU
VGG-16 [25]	✓	✓	×	83.24	81.46	87.24	83.55
	×	✓	×	80.88	78.05	85.27	79.01
	✓	×	✓	82.86	79.12	86.86	80.89
DensNet [15]	✓	✓	×	91.38	89.65	89.69	83.97
	×	✓	×	89.90	85.43	88.51	80.49
	✓	×	✓	88.43	86.70	88.13	81.82
ResNet-50 [11]	✓	✓	×	91.76	89.97	93.92	**88.80**
	×	✓	×	90.78	89.41	92.12	83.27
	✓	×	✓	91.06	88.95	91.97	86.64
Proposed	✓	✓	×	**96.56**	**93.13**	**95.16**	88.71
	×	✓	×	88.76	92.68	92.15	82.59
	✓	×	✓	93.52	91.47	91.10	85.46

Figure 2 (c) represents the confusion matrix for our model on the DDSM dataset, which indicates that the proposed model typically classifies mammogram images accurately, but some critical errors are notable. Specifically, it misclassifies six malignant images: four are mistaken for normal and two for benign. Such misclassifications may stem from several potential limitations, including the lack of diversity in the training data, the overlap in feature spaces across different classes, or the model's inadequacy to learn complex and subtle differences in breast cancer imaging. These errors can lead to serious consequences like treatment misdirection or delayed diagnosis. Consequently, we need further improvements such as enriching the dataset with diverse examples, refining feature extraction, or enhancing the model's discrimination ability to mitigate these limitations and improve overall performance. Table 3 compares the performance of various classification and detection techniques with 512×512 input images on the DDSM dataset. In the CF approach, Li et al. [18] obtained the highest accuracy at 91.86%. However, for the DT+CF approach, our proposed model outperformed others, achieving a top accuracy of 93.02%, an IoU score of 89.93%, a sensitivity of 92.55%, a specificity of 92.40%, and an AUC of 92.77%. Thus, the proposed model demonstrated an excellent ability to analyze high-resolution mammogram images. As seen in Tables 1 and 3, increasing the input image size from 224×224 to 512×512 provides more contextual information. However, it also introduces more background noise and irrelevant details, which makes it harder for the model to concentrate on significant features, possibly reducing its ability to capture key features, fine-grained details, and the overall context.

Table 3. Evaluating and comparing the proposed model existing methods on 512×512 input images in the same experimental setups on the DDSM dataset. Here, CF and DT depict the classification and detection.

Type	Approaches	Detection		Classification			
		Accuracy	IoU	Accuracy	Sensitivity	Specificity	AUC
CF	Chougrad et al. [5]	–	–	91.78%	90.98%	91.46%	90.66%
	Shen et al. [24]	–	–	85.12%	84.37%	85.89%	87.10%
	Rahman et al. [1]	–	–	84.19%	83.74%	85.05%	81.53%
	Li et al. [18]	–	–	91.86%	92.33%	91.76%	89.25%
DT+CF	Platania et al.'s [23]	92.71%	83.48	92.38%	92.37%	93.06%	89.25%
	Ueda et al. [26]	**92.87%**	85.29	92.98%	91.94%	92.38%	92.69%
	Proposed model	92.84%	**89.93**	**93.02%**	**92.55%**	**92.40%**	**92.77%**

Table 4. Comparative performance evaluation of the proposed model against established detection methods on the DDSM and INbreast Datasets at two different image resolutions in terms of accuracy and IoU.

Image Size	Model	DDSM		INbreast	
		Accuracy	IoU	Accuracy	IoU
224 × 224	SSD [19]	82.95	80.37	83.41	81.17
	Faster R-CNN [7]	86.47	81.60	86.89	81.85
	YOLOv5 [16]	87.15	82.64	86.72	82.09
	Proposed Model	**95.16**	**88.71**	**93.58**	**86.36**
512 × 512	SSD [19]	83.52	87.23	86.97	85.28
	Faster R-CNN [7]	85.66	88.74	88.32	87.95
	YOLOv5 [16]	89.85	**91.45**	**90.77**	89.16
	Proposed Model	**92.84**	89.93	90.24	**89.63**

Table 4 depicts a detailed comparative detection evaluation of our proposed model against established methods such as SSD [19], Faster R-CNN [7], and YOLOv5 [16], across two datasets-DDSM and INbreast-and two image sizes (224 × 224 and 512 × 512 pixels). Our model consistently outperforms the alternatives, achieving the highest accuracy and IoU scores in most scenarios. Notably, for the 224 × 224 image size, the proposed model reaches an Accuracy of 95.16% and an IoU of 88.71% on the DDSM dataset. However, at a 512 × 512 image size, YOLOv5 [16] exhibits a slightly higher IoU on the DDSM dataset, although our model still leads in accuracy. This underlines the robustness and efficacy of our proposed approach across different image dimensions and datasets.

5   Conclusion

This study introduced a novel automated hybrid model that detects and classifies breast cancer from mammogram images. The model consists of a hierarchical depthwise convolution backbone, an Enhancement Feature Block (EFB), a transformer with Scale-Denoising Attention (SDA), and two distinct heads for classification and detection. The backbone extracts multi-level features, the EFB aggregates these features, while the transformer uses SDA to simultaneously denoise and gather essential global contextual information for cancerous areas. Finally, the regression head predicts the coordinates of the bounding box for detection, and the classification head categorizes breast cancer from mammogram images. We tested our method on the DDSM database, where it achieved a detection accuracy of 95.16% and a classification accuracy of 96.56%. It demonstrated a strong ability to identify and classify malignant breast masses. For robust performance in mammograms, our future work aims to extend its capabilities to tumor localization, segmentation, and testing on more complex datasets, and ultimately, apply it to other healthcare areas for broader benefits.

References

1. Abdel Rahman, A.S., Belhaouari, S.B., Bouzerdoum, A., Baali, H., Alam, T., Eldaraa, A.M.: Breast mass tumor classification using deep learning. In: 2020 IEEE International Conference on Informatics, IoT, and Enabling Technologies (ICIoT), pp. 271–276 (2020). https://doi.org/10.1109/ICIoT48696.2020.9089535
2. Al-Antari, M.A., Al-Masni, M.A., Choi, M.T., Han, S.M., Kim, T.S.: A fully integrated computer-aided diagnosis system for digital x-ray mammograms via deep learning detection, segmentation, and classification. Int. J. Med. Inform. 117, 44–54 (2018)
3. Bai, J., Posner, R., Wang, T., Yang, C., Nabavi, S.: Applying deep learning in digital breast tomosynthesis for automatic breast cancer detection: A review. Med. Image Anal. 71, 102049 (2021)
4. Chen, H., et al.: Denoising self-attentive sequential recommendation. In: Proceedings of the 16th ACM Conference on Recommender Systems, pp. 92–101 (2022)
5. Chougrad, H., Zouaki, H., Alheyane, O.: Deep convolutional neural networks for breast cancer screening. Comput. Methods Programs Biomed. 157, 19–30 (2018)
6. Dhungel, N., Carneiro, G., Bradley, A.P.: Automated mass detection in mammograms using cascaded deep learning and random forests. In: 2015 International Conference on Digital Image Computing: Techniques and Applications (DICTA), pp. 1–8. IEEE (2015)
7. Girshick, R.: Fast R-CNN. In: Proceedings of the IEEE International Conference on Computer Vision, pp. 1440–1448 (2015)
8. Guan, B., et al.: Automatic detection and localization of thighbone fractures in x-ray based on improved deep learning method. Comput. Vis. Image Underst. 216, 103345 (2022)
9. Guo, Y., Li, Y., Wang, L., Rosing, T.: Depthwise convolution is all you need for learning multiple visual domains. In: Proceedings of the AAAI Conference on Artificial Intelligence, vol. 33, pp. 8368–8375 (2019)
10. He, J., Zhang, S., Yang, M., Shan, Y., Huang, T.: Bi-directional cascade network for perceptual edge detection. In: Proceedings of the IEEE/CVF Conference on Computer Vision and Pattern Recognition, pp. 3828–3837 (2019)
11. He, K., Zhang, X., Ren, S., Sun, J.: Deep residual learning for image recognition. In: Proceedings of the IEEE Conference on Computer Vision and Pattern Recognition, pp. 770–778 (2016)

12. health, L.: Breast Cancer Awareness (2023). https://www.leehealth.org/health-and-wellness/healthy-news-blog/cancer-care/breast-cancer-awareness-importance-of-early-detection#::text=According%20to%20the%20National%20Breast,so%20important%2C%E2%80%9D%20says%20Dr. Accessed 27 June 2023

13. Heath, M.D., Bowyer, K., Kopans, D.B., Moore, R.H.: The digital database for screening mammography (2007)

14. Hendrycks, D., Gimpel, K.: Gaussian error linear units (gelus). arXiv preprint arXiv:1606.08415 (2016)

15. Huang, G., Liu, Z., Van Der Maaten, L., Weinberger, K.Q.: Densely connected convolutional networks. In: Proceedings of the IEEE Conference on Computer Vision and Pattern Recognition, pp. 4700–4708 (2017)

16. Jocher, G., et al.: ultralytics/yolov5: v3. 0. Zenodo (2020)

17. Jubeen, M., et al.: An automatic breast cancer diagnostic system based on mammographic images using convolutional neural network classifier. J. Comput. Biomed. Inform. 4(01), 77–86 (2022)

18. Li, H., Niu, J., Li, D., Zhang, C.: Classification of breast mass in two-view mammograms via deep learning. IET Image Proc. 15(2), 454–467 (2021)

19. Liu, W., et al.: SSD: single shot multibox detector. In: Leibe, B., Matas, J., Sebe, N., Welling, M. (eds.) ECCV 2016. LNCS, vol. 9905, pp. 21–37. Springer, Cham (2016). https://doi.org/10.1007/978-3-319-46448-0_2

20. Liu, Z., et al.: Swin transformer: hierarchical vision transformer using shifted windows. In: Proceedings of the IEEE/CVF International Conference on Computer Vision, pp. 10012–10022 (2021)

21. Meng, L., et al.: Adavit: adaptive vision transformers for efficient image recognition. In: Proceedings of the IEEE/CVF Conference on Computer Vision and Pattern Recognition, pp. 12309–12318 (2022)

22. Moreira, I.C., Amaral, I., Domingues, I., Cardoso, A., Cardoso, M.J., Cardoso, J.S.: Inbreast: toward a full-field digital mammographic database. Acad. Radiol. 19(2), 236–248 (2012)

23. Platania, R., Shams, S., Yang, S., Zhang, J., Lee, K., Park, S.J.: Automated breast cancer diagnosis using deep learning and region of interest detection (bc-droid). In: Proceedings of the 8th ACM International Conference on Bioinformatics, Computational Biology, and Health Informatics, pp. 536–543 (2017)

24. Shen, L., Margolies, L.R., Rothstein, J.H., Fluder, E., McBride, R., Sieh, W.: Deep learning to improve breast cancer detection on screening mammography. Sci. Rep. 9(1), 1–12 (2019)

25. Simonyan, K., Zisserman, A.: Very deep convolutional networks for large-scale image recognition. arXiv preprint arXiv:1409.1556 (2014)

26. Ueda, D., et al.: Development and validation of a deep learning model for detection of breast cancers in mammography from multi-institutional datasets. PLoS ONE 17(3), e0265751 (2022)

27. Yao, C., Jin, S., Liu, M., Ban, X.: Dense residual transformer for image denoising. Electronics 11(3), 418 (2022)

Distilling Local Texture Features for Colorectal Tissue Classification in Low Data Regimes

Dmitry Demidov[✉], Roba Al Majzoub, Amandeep Kumar, and Fahad Khan

Mohamed Bin Zayed University of Artificial Intelligence, Abu Dhabi, UAE
{dmitry.demidov,roba.majzoub,amandeep.kumar,fahad.khan}@mbzuai.ac.ae

Abstract. Multi-class colorectal tissue classification is a challenging problem that is typically addressed in a setting, where it is assumed that ample amounts of training data is available. However, manual annotation of fine-grained colorectal tissue samples of multiple classes, especially the rare ones like stromal tumor and anal cancer is laborious and expensive. To address this, we propose a knowledge distillation-based approach, named KD-CTCNet, that effectively captures local texture information from few tissue samples, through a distillation loss, to improve the standard CNN features. The resulting enriched feature representation achieves improved classification performance specifically in low data regimes. Extensive experiments on two public datasets of colorectal tissues reveal the merits of the proposed contributions, with a consistent gain achieved over different approaches across low data settings. The code and models are publicly available on GitHub.

Keywords: Colorectal Tissue Classification · Low Data Regimes

1 Introduction

Colorectal cancer (CRC) remains a prominent global health concern, claiming over 10 million lives in 2020 [7] and ranking as the second leading cause of death. In response, significant research efforts have been dedicated to facilitating early diagnosis and treatment through histopathology tissue data classification. The automatic classification of such images in general presents a complex and vital task, enabling precise quantitative analysis of tumor tissue characteristics and contributing to enhanced medical decision-making.

Colorectal tissue classification is typically solved in a setting that assumes the availability of sufficient amounts of labelled training samples [11,13,25]. Therefore, most existing works on colorectal tissue classification follow this setting based on datasets with hundreds to thousands of annotated training images.

D. Demidov and R. Al Majzoub—Contributed equally to this work.

© The Author(s), under exclusive license to Springer Nature Switzerland AG 2024
X. Cao et al. (Eds.): MLMI 2023, LNCS 14349, pp. 357–366, 2024.
https://doi.org/10.1007/978-3-031-45676-3_36

However, this is not the case for rare instances such as gastrointestinal stromal tumors, anal cancers, and other uncommon CRC syndromes, where data availability is limited. Further, manual annotation of such fine-grained colorectal tissue data of multiple classes is laborious and expensive. To reduce the dependency on the training data, few works have explored other settings such as self-supervised [18], semi-supervised [2], and few-shot learning [20]. However, both self-supervised and semi-supervised colorectal tissue classification scenarios assume the availability of unlabeled colorectal tissue training data, whereas the few-shot counterpart setting relies on a relatively larger amount of labelled training samples belonging to base classes for general feature learning. In this work, we explore the problem of colorectal tissue classification in a more challenging setting, termed *low data regime*. In this setting, the number of labelled training samples for all the classes is significantly reduced, i.e., 1% to 10% *per* category.

The aforementioned challenging setting of low data regimes has been recently studied [21] in the context of natural image classification. The authors proposed an approach that improves the feature representation of standard vanilla convolutional neural networks (CNNs) by having a second branch that uses off-the-shelf class activation mapping (CAM) technique [19] to obtain heat maps of gradient-based feature class activations. The feature-level outputs of both branches are trained to match each other, which aids to refine the standard features of the pre-trained CNN in the low-data regimes. While effective on natural images having distinct structures of *semantic objects*, our empirical study shows that such an approach is sub-optimal for colorectal tissue classification likely due to CAM being ineffective in capturing *texture* feature representations in colorectal tissues. Therefore, we look into an alternative approach to address the problem of colorectal tissue classification in low data regimes.

When designing such an approach, effectively encoding local texture information is crucial to obtain improved classification performance [12]. One approach is to utilize a pre-trained model and fine-tune it to encode data-specific local textures. This has been previously endorsed in [4,23,25]. Model fine-tuning in standard settings with ample training samples allows it to encode local texture information. However, this strategy may struggle with extremely scarce training data. A straightforward alternative is random cropping, where smaller regions of the input image are resized for data augmentation to capture local texture patterns [3]. However, such data augmentations prove to be sub-optimal due to the large pixel-level input variations and the artificial generation of a large portion of the pixels due to data extrapolation. To address this issue, we adopt knowledge distillation, transferring knowledge between models without performance sacrifice. This approach has been previously used for both natural and medical image classification [3,9]. In this paper, we utilize all three approaches combined to mitigate the effect of data scarcity on model training and allow the model to learn better representative features even in low data regimes.

Contributions: We propose a knowledge distillation-based approach, named KD-CTCNet, to address the problem of colorectal tissue classification in low

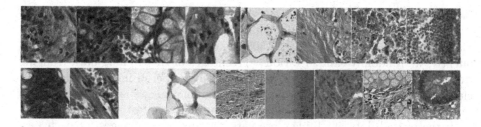

Fig. 1. Samples from Kather-2016 (top) and Kather-2019 (bottom) datasets

Table 1. Number of images per class for train sets of the Kather-2016 dataset in our low data sampling strategy.

Percentage	1%	3%	5%	10%	20%	30%	40%	50%	75%	100%
Samples/class	3	9	15	30	62	93	124	156	234	312

data regimes. Our proposed KD-CTCNet strives to improve the feature representation of the standard fine-tuned CNN by explicitly capturing the inherent local texture information from very few colorectal tissue samples. Our KD-CTCNet consists of a standard branch with a conventional CNN stream and a local image branch that performs local image sampling to encode local texture information. We further utilize a knowledge distillation technique, where the standard global branch serves as a teacher and the local image branch acts as a student. The corresponding output logits of the two branches are compared through a distillation loss to obtain enriched feature representations that are specifically effective for low data regimes. Extensive experiments conducted on two histopathological datasets of CRC including eight and nine different types of tissues respectively, reveal the merits of our proposed contributions. Our KD-CTCNet consistently outperforms standard pre-trained and fine-tuned CNN approaches on a variety of low data settings.

Datasets: Our main experiments are conducted on **Kather-2016**, a dataset of CRC images introduced by [11]. Image patches of size 150×150 are extracted from 10 anonymized whole slide images (WSI) obtained from Mannheim University Medical Center, Heidelberg University, Mannheim, Germany. The dataset includes 5,000 images equally distributed among eight classes, with 625 images per class. To further validate our approach we conduct additional experiments on **Kather-2019**, the NCT-CRC-HE-100K dataset [10]. This publicly available dataset contains nearly 100,000 H&E stained patches of size 224×224. These patches are extracted from Whole Slide Images (WSI) of 86 patients sourced from NCT Biobank and the UMM pathology archive. The dataset comprises 9 classes, and sample images are shown in Fig. 1.

Although these datasets were previously utilized for histopathology classification [1,5], most of the performed research utilizes either all the data or a very large percentage of it, which does not reflect real-life scenarios for many diseases.

Sampling Strategy. To address low data regimes, we employ a sampling app-roach on the dataset, mimicking limited data availability for rare cases and diseases. Each class is divided into 50% training and 50% testing sets. Subse-quently, we randomly sample various percentages of the training data (ranging from 1% to 100% of the train set with an equal number of images sampled per class) for model training. This sampling of training images is done three times randomly and the resulting model is tested on the same test set originally split from the overall dataset. Table 1 presents the number of images for each per-centage. To further validate our model's performance, we first randomly sample 625 images from each class of the Kather-2019 dataset (to match the number of images in the Kather-2016 dataset) and then repeat the sampling scheme.

2   Method

2.1   Baseline Framework

Transfer Learning. A ResNet model [8], pre-trained on ImageNet [6] is fine-tuned on the downstream colorectal tissue classification task. Typically, fine-tuning allows the transfer of generic features learned on large datasets across similar tasks for better model initialization in cases of small training datasets. Here, we are given source domain data D_S and target-domain data D_T, each consisting of a feature space χ and probability distribution $P(X)$, where $X = \{x_1, ..., x_n\} \in \chi$. Therefore, the task to solve is a prediction of the corresponding label, $f(x')$, of a new instance x' using label space Y and an objective function $f(\cdot)$, which should be learned from the labelled training data of pairs $\{x_i, y_i\}$, where $x_i \in X$ and $y_i \in Y$ represent samples and their assigned labels.

Limitations. The current scheme is sub-optimal due to two reasons. First, the scarcity of training samples in low data regimes leads to model overfitting. Sec-ond, previous works in texture recognition literature [14,22] have shown that local feature representations are more effective in encoding distinct texture pat-terns found in colorectal histopathology tissue samples. We empirically show that fine-tuning an ImageNet pre-trained model on colorectal tissue data in a low data regime is sub-optimal in capturing local texture patterns due to limited training data. Consequently, standard transfer learning is relatively ineffective in this context. To address these challenges, we propose an approach that pri-oritizes capturing local texture patterns to improve feature representations in low-data settings.

2.2   Local-Global Feature Enrichment via Knowledge Distillation

Overall Architecture. Figure 2 shows an overview of our proposed architec-ture of a knowledge distillation-based framework, named KD-CTCNet, for CRC tissue classification in low data regimes. The proposed KD-CTCNet comprises two branches: a standard global branch and a local image branch. The standard global branch, a pre-trained ResNet model, takes the entire image as input and

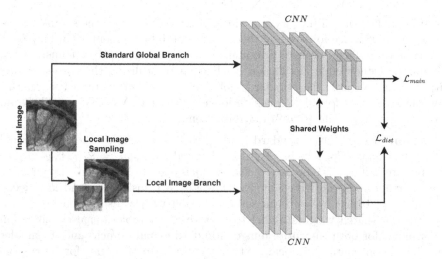

Fig. 2. Overview of our proposed knowledge distillation-based framework, named KD-CTCNet, for CRC tissue classification in low data regimes. The framework strives to obtain enriched feature representations by explicitly capturing the inherent local texture patterns within the CRC data. The proposed KD-CTCNet comprises a conventional CNN stream (top branch) encoding features from the full image content with a standard cross-entropy loss ($\mathcal{L}_{main}$) and a parallel branch that is specifically designed to capture local texture feature representations by performing local image sampling. Both branches share the weights and their corresponding output logits are compared using a self-distillation loss ($\mathcal{L}_{dist}$). The resulting enriched feature representations are beneficial to obtain improved classification performance, especially in low data regimes.

utilizes a standard cross-entropy loss. On the other hand, the local image branch first performs local image sampling by randomly cropping different-sized regions from the input image. The two branches in our framework use the shared weights and their corresponding output logits are compared using a self-distillation loss.

Standard Global Branch. This branch represents the standard CNN stream, which follows the conventional training procedure [17] including fine-tuning all the layers for colorectal tissue classification. To improve the model training, different standard data augmentation techniques are employed, such as horizontal and vertical flips. However, we do not employ colour and geometry-based augmentations as they are observed to be sub-optimal in low data regimes. The randomness of the image sampling leads to large relative data variations compared to the number of images and, therefore, may harm the learning process.

Local Image Branch. This branch strives to capture local texture features complementary to the features learned in the standard global branch. The branch includes the same shared weights as those of the pre-trained ResNet model used in the standard global branch, except that now random samples of different sizes are fed to the model instead of the entire image. Specifically, in order to capture diverse texture patterns, we randomly sample patches ranging between

10–50 % of the original image dimensions. Next, we resize all the sampled local regions to a fixed size followed by the same set of data augmentations employed in the standard global branch. Consequently, motivated by [24] in transformers, we employ a knowledge distillation mechanism by utilizing the corresponding output logits of both the standard global and local image branches. In this way, the network is forced to focus on local texture patterns that can likely aid colorectal tissue classification in low data scenarios.

Loss Function. For the standard global branch, we employ the conventional cross-entropy loss [26] for classification, denoted as $\mathcal{L}_{main} = \mathcal{L}_{CE}(\hat{y}, y)$ in our architecture, where $\hat{y}$ is a predicted class and y is a ground truth one-hot encoded label. Furthermore, motivated by recent works in transformers and knowledge distillation [3,16,24], we perform self-distillation with a distillation loss $\mathcal{L}_{dist}$. We utilize a shared weights model to achieve comparable image classification performance for both the entire image (standard global branch) and its smaller sampled region containing local texture patterns. To this end, for knowledge distillation, we use the standard global branch as the teacher and the local image branch as the student. The teacher's output, being based on more data, is expected to be more confident and can be treated as a label. In our approach, a "hard" label coming from the teacher branch is used as a target. Similar to [24], the label is obtained by feeding a full image into the standard global branch, then taking the teacher's prediction $y_t = argmax(Z_t)$, where Z_t is the output logits of the teacher after the *softmax* function ψ. This parameter-free type of label has previously been shown to provide better distillation performance. Similarly, for the local image branch (student), we obtain its output logits Z_s for the randomly sampled local image regions.

Next, depending on the amount of the available data per class $n_{im/c}$, we compute the distillation loss. We observe that under extremely low data regimes (fewer than 20 images per class), the randomly sampled images may be of inconsistent complexity since adequate image diversity is not preserved. This situation may cause class complexity imbalance, where some classes have varying enough images to cover most of the important patterns, while others do not have this property. Therefore, in order to mitigate this problem, for the sampled dataset portions with fewer than $n_{min} = 20$ images (usually $\leq 5\%$ of the original dataset), instead of the standard cross-entropy loss we utilize focal loss [15], which is specifically designed to deal with the data imbalance. In this way, the distillation loss can be described as follows:

$$\mathcal{L}_{dist} = \begin{cases} \mathcal{L}_{focal}(\psi(Z_s), y_t), & \text{if } n_{im/c} \leq n_{min} \\ \mathcal{L}_{CE}(\psi(Z_s), y_t), & \text{otherwise,} \end{cases} \tag{1}$$

Consequently, the final objective function $\mathcal{L}$ is defined as a combination of both classification and distillation losses:

$$\mathcal{L} = \frac{1}{2}\mathcal{L}_{main} + \alpha\frac{1}{2}\mathcal{L}_{dist}, \tag{2}$$

where $\alpha = 0.1$ is a controlling scaler for the self-distillation loss needed to manage the amount of complementary refinement information.

Table 2. Comparison of different approaches using different percentages of the data on the Kather-2016 dataset. We conduct the experiments with three different train sets (for 1 % to 50 % splits) and report the mean classification accuracy. Our proposed KD-CTCNet achieves consistent improvement in performance over other methods across different data settings. Best results are highlighted in bold.

Model	1%	3%	5%	10%	20%	50%	100%
ResNet-50	72.64	82.88	86.29	88.91	91.82	94.57	96.04
ResNet-50 + Sampling	72.75	82.74	84.98	90.35	92.87	95.12	96.65
SAM ResNet-50 [21]	71.85	82.91	86.18	90.06	91.85	94.65	95.41
KD-CTCNet (Ours)	$\mathbf{72.91}_{\pm 0.69}$	$\mathbf{83.76}_{\pm 2.35}$	$\mathbf{86.82}_{\pm 0.47}$	$\mathbf{91.58}_{\pm 0.65}$	$\mathbf{94.13}_{\pm 0.54}$	$\mathbf{95.76}_{\pm 0.15}$	**96.88**

Table 3. Performance of KD-CTCNet on NCT-CRC-HE-100K dataset.

Model	1%	3%	5%	10%	20%	50%	100%
ResNet-50	**73.1**	86.66	91.47	94.10	95.65	97.4	97.8
KD-CTCNet (Ours)	$72.25_{\pm 3.53}$	$\mathbf{87.21}_{\pm 1.98}$	$\mathbf{92.67}_{\pm 1.25}$	$\mathbf{94.94}_{\pm 0.74}$	$\mathbf{96.31}_{\pm 0.37}$	$\mathbf{97.91}_{\pm 0.22}$	**98.23**

3 Experiments and Results

3.1 Implementation Details

To adapt to low data regime settings, we apply the following changes. First, unlike the default ImageNet resizing dimensions, images are resized to 192×192. This choice of dimensions is a trade-off between the optimal 224×224 size the model was pre-trained on and the original 150×150 size of the input images. Considering the receptive field of the ResNet-50 last convolutional layer, we select the closest number divisible by 32 pixels size. We observe a slight deterioration in classification accuracy when using the 224×224 size. To preserve the learnable texture, we do not use any colour or geometry-disturbing augmentations and only utilize horizontal and vertical flips. For our local image branch, after performing random local image sampling from the input image, we resize the sampled regions to 96×96 pixels; a trade-off between the sizes of the standard ResNet-50 input and the average of our final truncated images. The final choice is also selected as the closest value divisible by 32. Lastly, we perform the same horizontal and vertical flip augmentations here. Our training setup includes the standard SGD optimizer with a momentum equal to 0.9, a learning rate of 0.01, and a training batch size of 32. We emphasise that the choice of the above-mentioned hyper-parameters is geometrically-inferred and not optimised based on the test set, while the value for hyper-parameter α in 2 is based on experiments performed on independent random train and test sets from the Kather-2016 dataset. All experiments were conducted on a single NVIDIA RTX 6000 GPU using the PyTorch framework and the APEX utility for mixed precision training.

3.2   Results and Analysis

Quantitative Results. Results in Table 2 demonstrate that our local image sampling strategy boosts the performance of the vanilla network across all sampled percentages, showing that the local sampling of the image introduces further local information to the network. Meanwhile, the SAM approach proposed for natural images does not perform well on CRC patches and its performance appears sub-optimal to the medical data domain. Training KD-CTCNet from scratch yields much lower accuracies than those obtained by the other approaches, since the initialization of the other networks is based on their pre-trained weights, allowing the model to reach better results even in cases where data is scarce. However, using a pre-trained model in KD-CTCNet along with our local image sampling strategy proves to be superior to all other approaches across all data regimes reaching up to **2.67%** improvement using only 10% of the data. This proves the ability of our approach to capture better texture information. To showcase the scalability of KD-CTCNet, we also conduct experiments on another CRC dataset [10] reported in Table 3. We observe that our KD-CTCNet improves over vanilla ResNet in almost all data percentages, proving its ability to enhance classification accuracy even in very low data regimes.

Qualitative Analysis. For qualitative analysis, two confusion matrices for the standard ResNet-50 model and our proposed KD-CTCNet are generated for one of the low data regimes (20%). As can be seen in Fig. 3, our KD-CTCNet outperforms the vanilla ResNet in correctly classifying positive samples into their correct labels for most classes. And while ResNet struggles more in the distinction between stroma, lymph, and tumor, our model shows a higher discriminative ability in such cases, demonstrating the effectiveness of our approach in the CRC classification setting. On the other hand, our model seems to be less discriminative for adipose classification compared to the vanilla ResNet.

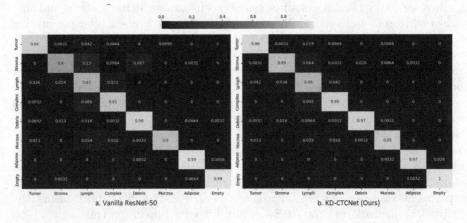

Fig. 3. Comparison of confusion matrices calculated on the test set with 20% of the available data for (a) vanilla ResNet-50 and (b) our KD-CTCNet approach.

4 Conclusion

CRC histopathological image classification poses a great challenge for deep learning models, specifically in scenarios where data is scarce. In this paper, we propose a dedicated architecture for the classification of CRC scans and demonstrate its superiority to similar approaches. More specifically, through extensive quantitative experiments, we showcase the ability of our model to reach advanced performance in the low data regime. This is achieved by enriching the standard CNN features with effectively captured local texture information from tissue samples through our specifically designed distillation loss. Moreover, from the qualitative experiments, we show that learning low-level texture information allows the model to achieve better per-class accuracy and decrease confusion of visually similar classes by identifying subtle but important details.

References

1. Anju, T., Vimala, S.: Tissue and tumor epithelium classification using fine-tuned deep CNN models. Int. J. Adv. Comput. Sci. Appl. **13**(9) (2022)
2. Bakht, A.B., Javed, S., Al Marzouqi, H., Khandoker, A., Werghi, N.: Colorectal cancer tissue classification using semi-supervised hypergraph convolutional network. In: 2021 IEEE 18th International Symposium on Biomedical Imaging (ISBI), pp. 1306–1309. IEEE (2021)
3. Caron, M., et al.: Emerging properties in self-supervised vision transformers. In: Proceedings of the IEEE/CVF International Conference on Computer Vision, pp. 9650–9660 (2021)
4. Chen, T., Wu, M., Li, H.: A general approach for improving deep learning-based medical relation extraction using a pre-trained model and fine-tuning. Database **2019**, baz116 (2019)
5. Dabass, M., Vashisth, S., Vig, R.: A convolution neural network with multi-level convolutional and attention learning for classification of cancer grades and tissue structures in colon histopathological images. Comput. Biol. Med. **147**, 105680 (2022)
6. Deng, J., Dong, W., Socher, R., Li, L.J., Li, K., Fei-Fei, L.: ImageNet: a large-scale hierarchical image database. In: 2009 IEEE Conference on Computer Vision and Pattern Recognition, pp. 248–255 (2009). https://doi.org/10.1109/CVPR.2009.5206848
7. Ferlay, J., Ervik, M., Lam, F., Colombet, M., Mery, L., Piñeros, M.: Global cancer observatory: cancer today. https://gco.iarc.fr/today. Accessed Mar 2023
8. He, K., Zhang, X., Ren, S., Sun, J.: Deep residual learning for image recognition. CoRR, abs/1512 3385, 2 (2015)
9. Javed, S., Mahmood, A., Qaiser, T., Werghi, N.: Knowledge distillation in histology landscape by multi-layer features supervision. IEEE J. Biomed. Health Inform. **27**(4), 2037–2046 (2023)
10. Kather, J.N., et al.: Predicting survival from colorectal cancer histology slides using deep learning: a retrospective multicenter study. PLoS Med. **16**(1), e1002730 (2019)
11. Kather, J.N., et al.: Multi-class texture analysis in colorectal cancer histology. Sci. Rep. **6**(1), 27988 (2016)

12. Khan, F.S., Anwer, R.M., van de Weijer, J., Felsberg, M., Laaksonen, J.: Compact color-texture description for texture classification. Pattern Recogn. Lett. **51**, 16–22 (2015). https://doi.org/10.1016/j.patrec.2014.07.020

13. Kumar, A., Vishwakarma, A., Bajaj, V.: CRCCN-Net: automated framework for classification of colorectal tissue using histopathological images. Biomed. Signal Process. Control **79**, 104172 (2023)

14. Lazebnik, S., Schmid, C., Ponce, J.: A sparse texture representation using local affine regions. IEEE Trans. Pattern Anal. Mach. Intell. **27**(8), 1265–1278 (2005)

15. Lin, T., Goyal, P., Girshick, R., He, K., Dollar, P.: Focal loss for dense object detection. In: 2017 IEEE International Conference on Computer Vision (ICCV), pp. 2999–3007. IEEE Computer Society, Los Alamitos, CA, USA, October 2017

16. Pham, M., Cho, M., Joshi, A., Hegde, C.: Revisiting self-distillation (2022). https://doi.org/10.48550/ARXIV.2206.08491. https://arxiv.org/abs/2206.08491

17. Ramdan, A., Heryana, A., Arisal, A., Kusumo, R.B.S., Pardede, H.F.: Transfer learning and fine-tuning for deep learning-based tea diseases detection on small datasets. In: 2020 International Conference on Radar, Antenna, Microwave, Electronics, and Telecommunications (ICRAMET), pp. 206–211 (2020)

18. Saillard, C., et al.: Self supervised learning improves dMMR/MSI detection from histology slides across multiple cancers. arXiv preprint arXiv:2109.05819 (2021)

19. Selvaraju, R., Cogswell, M., Das, A., Vedantam, R., Parikh, D., Batra, D.: Grad-CAM: visual explanations from deep networks via gradient-based localization (2016)

20. Shakeri, F., et al.: FHIST: a benchmark for few-shot classification of histological images. arXiv preprint arXiv:2206.00092 (2022)

21. Shu, Y., Yu, B., Xu, H., Liu, L.: Improving fine-grained visual recognition in low data regimes via self-boosting attention mechanism. In: Avidan, S., Brostow, G., Cissé, M., Farinella, G.M., Hassner, T. (eds.) Computer Vision-ECCV 2022: 17th European Conference, Tel Aviv, Israel, 23–27 October 2022, Proceedings, Part XXV, pp. 449–465. Springer, Cham (2022). https://doi.org/10.1007/978-3-031-19806-9_26

22. Simon, P., Uma, V.: Deep learning based feature extraction for texture classification. Procedia Comput. Sci. **171**, 1680–1687 (2020). Third International Conference on Computing and Network Communications (CoCoNet'19)

23. Sun, G., Cholakkal, H., Khan, S., Khan, F., Shao, L.: Fine-grained recognition: accounting for subtle differences between similar classes. In: Proceedings of the AAAI Conference on Artificial Intelligence, vol. 34, no. (07), pp. 12047–12054 (2020)

24. Touvron, H., Cord, M., Douze, M., Massa, F., Sablayrolles, A., Jegou, H.: Training data-efficient image transformers & distillation through attention. In: Meila, M., Zhang, T. (eds.) Proceedings of the 38th International Conference on Machine Learning. Proceedings of Machine Learning Research, vol. 139, pp. 10347–10357. PMLR, 18–24 July 2021. https://proceedings.mlr.press/v139/touvron21a.html

25. Tsai, M.J., Tao, Y.H.: Deep learning techniques for colorectal cancer tissue classification. In: 2020 14th International Conference on Signal Processing and Communication Systems (ICSPCS), pp. 1–8. IEEE (2020)

26. Zhang, Z., Sabuncu, M.: Generalized cross entropy loss for training deep neural networks with noisy labels. In: Bengio, S., Wallach, H., Larochelle, H., Grauman, K., Cesa-Bianchi, N., Garnett, R. (eds.) Advances in Neural Information Processing Systems, vol. 31. Curran Associates, Inc. (2018)

Delving into Ipsilateral Mammogram Assessment Under Multi-view Network

Toan T. N. Truong[1], Huy T. Nguyen[2(✉)], Thinh B. Lam[3],
Duy V. M. Nguyen[4], and Phuc H. Nguyen[5]

[1] Ho Chi Minh City International University, Ho Chi Minh City, Vietnam
4401101074@student.hcmue.edu.vn
[2] National Cheng Kung University, Tainan, Taiwan
[3] Ho Chi Minh City University of Science, Ho Chi Minh City, Vietnam
[4] Military Hospital 175, Ho Chi Minh City, Vietnam
[5] Eastern International University, Thu Dau Mot, Vietnam
phuc.nguyenhong@eiu.edu.vn

Abstract. In many recent years, multi-view mammogram analysis has been focused widely on AI-based cancer assessment. In this work, we aim to explore diverse fusion strategies (average and concatenate) and examine the model's learning behavior with varying individuals and fusion pathways, involving Coarse Layer and Fine Layer. The Ipsilateral Multi-View Network, comprising five fusion types (Pre, Early, Middle, Last, and Post Fusion) in ResNet-18, is employed. Notably, the Middle Fusion emerges as the most balanced and effective approach, enhancing deep-learning models' generalization performance by +5.29% (concatenate) and +5.9% (average) in VinDr-Mammo dataset and +2.03% (concatenate) and +3% (average) in CMMD dataset on macro F1-Score. The paper emphasizes the crucial role of layer assignment in multi-view network extraction with various strategies.

Keywords: Mammogram Analysis · Multi-view · Fusion Network

1 Introduction

In recent years, many machine learning methods based on texture descriptors or deep learning networks have been proposed for classification using ipsilateral views. A mammographic screening normally consists of two views: craniocaudal view (CC), a top-down view of the breast, and mediolateral oblique (MLO), a side view of the breast taken at a certain angle. Radiologists examine the two views from both patient's breasts left and right, which are both views of the same breast (ipsilateral views) and the same view of both breasts (bilateral views). *Y. Chen et.al.* [2] proposed two pathways to extract the global feature and local feature between two ipsilateral views (CC and MLO). This methodology achieved

---

Toan T. N. Truong and Huy T. Nguyen—Equally contribute to this paper.

© The Author(s), under exclusive license to Springer Nature Switzerland AG 2024
X. Cao et al. (Eds.): MLMI 2023, LNCS 14349, pp. 367–376, 2024.
https://doi.org/10.1007/978-3-031-45676-3_37

good and desirable results. Continuously, *Liu et.al.* [3,4] successfully applied the well-known graph convolutional network (GCN) to the mammographic field which processes the bipartite GCN and inception GCN individually. Then, they fused both of them together in a correspondence reasoning enhancement stage to procedure the prediction. There are many other multi-view-based approaches [2,5,6,8,12] using from two to four images as inputs.

Recent research has demonstrated that multi-perspective approaches [9–11] enhance breast cancer diagnosis. The main technique underpinning these approaches is the development of an end-to-end deep learning classification model for mammographic pathology. Before combining four screening mammogram views for prediction, this strategy extracts features from each view separately. Also, the majority of current breast cancer diagnostic research is devoted to determining whether a mammogram is malignant or benign. However, those frameworks did not consider how hidden layers affect the fusion layer before, after, or between them, which can be inaccurate and lead to a poor learning process. Thus, learning two examined-auxiliary (EA) low-dimensional and high-dimensional spaces can more effectively exploit the complex correlations between EA image pairings, producing higher-quality reconstruction results.

In summary, the main contributions of our work are as follows:

1. We introduce various mutations of multi-view networks to understand how the effectiveness of fusion at multiple positions in architecture. We proposed five strategies for fusion types: Pre Fusion, Early Fusion, Middle Fusion, Last Fusion, and Post Fusion. Additionally, we use ResNet-18 in our instance as a backbone and then divide it into different strategies to assess its efficacy.
2. We proposed a robust fusion block using the two well-known functions: average and concatenate as the aggregation operations and ablate the effect of skip connection between different views and fused features.

2   Methodology

2.1   Fusion Type

Previously, many mammogram classification approaches have simply let the images go through the common-shared feature extractors to learn their information. Then, they continuously extend their methodology to increase performance. Afterward, they fused those features and fed them into various multiple perceptron layers (MLP) at last to produce the prediction. However, this can cause poor learning processing. Because of two different views of information, fused features can have specific noise on each view when combined together. This motivates us to propose new strategies of multi-view networks that empirically extracted before, after, or between fused features. Those approaches separately cut down the original backbone architecture into two parts: Coarse Layer (low-level individual extractor) and Fine Layer (high-level fusion extractor), which are the layers before the fusion block and the layers after the fusion block, respectively.

To be used as a building block of our network, ResNet-18 [20] was used to be the main architecture, which consists of a pre-trained convolutional neural network. According to *He et al. (2016)* [20], ResNet was originally built with many blocks in many different channel dimension sizes, which enable it to split apart. As shown in Fig. 2, the input mammograms are first fed into the Coarse Layer. Following that, the fusion block, average or concatenate, is subsequently applied to make the unchanging flow and combine in the backbone. The combined feature is then passed to the Fine Layer to extract the high-level features for distinguishing among the classes at the last classification layer. Therefore, there are five positions to separate and two aggregation functions which a total of ten models are discussed and evaluated in Sect. 4.

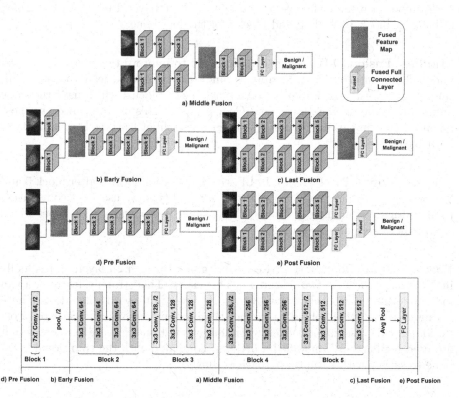

Fig. 1. Five proposed fusion types (top) and ResNet-18 architecture with the fusion separated line (bottom). (a) Pre Fusion put all layers of the backbone into the Fine Layer, nothing for the Coarse Layer. (b) Early Fusion separates between block 1 and block 2 of the backbone. (c) Middle Fusion separates between block 3 and block 4 which contributes equally in both layers. (d) Last Fusion, in contrast with Early Fusion, separates most of the blocks in the backbone to the Coarse Layer, which cuts between block 5 and the average pooling function. (e) Post Fusion, in contrast with Pre Fusion, put all layers of the backbone into the Coarse Layer, nothing for the Fine Layer.

Pre - Fusion (PreF) (Fig. 1.d). This is the simplest and fewest computational resources model. Pre Fusion works as two screening mammography views individually not going through any layers. Instead, they are firstly aggregated together at the beginning without learning the mapping from low-dimensional and then passed the combined feature to the fully Fine Layer. This approach focuses mostly on the Fine Layer and removes totally process in Coarse Layer.

Early - Fusion (EF) (Fig. 1.b). The separation now shifts the position to the right a little bit compared to PreF and locates between block 1 and block 2 in the feature extractor. The two views input now can be learned individually from all layers in block 1 in the Coarse Layer. This enhances the combined feature to perform better results when going through the Fine Layer. Fine Layer consisting of block 2, block 3, block 4, and block 5 of the backbone.

Middle - Fusion (MF) (Fig. 1.a). In order to improve the performance and make full advantage of deep learning technologies to automatically boost resolution. Middle Fusion is shifted between block 3 and block 4 which have the same number of layers in Fine Layer and Coarse Layer. This can be called the most balanced type in various proposed strategies, which let features learn equally in both Layers.

Last - Fusion (LF) (Fig. 1.c). In LF, the aggregation locates after block 5 and before the average pooling layer. This is the final type we use 1×1 convolution to make an equal number of channel dimensions in two aggregation types which will be discussed later in Sect. 2.2.

Post - Fusion (PostF) (Fig. 1.e). After going through all layers individually in ResNet-18, we fuse two features of two views. This strategy learns a lot of individual information about two views. However, afterward, they do not have learnable parameters behind to extract the high-dimensional information. Post Fusion, in this case, is most likely with many traditional feature extractors that are used regularly. We modified a little bit compared with the traditional backbone to satisfy with various approaches above.

2.2   Fusion Block

There are two main aggregation types that are mainly used in several methods nowadays: concatenate and average. Because two individual ipsilateral views are learned from two shared weight Fine Layer. Therefore, they need the combination function to continuously adapt the remained Coarse Layer. In the average fusion block, the average function uses the pixel-wise additional and then pixel-wise divided by two to perform the combined feature. In the concatenate fusion block, the concatenation function combines two features in the Coarse Layer along the channel dimension if there are feature maps that occur in PreF, EF, MF, and

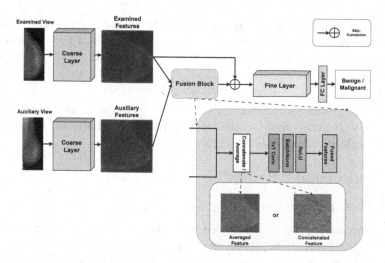

Fig. 2. The overview of Ipsilateral Multi-View Network architecture. The two examine and auxiliary views are fed into the Coarse Layer, then fused together in the Fusion Block. Afterward, it goes through the Fine Layer and finally classification in the FC Layer.

LF approaches. If there are flattened features in the PostF strategy, the fused feature will be obtained along the node size in a dense layer. These functions can be formulated in the following form:

$$a_{avg-fused} = \frac{C(x_{CC}) + C(x_{MLO})}{2} \tag{1}$$

$$a_{concat-fused} = [C(x_{CC}), C(x_{MLO})] \tag{2}$$

where $C(.)$ is the Coarse Layer network, $a_{avg-fused}$, $a_{concat-fused}$ is the averaged feature and concatenated feature in the fusion block, respectively, and [.] is the concatenation function.

To alleviate the inconsistency in our model, the 1×1 convolution layer was added to keep the stabilization of the feature extractor in depth dimension flow. Because of the double number of channels in concatenation, we need 1×1 convolution to resize its channel dimension by half and continuously go through the remaining layers. In average block fusion, we do not need halving because it used pixel-wise calculation. Therefore, we keep the same dimensional in the 1×1 convolution layer to stabilize the flow. Furthermore, we used batch normalization [16] and ReLU activation layers [17] in the fusion block after 1×1 conv2d was used for normalizing the fused features.

2.3   Class Imbalanced

Dealing with unbalanced data is one of the biggest difficulties in medical image analysis. The issue is considerably more obvious in the area of medical imaging.

We present the Focal Loss [15], which pushes the model to down-weight simple instances in order to learn about hard samples and is defined by adding a modulating component to the cross-entropy loss and a class balancing parameter.

$$FL(p_t) = -\alpha_t(1 - p_t)^\gamma \log(p_t), \tag{3}$$

where p_t is the predicted probability of the true class, α_t is the class balancing factor for the true class, and γ is the focusing parameter that controls the degree of down-weighting for well-classified examples.

3   Experimental Settings

3.1   Dataset Preparation

VinDr-Mammo: [19] is a large-scale full-field digital mammography dataset with 20000 images from 5000 patients and 4 views per patient. From 2018 to 2020, the dataset was collected and labeled by three radiologists with an average of 19 years of clinical experience. Each image was annotated using a BI-RADS scale ranging from 1 to 5. Due to the heavy imbalance from BI-RADS 1 and inconsistent annotation from BI-RADS 3, a subset of VinDr-Mammo with BI-RADS 2, 4, and 5 was brought for assessment. We divided them into two classes: benign is a set of BI-RADS 2 samples, and suspicious for malignancy is a set of BI-RADS 4 and 5 samples. According to the original train-test splitting information from a metadata file, we used 4532 images for training and validating and 1132 images for testing, with 4676 benign and 988 suspicious samples. For multi-view training settings, each patient from two sets remains four views, with two views of each breast. The training stage and Inference stage require two views each for performing classification.

The Chinese Mammography Database (CMMD): [7] includes 5.202 screening mammogram images from 1.775 studies. We trained on 1.172 non-malignant mammograms and 2.728 malignant screening images with 85%:15% ratio splitting on the training set and test set. Furthermore, we employ stratified sampling, resulting in 498 benign and 1157 malignancy ipsilateral view samples on the training set and 88 benign and 205 malignancy ipsilateral view samples on the testing set.

3.2   Detailed Training and Evaluation Metrics

In our settings, we used the same architecture (ResNet-18 [20]) for the feature extractor part of the framework. In the data loading part, the images are loaded with a batch size of 32 (two views for each breast with a total of 16 breasts on one side). The model was trained for 200 epochs using SGD optimizer [21] with an initial learning rate 1×10^{-3} and decays by 0.1 after $20, 40, 60,$ and 80 epochs. Our images are preprocessed by resizing them into 800 for both training and testing. In our case, macro F1 - Score is the appropriate evaluation metric. Furthermore, the area under the ROC Curve (AUC ROC) also is used for measuring the models' performance under slightly imbalanced dataset training.

4 Results and Ablation Studies

4.1 Ipsilateral Multi-view Network

Table 1. Quantitative results (%) using AUC-ROC and macro F1-Score with ResNet-18 on the VinDr-Mammo and CMMD datasets.

		Average		Concatenate	
DataSet	Fusion Type	F1-Score	AUC-ROC	F1-Score	AUC-ROC
VinDr-Mammo	PreF	68.71	68.49	73.28	73.46
	EF	71.49	71.92	69.44	68.67
	MF	74.00	72.15	**75.34**	74.24
	LF	**74.09**	74.47	74.91	74.61
	PostF	74.08	75.35	74.48	73.23
CMMD	PreF	78.45	81.82	75.74	76.92
	EF	79.92	80.02	72.23	76.64
	MF	**81.45**	84.16	**77.77**	80.42
	LF	80.67	82.52	77.02	79.92
	PostF	81.32	83.70	76.98	79.12

Table 1 illustrates the experimental results of the five proposed methods with two fusion block types: average and concatenate on VinDr-Mammo and CMMD datasets. In the comparison in fusion type, it is also shown that the Middle Fusion achieves high results on the whole experiments. It shows a significant improvement on macro F1-Score by (+5.29%) on average and (+5.9%) on concatenate in VinDr-Mammo compared with the conventional technique, Pre Fusion. In CMMD, it also improves with around (+3%) on average and (+2.03%) on concatenate. In particular, Middle Fusion outperforms both fusion block types with two datasets, but the average Last Fusion achieves a little improvement on VinDr-Mammo with a lower only 0.09% than Middle Fusion. This loosening can be ignored because the average Middle Fusion also gets a good result. As a result, Middle Fusion can produce better performance in both fusion block types. We conjecture that claim is due to the equity in layers flow in the Middle Fusion. Since the Coarse Layer and the Fine Layer contains the balance separation in the feature extractor ResNet-18, the low-level features can adequately learn individual two examined-auxiliary (EA) images before fusing them together. Later, the fused feature might include noise inside needed the completely Fine Layer with various layers to procedure them into useful knowledge before the classifier layers.

4.2 Skip Connection

Table 2 shows the ablation studies on the skip connection in various ways: examined, auxiliary, and examined-auxiliary (EA) in two backbones: ResNet-18 and

RestNet-34. The skip connection with examined sample outperformed all of those strategies which rounded (2.12%–1.62%) on VinDr-Mammo and (1.91%–1.39%) on CMMD compared to the baseline, no skip connection. The skip connection with two examined-auxiliary views resulted in poor performance, even with baseline no skip connection.

Table 2. Ablation studies of skip connection on different views with ResNet-18 and ResNet-34 on VinDr-Mammo and CMMD datasets

VinDr-Mammo			ResNet-18 MF		ResNet-34 MF	
Concatenate	Skip Connection (Examined)	Skip Connection (Auxiliary)	Macro F1-Score	AUC ROC	Macro F1-Score	AUC ROC
✓			73.22	70.66	74.63	72.18
✓	✓		**75.34**	74.24	**75.98**	74.86
✓	✓	✓	71.68	72.66	73.27	68.92
CMMD			ResNet-18 MF		ResNet-34 MF	
Concatenate	Skip Connection (Examined)	Skip Connection (Auxiliary)	Macro F1-Score	AUC ROC	Macro F1-Score	AUC ROC
✓			75.86	77.10	78.12	77.67
✓	✓		**77.77**	80.42	**79.51**	81.97
✓	✓	✓	74.87	79.21	77.86	79.54

5   Conclusion

In this paper, we delved into many variants of CNN-based multi-view networks for ipsilateral breast cancer analysis. We empirically validate various fusion strategies and two fusion blocks in the same configuration. The experimental results show that the Middle Fusion significantly outperformed the remaining methods on large-scale clinical datasets. The intuition is that Middle Fusion has balance layers in both Coarse Layer and Fine Layer which extract enough low-level individual dimension and high-level fusion dimension, respectively. In addition, concatenate fusion block also keeps the information in two EA features without alleviating its as the average function does. In future work, we plan to develop this simple methodology to address the issue of dependency on multi-view data, and partial multi-view learning via the optimizing domain in EA views.

Acknowledgement. This paper is partially supported by AI VIETNAM. We thank the lab from the Biomedical Engineering Department of National Cheng Kung University - Integrated MechanoBioSystems Lab (IMBSL) for providing the GPU to support the numerical calculations in this paper.

References

1. World Health Organization (WHO): Latest global cancer data: Cancer burden rises to 19.3 million new cases and 10.0 million cancer deaths in 2020. International Agency for Research on Cancer (IARC) (2020)
2. Chen, Y., et al.: Multi-view local co-occurrence and global consistency learning improve mammogram classification generalisation. In: Wang, L., Dou, Q., Fletcher, P.T., Speidel, S., Li, S. (eds.) Medical Image Computing and Computer Assisted Intervention – MICCAI 2022. MICCAI 2022. LNCS, vol. 13433, pp. 3–13. Springer, Cham (2022). https://doi.org/10.1007/978-3-031-16437-8_1
3. Liu, Y., Zhang, F., Chen, C., Wang, S., Wang, Y., Yu, Y.: Act like a radiologist: towards reliable multi-view correspondence reasoning for mammogram mass detection. IEEE Trans. Pattern Anal. Mach. Intell. 44(10), 5947–59612 (2022). https://doi.org/10.1109/TPAMI.2021.3085783
4. Liu, Y., Zhang, F., Zhang, Q., Wang, S., Wang, Y., Yu, Y.: Cross-view correspondence reasoning based on bipartite graph convolutional network for mammogram mass detection. In: 2020 IEEE/CVF Conference on Computer Vision and Pattern Recognition (CVPR), pp. 3811–3821 (2020). https://doi.org/10.1109/CVPR42600.2020.00387
5. Wang, H., et al.: Breast mass classification via deeply integrating the contextual information from multi-view data. Pattern Recognit. 80, 42–52 (2018)
6. Carneiro, G., Nascimento, J., Bradley, A.P.: Automated analysis of unregistered multi-view mammograms with deep learning. IEEE Trans. Med. Imag. 36(11), 2355–2365 (2017)
7. Nolan, T.: The Chinese mammography database (CMMD) (2023). https://wiki.ancerimagingarchive.net/pages/viewpage.action?pageId=70230508
8. Li, Y., Chen, H., Cao, L., Ma, J.: A survey of computer-aided detection of breast cancer with mammography. J. Health Med. Inform. 4(7), 1–6 (2016)
9. Wu, N., et al.: Deep neural networks improve radiologists' performance in breast cancer screening. IEEE Trans. Med. Imaging 39(4), 1184–1194 (2020)
10. Khan, H.N., Shahid, A.R., Raza, B., Dar, A.H., Alquhayz, H.: Multiview feature fusion based four views model for mammogram classification using convolutional neural network. IEEE Access 7, 165724–165733 (2019)
11. Geras K.J., Wolfson, S., Kim, S.G., Moy, L., Cho, K.: High-resolution breast cancer screening with multiview deep convolutional neural networks. ArXiv, vol.abs/1703.07047 (2017)
12. Nguyen, H.T.X., Tran, S.B., Nguyen, D.B., Pham, H.H., Nguyen, H.Q.: A novel multi-view deep learning approach for BI-RADS and density assessment of mammograms. In: 2022 44th Annual International Conference of the IEEE Engineering in Medicine & Biology Society (EMBC), pp. 2144–2148 (2022). https://doi.org/10.1109/EMBC48229.2022.9871564
13. Ke, G., et al.: LightGBM: a highly efficient gradient boosting decision tree. In: Advances in Neural Information Processing System 30 (NIPS 2017), vol. 30, pp. 3146–3154 (2022)
14. Wang, C.Y., Bochkovskiy, A., Liao, H.Y.M.: YOLOv7: trainable bag-of-freebies sets new state-of-the-art for real-time object detectors. In: Proceedings of the IEEE/CVF Conference on Computer Vision and Pattern Recognition (CVPR), pp. 7464–7475 (2023)
15. Lin, T.Y., Goyal, P., Girshick, R., He, K., Dollár, P.: Focal loss for dense object detection. In: 2017 IEEE International Conference on Computer Vision (ICCV), Venice, Italy, pp. 2999–3007 (2017)

16. Ioffe, S., Szegedy, C.: Batch normalization: accelerating deep network training by reducing internal covariate shift. In: Proceedings of the 32nd International Conference on International Conference on Machine Learning, vol. 37, pp. 448–456 (2015)
17. Agarap A.F.: Deep learning using rectified linear units (ReLU). arXiv preprint arXiv:1803.08375 (2019)
18. Nguyen, H.T., et al.: In-context cross-density adaptation on noisy mammogram abnormalities detection. arXiv preprint arXiv:2306.06893 (2023)
19. Nguyen, H.T., Nguyen, H.Q., Pham, H.H. et al.: VinDr-Mammo: a large-scale benchmark dataset for computer-aided diagnosis in full-field digital mammography. Sci. Data **10**, 277 (2023). https://doi.org/10.1038/s41597-023-02100-7
20. He, K., Zhang, X., Ren, S., Sun, J.: Deep residual learning for image recognition. In: IEEE Conference on Computer Vision and Pattern Recognition (2016)
21. Ruder, S.: An overview of gradient descent optimization algorithms. arXiv preprint arXiv:1609.04747 (2016)

ARHNet: Adaptive Region Harmonization for Lesion-Aware Augmentation to Improve Segmentation Performance

Jiayu Huo[1](✉), Yang Liu[1], Xi Ouyang[2], Alejandro Granados[1], Sébastien Ourselin[1], and Rachel Sparks[1]

[1] School of Biomedical Engineering and Imaging Sciences (BMEIS), King's College London, London, UK
jiayu.huo@kcl.ac.uk
[2] Shanghai United Imaging Intelligence Co., Ltd., Shanghai, China

Abstract. Accurately segmenting brain lesions in MRI scans is critical for providing patients with prognoses and neurological monitoring. However, the performance of CNN-based segmentation methods is constrained by the limited training set size. Advanced data augmentation is an effective strategy to improve the model's robustness. However, they often introduce intensity disparities between foreground and background areas and boundary artifacts, which weakens the effectiveness of such strategies. In this paper, we propose a foreground harmonization framework (ARHNet) to tackle intensity disparities and make synthetic images look more realistic. In particular, we propose an Adaptive Region Harmonization (ARH) module to dynamically align foreground feature maps to the background with an attention mechanism. We demonstrate the efficacy of our method in improving the segmentation performance using real and synthetic images. Experimental results on the ATLAS 2.0 dataset show that ARHNet outperforms other methods for image harmonization tasks, and boosts the down-stream segmentation performance. Our code is publicly available at https://github.com/King-HAW/ARHNet.

Keywords: Stroke segmentation · Lesion-aware augmentation · Adaptive image harmonization

1 Introduction

Accurate brain lesion segmentation is essential for understanding the prognoses of neurological disorders and quantifying affected brain areas by providing information on the location and shape of lesions [8]. With advanced deep learning techniques, various brain lesion segmentation methods based on Convolutional Neural Networks (CNNs) have been proposed [11,23]. However, a noteworthy hurdle is the prerequisite of an adequate number of training samples to ensure the model's generalization ability. Utilizing small-scale datasets for the segmentation model training can result in over-fitting, thereby limiting its robustness to unseen samples. Due to the variance of lesion appearance and size, as well as the

X. Cao et al. (Eds.): MLMI 2023, LNCS 14349, pp. 377–386, 2024.
https://doi.org/10.1007/978-3-031-45676-3_38

extreme data imbalance between foreground and background voxels, many deep learning models also struggle to perform the small lesion segmentation task.

To this end, some data augmentation techniques have been proposed that aim to increase the diversity of the training set, which helps to boost the performance of the segmentation model for unseen images [2,20]. Often data augmentation is realized by basic image transformations such as rotation and flipping. As the diversity of the data generated through basic image transformations is deficient, advanced data augmentation approaches have been developed. For instance, Huo *et al.* [6] designed a progressive generative framework to synthesize brain lesions that can be inserted into normal brain scans to create new training instances. Zhang *et al.* [22] proposed a lesion-aware data augmentation strategy to increase the sample diversity. However, these methods often inevitably introduce boundary artifacts that may cause the intensity distribution to shift, resulting in segmentation performance degradation [21]. Recently, some image harmonization frameworks [3,10] and watermark removal models [12] have been developed to solve the boundary and style discontinuities between the foreground and background for natural images. However, these frameworks have limitations when applied to brain MRI scans, where the smooth transition between the lesion and surrounding tissues is more critical than natural images.

In this paper, we tackle the problem of foreground intensity and style mismatch created by data augmentation, so that plausible images can be generated. As we do not have paired real and simulated images, we create simulated images by taking real images and introducing foreground disparities to use for training the image harmonization network (ARHNet). We further present an Adaptive Region Harmonization (ARH) module to align foreground feature maps guided by the background style information. Finally, we train a segmentation model based on the mixture of real and synthetic images produced by ARHNet to demonstrate its effectiveness for improving down-stream segmentation performance.

2   Methodology

The purpose of ARHNet is to harmonize the foreground in augmented images created by a data augmentation technique such as Copy-Paste [4], to further serve down-stream tasks like segmentation. We try to find a function f such that $f_\theta(\tilde{I}_a, M_a) \approx I_a$. Here, $\tilde{I}_a$ is the augmented image, I_a is the corresponding real image, and M_a is the foreground mask of $\tilde{I}_a$. θ refers to the parameter vector of f, a.k.a., ARHNet. However, since the augmented image $\tilde{I}_a$ does not have a corresponding real image I_a, we perform foreground intensity perturbation using a real brain MRI scan I with stroke lesions and its foreground mask M to create an image $\tilde{I}$ that simulates $\tilde{I}_a$ with a disharmonious foreground. We train ARHNet using the pairs $(\tilde{I}, M) \rightarrow I$ to learn the parameter vector θ.

2.1   Overview of ARHNet

Figure 1 represents our framework (ARHNet) for foreground harmonization, which comprises four components: a foreground intensity perturbation unit, a boundary extractor, a generator G, and a discriminator D. Given I and M, I is first scaled from 0 to 1. Next, the foreground intensity perturbation unit generates a foreground intensity-perturbed image $\tilde{I}$. Intensity perturbation is performed as follows:

$$\tilde{I} = [(1 + \alpha) \cdot I + \lambda] \odot M + I \odot (1 - M), \tag{1}$$

where $\alpha \sim \mathcal{U}(-0.3, 0.3)$, $\lambda \sim \mathcal{U}(-0.3, 0.3)$. Here α and λ can simulate large intensity variance in augmented images $\tilde{I}_a$ generated by advanced data augmentation approaches like Copy-Paste [4]. "$\odot$" denotes element-wise multiplication. After the foreground intensity perturbation, the stroke area is either brighter or darker compared to the background tissue, which is a boundary mismatch. Next, $\tilde{I}$ is passed through G to obtain the intensity difference map. The foreground region of the intensity difference map is then extracted using M and further added by $\tilde{I}$ to get a harmonized image $\hat{I}$. Inspired by [15], we concatenate $\hat{I}$ with $\tilde{I}$ and M to create the input image pair for D. Here $\tilde{I}$ and M provide location information of the foreground, which benefits the adversarial training process and ensures $\hat{I}$ have high fidelity to the ground truth image.

To optimize G and D so that harmonized images $\hat{I}$ have realistic texture and harmonized boundary intensities, three loss functions are deployed during model training: reconstruction loss $\mathcal{L}_{rec}$, boundary-aware total variation loss $\mathcal{L}_{btv}$, and adversarial loss $\mathcal{L}_{adv}$. The reconstruction loss implemented in our framework is defined as:

$$\mathcal{L}_{rec} = \|I - \hat{I}\|_1. \tag{2}$$

Reconstruction L1 loss makes the output and ground truth have similar appearances but may cause over-smoothing of images. Therefore, the model tends to output images with low mean square error but with relatively blurred texture. To prevent texture blurring we add a discriminator so that the generator will produce distinct and realistic images. The adversarial loss $\mathcal{L}_{adv}$ is added as additional supervision to the training process. In particular, we use hinge loss [9] instead of the cross-entropy loss to stabilize the training process and prevent the gradient from vanishing. The $\mathcal{L}_{adv}$ is formulated as follows:

$$\mathcal{L}_{adv}(D) = \mathbb{E}_{\hat{I},\tilde{I},M}[max(0, 1 - D(\hat{I}, \tilde{I}, M))] + \mathbb{E}_{I,\tilde{I},M}[max(0, 1 + D(I, \tilde{I}, M))], \tag{3}$$

$$\mathcal{L}_{adv}(G) = -\mathbb{E}_{\hat{I},\tilde{I},M}[D(\hat{I}, \tilde{I}, M)]. \tag{4}$$

A loss with only $\mathcal{L}_{rec}$ and $\mathcal{L}_{adv}$ leads to an abrupt boundary between the foreground and background. To encourage the network to give low gradients on the border area of $\hat{I}$ and make the transition from background to foreground smoother, we present a boundary-aware total variation loss $\mathcal{L}_{btv}$. If $\tilde{M}$ is the set of boundary voxels extracted by the boundary extractor, $\mathcal{L}_{btv}$ can be defined as:

$$\mathcal{L}_{btv} = \sum_{(i,j,k)\in\tilde{M}} \|\hat{I}_{i+1,j,k} - \hat{I}_{i,j,k}\|_1 + \|\hat{I}_{i,j+1,k} - \hat{I}_{i,j,k}\|_1 + \|\hat{I}_{i,j,k+1} - \hat{I}_{i,j,k}\|_1, \tag{5}$$

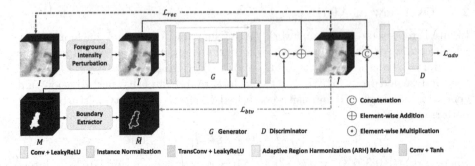

Fig. 1. The pipeline of ARHNet for adaptive image harmonization for simulated brain MRI with stroke lesions.

where i, j and k represent the $(i, j, k)^{th}$ voxel in $\tilde{M}$. By adding the boundary-aware loss, our framework makes the boundary transition smoother compared to other methods (see Figs. 3 and 4), which makes the harmonized images more like those observed on real MRI. Overall, our total loss function is defined as:

$$\mathcal{L}_{total} = \lambda_{rec}\mathcal{L}_{rec} + \lambda_{btv}\mathcal{L}_{btv} + \lambda_{adv}\mathcal{L}_{adv}, \tag{6}$$

where λ_{rec}, λ_{btv} and λ_{adv} are weighting factors for each term.

2.2   Adaptive Region Harmonization (ARH) Module

To better align the foreground and background feature maps obtained from $\tilde{I}$, we design a new feature normalization paradigm called Adaptive Region Harmonization (ARH) module. As depicted in Fig. 2, the ARH module takes the resized foreground mask M and the feature maps F as input. Here $F \in \mathbb{R}^{C \times H \times W \times D}$ and $M \in \mathbb{R}^{1 \times H \times W \times D}$, where C, H, W, D indicate the number of feature channels, height, width, and depth of F, respectively. We first divide the feature maps into foreground $F_f = F \odot M$ and background $F_b = F \odot (1 - M)$ according to M. Then we normalize F_f and F_b using Instance Normalization (IN) [19], and calculate the channel-wise background mean value $\mu \in \mathbb{R}^C$ and standard deviation $\sigma \in \mathbb{R}^C$ as follows:

$$\mu = \frac{1}{sum(1 - M)} \sum_{h,w,d} F_{c,h,w,d} \odot (1 - M_{h,w,d}), \tag{7}$$

$$\sigma = \sqrt{\frac{1}{sum(1 - M)} \sum_{h,w,d} [F_{c,h,w,d} \odot (1 - M_{h,w,d}) - \mu]^2}, \tag{8}$$

where $sum(\cdot)$ indicates summing all elements in the map. Different from the RAIN module [10] that directly applies μ and σ to F_f to align the foreground to the background, we present a learned scaling parameter strategy, with an attention mechanism so that the network focuses more on task-relevant areas

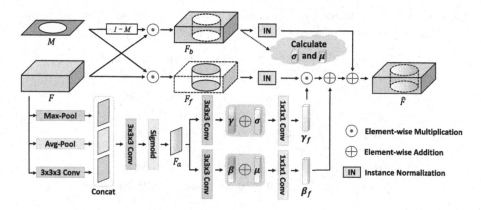

Fig. 2. The structure of our Adaptive Region Harmonization (ARH) module. μ and σ represent the channel-wise mean value and standard deviation calculated from F_b.

to better learn the consistent feature representation for both foreground and background.

Specifically, we calculate an attention map $F_a \in \mathbb{R}^{1 \times H \times W \times D}$ based on the entire feature maps in the ARH module, to let the module adaptively extract style information from important areas. F_a is formulated as:

$$F_a = S(Conv([F_{max}, F_{avg}, F_{Conv}])), \tag{9}$$

where S denotes the sigmoid function and $Conv$ denotes the convolution operation. Additionally, we calculate two channel-wised scaling parameters $\gamma \in \mathbb{R}^{C \times H \times W \times D}$ and $\beta \in \mathbb{R}^{C \times H \times W \times D}$ as:

$$\gamma = Conv(F_a), \beta = Conv(F_a). \tag{10}$$

γ and β allow element-wise adjustments on σ and μ which represent the global intensity information extracted from the background feature maps. We fuse γ and β with σ and μ with two convolutional layers to obtain the foreground scaling factors γ_f and β_f, which can be calculated as:

$$\gamma_f = Conv(\gamma + \sigma), \beta_f = Conv(\beta + \mu). \tag{11}$$

By applying γ_f and β_f to the foreground feature maps F_f, we finally attain the aligned feature maps via $\hat{F} = F_f \odot (1 + \gamma_f) + \beta_f + F_b$.

3   Experiments

3.1   Experiment Settings

Dataset. We use the ATLAS v2.0 dataset [8] to evaluate the performance of ARHNet. ATLAS (short for ATLAS v2.0) is a large stroke dataset, which contains 655 T1-weighted brain MRIs with publicly available voxel-wise annotations. All images were registered to the MNI-152 template with a voxel spacing

Table 1. Metrics for image harmonization on the ATLAS test set. The best results are highlighted in bold. fMAE and fPSNR are computed in the foreground.

Method	MAE↓	fMAE↓	PSNR↑	fPSNR↑
Composite	0.0014	0.23	39.70	14.62
HM	0.0010	0.14	41.38	16.30
UNet [18]	0.0009	0.11	43.94	18.88
Hinge-GAN [9]	0.0011	0.14	43.44	18.38
UNet-GAN [10]	0.0009	0.12	44.16	19.11
RainNet [10]	0.0007	0.09	45.33	20.30
Ours	**0.0006**	**0.07**	**46.74**	**21.74**

of $1\,mm \times 1\,mm \times 1\,mm$. According to [5], about half of the images are characterized as small lesion images (foreground voxels $\leq 5,000$). In this work we focus on only these images, corresponding to 320 MRIs. We split the dataset into five folds, stratified by lesion size to ensure both training and testing sets have the same data distribution. We randomly select one fold (20%) as the test set and the remaining four folds are the training set.

Implementation Details. ARHNet is implemented within PyTorch [16] and uses TorchIO [17] for loading data and creating intensity perturbations. To optimize the generator and discriminator, we use two AdamW optimizers [13]. The initial learning rates for G and D are set to $1e-4$, and $5e-5$, respectively. The batch size is set to 16 and total training epochs are 200 for each model. For input images, we randomly extract a $64 \times 64 \times 64$ patch from the MRI scans corresponding to the region that contains the stroke annotation(s). The loss weight factors λ_{rec}, λ_{btv}, and λ_{adv} are set to 100, 10, 1, respectively. For the down-stream segmentation task that is used to evaluate our framework, we implement a segmentation model based on Attention UNet [14] in the MONAI framework [1]. The initial learning rate is set to $1e-3$, and the batch size is 4. For a fair comparison, we train each setting for 30,000 iterations.

Evaluation Metrics. We evaluate the performance of ARHNet on the image harmonization task and also a down-stream stroke segmentation task. For the image harmonization task, we use four metrics to measure the fidelity of the output, *i.e.*, mean absolute error (MAE), mean absolute error of the foreground region (fMAE), peak signal-to-noise ratio (PSNR), and signal-to-noise ratio of the foreground region (fPSNR). For the down-stream stroke segmentation task, we use three metrics to evaluate the segmentation performance: the Dice coefficient, 95% Hausdorff Distance (95HD), and average surface distance (ASD).

3.2   Experimental Results

Comparison of Image Harmonization Results. We quantitatively compare the foreground image harmonization results of ARHNet on the ATLAS test

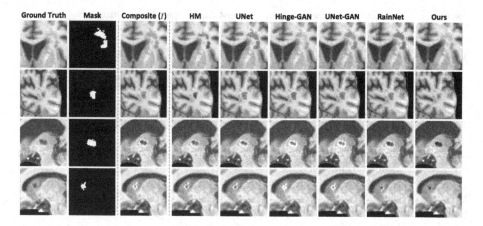

Fig. 3. Qualitative comparison between different harmonization methods.

set with other non-learning- and learning-based methods. Results are shown in Table 1 where "Composite" means we do not use any image harmonization method but directly calculating the metrics based on the images with foreground disparities which are inputs for all other methods. It gives the worst results as expected. If we adapt the foreground intensity to be consistent with the background based on Histogram Matching ("HM" in Table 1), we can achieve better results, but still worse than all of the learning-based methods evaluated.

Four learning-based methods are implemented as comparisons. Here "UNet" refers to the UNet model trained with only the reconstruction loss $\mathcal{L}_{rec}$. "Hinge-GAN" means the UNet model trained with only the adversarial loss $\mathcal{L}_{adv}$. "UNet-GAN" denotes the UNet model is trained under the supervision of $\mathcal{L}_{rec}$ and $\mathcal{L}_{adv}$. "RainNet" is a generator that consists of the RAIN module [10], also only $\mathcal{L}_{rec}$ and $\mathcal{L}_{adv}$ are used for backpropagation. From Table 1, we can find that our method outperforms other methods on all metrics, proving the efficacy and rationality of ARHNet. Furthermore, compared with RainNet, our method achieve a big improvement of 1.41 dB in PSNR and 1.44 dB in fPSNR.

We present qualitative results in Figs. 3 and 4. In Fig. 3 we can observe that ARHNet can achieve realistic harmonization images no matter if the foreground is brighter or darker than the background (top two rows: darker, bottom two rows: brighter). Also, the boundaries in our results are smoother than other methods. Additionally, we show the image harmonization results on composite brain MRI scans in Fig. 4. By zooming in on the boundary area, it is easy to observe that composite images harmonized by ARHNet are more realistic than RainNet, which demonstrates the superiority of our method again.

Comparison of Down-Stream Segmentation Performance. We report quantitative measures of the down-stream lesion segmentation performance for different training sets in Table 2. For each setting, we keep the batch size the same and train for 30,000 iterations for a fair comparison. "-" denotes no additional

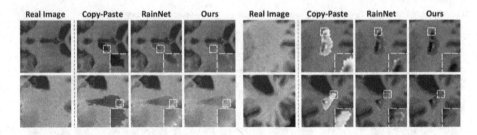

Fig. 4. Visualization results on composite brain MRI scans which are used for the down-steam segmentation task.

Table 2. Segmentation performances under different training data settings.

Additional Data	Dice↑	ASD↓	95HD↓
–	23.84	48.85	85.67
200 real	25.05	48.93	88.08
200 by [22]	32.38	40.11	77.78
200 by Ours	**36.41**	**25.14**	**49.30**

Table 3. Ablation studies on different feature normalization methods.

Method	MAE↓	fMAE↓	PSNR↑	fPSNR↑
BN [7]	0.0007	0.09	45.80	20.79
IN [19]	0.0008	0.09	45.68	20.66
RAIN [10]	0.0008	0.10	43.89	18.87
Ours	**0.0006**	**0.07**	**46.74**	**21.74**

data is used for model training. "200 real" means 200 images with big lesions (foreground voxels > 5,000) from the original ATLAS v2.0 dataset are utilized as additional training samples. "200 by [22]" refers to using CarveMix to generate additional 200 images for model training. "200 by Ours" means we first use Copy-Paste [4] strategy to create 200 composite images, then we use ARHNet to adjust the foreground intensity to harmonize the images. As shown in Table 2, our method achieves the best segmentation result and brings a large performance gain of 12.57% in Dice compared to not using any additional data.

Ablation Study. We also investigate the performance gain achieved by our ARH module, results are shown in Table 3. We can find that if we keep all other settings unchanged and only replace the ARH module with InstanceNorm or BatchNorm, higher PSNR is reached compared to RainNet (see in Table 1). This demonstrates the effectiveness of some of the additional elements we presented in this work, such as boundary-aware total variation loss and the foreground intensity perturbation unit. However, if we replace the ARH module with the RAIN module, the result is the worst among all normalization methods. This is likely because the RAIN module only considers the entire style of the background, and therefore cannot align the foreground feature maps properly.

4 Conclusion

In this paper, we propose an Adaptive Region Harmonization Network (ARHNet) that can effectively harmonize a target area and make the style of foreground and background consistent in this region. This framework can be utilized to harmonize synthetic samples generated by other data augmentation methods, and make these images more realistic and natural. Harmonized augmented samples can be further utilized in down-stream segmentation tasks to improve the segmentation model's generalization ability. Extensive experimental results demonstrate that our proposed method can generate style-consistent images and is effective for segmenting small stroke lesions on T1-weighted MRI.

Acknowledgement. This work was supported by Centre for Doctoral Training in Surgical and Interventional Engineering at King's College London; King's-China Scholarship Council PhD Scholarship programme (K-CSC); and the Engineering and Physical Sciences Research Council Doctoral Training Partnership (EPSRC DTP) grant EP/T517963/1. This publication represents, in part, independent research commissioned by the Wellcome Innovator Award [218380/Z/19/Z]. The views expressed in this publication are those of the authors and not necessarily those of the Wellcome Trust.

References

1. Cardoso, M.J., et al.: MONAI: an open-source framework for deep learning in healthcare. arXiv preprint arXiv:2211.02701 (2022)
2. Chen, S., Dobriban, E., Lee, J.H.: A group-theoretic framework for data augmentation. J. Mach. Learn. Res. **21**(1), 9885–9955 (2020)
3. Cong, W., et al.: Dovenet: deep image harmonization via domain verification. In: Proceedings of the IEEE/CVF Conference on Computer Vision and Pattern Recognition, pp. 8394–8403 (2020)
4. Ghiasi, G., et al.: Simple copy-paste is a strong data augmentation method for instance segmentation. In: Proceedings of the IEEE/CVF Conference on Computer Vision and Pattern Recognition, pp. 2918–2928 (2021)
5. Huo, J., et al.: Mapping: model average with post-processing for stroke lesion segmentation. arXiv preprint arXiv:2211.15486 (2022)
6. Huo, J., et al.: Brain lesion synthesis via progressive adversarial variational auto-encoder. In: Zhao, C., Svoboda, D., Wolterink, J.M., Escobar, M. (eds.) SASHIMI 2022. LNCS, vol. 13570, pp. 101–111. Springer, Cham (2022). https://doi.org/10.1007/978-3-031-16980-9_10
7. Ioffe, S., Szegedy, C.: Batch normalization: accelerating deep network training by reducing internal covariate shift. In: International Conference on Machine Learning, pp. 448–456. PMLR (2015)
8. Liew, S.L., et al.: A large, curated, open-source stroke neuroimaging dataset to improve lesion segmentation algorithms. Sci. Data **9**(1), 320 (2022)
9. Lim, J.H., Ye, J.C.: Geometric GAN. arXiv preprint arXiv:1705.02894 (2017)
10. Ling, J., Xue, H., Song, L., Xie, R., Gu, X.: Region-aware adaptive instance normalization for image harmonization. In: Proceedings of the IEEE/CVF Conference on Computer Vision and Pattern Recognition, pp. 9361–9370 (2021)
11. Liu, X., et al.: MSDF-net: multi-scale deep fusion network for stroke lesion segmentation. IEEE Access **7**, 178486–178495 (2019)
12. Liu, Y., Zhu, Z., Bai, X.: Wdnet: watermark-decomposition network for visible watermark removal. In: Proceedings of the IEEE/CVF Winter Conference on Applications of Computer Vision, pp. 3685–3693 (2021)

13. Loshchilov, I., Hutter, F.: Decoupled weight decay regularization. arXiv preprint arXiv:1711.05101 (2017)
14. Oktay, O., et al.: Attention U-net: learning where to look for the pancreas. arXiv preprint arXiv:1804.03999 (2018)
15. Ouyang, X., Cheng, Y., Jiang, Y., Li, C.L., Zhou, P.: Pedestrian-synthesis-GAN: generating pedestrian data in real scene and beyond. arXiv preprint arXiv:1804.02047 (2018)
16. Paszke, A., et al.: Pytorch: an imperative style, high-performance deep learning library. In: Advances in Neural Information Processing Systems, vol. 32 (2019)
17. Pérez-García, F., Sparks, R., Ourselin, S.: Torchio: a python library for efficient loading, preprocessing, augmentation and patch-based sampling of medical images in deep learning. Comput. Methods Programs Biomed. **208**, 106236 (2021)
18. Ronneberger, O., Fischer, P., Brox, T.: U-net: convolutional networks for biomedical image segmentation. In: Navab, N., Hornegger, J., Wells, W.M., Frangi, A.F. (eds.) MICCAI 2015. LNCS, vol. 9351, pp. 234–241. Springer, Cham (2015). https://doi.org/10.1007/978-3-319-24574-4_28
19. Ulyanov, D., Vedaldi, A., Lempitsky, V.: Instance normalization: the missing ingredient for fast stylization. arXiv preprint arXiv:1607.08022 (2016)
20. Wan, J., Liu, Y., Wei, D., Bai, X., Xu, Y.: Super-BPD: super boundary-to-pixel direction for fast image segmentation. In: Proceedings of the IEEE/CVF Conference on Computer Vision and Pattern Recognition, pp. 9253–9262 (2020)
21. Wei, W., et al.: Adversarial examples in deep learning: characterization and divergence. arXiv preprint arXiv:1807.00051 (2018)
22. Zhang, X., et al.: CarveMix: a simple data augmentation method for brain lesion segmentation. In: de Bruijne, M., et al. (eds.) MICCAI 2021. LNCS, vol. 12901, pp. 196–205. Springer, Cham (2021). https://doi.org/10.1007/978-3-030-87193-2_19
23. Zhang, Y., Wu, J., Liu, Y., Chen, Y., Wu, E.X., Tang, X.: Mi-unet: multi-inputs unet incorporating brain parcellation for stroke lesion segmentation from t1-weighted magnetic resonance images. IEEE J. Biomed. Health Inform. **25**(2), 526–535 (2020)

Normative Aging for an Individual's Full Brain MRI Using Style GANs to Detect Localized Neurodegeneration

Shruti P. Gadewar[✉], Alyssa H. Zhu, Sunanda Somu, Abhinaav Ramesh,
Iyad Ba Gari, Sophia I. Thomopoulos, Paul M. Thompson, Talia M. Nir,
and Neda Jahanshad

Imaging Genetics Center, Mark and Mary Stevens Neuroimaging and Informatics Institute,
University of Southern California, Marina del Rey, CA, USA
{gadewar,njahansh}@usc.edu

Abstract. In older adults, changes in brain structure can be used to identify and predict the risk of neurodegenerative disorders and dementias. Traditional 'brainAge' methods seek to identify differences between chronological age and biological brain age predicted from MRI that can indicate deviations from normative aging trajectories. These methods provide one metric for the entire brain and lack anatomical specificity and interpretability. By predicting an individual's healthy brain at a specific age, one may be able to identify regional deviations and abnormalities in a true brain scan relative to healthy aging. This study aims to address the problem of domain transfer in estimating age-related brain differences. We develop a fully unsupervised generative adversarial network (GAN) with cycle consistency reconstruction losses, trained on 4,000 cross-sectional brain MRI data from UK Biobank participants aged 60 to 80. By converting the individual anatomic information from their T1-weighted MRI as "content" and adding the "style" information related to age and sex from a reference group, we demonstrate that brain MRIs for healthy males and females at any given age can be predicted from one cross-sectional scan. Paired with a full brain T1w harmonization method, this new MRI can also be generated for any image from any scanner. Results in the ADNI cohort showed that without relying on longitudinal data from the participants, our style-encoding domain transfer model might successfully predict cognitively normal follow-up brain MRIs. We demonstrate how variations from the expected structure are a sign of a potential risk for neurodegenerative diseases.

Keywords: domain transfer · GAN · aging patterns · brain MRI

1 Introduction

The structure of the human brain undergoes dynamic changes as individuals age [1]. In brain mapping studies, we often seek to identify patterns of changes that predict underlying risk for disease. For instance, atrophy in the temporal lobe may indicate a higher risk of late-onset Alzheimer's disease (AD) and related dementias [2]. However, to determine whether atrophy patterns are as expected with age, or accelerated compared to typical aging patterns, a reference aging model is required.

X. Cao et al. (Eds.): MLMI 2023, LNCS 14349, pp. 387–395, 2024.
https://doi.org/10.1007/978-3-031-45676-3_39

"BrainAge" has been introduced to provide an estimate for how old an individual brain appears, most often providing a single metric for the entire brain [3]. BrainAge models are generally built using machine learning or deep learning methods on large-scale normative datasets and are applied to out of sample testing data to estimate a new individual's brainAge, and the brainAge gap. In older adult populations, a brainAge older than the chronological age of the person implies that the individual has undergone accelerated brain aging and may be at risk for, or have, a neurodegenerative condition [4]. BrainAge has several limitations and constraints described in [5], including a loss of anatomical specificity and interpretability as 1) brain regions may "age" differently and 2) specific patterns of neuroanatomical aging may reflect distinct disease processes. In contrast, a detailed 3D brain map highlighting the specific areas where deviations occur, could offer a more comprehensive understanding of underlying anomalies. While brainAge saliency maps [4] provide a map of the relative importance of features across the brain, they do not give insights on the individual level as to whether these specific regions contribute to greater or lower brainAge estimates, as directionality is not provided.

Studies of structural brain changes typically involve mapping longitudinally collected brain MRIs to each other to quantify change over time. Deformation-based methods allow us to map regional changes, but a reference population is needed to give any insight into how individual patterns of neuroanatomic changes deviate from their typical aging trajectories. The reference populations are generally collected on the same scanner and with the same acquisition protocols, which can be challenging to maintain in longitudinal studies as scanners and protocols are upgraded. Creating a normative population aging full brain MRI may help provide a reference to guide the typical brain aging pattern. Building on a normative framework, the generation of brain MRI images at future ages could potentially offer insights into how the brain would age under healthy conditions. These generated images could serve as a benchmark for an individual's brain structure at any given age. Similar to brainAge, any deviations or differences detected between the true image and the generated image could potentially indicate anatomical abnormalities and risk factors for cognitive decline.

Deep learning techniques can generate brain MRIs in order to simulate and predict neurodegenerative trajectories [6–8]. Deep learning models can also perform various image translations by separating the image into distinct 'content' and 'style' spaces [9]. The 'content' space captures low-level information in images, such as contours and orientations, while the 'style' space represents high-level information like colors and textures. Different domains of images may share the same content space but exhibit different styles.

Here, we train a style GAN model on a subset of images from UK Biobank and predict an individual's healthy brain MRI at any age between 60 and 79 years, map deviations in brain regions between the predictions and the actual follow-up scans for different diagnoses like stable mild cognitive impairment (MCI), MCI to dementia convertors and stable dementia.

2   Methods and Materials

2.1   The Architecture of the Style-Encoding GAN

The StarGAN-v2 framework [10] used in this study included a single encoder (E), generator (G), and discriminator (D), as illustrated in Fig. 1. We defined X as a collection of brain MR images, and Y as a set of domains specific to age and sex, totaling 40 domains - twenty age bins for each of male and female sex. The goal was to train this network to create a variety of output images across all age domains, corresponding to the indicated sex of the subject, given an image x from X and the original age group y from Y. To achieve this, the style encoding block E was used to extract age data, which in turn generated age/sex specific styles using training images sorted into different domains. This enabled G to transform an input image x into an output image $G(x,t)$, mirroring the aging pattern of the target age group y_t. In this context, $t = M(z,y_t)$ represents the age factor for input x, created using a mapping network M that samples from a latent vector z and the target domain y_t. The resulting target age group t is then associated with both the patterns of the target domain and the patterns irrelevant to the domain from the input image. We opted to use t instead of z because the intermediate latent space, derived from the mapping network, doesn't need to support the sampling of the latent vector z based on a fixed distribution. This approach enforces a disentangled representation, leading to the generation of more realistic images [11]. Adaptive instance normalization (AdIN) [12] was employed to incorporate t into G. The discriminator D then aims to differentiate between the original input image x and the generated output image $G(x,t)$ produced by G. During the training process, E is trained to ensure that if an image is generated based on age group t, then t will also be encoded when this generated image is input to E with the target domain, resulting in $E(G(x, t), y_t) = t$.

2.2   Network Losses

The following loss functions were used to train our model:

Adversarial Loss. The generator, denoted as G, yields $G(x, s)$, where 'x' is the input image and 's' is the target style code, represented as $M(z, y)$. This target style code is generated during the training process of the network, where 'M' stands for the mapping network and 'z' is a randomly generated latent code. The discriminator, referred to as D, differentiates the output image $G(x, s)$ from the input image by using the adversarial loss:

$$L_{GAN} = E_x\big[logD(x)\big] + E_{x,z}[log(1 - D(G(x, s)))]$$

Cycle-consistency Loss [13]. This term signifies the variance between the original and the generated images. This loss metric is applied to ensure the images created retain the anatomical patterns inherent in the input image x:

$$L_{cyc} = E_{x,z}[\|x - G(G(x, s), s_x)\|_1]$$

Here, $s_x = E(x, y)$ is the estimated style code of the input image x. The generator G undergoes extensive training to preserve the characteristics of the original image x while incorporating the desired style s_x.

Style Reconstruction Loss. This loss function compels the generator, denoted as G, to utilize a style code s extracted from a reference set. The objective is to generate an output image, represented as $G(x, s)$, in conjunction with the encoder:

$$L_{sty} = E_{x,z}[\|s - E(G(x, s))\|_1]$$

Style Diversification Loss [14]. This regularization technique encourages the generator, G, to explore the space of images and generate a wide range of diverse outputs. The goal is to maintain the unique characteristics and individuality of each input image:

$$L_{div} = E_{x,z_1,z_2}[\|G(x, s_1) - G(x, s_2)\|_1],$$

where s_1 and s_2 are the target style codes generated using the mapping network M and two random latent codes z_1 and z_2 (i.e., $s_i = M(z_i)$ for i $\in \{1, 2\}$).

Complete Loss. All the losses combined - with λ_{cyc}, λ_{sty} and λ_{div}, being the hyper-parameters for each of the loss functions - the final objective function can be written as:

$$L(G, M, E, D) = L_{GAN} + \lambda_{cyc}L_{cyc} + \lambda_{sty}L_{sty} - \lambda_{div}L_{div}$$

2.3   2D Implementation on 3D Images

Our model implementation is based on the PyTorch framework, which can be found at this GitHub repository: (https://github.com/clovaai/stargan-v2.git). To train the model, we used a single GPU node on a Cisco C480 ML server equipped with 8 SXM2 V100 32 GB GPUs, NVLink interconnect, and 2 x 6140 Intel Skylake processors. The training process lasted for 150,000 epochs over a period of 6 days. We used a batch size of 4 and adjusted the lambda values (λ_{cyc}=3, λ_{sty}=2, λ_{div} = 2) until we observed desired anatomical variations in the generated images.

Although brain MRI scans consist of 3D volumes, we designed our network architecture to handle 2D images due to limitations in the GPU memory. We transformed the 3D scans into a series of 3-slice or 3-channel images along the brain's sagittal plane. Each channel represented one slice, and these transformed images were saved as conventional PNG files. The model was then trained on these PNG images. The generated 3-channel image series from the model were stacked and then transformed back into a 3D volume.

In the generative adversarial architecture for style transfer, we loaded all slices in a manner that allowed the model to learn the transformation across all bins for a specific slice number within each pair of age bins. For example, if we wanted to transform an image from age bin 65 to age bin 75, we would first harmonize the image to an image from UK Biobank dataset using [15] and then choose a reference subject from age bin 75, and the harmonized images from age bin 65 would be transformed to age bin 75 using the trained generator. The error in the prediction was controlled during training by the discriminator, and the generator weights were updated to minimize the discrepancy from the desired target population.

2.4 Data Processing for Model Training and Testing

A subset of 5,000 T1-weighted brain MR images (T1w) from the UK Biobank dataset [16], specifically focusing on subjects aged 60–79 was used. We used 20 one-year age bins per sex to improve the accuracy of the model. The images were selected such that each age group stratified by sex consisted of 125 subject images. 100 images from each group were used for training while the rest of them were set aside for validation. Prior to training, all the images underwent non-uniformity correction and skull stripping using HD-BET [17]. Images were then linearly registered to the MNI template using FSL's *flirt* [18] with 9 degrees-of-freedom and zero-padding was performed to obtain an image dimension of $256 \times 256 \times 256$.

For evaluation, we used an independent longitudinal test set consisting of 500 subjects from the ADNI dataset [19]. This test set included two scans per subject captured at varying time intervals (1, 2, 3, 4, 5 or 6 years), allowing for assessment of the model's effectiveness. All scans were first harmonized to a reference image from UK Biobank dataset using [15]. The data distribution for training and test set is summarized in Table 1.

Table 1. Demographic breakdown of subjects in the two datasets used.

Dataset	Group	BL Age (years)	FU Age (years)	N Male/Female	N Train/Val
UK Biobank	–	(60–79)	–	2500/2500	4000/1000
ADNI	Cognitive Normal	70.52 ± 3.93	73 ± 3.82	54/90	–
	MCI Stable	69.68 ± 4.77	72.52 ± 4.61	115/78	–
	MCI to Dementia Convertors	70.81 ± 4.29	73.48 ± 4.15	20/23	–
	Dementia stable	71.04 ± 4.65	72.31 ± 4.61	25/15	–

2.5 Evaluation: Hippocampal Volumes and Tensor Based Morphometry

Hippocampal volumes were extracted for all the true and generated MRIs using Hippodeep [20]. The differences between the average hippocampal volumes for predicted and true follow-up (FU) scans were calculated for each test subject. Full brain tensor based morphometry (TBM) was performed to map local volumetric differences between predicted and true FU T1w scans. Median normalization was performed on all FU images. The predicted FUs were warped to the true FU images and the log Jacobian (Jac) maps were generated using ANTS's Syn registration [21]. To spatially normalize the Jacs for group statistics, the true FU T1w were linearly registered (12 dof) using FSL's *flirt* tool [18] and then warped to the *mni_icbm152_nlin_sym_09a* template. The resulting transforms were applied to the Jac maps.

2.6   Statistical Analysis

Linear regressions were performed to evaluate differences in predicted and true FU hippocampal volumes (i.e., dependent variable was predicted minus FU volume) between 1) stable CN and stable AD participants or 2) stable MCI and MCI to dementia converters. Voxel wise linear regressions were also performed to test for differences in log Jacobian (i.e., regional volumetric differences between predicted and true FU T1w images) between stable MCI and MCI to dementia converters. All models were adjusted for baseline age, sex, age-by-sex interaction, education (years), intracranial volume, and the time interval between and FU and baseline scans (years).

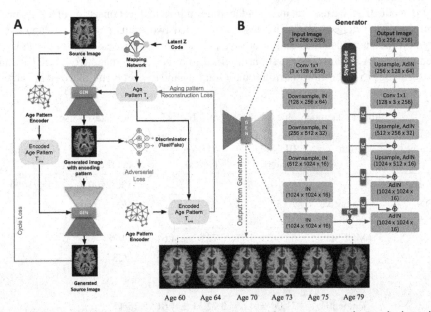

Fig. 1. A. Style-encoding GAN architecture. The generators learn to generate images by inputting a source image and a style code, the discriminator tries to differentiate between the real image and one generated from a generator block. This whole network is guided using an adversarial loss, cycle loss, and an aging pattern reconstruction loss. **B.** Generator architecture: The blue blocks represent the image with its dimensions. The three numbers in each green block represent the input channels, output channels, and the image size respectively; where Conv is a convolution layer, IN: Instance Normalization; AdIN: Adaptive Instance Normalization.

3   Results

Hippocampal volume differences between predicted and the true FU scans were close to zero for stable CN participants; these differences got progressively larger with cognitive impairment (Fig. 2A). Compared to CN participants, those with dementia showed significantly greater differences, suggesting greater hippocampal atrophy in dementia than expected with aging ($p = 7.95 \times 10^{-17}$; $\beta = 359.5$, se $= 38.9$). Compared to participants with stable MCI, those that converted from MCI to dementia also showed greater hippocampal volume differences ($p = 3.06 \times 10^{-8}$, $\beta = 596.9$, se $= 104.1$). Figure 2B shows an example of qualitative differences between true and predicted FU T1w in one CN participant; compared to true FU, the predicted image shows smaller ventricular and sulcal volumes which is confirmed in the voxelwise log Jacobian map. Significant differences in log Jacobians (i.e., voxelwise differences between predicted and true FU T1w MRI) were detected between CN and AD ($p < 0.0028$; Fig. 2C).

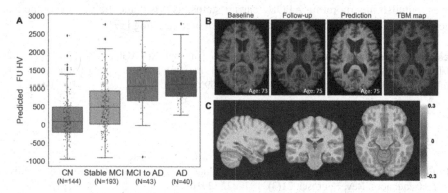

Fig. 2. A) Average hippocampal volume difference between the predicted and real FU scans for different diagnostic groups. **B)** T1w MRI of a CN female at baseline age 73, true FU at 75, predicted image at 75, and TBM map between predicted and true FU; blue voxels (e.g., sulci and ventricles) indicate larger volume in true compared to predicted FU, while red (e.g., white matter) indicates smaller volume. **C)** Regional β-values for differences in log Jacobians between CN vs. dementia groups with $p < 0.0028$ are highlighted. The red color shows dementia participants showed relatively greater tissue volume in the predicted scans compared to the true follow up T1ws (e.g., in the hippocampus, suggesting greater atrophy in the hippocampus relative to healthy aging in dementia) while blue shows regions where dementia participants showed relatively smaller volumes in predicted compared to true T1w (e.g., ventricles, suggesting greater expansion in individuals with dementia relative to healthy aging).

4   Discussion and Conclusion

We demonstrate that GANs trained on a large cross-sectional database of MR images, specifically domain and style-encoding GANs, offer the potential to predict how an individual's brain may change over time under normal and healthy conditions. Notably, our model does not rely on longitudinal images from the same subjects for training. Other generative approaches, such as the one presented in [22], generate 3D brain volumes for data augmentation purposes rather than individualized predictions. In work similar to what we present here, [23] predicts aged images using cross-sectional data; our model builds on this by allowing sex specific models, and a framework for other covariates to be incorporated.

The proposed method holds potential for charting the brain-wide trajectory of neurodegenerative diseases. A model trained on healthy subjects can establish individual normative models. These models enable the monitoring of individual brain changes in relation to normative trajectories, offering increased sensitivity in tracking the early progression of abnormalities associated with brain diseases.

One limitation of our work is that the model is currently trained only on UK Biobank data, using largely healthy individuals; population biases in the training model may influence the outcome, as many brain wide applications have been shown to be biased by templates [24–26]. Expansions of our work may include not only sex-specific trajectories, but also incorporating APOE4 carrier status, as this dementia risk genotype has been associated with the rate of brain volume change in aging [27].

Acknowledgments. This work was supported in part by: R01AG059874, RF1AG057892, R01AG058854, P41EB015922, U01AG068057. This research has been conducted using the UK Biobank Resource under Application Number '11559'.

References

1. Fjell, A.M., Walhovd, K.B.: Structural brain changes in aging: courses, causes and cognitive consequences. Rev. Neurosci. **21**(3), 187–222 (2010)
2. Rabinovici, G.D.: Late-onset Alzheimer disease. CONTINUUM: Lifelong Learn. Neurol. **25**(1), 14–33 (2019)
3. Cole, J.H., Franke, K.: Predicting age using neuroimaging: innovative brain ageing biomarkers. Trends Neurosci. **40**(12), 681–690 (2017)
4. Yin, C., et al.: Anatomically interpretable deep learning of brain age captures domain-specific cognitive impairment. In: Proceedings of the National Academy of Sciences 120, no. 2, e2214634120 (2023)
5. Butler, E.R., et al. Pitfalls in brain age analyses, vol. 42, no. 13. Hoboken, USA: John Wiley & Sons, Inc. (2021)
6. Rachmadi, M.F., Valdés-Hernández, M.C., Makin, S., Wardlaw, J., Komura, T.: Automatic spatial estimation of white matter hyperintensities evolution in brain MRI using disease evolution predictor deep neural networks. Medical image analysis 63, 101712 (2020)
7. Rachmadi, M.F., Valdés-Hernández, M.C., Makin, S., Wardlaw, J.M., Komura, T.: Predicting the evolution of white matter Hyperintensities in brain MRI using generative adversarial networks and irregularity map. In: Shen, D., Liu, T., Peters, T.M., Staib, L.H., Essert, C., Zhou, S., Yap, P.-T., Khan, A. (eds.) MICCAI 2019. LNCS, vol. 11766, pp. 146–154. Springer, Cham (2019). https://doi.org/10.1007/978-3-030-32248-9_17
8. Ravi, D., Alexander, D.C., Oxtoby, N.P.: Degenerative adversarial neuroimage nets: generating images that mimic disease progression. In: Shen, D., Liu, T., Peters, T.M., Staib, L.H., Essert, C., Zhou, S., Yap, P.-T., Khan, A. (eds.) MICCAI 2019. LNCS, vol. 11766, pp. 164–172. Springer, Cham (2019). https://doi.org/10.1007/978-3-030-32248-9_19
9. Huang, X., Liu, M.-Y., Belongie, S., Kautz, J.: Multimodal unsupervised image-to-image translation. In: Proceedings of the European Conference on Computer Vision (ECCV), pp. 172–189 (2018)
10. Choi, Y., Uh, Y., Yoo, J., Ha, J.-W.: Stargan v2: diverse image synthesis for multiple domains. In: Proceedings of the IEEE/CVF Conference on Computer Vision and Pattern Recognition, pp. 8188–8197 (2020)

11. Karras, T., Laine, S., Aila, T.: A style-based generator architecture for generative adversarial networks. In: Proceedings of the IEEE/CVF Conference on Computer Vision and Pattern Recognition, pp. 4401–4410 (2019)
12. Huang, X., Belongie, S.: Arbitrary style transfer in real-time with adaptive instance normalization. In: Proceedings of the IEEE International Conference on Computer Vision, pp. 1501–1510 (2017)
13. Zhao, F., et al.: Harmonization of infant cortical thickness using surface-to-surface cycle-consistent adversarial networks. In: Shen, D., Liu, T., Peters, T.M., Staib, L.H., Essert, C., Zhou, S., Yap, P.-T., Khan, A. (eds.) Medical Image Computing and Computer Assisted Intervention – MICCAI 2019: 22nd International Conference, Shenzhen, China, October 13–17, 2019, Proceedings, Part IV, pp. 475–483. Springer International Publishing, Cham (2019). https://doi.org/10.1007/978-3-030-32251-9_52
14. Wang, X., et al.: ESRGAN: enhanced super-resolution generative adversarial networks. In: Proceedings of the European Conference on Computer Vision (ECCV) Workshops (2018)
15. Liu, M., et al.: Style transfer generative adversarial networks to harmonize multi-site MRI to a single reference image to avoid over-correction. bioRxiv (2022)
16. Miller, K.L.: Multimodal population brain imaging in the UK Biobank prospective epidemiological study. Nature Neuroscience 19(11), 1523–1536 (2016)
17. Isensee, F., et al.: Automated brain extraction of multisequence MRI using artificial neural networks. Hum. Brain Mapp. 40(17), 4952–4964 (2019). https://doi.org/10.1002/hbm.24750
18. Jenkinson, M., Bannister, P., Brady, M., Smith, S.: Improved optimization for the robust and accurate linear registration and motion correction of brain images. Neuroimage 17(2), 825–841 (2002)
19. Mueller, S.G., et al.: Ways toward an early diagnosis in Alzheimer's disease: the Alzheimer's Disease Neuroimaging Initiative (ADNI). Alzheimer's Dement 1(1), 55–66 (2005)
20. Thyreau, B., Sato, K., Fukuda, H., Taki, Y.: Segmentation of the hippocampus by transferring algorithmic knowledge for large cohort processing. Med. Image Anal. 43, 214–228 (2018)
21. Avants, B.B., Tustison, N., Song, G.: Advanced normalization tools (ANTS). Insight J 2(365), 1–35 (2009)
22. Kwon, G., Han, C., Kim, D.-S.: Generation of 3D brain MRI using auto-encoding generative adversarial networks. In: Shen, D., Liu, T., Peters, T.M., Staib, L.H., Essert, C., Zhou, S., Yap, P.-T., Khan, A. (eds.) Medical Image Computing and Computer Assisted Intervention – MICCAI 2019: 22nd International Conference, Shenzhen, China, October – 13–17, 2019, Proceedings, Part III, pp. 118–126. Springer International Publishing, Cham (2019). https://doi.org/10.1007/978-3-030-32248-9_14
23. Xia, T, Chartsias, A., Wang, C., Tsaftaris, S.A., Alzheimer's Disease Neuroimaging Initiative: Learning to synthesise the ageing brain without longitudinal data. Med. Image Anal. 73, 102169 (2021)
24. Dey, N., Ren, M., Dalca, A.V., Gerig, G.: Generative adversarial registration for improved conditional deformable templates. In: Proceedings of the IEEE/CVF International Conference on Computer Vision, pp. 3929–3941 (2021)
25. Zhu, A.H., Thompson, P.M., Jahanshad, N.: Age-related heterochronicity of brain morphometry may bias voxelwise findings. In: 2021 IEEE 18th International Symposium on Biomedical Imaging (ISBI), pp. 836–839. IEEE (2021)
26. Fonov, V., Evans, A.C., Kelly Botteron, C., Almli, R., McKinstry, R.C., Louis Collins, D.: Unbiased average age-appropriate atlases for pediatric studies. Neuroimage 54(1), 313–327 (2011). https://doi.org/10.1016/j.neuroimage.2010.07.033
27. Brouwer, R.M., et al.: Genetic variants associated with longitudinal changes in brain structure across the lifespan. Nat. Neurosci. 25(4), 421–432 (2022)

Deep Bayesian Quantization
for Supervised Neuroimage Search

Erkun Yang[1,2], Cheng Deng[2], and Mingxia Liu[1(✉)]

[1] Department of Radiology and BRIC, University of North Carolina at Chapel Hill,
Chapel Hill, NC 27599, USA
mingxia_liu@med.unc.edu
[2] Xidian University, Xi'an, China

Abstract. Neuroimage retrieval plays a crucial role in providing physicians with access to previous similar cases, which is essential for case-based reasoning and evidence-based medicine. Due to low computation and storage costs, hashing-based search techniques have been widely adopted for establishing image retrieval systems. However, these methods often suffer from nonnegligible quantization loss, which can degrade the overall search performance. To address this issue, this paper presents a compact coding solution namely *Deep Bayesian Quantization* (DBQ), which focuses on deep compact quantization that can estimate continuous neuroimage representations and achieve superior performance over existing hashing solutions. Specifically, DBQ seamlessly combines the deep representation learning and the representation compact quantization within a novel Bayesian learning framework, where a proxy embedding-based likelihood function is developed to alleviate the sampling issue for traditional similarity supervision. Additionally, a Gaussian prior is employed to reduce the quantization losses. By utilizing pre-computed lookup tables, the proposed DBQ can enable efficient and effective similarity search. Extensive experiments conducted on $2,008$ structural MRI scans from three benchmark neuroimage datasets demonstrate that our method outperforms previous state-of-the-arts.

Keywords: Neuroimage search · Deep Bayesian learning · Quantization

1 Introduction

Neuroimage analysis plays a fundamental role in modern medicine and has gained significant usage in automated disease diagnosis [1–3], clinical analysis [4,5], and treatment planning [6]. As the utilization of digital imaging data becomes widespread in hospitals and research centers, the size of neuroimage repositories is experiencing rapid growth. In the medical domain, the question of neuroimage similarity has gained greater importance since clinical decision-making traditionally relies on evidence from patients' data along with physicians'

Supplementary Information The online version contains supplementary material available at https://doi.org/10.1007/978-3-031-45676-3_40.

X. Cao et al. (Eds.): MLMI 2023, LNCS 14349, pp. 396–406, 2024.
https://doi.org/10.1007/978-3-031-45676-3_40

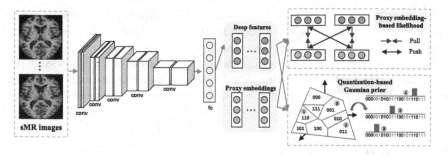

Fig. 1. Framework of the Deep Bayesian Quantization (DBQ)

prior experience with similar cases [7]. A recent study [8] has shown that physicians select similar cases primarily based on visual characteristics. Consequently, the search for approximate nearest neighbors (ANN) in neuroimages, which facilitates the retrieval of similar objects from existing neuroimage datasets, has become more critical than ever before.

Current promising solutions for ANN search include hashing [9–12] and quantization [13–16] approaches, both of which can transform high-dimensional input data into compact binary codes. Hashing-based methods generally first learn a family of hash functions to encode inputs to binary codes, and then conduct similarity search by computing the Hamming distance. However, since binary codes are highly compressed, existing hashing approaches usually suffer from the nonnegligible quantization loss and fail to output informative codes without exploring any quantization techniques.

Quantization-based methods [14, 17–19] can estimate the similarities of continuous features and have shown superior performance over hashing solutions. These methods first learn continuous image representations and then quantize them with advanced quantization techniques [20, 21]. The ANN search are realized by approximating the similarities of continuous representations with precomputed lookup tables. As the quantization procedure is based on continuous representations, learning high quality and quantizable continuous representations is critical for these methods. Existing studies [22–25] usually learn representations based on similarity-style supervision (*e.g.*, pairwise or triplet examples) but empirically suffer from the sampling issue, *i.e.*, it is challenging to select informative training pairs/triplets for effective learning and algorithm convergence [26, 27]. Besides, how to learn quantizable continuous representations is also a challenge problem.

To address the above problem, we propose a *Deep Bayesian Quantization* (DBQ) framework, which can effectively learn compact representations for neuroimage search. As illustrated in Fig. 1, the proposed DBQ consists of three key components: 1) a 3D CNN, which serves as a representation learning mapping function and can output deep representations directly from raw structural MR (sMRI) images; 2) a proxy embedding-based likelihood function, which can effectively solve the sampling issue and learn discriminative representations; 3) a quantization-based Gaussian prior, which can quantize deep representations with several codebooks and indicator matrices and enable efficient similarity search. These three components are seamlessly incorporated into an end-to-end trainable deep Bayesian learning framework. To the best of our knowledge, this is among the first attempts to design a deep quantization framework for medical image search. Experiments on three neuroimage datasets demonstrate that the DBQ outperforms several state-of-the-art methods in sMR image search.

2   Method

Given N neuroimages in a dataset $X = \{x_i\}_{i=1}^N$ with an associated label set $L = \{l_i\}_{i=1}^N$, where l_i is the class label for the i-th input image $x_i \in \mathbb{R}^D$. For image search, the proposed DBQ aims to learn a deep quantizer $f(x) : \mathbb{R}^D \to \{0,1\}^M$ that employs deep neural networks to encode each input image x into a M-bit binary code coupled with C learned codebooks $D = [D_1, D_2, \cdots, D_C]$, so that the representations reconstructed by the binary code and learned codebooks can maximally preserve the semantic information of images.

2.1   Proposed Framework

For effective neuroimage quantization, we aim to learn representations that are jointly optimal for *semantic preserving* and *compact quantization*. This goal is achieved through a Bayesian learning framework, with details given as follows. Given an image x_i and its label l_i, we first design a 3D CNN as a mapping function $\nu(\cdot, \theta)$ to learn the vector representation $z_i = \nu(x_i, \theta)$, where θ is the parameter set of the deep mapping function. Then the logarithm Maximum a Posteriori (MAP) estimation of $Z = [z_1, \cdots, z_n]$ is formulated as:

$$\log(p(Z|L)) \propto \log(p(L|Z)p(Z)), \tag{1}$$

where $p(L|Z)$ is the likelihood function and $p(Z)$ is the prior distribution. The $p(L|Z)$ is often designed to preserve the semantic information and is achieved with similarity-style supervision, e.g., triplets of similar and dissimilar examples (z_a, z_p, z_n), where z_a is an anchor point, z_p is a positive point from the same category as z_a, and z_n is a negative point that belongs to different categories from z_a. When a set of points are mentioned, we use Z_p and Z_n to denote the set of positive points and the set of negative points, respectively.

To preserve semantic similarity among input images, we try to make z_a closer to z_p than to any points in the negative point set Z_n. Therefore, given a triplet set $T = \{(z_a, z_p, Z_n)\}$, the likelihood function $p(L|Z)$ can be explicitly defined using exponential weighting [28] as follows:

$$p(L|Z) = \prod_{(z_a, z_p, Z_n) \in T} \frac{\exp(s(z_a, z_p))}{\exp(s(z_a, z_p)) + \sum_{z_n \in Z_n} \exp(s(z_a, z_n))}, \tag{2}$$

where $s(\cdot, \cdot)$ measures the similarity between two input images. Although maximizing Eq. (2) can enhance the triplet similarities, the likelihood function in Eq. (2) is highly dependent on triplet samples, while selecting informative training triplets is usually challenging in practice especially for neuroimage data. To address this issue, we resort to the proxy embedding technique [26], which learns a small set of data points P to approximate the set of all training points. Specifically, we learn one proxy embedding for each class to ensure that there is one positive element and several negative elements in P for any input point. Denote the proxy embedding for the j-th class as $\boldsymbol{p}_j$. We can reformulate the likelihood function in Eq. (2) as:

$$p(\boldsymbol{L}|\boldsymbol{Z}) = \prod_{\boldsymbol{z} \in Z} \frac{\exp(s(\boldsymbol{z}, \boldsymbol{p}_y))}{\sum_i \exp(s(\boldsymbol{z}, \boldsymbol{p}_i))}. \tag{3}$$

Since the proxy set P is smaller than the original training set, the number of triplets can be significantly reduced. Besides, Eq. (3) can be considered as an extension of a softmax classification loss, with $\boldsymbol{p}_y$ as the mean vector for the y-th category and $s(\cdot, \cdot)$ as the negative Bregman divergence [29].

For effective compact coding, we need to enhance the *quantizability* of the learned vectors (i.e., the possibility of being quantized with infinitesimal error). To achieve this goal, we design a novel Gaussian prior on the vector representation $\boldsymbol{Z}$ as

$$p(\boldsymbol{Z}) = \prod_{\boldsymbol{z} \in Z} \frac{1}{\sqrt{2\pi}\sigma} \exp(-||\boldsymbol{z} - \hat{\boldsymbol{z}}||^2/2\sigma^2), \tag{4}$$

where σ and $\hat{\boldsymbol{z}}$ are the covariance and mean values for a Gaussian distribution. Furthermore, $\hat{\boldsymbol{z}}$ is assumed to be a vector that can be perfectly quantized via existing quantization techniques. It can be seen that maximizing the prior in Eq. (4) will minimize the squared loss between $\boldsymbol{z}$ and the perfectly quantizable representation $\hat{\boldsymbol{z}}$, thus improving the quantizability of the learned representation.

With the likelihood function in Eq. (3) and the Gaussian prior in Eq. (4), we still need to define the perfectly quantizable representation $\hat{\boldsymbol{z}}$ to enable the learning with Eq. (1). In the following, we will elaborate how to obtain $\hat{\boldsymbol{z}}$ by using the state-of-the-art additive quantization [20] as the quantizer.

2.2 Compact Quantization

Additive quantization [20] aims to approximate representations as the sum of several codewords with each selected from a codebook. Motivated by this idea, we assume that $\hat{\boldsymbol{z}}_i$ can be perfectly quantized as

$$\hat{\boldsymbol{z}}_i = \boldsymbol{D}\boldsymbol{B}_i = \sum_{c=1}^{C} \boldsymbol{D}_c \boldsymbol{b}_{ic}, \tag{5}$$

where $\boldsymbol{D}$ denotes C codebooks, $\boldsymbol{D}_c = [\boldsymbol{d}_{c1}, \boldsymbol{d}_{c2}, \cdots, \boldsymbol{d}_{cK}]$ is the c-th codebook with K elements, $\boldsymbol{B}_i = [\boldsymbol{b}_{i1}, \boldsymbol{b}_{i2}, \cdots, \boldsymbol{b}_{iC}]$ is an indicator matrix, and $||\boldsymbol{b}_{ic}||_0 = 1$ with $\boldsymbol{b}_{ic} \in \{0, 1\}^K$ is a 1-of-K binary vector indicating which one (and only one) of the K codebook elements is selected to approximate the data point. Here, we adopt C codebooks instead of one codebook to further reduce the quantization loss, since using only one codebook will produce significant quantization errors and cause significant performance degradation [14].

Compact quantization is a powerful technique for ANN search. Given a query point x_q, we first compute its deep continuous representation z_q using the trained mapping function $\nu(\cdot, \theta)$. We then use the *Asymmetric Quantizer Distance* (AQD) [18] to calculate the similarity between the query point x_q and a database point x_i as

$$AQD(x_q, x_i) = z_q \cdot \left(\sum_{c=1}^{C} D_c b_{ic}\right). \tag{6}$$

The inner products for all C codebooks $\{D_c\}_{c=1}^{C}$ and all K possible values of b_{ic} can be pre-calculated and stored in an $M \times K$ lookup table, which can then be used to compute AQD between any given query and all database points. The AQD calculation requires M table lookups and M additions, and its calculation cost is only slightly higher than the Hamming distance calculation. The approximate error analysis of using AQD to estimate the similarities of continuous representations is provided in the *Supplementary Materials*.

2.3   Optimization

The optimization of DBQ involves two sub-problems: 1) learning continuous representations via optimizing the deep mapping function and 2) representation quantization. We propose to alternatively solve each sub-problem as follows.

1) Learning Continuous Representations. The parameter set θ for the deep mapping function $\nu(\cdot, \theta)$ can be readily optimized with stochastic gradient descent (SGD) algorithm. The details for this part is omitted in this paper, since most deep learning platforms support automatic gradient back-propagation.

2) Representation Quantization. The quantization part contains two sets of variables, i.e., the codebook $D = [D_1, \cdots, D_C]$ and the indicator matrix $B = [B_1, \cdots, B_N]$. To optimize these two parameter sets, we employ an alternating optimization paradigm. With θ and B fixed, Given $Z = [z_1, ..., z_N]$, the optimization for D can be reformulated as

$$\min_{D} \|Z - DB\|_2^2, \tag{7}$$

which is an unconstrained quadratic problem with an analytic solution. Thus, we can update D by $D = [ZB^T][BB^T]^{-1}$. With θ and D fixed, the optimization of B can be decomposed to N sub-problems

$$\min_{B_i} \ \left\|z_i - \sum_{c=1}^{C} D_c b_{ic}\right\|_2^2 \tag{8}$$
$$s.t. \ \|b_{ic}\|_0 = 1, b_{ic} \in \{0, 1\}^K.$$

This optimization problem is generally NP-hard. As shown in [30], it is essentially a high-order Markov Random Field problem, which can be approximately solved through the Iterated Conditional Modes (ICM) algorithm [31]. Specifically, we first fix $\{b_{ic'}\}_{c' \neq c}$. Then, b_{ic} can be updated by exhaustively checking each element in D_c, finding the element such that the objective function in Eq. (8) is minimized, and setting the corresponding element of b_{ic} to be 1 and the rest to be 0. This iterative updating algorithm is guaranteed to converge and can be terminated if maximum iteration numbers are reached.

3 Experiments

3.1 Experimental Setup

1) Datasets. We validate our DBQ on 3 datasets: Alzheimer's Disease Neuroimaging Initiative (ADNI1), ANDI2 [32], and Open Access Series of Imaging Studies (OASIS3) [33]. **ADNI1** has 748 subjects with 1.5T T1-weighted structural magnetic resonance imaging (sMRI) scans. Each subject was annotated by a class-level label, *i.e.*, Alzheimer's disease (AD), normal control (NC), or mild cognitive impairment (MCI). These labels were obtained based on the standard clinical principles (*e.g.*, mini-mental state examination and clinical dementia rating). There are 231 NC, 312 MCI, and 205 AD subjects in ADNI1. **ADNI2** consists of 708 subjects with 3T T1-weighted sMRIs, including 205 NC, 341 MCI, and 162 AD subjects. For an independent test, subjects who participated in both ADNI1 and ADNI2 were removed from ADNI2. **OASIS3** has 552 subjects with T1-weighted sMRI, with 426 NC and 126 AD subjects. The demographic and clinical information of the subjects are provided in the *Supplementary Materials*.

For all the three datasets, we pre-process sMRI scans using a standard pipeline. We select 10% subjects from each class to form a test set, and use the remaining subjects as the training set as well as the retrieval set. For ADNI1 and ADNI2, we select training and test samples randomly. For OASIS3, since there may contain multiple sMRI scans for one subject at multiple time points, we apply a *subject-level partition* strategy to avoid that the training and the test sets contain sMRI scans from the same subject.

2) Competing Methods. We compare the proposed DBQ with two non-deep hashing methods, *i.e.*, **DSH** [34] and **SKLSH** [35], and five deep hashing methods, *i.e.*, **DPSH** [36], **DHN** [37], **HashNet** [38], **DCH** [39], and **DBDH** [40]. For two non-deep hashing methods, we use volumes of gray matter tissue inside 90 regions-of-interest (defined in the AAL atlas) as sMRI features. For five deep hashing methods, we use the same 3D CNN as the backbone network (with sMR images as input) for the fair comparison. The parameters for all the competing methods are set according to the original papers.

3) Implementation and Evaluation. We design a 3D CNN as the deep mapping function and train it from scratch with the Adam optimizer. The batch size is set to 2. The learning rate is set to 10^{-4}. The architecture of this 3D CNN are provided in the *Supplementary Materials*. For each sample, the binary codes for all C subspaces require $M = C \log_2 K$ bits. Empirically, we set $K = 256$, and C can be set to $M/8$. The inner product is used to measure the similarity between two inputs. Following existing methods, three metrics are used to evaluate the search performance, including mean of average precision (MAP), topN-precision, and recall@k curves.

3.2 Results and Analysis

We first report MAP values of all methods with different lengths of hash code (*i.e.*, M) to provide a global evaluation. Then we report the topN-precision and recall@k curves with $M = 64$ for comprehensive contrastive study.

Table 1. MAP results of eight methods on three neuroimage datasets with different lengths of binary codes. Best results are shown in boldface.

Method	ADNI1			ADNI2			OASIS3		
	64 bits	128 bits	256 bits	64 bits	128 bits	256 bits	64 bits	128 bits	256 bits
DSH	0.394	0.398	0.401	0.402	0.395	0.397	0.695	0.688	0.692
SKLSH	0.492	0.491	0.410	0.492	0.430	0.395	0.661	0.661	0.656
DHN	0.478	0.429	0.398	0.547	0.542	0.554	0.938	0.935	0.935
DPSH	0.512	0.433	0.427	0.557	0.550	0.550	0.920	0.946	0.937
HashNet	0.430	0.441	0.419	0.600	0.526	0.550	0.938	0.952	0.930
DCH	0.619	0.592	0.605	0.562	0.630	0.602	0.936	0.946	0.926
DBDH	0.548	0.539	0.582	0.558	0.606	0.603	0.939	0.940	**0.952**
DBQ (Ours)	**0.622**	**0.612**	**0.625**	**0.649**	**0.641**	**0.622**	**0.948**	**0.964**	0.946

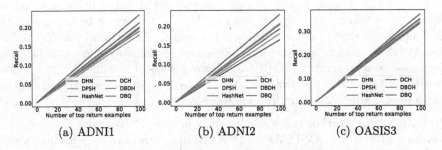

(a) ADNI1   (b) ADNI2   (c) OASIS3

Fig. 2. Recall@k for ADNI1, ADNI2, and OASIS3 datasets.

1) MAP. The MAP results are reported in Table 1. We can observe that our DBQ substantially outperforms these state-of-the-art methods in most cases. Besides, our DBQ with $M = 128$ generally outperforms DBDH with 256-bit code on three datasets. This implies that, by effectively quantizing continuous representations, the DBQ can achieve better performance with shorter hash code length.

2) Recall@k. The recall@k results of different methods are reported in Fig. 2. One can see from Fig. 2 that, compared to seven competing methods, our DBQ consistently yields higher recall values when using different numbers of returned examples on three datasets. These results further prove that the proposed DBQ can learn high-quality binary codes for effective neuroimage search.

3) TopN-precision. The topN-precision results are illustrated in Fig. 3, which shows that the proposed DBQ generally produces higher precision results. This is consistent with the MAP results. Note that physicians usually pay more attention to those top returned instances. Therefore, it is critical to provide users with top ranked instances that are highly relevant to the given query. In addition, one can observe from Fig. 3 that the proposed DBQ outperforms seven competing methods by a large margin, especially when the number of returned objects is small (e.g., 20). This implies that our DBQ can effectively retrieve semantic similar examples and is suitable for neuroimage search task.

Table 2. Quantization loss of different methods on the three datasets.

Method	ADNI1			ADNI2			OASIS3		
	64 bits	128 bits	256 bits	64 bits	128 bits	256 bits	64 bits	128 bits	256 bits
DCH	33.32	65.186	145.59	33.57	62.47	129.65	29.60	57.99	124.79
DBQ (Ours)	**0.05**	**0.08**	**0.07**	**0.06**	**0.14**	**0.02**	**0.14**	**0.13**	**0.18**

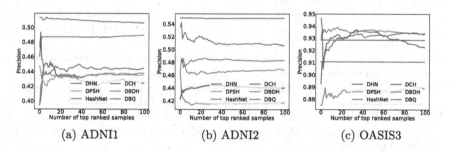

(a) ADNI1   (b) ADNI2   (c) OASIS3

Fig. 3. TopN-precision for ADNI1, ADNI2, and OASIS3 datasets.

4) Quantization Loss. To further demonstrate the superiority of the proposed DBQ, we evaluate the quantization loss of DBQ and the best deep hashing method (i.e. DCH). The average of squared Euclidean distances between continuous and quantized representations are reported in Table 2. We can see that, compared with DCH, our DBQ can effectively reduce the quantization loss by a factor of 4. This clearly demonstrates that the proposed quantization-based Gaussian prior in Eq. (4) can significantly reduce the quantization loss.

4 Conclusion and Future Work

This paper introduces a deep Bayesian quantization (DBQ) method for neuroimage search. The DBQ approach utilizes a Bayesian learning framework that simultaneously optimizes a likelihood function based on proxy embeddings for effective representation learning and a Gaussian prior function for efficient and compact quantization. Experimental results on three neuroimage datasets demonstrate that DBQ outperforms previous state-of-the-art approaches.

In the present study, our DBQ framework is specifically designed for neuroimages from a single modality. However, an intriguing avenue for future research involves adapting the proposed DBQ to enable multi-modality ANN search. Additionally, we acknowledge the presence of heterogeneity between datasets arising from the utilization of different scanners and/or scanning protocols at various imaging sites. Therefore, we aim to develop advanced learning techniques to address cross-dataset/domain neuroimage search in our future work.

Acknowledgements. This work was partly supported by NIH grant AG073297. This work was finished when E. Yang worked at the University of North Carolina at Chapel Hill.

References

1. Owais, M., Arsalan, M., Choi, J., Park, K.R.: Effective diagnosis and treatment through content-based medical image retrieval (CBMIR) by using artificial intelligence. J. Clin. Med. **8**(4), 462 (2019)
2. Liu, M., Zhang, D., Shen, D.: Relationship induced multi-template learning for diagnosis of Alzheimer's disease and mild cognitive impairment. IEEE Trans. Med. Imag. **35**(6), 1463–1474 (2016)
3. Liu, M., Zhang, J., Adeli, E., Shen, D.: Landmark-based deep multi-instance learning for brain disease diagnosis. Med. Image Anal. **43**, 157–168 (2018)
4. Cheng, C.H., Liu, W.X.: Identifying degenerative brain disease using rough set classifier based on wavelet packet method. J. Clin. Med. **7**(6), 124 (2018)
5. Graham, R.N., Perriss, R., Scarsbrook, A.F.: DICOM demystified: a review of digital file formats and their use in radiological practice. Clin. Radiol. **60**(11), 1133–1140 (2005)
6. Zaidi, H., Vees, H., Wissmeyer, M.: Molecular PET/CT imaging-guided radiation therapy treatment planning. Acad. Radiol. **16**(9), 1108–1133 (2009)
7. Holt, A., Bichindaritz, I., Schmidt, R., Perner, P.: Medical applications in case-based reasoning. Knowl. Eng. Rev. **20**(3), 289–292 (2005)
8. Sedghi, S., Sanderson, M., Clough, P.: How do health care professionals select medical images they need? In: Aslib Proceedings. Emerald Group Publishing Limited (2012)
9. Dong, J., et al.: Dual encoding for video retrieval by text. IEEE Trans. Pattern Anal. Mach. Intell. **44**(8), 4065–4080 (2022)
10. Yang, E., Deng, C., Liu, T., Liu, W., Tao, D.: Semantic structure-based unsupervised deep hashing. IJCA **I**, 1064–1070 (2018)
11. Dong, J., Li, X., Snoek, C.G.: Predicting visual features from text for image and video caption retrieval. IEEE Trans. Multimedia **20**(12), 3377–3388 (2018)
12. Yang, E., Deng, C., Liu, W., Liu, X., Tao, D., Gao, X.: Pairwise relationship guided deep hashing for cross-modal retrieval. AAA **I**, 1618–1625 (2017)
13. Yang, E., Deng, C., Li, C., Liu, W., Li, J., Tao, D.: Shared predictive cross-modal deep quantization. IEEE Trans. Neural Netw. Learn. Syst. (2018)

14. Zhang, T., Wang, J.: Collaborative quantization for cross-modal similarity search. In: CVPR, pp. 2036–2045 (2016)
15. Yang, X., Feng, F., Ji, W., Wang, M., Chua, T.S.: Deconfounded video moment retrieval with causal intervention. In: SIGIR (2021)
16. Yang, X., Dong, J., Cao, Y., Wang, X., Wang, M., Chua, T.S.: Tree-augmented cross-modal encoding for complex-query video retrieval. In: SIGIR (2020)
17. Ge, T., He, K., Ke, Q., Sun, J.: Optimized product quantization. IEEE Trans. Pattern Anal. Mach. Intell. **36**(4), 744–755 (2014)
18. Long, M., Cao, Y., Wang, J., Yu, P.S.: Composite correlation quantization for efficient multimodal retrieval. In: SIGIR, pp. 579–588 (2016)
19. Yang, E., Liu, T., Deng, C., Tao, D.: Adversarial examples for hamming space search. IEEE Trans. Cybern. **50**(4), 1473–1484 (2018)
20. Babenko, A., Lempitsky, V.: Additive quantization for extreme vector compression. In: CVPR, pp. 931–938 (2014)
21. Yang, E., et al.: Deep Bayesian hashing with center prior for multi-modal neuroimage retrieval. IEEE Trans. Med. Imaging **40**(2), 503–513 (2020)
22. Hadsell, R., Chopra, S., LeCun, Y.: Dimensionality reduction by learning an invariant mapping. In: CVPR, pp. 1735–1742 (2006)
23. Weinberger, K.Q., Saul, L.K.: Distance metric learning for large margin nearest neighbor classification. J. Mach. Learn. Res. **10**(2) (2009)
24. Deng, C., Yang, E., Liu, T., Tao, D.: Two-stream deep hashing with class-specific centers for supervised image search. IEEE Trans. Neural Netw. Learn. Syst. **31**(6), 2189–2201 (2019)
25. Yang, E., Yao, D., Liu, T., Deng, C.: Mutual quantization for cross-modal search with noisy labels. In: CVPR, pp. 7551–7560 (2022)
26. Movshovitz-Attias, Y., Toshev, A., Leung, T.K., Ioffe, S., Singh, S.: No fuss distance metric learning using proxies. In: ICCV, pp. 360–368 (2017)
27. Yang, E., et al.: Deep disentangled hashing with momentum triplets for neuroimage search. In: Martel, A.L., et al. (eds.) MICCAI 2020. LNCS, vol. 12261, pp. 191–201. Springer, Cham (2020). https://doi.org/10.1007/978-3-030-59710-8_19
28. Roweis, S., Hinton, G., Salakhutdinov, R.: Neighbourhood component analysis. In: NeurIPS, vol. 17, pp. 513–520 (2004)
29. Bregman, L.M.: The relaxation method of finding the common point of convex sets and its application to the solution of problems in convex programming. USSR Comput. Math. Math. Phys. **7**(3), 200–217 (1967)
30. Zhang, T., Du, C., Wang, J.: Composite quantization for approximate nearest neighbor search. ICML. Number **2**, 838–846 (2014)
31. Besag, J.: On the statistical analysis of dirty pictures. J. Roy. Stat. Soc.: Ser. B (Methodol.) **48**(3), 259–279 (1986)
32. Jack Jr, C.R., et al.: The Alzheimer's disease neuroimaging initiative (ADNI): MRI methods. J. Magn. Resonan. Imaging Off. J. Int. Soc. Magn. Resonan. Med. **27**(4), 685–691 (2008)
33. LaMontagne, P.J., et al.: Oasis-3: longitudinal neuroimaging, clinical, and cognitive dataset for normal aging and Alzheimer disease. medRxiv (2019)
34. Jin, Z., Li, C., Lin, Y., Cai, D.: Density sensitive hashing. IEEE Trans. Cybern. **44**(8), 1362–1371 (2014)
35. Raginsky, M., Lazebnik, S.: Locality-sensitive binary codes from shift-invariant kernels. In: NeurIPS, pp. 1509–1517 (2009)
36. Li, W.J., Wang, S., Kang, W.C.: Feature learning based deep supervised hashing with pairwise labels. IJCA **I**, 1711–1717 (2016)

37. Zhu, H., Long, M., Wang, J., Cao, Y.: Deep hashing network for efficient similarity retrieval. AAAI, 2415–2421 (2016)
38. Cao, Z., Long, M., Wang, J., Yu, P.S.: Hashnet: deep learning to hash by continuation. In: ICCV, pp. 5608–5617 (2017)
39. Cao, Y., Long, M., Liu, B., Wang, J., Kliss, M.: Deep cauchy hashing for hamming space retrieval. In: CVPR, pp. 1229–1237 (2018)
40. Zheng, X., Zhang, Y., Lu, X.: Deep balanced discrete hashing for image retrieval. Neurocomputing **403**, 224–236 (2020)

Triplet Learning for Chest X-Ray Image Search in Automated COVID-19 Analysis

Linmin Wang[1], Qianqian Wang[2], Xiaochuan Wang[1], Yunling Ma[1], Lishan Qiao[1,3(✉)], and Mingxia Liu[2(✉)]

[1] School of Mathematics Science, Liaocheng University, Liaocheng 252000, Shandong, China
qiaolishan@lcu.edu.cn
[2] Department of Radiology and BRIC, University of North Carolina at Chapel Hill, Chapel Hill, NC 27599, USA
mingxia_liu@med.unc.edu
[3] School of Computer Science and Technology, Shandong Jianzhu University, Jinan 250101, Shandong, China

Abstract. Chest radiology images such as CT scans and X-ray images have been extensively employed in computer-assisted analysis of COVID-19, utilizing various learning-based techniques. As a trending topic, image retrieval is a practical solution by providing users with a selection of remarkably similar images from a retrospective database, thereby assisting in timely diagnosis and intervention. Many existing studies utilize deep learning algorithms for chest radiology image retrieval by extracting features from images and searching the most similar images based on the extracted features. However, these methods seldom consider the complex relationship among images (*e.g.*, images belonging to the same category tend to share similar representations, and vice versa), which may result in sub-optimal retrieval accuracy. In this paper, we develop a triplet-constrained image retrieval (TIR) framework for chest radiology image search to aid in COVID-19 diagnosis. The TIR contains two components: (a) feature extraction and (b) image retrieval, where a *triplet constraint* and an *image reconstruction constraint* are embedded to enhance the discriminative ability of learned features. In particular, the triplet constraint is designed to minimize the distances between images belonging to the same category and maximize the distances between images from different categories. Based on the extracted features, we further perform chest X-ray (CXR) image search. Experimental results on a total of $29,986$ CXR images from a public COVIDx dataset with $16,648$ subjects demonstrate the effectiveness of the proposed method compared with several state-of-the-art approaches.

Keywords: COVID-19 · Image retrieval · Triplet constraint · Chest X-ray

1 Introduction

Chest radiology images, such as CT scans and X-ray images [1], can help capture pulmonary changes caused by COVID-19 and play a significant role in computer-assisted COVID-19 analysis [2]. In particular, chest X-ray images are often the preferable medical imaging modality for COVID-19 detection due to their availability, low cost, and

X. Cao et al. (Eds.): MLMI 2023, LNCS 14349, pp. 407–416, 2024.
https://doi.org/10.1007/978-3-031-45676-3_41

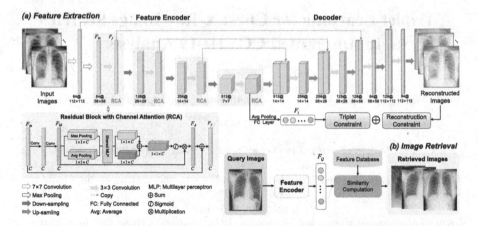

Fig. 1. Illustration of the proposed triplet-constrained image retrieval (TIR) framework, including (a) feature extraction and (b) image retrieval. The feature extraction component consists of (1) a *feature encoder* that takes CXR images as input and maps them into a lower-dimensional feature space, and (2) a *decoder* that takes the learned features as input and reconstructs the original CXR images. Especially, a triplet constraint is designed to enhance the discriminative ability of learned features in feature extraction. For image retrieval, we will search for the most similar images based on learned feature embeddings with the query CXR image as input.

fast acquisition time [3]. They are two-dimensional images obtained by passing X-ray through the chest and capturing the transmitted radiation using a sensor [4]. The analysis of COVID-19 often requires experts' clinical experiences and professional knowledge, which is labor-intensive and time-consuming. Thus, it is clinically meaningful to design an automated COVID-19 diagnosis system based on previous chest X-ray images and related diagnostic records, facilitating clinical decision-making [5].

Medical image retrieval technique provides an effective solution to search and access similar images associated with COVID-19 cases from large-scale retrospective databases [6,7]. Researchers have proposed various machine/deep learning algorithms to retrieve chest X-ray images for assisting in timely diagnosis and intervention of COVID-19 [8,9]. These learning-based retrieval methods generally first extract image features, followed by converting the extracted features into binary hash codes, and then utilize the extracted features or binary hash codes to search for the most similar images or cases [10]. There exists some complex relationship among images; for example, images belonging to the same category tend to share similar representations, while images belonging to the same category from different categories would have different features [11–14]. However, existing image retrieval methods usually ignore such intrinsic relationships, thus possibly resulting in suboptimal accuracy [15].

To this end, we propose a triplet-constrained image retrieval (**TIR**) framework for automated COVID-19 analysis based on chest X-ray images. As illustrated in Fig. 1, the proposed TIR framework consists of two key components: (a) feature extraction, which consists of a feature encoder and a decoder to learn latent feature embeddings of chest X-ray images, and (b) image retrieval, which searches for the most similar images based on the extracted feature embeddings. During feature extraction, we use a triplet constraint and an image reconstruction constraint to enhance the discriminative power of the learned features. In particular, the triplet constraint is designed to minimize the distances between images belonging to the same category and maximize the distances between images from different categories. Experimental results on the public COVIDx dataset with a total of 29,986 CXR images from 16,648 subjects demonstrate the effectiveness of the proposed TIR method in CXR image search [2].

2   Proposed Method

As illustrated in Fig. 1, the proposed triplet-constrained image retrieval (TIR) framework is comprised of a *feature extraction module* and a *image retrieval module*. More details are introduced as follows.

Feature Extraction. As a fundamental step, feature extraction plays a crucial role in medical image retrieval tasks, which aims to learn discriminative image embeddings for efficient retrieval. In this work, we develop a UNet-based model for CXR image feature extraction, where a *triplet constraint* and an *image reconstruction constraint* are explicitly incorporated to enhance the discriminative power of the learned features.

(1) Feature Encoder. Due to the high dimensionality and complexity of medical image data, direct retrieving similar images from a large-scale retrospective database can be challenging in clinical practice. Thus, we employ a feature encoder to map high-dimensional CXR images into a lower-dimensional feature space [16, 17].

As shown in Fig. 1 (a), the feature encoder consists of 2D convolution, max pooling, downsampling, residual blocks with channel attention (RCAs) [18], average pooling and fully connected layers. Specifically, with each CXR image X_{in} as input (size: $3 \times 224 \times 224$), we first use 2D convolution operation with larger kernel size to capture local patterns of CXR images, and use max pooling to reduce spatial dimensions, obtaining the feature representation F_{in}. Then, we employ a residual block with a skip connection to alleviate the degradation problem in deep neural networks. Furthermore, a channel attention technique is incorporated into each residual block to adaptively learn the importance of different feature maps [19].

Our goal is to perform a transformation from $F_M \rightarrow F_A$ through the channel attention technique. Specifically, we first simultaneously apply max pooling and average pooling to F_M, obtaining two vectorized representations. Then, these two representations are fed into a multilayer perceptron (MLP) that shares parameters within each residual operation. The output of the MLP is summed, followed by multiplication with F_M to obtain a feature vector with channel attention, denoted as F_A. Mathematically, the above channel attention can be represented as follows:

$$F_A = \sigma(W(\phi_1(F_M)) \oplus W(\phi_2(F_M))) \otimes F_M, \tag{1}$$

where σ is the sigmoid function, ϕ_2 and ϕ_2 denote max pooling and average pooling respectively, $\oplus$ and $\otimes$ represent element-wise sum and multiplication, respectively, and

W is a learnable parameter matrix. After that, we perform skip connection to obtain the output of the residual block, formulated as:

$$F_f = F_{in} \oplus F_A, \tag{2}$$

In this work, we use four residual blocks and the size of the feature map remains unchanged throughout each channel attention. The output of each residual block is then downsampled and then fed into the next residual block. For each input image, we can obtain a feature map (size: $512 \times 7 \times 7$) via the feature encoder, followed by global average pooling and a fully connected layer to yield a $1,000$-dimensional vector F_x. The feature F_x is stored in a feature database for subsequent image retrieval.

(2) Triplet Constraint. Considering the complex relationship among input samples, we introduce a triplet constraint in the TIR to improve the discriminative power of learned features. Specifically, the triplet constraint aims to enhance the similarity between features of samples from the same category, while increasing the distance between features of samples from different categories [11]. Mathematically, the triplet constraint can be formulated as follows:

$$d(a, p) < d(a, n), \tag{3}$$

where $d(a, p)$ represents the Euclidean distance between the anchor sample a and the positive sample p, $d(a, n)$ represents the Euclidean distance between the anchor sample a and the negative sample n.

An MN sampling strategy is used in the training phase, where we first select M categories and then randomly select N images of each category for model training [11]. Mathematically, given $M \times N$ medical triplet units $X = \{X_i^a, X_i^p, X_i^n\}_{i=1}^{M \times N}$ and corresponding triplet labels $L = \{L_i^a, L_i^p, L_i^n\}_{i=1}^{M \times N}$ where X_i^a, X_i^p and X_i^n represents the i-th anchor medical sample, positive medical sample and negative medical sample respectively, L_i^a, L_i^p and L_i^n represents the label of the anchor medical sample, positive medical sample and negative medical sample respectively. Based on this sampling strategy, the triplet constraint in a mini-batch is calculated as follows:

$$L_T = \sum_{i=1}^{M \times N} [m + d(F(x_i^a), F(x_i^p)) - d(F(x_i^a), F(x_j^n))]_+ \tag{4}$$

where $F(x_i^a)$, $F(x_i^p)$ and $F(x_j^n)$ represent the features of a anchor sample, a positive sample, and a negative sample respectively. Also, m is a pre-defined margin value used to control the distance between the positive and negative samples, $d(\cdot)$ is the Euclidean distance, and $[x]_+ = \max(x, 0)$.

(3) Image Reconstruction Constraint. To further enhance the representation ability of the features extracted by the encoder, we introduce an image reconstruction constraint that emphasizes the useful information in the feature space while filtering out irrelevant noise and redundant features [20]. The objective of creating this constraint is to aid in acquiring more distinguishing features for downstream image retrieval tasks.

In this work, we use a UNet-based feature decoder to reconstruct input images, which facilitates precise localization and reconstruction [16, 17]. In this decoder, the feature obtained from the encoder is first upsampled and concatenated with the corresponding output from the encoder. Then, the concatenated feature is further processed through convolution. The above process is repeated four times and the output is finally upsampled to obtain the reconstructed image X_{Re}. The reconstruction constraint in a mini-batch is formulated as follows:

$$L_R = \frac{1}{M \times N} \sum_{i=1}^{M \times N} (X_{Re} - X_{in})^2, \tag{5}$$

where $M \times N$ represents the number of samples in a min-batch, X_{in} and X_{Re} represent the input image and the reconstructed image, respectively.

(4) Objective Function. As illustrated in the bottom right of Fig. 1 (a), our TIR consists of a triplet constraint and an image reconstruction constraint. The triplet constraint ensures that the extracted features can reflect complex relationships among images, and the reconstruction constraint ensures that the extracted features can preserve image semantic information. With Eqs. (4)–(5), the objective function of TIR is formulated as:

$$L = L_T + \alpha L_R, \tag{6}$$

where $\alpha \in [0, 1]$ is a hyperparameter.

Image Retrieval. Image features produced by the feature encoder are stored in a feature database, and are then used for image retrieval. As shown in Fig. 1 (b), given a query image X_Q, the pre-trained feature encoder is first used to generate feature representation F_Q. Denote the features of all training images in the database as $F^n = [F_1, \cdots, F_n]$, where F_i corresponds to the embedded feature learned from the i-th image and n is the number of features in the database. Then, we calculate the Euclidean distance between feature F_Q of query image and each training image feature F_i ($i = 1, \cdots, n$) in the database to retrieve images similar to query images.

Implementation Details. The feature extraction phase in our proposed TIR consists of a *feature encoder* and a *decoder*. In the *feature encoder*, the first convolutional layer has a kernel size of 7×7, the max pooling operation is a convolution operation with a kernel size of 3×3 and a stride of 2, and the convolutional layers in each residual block have a kernel size of 3×3. The MLP in the channel attention module has 3 layers with 64, 4, and 64 neurons, respectively. After the first, second, and third residual block, downsampling operations are performed with a kernel size of 3×3 and a stride of 2 to double the number of channels and halve the size. In the *decoder*, the first upsampling operation has a kernel size of 2×2 and a stride of 2, while others have a kernel size of 4×4 and a stride of 2. The convolutional layers have two convolution kernels with the size of 3×3 and a stride of 1. The margin m in triplet constraint is set at 0.2 and the parameter α is fixed to 0.2. We sampled triplets for each mini-batch using a MN sampling strategy (N samples randomly from each of M categories). We set $M = 2$ and $N = 16$ in this work. The objective function is optimized via Adam, with the learning rate $1e - 4$, the batch size of 32, and the training epoch of 30.

3   Experiments

Materials. We evaluate the proposed method on the COVIDx dataset [2], a publicly available COVID-19 CXR dataset. For a fair comparison, we follow the same data training and test data partition strategy as [2]. The training set consists of 29,986 CXR images from 16,648 subjects, including 15,994 COVID-19 positive images from 2,808 subjects and 13,992 COVID-19 negative images from 13,850 subjects. The test set contains of 400 CXR images from 378 subjects, including 200 COVID-19 positive images from 200 subjects and 200 COVID-19 negative images from 178 subjects. Standard data normalization techniques are applied for the CXR images, including (1) resizing the images to 256×256 pixels, (2) center or random resized cropping to 224×224 pixels, and (3) random horizontal flipping.

Experimental Setting. The retrieval performance of each method is evaluated using two metrics, including mean average precision (mAP) and mean precision (P@K), where $K = 1, 5, 10$. Specifically, the mAP represents the average precision across all queries in image retrieval. It measures the overall performance of an image retrieval system by considering the proportion of relevant images among the retrieved images for each query. The P@K denotes the precision of a retrieval system at a specified value of K, indicating the proportion of relevant images among the top K retrieved images. To reduce the bias caused by model initialization, we repeat experiments five times and record the mean and standard deviation results of each method.

Competing Methods. We first compare the proposed TIR with four state-of-the-art (SOTA) methods for CXR image retrieval. The SOTA competing methods include three image retrieval methods based on hash codes and a method based on features learned from a deep convolutional network. (1) Deep Hash Network (**DHN**) [21] employs an unsupervised learning approach to directly learn compact binary codes from image features, where a pair-wise similarity loss and a quantization loss are used to optimize the binary codes. (2) Deep Supervised Hashing (**DSH**) [22] learns binary codes using a deep neural network architecture. It optimizes the binary codes using a hashing loss function, which minimizes the Hamming distance between similar images and maximizes the Hamming distance between dissimilar images. (3) Deep Cauchy Hash (**DCH**) [23] employs a supervised hashing loss function to learn discriminative binary codes. These three hash methods (*i.e.*, DHN, DSH, and DCH) utilize their generated binary hash codes as feature representations for image retrieval. (4) Deep Convolutional Neural Network (**DCNN**) [24] utilizes discriminative features learned from a convolutional neural network for image retrieval.

We further compare our TIR with its 3 variants that use different constraints for image retrieval. (1) **TIR-CE** is used to optimize the model by performing a classification task, which minimizes the difference between the prediction and the label via a cross-entropy loss. It can be written as: $L_E = -(y \log(y_1) + (1 - y) \log(1 - y_1))$, where y and y_1 represent label and prediction from F_i respectively. (2) **TIR-CS** [25] uses the cosine similarity $c(\cdot)$ to describe the similarity between samples instead of the Euclidean distance used in our triplet loss. The loss can be represented as:

Query Image   TOP-10 Retrieved Images

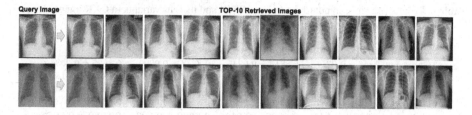

Fig. 2. Top 10 most similar CXR images retrieved by the TIR on COVIDx dataset. Images that are retrieved and have a different category from the query image are highlighted with a red box. (Color figure online)

Table 1. Results (mean ± standard deviation) of eight different methods in CXR image retrieval on the COVIDx dataset. Best results are shown in bold.

Method	mAP	P@1↓	P@5↓	P@10↓
DHN	0.7782 ± 0.0163	0.8350 ± 0.0143	0.8207 ± 0.0183	0.8175 ± 0.0168
DSH	0.8152 ± 0.0292	0.8306 ± 0.0276	0.8285 ± 0.0236	0.8261 ± 0.0227
DCH	0.7828 ± 0.0391	0.8825 ± 0.0367	0.8630 ± 0.0363	0.8735 ± 0.0349
DCNN	0.7908 ± 0.0016	0.8345 ± 0.1494	0.8336 ± 0.0174	0.8328 ± 0.0182
TIR-CE	0.6971 ± 0.0583	0.8726 ± 0.0340	0.8450 ± 0.0018	0.8163 ± 0.0278
TIR-CS	0.7816 ± 0.0056	**0.9005 ± 0.0293**	0.8863 ± 0.0373	0.8772 ± 0.0040
TIR-CL	0.7321 ± 0.0593	0.8725 ± 0.0302	0.8492 ± 0.0364	0.8312 ± 0.0364
TIR (Ours)	**0.8786 ± 0.0177**	0.8930 ± 0.0119	**0.8999 ± 0.0024**	**0.8989 ± 0.0035**

$L_S = max(0, c(F_{\hat{x}}^a, F_x^n) - c(F_x^a, F_x^p) + m)$, where the margin m is the same as Eq. (4). (3) **TIR-CL** [26] penalizes the Euclidean distance between each sample and its corresponding class center in the feature space, with the loss function expressed as $L_L = \frac{1}{2}\sum_{i=1}^{n}|F_i - c_i|^2$, where c_i represents the feature center point of all samples in the class corresponding to the i-th sample.

Experimental Result. As shown in Table 1, we report the experimental results of the proposed TIR and seven competing methods in CXR image retrieval. From Table 1, it can be observed that our TIR achieves superior performance than four SOTA retrieval methods (*i.e.*, DHN, DSH, DCH and DCNN). For instance, the proposed TIR achieves an improvement of 6.34% compared to the best SOTA method (*i.e.*, DSH) in terms of mAP. In addition, it can be seen from Table 1 that our proposed TIR outperforms its three variants that use other constraints (*i.e.*, TIR-CE, TIR-CS, and TIR-CL) instead of the triplet constraint in Eq. 4. The possible reason is that the triplet constraint used in our TIR can enhance the discriminative power of the learned features via modeling the complex relationships between samples, thus improving the retrieval performance. This validates the effectiveness of the proposed TIR framework in CXR image retrieval. In Fig. 2, we also visualize the top 10 most similar CXR images retrieved by the TIR on the COVIDx dataset, from which one can see that our method achieves good performance.

4   Discussion

Ablation Study. To explore the effectiveness of each component in the TIR, we compare TIR and its four variants, called **TIRw/oR, TIRw/oA, TIRw/oT** and **TIRw/oRA,**

Table 2. Results (mean ± standard deviation) of TIR and its four variants in CXR image retrieval on the COVIDx dataset. Best results are shown in bold.

Method	mAP	P@1↓	P@5↓	P@10↓
TIRw/oR	0.7856 ± 0.0711	0.8815 ± 0.0245	0.8718 ± 0.0296	0.8625 ± 0.0036
TIRw/oA	0.8034 ± 0.0498	0.8875 ± 0.0060	0.8884 ± 0.0138	0.8833 ± 0.0187
TIRw/oT	0.5276 ± 0.0060	0.6340 ± 0.0288	0.6203 ± 0.0233	0.6000 ± 0.0266
TIRw/oRA	0.7743 ± 0.0487	0.8595 ± 0.0309	0.8367 ± 0.0316	0.8367 ± 0.0339
TIR (Ours)	**0.8786 ± 0.0177**	**0.8930 ± 0.0119**	**0.8999 ± 0.0024**	**0.8989 ± 0.0035**

Table 3. Results (mean ± standard deviation) of different parameters α of TIR for CXR image retrieval on the COVIDx dataset. The best results are shown in bold.

Parameter α	mAP	P@1↓	P@5↓	P@10↓
0.2	**0.8786 ± 0.0177**	0.8930 ± 0.0119	**0.8999 ± 0.0024**	**0.8989 ± 0.0035**
0.4	0.8412 ± 0.0503	0.8763 ± 0.0240	0.8623 ± 0.0301	0.8605 ± 0.0370
0.6	0.8326 ± 0.0326	0.8985 ± 0.0140	0.8807 ± 0.0218	0.8768 ± 0.0211
0.8	0.8252 ± 0.0467	0.8775 ± 0.0086	0.8753 ± 0.0218	0.8700 ± 0.0253
1.0	0.8670 ± 0.0209	**0.9005 ± 0.0174**	0.8989 ± 0.0203	0.8954 ± 0.0193

respectively. The **TIRw/oR** has no reconstruction constraint. The **TIRw/oA** does not use the channel attention module during feature extraction. The **TIRw/oT** removes the triplet constraint in Eq. (6). The **TIRw/oRA** only uses ResNet-18 architecture for feature extraction, without using the channel attention module and the decoder. The experimental results are reported in Table 2. It can be seen that our TIR achieves better performance compared to its four variants in most cases. This indicates that the added channel attention module, our designed triplet constraint, and the proposed image reconstruction constraint (through the decoder) in our TIR can help learn discriminative features, thus enhancing image retrieval performance.

Parameter Analysis. To investigate the effect of the key parameter α in Eq. (6) on the CXR image retrieval performance of the proposed TIR. Specifically, we use different α values within the domain of $[0.2, 0.4, \cdots, 1]$ for TIR model training, followed by the downstream CXR image retrieval task. The results are shown in Table 3. From Table 3, we can see that the proposed TIR can obtain better results on most metrics when the parameter is set to 0.2, These results imply that the triplet constraint contributes more to the proposed TIR model than the image reconstruction constraint.

5 Conclusion and Future Work

In this paper, we propose a triplet-constrained image retrieval (TIR) framework for CXR image retrieval. Specifically, we employ a feature encoder and a decoder for feature abstraction, where a triplet constraint and an image reconstruct constraint are explicitly incorporated to enhance discriminative power of the learned features. Then, with all learned features stored in a feature database, we calculate the similarity between the feature of the query image and the features of training images in the database for image retrieval. Extensive experiments demonstrate that our TIR achieves state-of-the-art retrieval performance on the public COVIDx dataset.

In the present study, we only test the efficacy of our method using a CXR image dataset for retrieval. As part of our future endeavors, we plan to incorporate additional image modalities such as MRI and PET [27, 28] into the retrieval system, with the aim of enhancing the diagnosis of COVID-19. Furthermore, investigating deep hashing techniques for efficient medical image retrieval is intriguing. This can be done by executing feature extraction and image retrieval in a streamlined end-to-end approach.

Acknowledgements. L. Qiao was supported in part by National Natural Science Foundation of China (Nos. 61976110, 62176112, 11931008) and Natural Science Foundation of Shandong Province (No. ZR202102270451).

References

1. Apostolopoulos, I.D., Mpesiana, T.A.: Covid-19: automatic detection from X-ray images utilizing transfer learning with convolutional neural networks. Phys. Eng. Sci. Med. **43**, 635–640 (2020)
2. Wang, L., Lin, Z.Q., Wong, A.: Covid-Net: a tailored deep convolutional neural network design for detection of COVID-19 cases from chest X-ray images. Sci. Rep. **10**(1), 1–12 (2020)
3. Hu, B., Vasu, B., Hoogs, A.: X-MRI: explainable medical image retrieval. In: Proceedings of the IEEE/CVF Winter Conference on Applications of Computer Vision, pp. 440–450 (2022)
4. Qi, A., et al.: Directional mutation and crossover boosted ant colony optimization with application to covid-19 x-ray image segmentation. Comput. Biol. Med. **148**, 105810 (2022)
5. Covid, C., et al.: Severe outcomes among patients with coronavirus disease 2019 (COVID-19)-United States, February 12–March 16 2020. Morb. Mortal. Wkly Rep. **69**(12), 343 (2020)
6. Akgül, C.B., Rubin, D.L., Napel, S., Beaulieu, C.F., Greenspan, H., Acar, B.: Content-based image retrieval in radiology: current status and future directions. J. Digit. Imaging **24**, 208–222 (2011)
7. Das, P., Neelima, A.: An overview of approaches for content-based medical image retrieval. Int. J. Multimedia Inf. Retrieval **6**(4), 271–280 (2017). https://doi.org/10.1007/s13735-017-0135-x
8. Jain, R., Gupta, M., Taneja, S., Hemanth, D.J.: Deep learning based detection and analysis of COVID-19 on chest X-ray images. Appl. Intell. **51**, 1690–1700 (2021)
9. Minaee, S., Kafieh, R., Sonka, M., Yazdani, S., Soufi, G.J.: Deep-COVID: predicting COVID-19 from chest X-ray images using deep transfer learning. Med. Image Anal. **65**, 101794 (2020)
10. Zhong, A., et al.: Deep metric learning-based image retrieval system for chest radiograph and its clinical applications in COVID-19. Med. Image Anal. **70**, 101993 (2021)
11. Schroff, F., Kalenichenko, D., Philbin, J.: FaceNet: a unified embedding for face recognition and clustering. In: Proceedings of the IEEE Conference on Computer Vision and Pattern Recognition, pp. 815–823 (2015)
12. Jégou, H., Douze, M., Schmid, C.: Improving bag-of-features for large scale image search. Int. J. Comput. Vision **87**, 316–336 (2010)
13. Yao, D., et al.: A mutual multi-scale triplet graph convolutional network for classification of brain disorders using functional or structural connectivity. IEEE Trans. Med. Imaging **40**(4), 1279–1289 (2021)

14. Liu, M., Zhang, D., Shen, D.: Relationship induced multi-template learning for diagnosis of Alzheimer's disease and mild cognitive impairment. IEEE Trans. Med. Imaging **35**(6), 1463–1474 (2016)

15. Smeulders, A.W., Worring, M., Santini, S., Gupta, A., Jain, R.: Content-based image retrieval at the end of the early years. IEEE Trans. Pattern Anal. Mach. Intell. **22**(12), 1349–1380 (2000)

16. Ronneberger, O., Fischer, P., Brox, T.: U-Net: convolutional networks for biomedical image segmentation. In: Navab, N., Hornegger, J., Wells, W.M., Frangi, A.F. (eds.) MICCAI 2015. LNCS, vol. 9351, pp. 234–241. Springer, Cham (2015). https://doi.org/10.1007/978-3-319-24574-4_28

17. Rundo, L., et al.: USE-Net: incorporating squeeze-and-excitation blocks into U-Net for prostate zonal segmentation of multi-institutional MRI datasets. Neurocomputing **365**, 31–43 (2019)

18. Kaiming, H., Shaoqing, R., Jian, S.: Deep residual learning for image recognition. In: Proceedings of the IEEE Conference on Computer Vision and Pattern Recognition (2016)

19. Hu, J., Shen, L., Sun, G.: Squeeze-and-excitation networks. In: Proceedings of the IEEE Conference on Computer Vision and Pattern Recognition, pp. 7132–7141 (2018)

20. Liu, G., Reda, F.A., Shih, K.J., Wang, T.-C., Tao, A., Catanzaro, B.: Image inpainting for irregular holes using partial convolutions. In: Ferrari, V., Hebert, M., Sminchisescu, C., Weiss, Y. (eds.) ECCV 2018. LNCS, vol. 11215, pp. 89–105. Springer, Cham (2018). https://doi.org/10.1007/978-3-030-01252-6_6

21. Zhu, H., Long, M., Wang, J., Cao, Y.: Deep hashing network for efficient similarity retrieval. In: Proceedings of the AAAI Conference on Artificial Intelligence, vol. 30 (2016)

22. Liu, H., Wang, R., Shan, S., Chen, X.: Deep supervised hashing for fast image retrieval. In: Proceedings of the IEEE Conference on Computer Vision and Pattern Recognition, pp. 2064–2072 (2016)

23. Cao, Y., Long, M., Liu, B., Wang, J.: Deep cauchy hashing for hamming space retrieval. In: Proceedings of the IEEE Conference on Computer Vision and Pattern Recognition, pp. 1229–1237 (2018)

24. Mohite, N.B., Gonde, A.B.: Deep features based medical image retrieval. Multimedia Tools Appl. **81**(8), 11379–11392 (2022). https://doi.org/10.1007/s11042-022-12085-x

25. Shen, Y., Li, H., Yi, S., Chen, D., Wang, X.: Person re-identification with deep similarity-guided graph neural network. In: Ferrari, V., Hebert, M., Sminchisescu, C., Weiss, Y. (eds.) ECCV 2018. LNCS, vol. 11219, pp. 508–526. Springer, Cham (2018). https://doi.org/10.1007/978-3-030-01267-0_30

26. Wen, Y., Zhang, K., Li, Z., Qiao, Yu.: A discriminative feature learning approach for deep face recognition. In: Leibe, B., Matas, J., Sebe, N., Welling, M. (eds.) ECCV 2016. LNCS, vol. 9911, pp. 499–515. Springer, Cham (2016). https://doi.org/10.1007/978-3-319-46478-7_31

27. Liu, Y., Yue, L., Xiao, S., Yang, W., Shen, D., Liu, M.: Assessing clinical progression from subjective cognitive decline to mild cognitive impairment with incomplete multi-modal neuroimages. Med. Image Anal. **75**, 102266 (2022)

28. Liu, M., Gao, Y., Yap, P.T., Shen, D.: Multi-hypergraph learning for incomplete multimodality data. IEEE J. Biomed. Health Inf. **22**(4), 1197–1208 (2017)

Cascaded Cross-Attention Networks for Data-Efficient Whole-Slide Image Classification Using Transformers

Firas Khader[1(✉)], Jakob Nikolas Kather[2], Tianyu Han[3], Sven Nebelung[1], Christiane Kuhl[1], Johannes Stegmaier[4], and Daniel Truhn[1]

[1] Department of Diagnostic and Interventional Radiology, University Hospital Aachen, Aachen, Germany
fkhader@ukaachen.de
[2] Physics of Molecular Imaging Systems, Experimental Molecular Imaging, RWTH Aachen University, Aachen, Germany
[3] Else Kroener Fresenius Center for Digital Health, Medical Faculty Carl Gustav Carus, Technical University Dresden, Dresden, Germany
[4] Institute of Imaging and Computer Vision, RWTH Aachen University, Aachen, Germany

Abstract. Whole-Slide Imaging allows for the capturing and digitization of high-resolution images of histological specimen. An automated analysis of such images using deep learning models is therefore of high demand. The transformer architecture has been proposed as a possible candidate for effectively leveraging the high-resolution information. Here, the whole-slide image is partitioned into smaller image patches and feature tokens are extracted from these image patches. However, while the conventional transformer allows for a simultaneous processing of a large set of input tokens, the computational demand scales quadratically with the number of input tokens and thus quadratically with the number of image patches. To address this problem we propose a novel cascaded cross-attention network (CCAN) based on the cross-attention mechanism that scales linearly with the number of extracted patches. Our experiments demonstrate that this architecture is at least on-par with and even outperforms other attention-based state-of-the-art methods on two public datasets: On the use-case of lung cancer (TCGA NSCLC) our model reaches a mean area under the receiver operating characteristic (AUC) of 0.970 ± 0.008 and on renal cancer (TCGA RCC) reaches a mean AUC of 0.985 ± 0.004. Furthermore, we show that our proposed model is efficient in low-data regimes, making it a promising approach for analyzing whole-slide images in resource-limited settings. To foster research in this direction, we make our code publicly available on GitHub: https://github.com/FirasGit/cascaded_cross_attention.

Keywords: Computational Pathology · Transformers · Whole-Slide Images

Supplementary Information The online version contains supplementary material available at https://doi.org/10.1007/978-3-031-45676-3_42.

1   Introduction

Computational pathology is an emerging interdisciplinary field that synergizes knowledge from pathology and computer science to aid the diagnoses and treatment of diseases [3]. With the number of digitized pathology images increasing over the years, novel machine learning algorithms are required to analyse the images in a timely manner [13]. Deep learning-based methods have demonstrated a promising potential in handling image classification tasks in computational pathology [10]. A prevalent procedure for employing deep learning techniques to the analysis of whole-slide images (WSI) is multiple instance learning (MIL), where a training sample comprises a set of instances and a label for the whole set. The set of instances is commonly chosen to be the set of feature tokens pertaining to one WSI. These feature tokens are constructed by an initial extraction of patches contained in a single WSI and a subsequent utilization of pre-trained feature extractors to capture meaningful feature representations of these patches. This is followed by a feature aggregation, in which the feature representations of each WSI are combined to arrive at a final prediction. Naive approaches for aggregating the features consist of mean or max pooling but only achieve limited performances [9,20]. This has given rise to more sophisticated feature aggregation techniques specifically tuned to the field of computational pathology [5,7,12,17]. These are largely based on an attention operation, in which trainable parameters are used to compute the contribution of each instance to the final prediction [7,17]. Other approaches employ graph neural networks to the task of MSI classification [11,18] or frameworks that use clustering approaches on the patches of WSI [16].

More recently, the transformer architecture [19] has been introduced to the field of computational pathology [15]. Transformer architectures have demonstrated state-of-the-art performance in the fields of natural language processing and computer vision [4]. The transformer model can be seen as an input agnostic method, that leverages the self-attention mechanism in order to aggregate its input tokens. However, one notable short-coming is the quadratic scaling of the self-attention mechanism with respect to the sequence length [8]. This becomes particularly problematic in the context of computational pathology, where whole-slide images are often several gigapixels large [13], therefore resulting in thousands of image patches that have to be input into the transformer model. As a result, bigger GPUs are necessary to store the attention matrix in memory and the increasing number of compute operations limits training and inference speeds as well as model complexity. To overcome this problem, we present a novel neural network architecture that is based on the cross-attention mechanism [8,19] and show that it is capable of efficiently aggregating feature tokens of whole-slide images while demonstrating state-of-the-art performance on two publicly available datasets. Furthermore, we show how attention maps can be extracted for our network and therefore allow for an increased interpretability. Finally, we demonstrate that our model is superior to the previous state of the art in situations where only a small number of training samples is available.

2   Materials and Methods

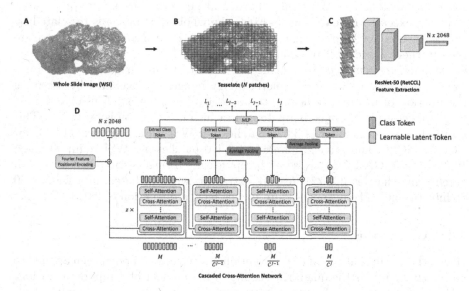

Fig. 1. Model architecture and pre-processing steps. In a first step, we extract a set of N patches from each WSI and subsequently derive feature tokens based on a pretrained ResNet-50 model [21] The resulting feature tokens are used as input to our model, where the cross attention mechanism is used to distil the information at each stage into a compressed representation that allows for efficient attention computation.

2.1   Dataset

We demonstrate the performance of our model on two publicly available datasets: (1) First, the TCGA-NSCLC (non-small-cell lung cancer) dataset, which is constructed by merging the lung cancer datasets TCGA-LUSC and TCGA-LUAD. This results in a total of n=1,042 whole-slide images, of which 530 are labeled as LUAD (Lung adenocarcinoma) and 512 are labeled as LUSC (Lung squamous cell carcinoma). We proceed by splitting the dataset into training (60%, n=627), validation (15%, n=145) and testing sets (25%, n=270) using a 4-fold cross validation scheme, while ensuring that images of the same patient only occur in the same set. (2) Additionally, we evaluate our model on the TCGA-RCC (renal cell carcinoma) dataset, which is composed of three other datasets: TCGA-KIRC, TCGA-KIRP and TCGA-KICH. This results in a dataset comprising n=937 whole-slide images, out of which 519 are labeled as KIRC (Kidney renal clear cell carcinoma), 297 labeled as KIRP (Kidney renal papillary cell carcinoma) and 121 labeled as KICH (Kidney Chromophobe). Similarly, we split the data into training (60%, n=561), validation (15%, n=141), and test sets (25%, n=235) following a 4-fold cross validation scheme.

2.2   Preprocessing

Prior to inputting the WSIs into the neural network, various pre-processings steps are executed on each image: First, the whole-slide image is tesselated by extracting non-overlapping square patches with a 256 μm edge length. These are then further resized to 256 × 256 pixels. Patches that possess a predominantly white color are removed by filtering out patches with a mean grayscale pixel value greater than 224. Furthermore, blurry images are excluded by assessing the fraction of pixels that belong to an edge within the patch using a Canny Edge Detector [2] and rejecting patches in which this fraction is below 2%. This results in a mean number of 3,091 patches per WSI (min: 38; max: 11,039) for TCGA-NSCLC, and a mean number of 3,400 patches per WSI (min: 90; max: 10,037) for TCGA-RCC. In a second step, feature tokens are obtained from each patch using the RetCCL [21] feature extractor, which is based on a ResNet-50 architecture [6]. The output feature tokens have a dimension of 2,048.

2.3   Architecture

One of the key limitations of the conventional transformer model when applied to pathology images is the quadratic scaling of the number of compute operations and the GPU memory footprint with respect to the number of input tokens [19]. In order to overcome this problem of quadratic scaling, we base our CCAN architecture (see Fig. 1) on the cross-attention mechanism proposed by Jaegle et al. [8]. The idea is to distil the information contained in the feature tokens of each MSI into a smaller set of latent tokens. More precisely, let $N \in \mathbb{R}^{D_f}$, denote the number of feature tokens of dimension D_f extracted from each patch of the WSI and let $M \in \mathbb{R}^{D_l}$ denote a pre-defined set of learn able latent tokens with dimensionality D_l. To provide the neural network with information about the position of each patch in the WSI, we first concatenate to each feacture token a positional encoding based on Fourier features [8]:

$$p = [\sin(f_i \pi \hat{D}), \cos(f_i \pi \hat{D})] \tag{1}$$

Here f_i denotes the ith frequency term contained in a set of I equidistant frequencies between 1 and f_{max} and $\hat{D}$ denotes the location of the feature token, normalized between -1 and 1, thereby mapping the top left feature token in the WSI to -1 and the bottom right token to 1. We motivate the use of Fourier features for positional encoding over the use of learn able encodings because they enable scaling to arbitrary sequence lengths. This property is particularly useful when dealing with WSI, which often present in varying sizes. We proceed by computing a key $K_f \in \mathbb{R}^{N \times D_l}$ and value $V_f \in \mathbb{R}^{N \times D_l}$ vector for each of the N input tokens, and a query $Q_l \in \mathbb{R}^{M \times D_l}$ vector for each of the M latent tokens, such that each latent token can attend to all of the feature tokens of the input image:

$$f_{\text{Cross-Attention}}(Q_l, K_f, V_f) = softmax\left(\frac{Q_l K_f^T}{\sqrt{M}}\right)V_f \tag{2}$$

Provided that $M \ll N$, the network learns a compressed representation of the information contained in the large set of input tokens. In addition the number of computation steps reduces to $O(MN)$, compared to $O(N^2)$ as is the case when self-attention is applied directly to the input tokens. Thus, by choosing M to be much smaller than N, the computational requirements for processing a whole-slide image can be significantly reduced. The information in the M latent tokens is then further processed through self-attention layers that scale with $O(M^2)$. The cross-attention and self-attention blocks are then repeated Z times.

When training neural networks, it is often beneficial to gradually change the input dimension opposed to abruptly reducing it (in convolutional neural networks the kernel size and stride are chosen in a way that gradually reduces the feature map dimensions at each stage). Based on this intuition, we propose to use a multi-stage approach in which the set of M tokens is further distilled by adding additional stages with a reduced set of $\frac{M}{C}$ latent tokens. C denotes an arbitrary compression factor and is chosen to be 2 in our model. Similar to the previous stage, cross-attention is used to distil the information, followed by self-attention blocks to process the compressed representation. Subsequently, we add the output of stage 1 to the output of stage 2, thereby serving as a skip-connection which allows for improved gradient flow when the number of stages J is chosen to be large [6]. To overcome the dimensionality mismatch that occurs when adding the M tokens of stage 1 to the $\frac{M}{C}$ tokens of stage 2, we aggregate the output tokens of stage 1 by means of an average pooling that is performed on each set of C consecutive tokens. This multi-stage approach is repeated for J stages, whereby in each stage the number of tokens of the previous stage is compressed by the factor C.

One limitation of the multi-stage approach and the repeated cross- and self attention blocks is that the computation of attention maps (using e.g. the attention rollout mechanism) becomes unintuitive. To overcome this limitation, we add a class token to each of the stages $j \in J$ (resuling in $\frac{M}{C^j}$ latent tokens and 1 class token), and feed it into a shared multi-layer perceptron, resulting in a prediction p_j and loss term L_j for each of the J stages. Additionally, in each stage we add a final cross-attention layer that takes as input the original set of N feature tokens, followed by a self-attention block. This allows us to visualize what regions in the WSI the class token is attending to at each stage. The final prediction is computed by averaging all the individual contributions p_j of each stage. Similarly, we back propagate and compute the total loss L_{total} by summing over all individual loss terms:

$$L_{\text{total}} = \Sigma_{j=1}^{J} L_j \qquad (3)$$

In our experiments, we chose the binary cross entropy loss as our loss function. Furthermore, to prevent the neural networks from overfitting, we randomly dropout a fraction p_{do} of the input feature tokens. All models are trained on an NVIDIA RTX A6000 GPU for a total of 100 epochs and were implemented using PyTorch 1.13.1. We chose the best model of each run in terms of the highest area under the receiver operating characteristic (AUC) reached on the validation set. Further details regarding the hyperparameters used can be found in Supplementary Table S1.

3   Results

3.1   Baseline

To assess the performance of our model, we compare its performance to Trans-MIL [15], a popular transformed-based method proposed for aggregating feature tokens in a MIL setting. TransMIL has demonstrated state-of-the-art performance on a number of datasets, outperforming other popular MIL techniques by a notable margin (see Supplementary Table S2 for a comparison to other methods). In essence, feature tokens are extracted from a set of patches pertaining to each WSI, and then fed through two transformer encoder layers. To handle the quadratic scaling of the self-attention mechanism with respect to the number of input tokens, they make use of Nyström attention [22] in each transformer layer. The low-rank matrix approximation thereby allows for a linear scaling of the attention mechanism. In addition, the authors of the architecture propose the use of a convolutional neural network-based pyramid position encoding generator to encode the positional information of each feature token.

3.2   WSI Classification

We compare the performance of the models on two publicly available datasets, comprising cancer subtype classification tasks, TCGA NSCLC and TCGA RCC (https://portal.gdc.cancer.gov/projects/). For the binary classification task (i.e., LUSC vs. LUAD in TCGA NSCLC), we use the AUC, to assess the model performance. For the multiclass classification problem (i.e., KICH vs. KIRC vs. KIRP in TCGA RCC), we assess the performance by computing the (macro-averaged) one vs. rest AUC. To guarantee a fair comparison, we train all models on the exact same data splits of the 4-fold cross validation. For TCGA NSCLC, we find that our model outperforms the TransMIL baseline (AUC: 0.970 ± 0.008 [standard deviation - SD] vs. 0.957 ± 0.013 [SD]) thereby demonstrating a strong performance. Similarly, we find that on the TCGA RCC dataset our model is on par with the state-of-the-art results set by TransMIL (both reach an AUC of 0.985 [SD CCAN: 0.004, SD TransMIL: 0.002], see Fig. 2 for a more detailed comparison).

3.3   Data Efficiency

A key obstacle in the field of medical artificial intelligence is the restricted access to large datasets. Therefore, neural network architectures need to be developed that can attain a high performance, even under conditions of limited available data. To simulate the performance in limited data settings, we train our model multiple times and restrict the amount of data seen during training to 2%, 5%,

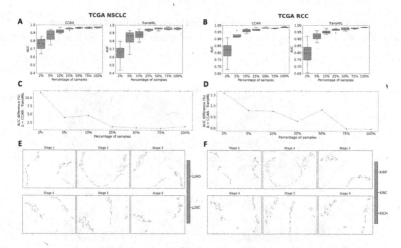

Fig. 2. Results of the CCAN and TransMIL models when trained on the TCGA NSCLC (A) and TCGA RCC (B) datasets. Boxplots show the results of the 4-fold cross validation for each fraction of the used training dataset. (C) and (D) thereby visualize the difference in the mean AUC between the two models, showing that CCAN is more data efficient when a small portion of training data is used. Finally, UMAP [14] projections of the class token of CCAN are visualized at each stage for TCGA NSCLC (E) and TCGA RCC (F), demonstrating separable clusters for all stages that contribute to the final prediction of the model.

10%, 25%, 50%, 75% and 100%. For comparison, we similarly train the baseline method on the same training data. For the TCGA NSCLC dataset, we find that with only 2% of data, our model (AUC: 0.756 ± 0.095 [SD]) outperforms the previous state-of-the-art method by 12% (AUC: 0.639 ± 0.149 [SD]). Similar results were found for other dataset sizes (see Fig. 2). Accordingly, for the TCGA RCC dataset, our model (AUC: 0.808 ± 0.096 [SD]) outperforms the baseline (AUC: 0.792 ± 0.109 [SD]) at only 2% of training data by 1.6%. Again, this trend of a superior performance can be seen for all other percentages of the training data.

3.4 Explainability

In order to gain a deeper understanding of the inner workings of the neural network, attention maps were extracted. Since the class tokens are used as input in the final MLP and the output thereof is used to arrive at the final prediction, we propose to compute the attention map of the class token with respect to the input patches. More precisely, in each stage we perform attention-rollout [1] using the last self-attention layers and the last cross-attention layer. We then use computed attention values to visualize the attention map of each class token, thereby providing insights into the decision making process of the model at each stage (see Supplementary Figure S1). To arrive at a single attention map for the whole model, we take the mean attention over all stages. We find that these attention maps largely coincide with regions containing the tumor (see Fig. 3). Additionally, we display regions of the model with low and high attention values.

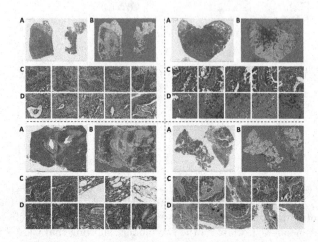

Fig. 3. Attention visualization for four different WSI contained in the TCGA NSCLC dataset. (A) shows the original WSI and (B) displays the aggregated attention map over all stages. Additionally, the top-5 patches with the lowest attention value (C) and top-5 patches with highest attention value (D) are visualized.

4   Conclusion

In this work we developed and presented a novel neural network architecture, based on the cross-attention mechanism, that is capable of aggregating a large set of input feature vectors previously extracted from whole-slide images. We demonstrate that this method outperforms the previous state-of-the-art methods on two publicly available datasets. In particular we show that our model architecture is more data efficient when training data is limited. Furthermore, we provide insights into the models decision-making process by showing how attention maps of each stage in the model can be aggregated. This allows for an interpretable visualization of which regions of the image the neural networks looks at to arrive at a final prediction. The model design lends itself to be extended to support multimodal inputs such as WSI in combination with genomics, which we will explore in future work.

Acknowledgements. The results published here are in whole or part based upon data generated by the TCGA Research Network: https://www.cancer.gov/tcga.

References

1. Abnar, S., Zuidema, W.: Quantifying attention flow in transformers (2020). arXiv:2005.00928
2. Canny, J.: A Computational approach to edge detection. In: IEEE Transactions on Pattern Analysis and Machine Intelligence, PAMI, vol. 8(6), pp. 679–698 (1986)
3. Cui, M., Zhang, D.Y.: Artificial intelligence and computational pathology. Lab. Invest. **101**(4), 412–422 (2021). https://www.nature.com/articles/s41374-020-00514-0
4. Dosovitskiy, A., et al.: An image is worth 16×16 words: transformers for image recognition at scale (2020). arXiv:2010.11929
5. Hashimoto, N., et al.: Multi-scale domain-adversarial multiple-instance CNN for cancer subtype classification with unannotated histopathological images. In: 2020 IEEE/CVF Conference on Computer Vision and Pattern Recognition (CVPR), pp. 3851–3860. IEEE, Seattle, WA, USA (2020). https://ieeexplore.ieee.org/document/9157776/
6. He, K., Zhang, X., Ren, S., Sun, J.: Deep residual learning for image recognition. In: 2016 IEEE Conference on Computer Vision and Pattern Recognition (CVPR), pp. 770–778 (2016). https://doi.org/10.1109/CVPR.2016.90. ISSN: 1063-6919
7. Ilse, M., Tomczak, J., Welling, M.: Attention-based deep multiple instance learning. In: Proceedings of the 35th International Conference on Machine Learning, pp. 2127–2136. PMLR (2018). https://proceedings.mlr.press/v80/ilse18a.html. ISSN: 2640-3498
8. Jaegle, A., Gimeno, F., Brock, A., Zisserman, A., Vinyals, O., Carreira, J.: Perceiver: general perception with iterative attention (2021). https://doi.org/10.48550/arXiv.2103.03206, arXiv:2103.03206
9. Kanavati, F., et al.: Weakly-supervised learning for lung carcinoma classification using deep learning. Sci. Rep. **10**(1), 9297 (2020). https://www.nature.com/articles/s41598-020-66333-x
10. Kather, J.N., et al.: Deep learning can predict microsatellite instability directly from histology in gastrointestinal cancer. Nat. Med. **25**(7), 1054–1056 (2019). https://www.nature.com/articles/s41591-019-0462-y
11. Konda, R., Wu, H., Wang, M.D.: Graph convolutional neural networks to classify whole slide images. In: ICASSP 2020–2020 IEEE International Conference on Acoustics, Speech and Signal Processing (ICASSP), pp. 1334–1338 (2020)
12. Lu, M.Y., Williamson, D.F.K., Chen, T.Y., Chen, R.J., Barbieri, M., Mahmood, F.: Data-efficient and weakly supervised computational pathology on whole-slide images. Nat. Biomed. Eng. **5**(6), 555–570 (2021). https://www.nature.com/articles/s41551-020-00682-w
13. Marini, N., et al.: Unleashing the potential of digital pathology data by training computer-aided diagnosis models without human annotations. NPJ Digit. Med. **5**(1), 1–18 (2022). https://www.nature.com/articles/s41746-022-00635-4
14. McInnes, L., Healy, J., Saul, N., Großberger, L.: UMAP: uniform manifold approximation and projection. J. Open Source Softw. 3(29), 861 (2018). https://doi.org/10.21105/joss.00861 https://doi.org/10.21105/joss.00861
15. Shao, Z., et al.: TransMIL: transformer based correlated multiple instance learning for whole slide image classification. In: Advances in Neural Information Processing Systems, vol. 34, pp. 2136–2147. Curran Associates, Inc. (2021). https://proceedings.neurips.cc/paper/2021/hash/10c272d06794d3e5785d5e7c5356e9ff-Abstract.html

16. Sharma, Y., Shrivastava, A., Ehsan, L., Moskaluk, C.A., Syed, S., Brown, D.: Cluster-to-Conquer: a framework for end-to-end multi-instance learning for whole slide image classification. In: Proceedings of the Fourth Conference on Medical Imaging with Deep Learning, pp. 682–698. PMLR (2021). https://proceedings. mlr.press/v143/sharma21a.html ISSN: 2640-3498

17. Tomita, N., Abdollahi, B., Wei, J., Ren, B., Suriawinata, A., Hassanpour, S.: Attention-based deep neural networks for detection of cancerous and precancerous esophagus tissue on histopathological slides. JAMA Netw. Open 2(11), e1914645 (2019)

18. Tu, M., Huang, J., He, X., Zhou, B.: Multiple instance learning with graph neural networks (2019). arXiv:1906.04881

19. Vaswani, A., et al.: Attention is all you need. In: Advances in Neural Information Processing Systems, vol. 30. Curran Associates, Inc. (2017). https://proceedings. neurips.cc/paper/2017/hash/3f5ee243547dee91fbd053c1c4a845aa-Abstract.html

20. Wang, X., Yan, Y., Tang, P., Bai, X., Liu, W.: Revisiting multiple instance neural networks. Pattern Recogn. 74, 15–24 (2018). arXiv:1610.02501

21. Wang, X., et al.: RetCCL: clustering-guided contrastive learning for whole-slide image retrieval. Med. Image Anal. 83, 102645 (2023). https://www.sciencedirect. com/science/article/pii/S1361841522002730

22. Xiong, Y., et al.: A Nyström-based algorithm for approximating self-attention (2021). arXiv:2102.03902

Enhanced Diagnostic Fidelity in Pathology Whole Slide Image Compression via Deep Learning

Maximilian Fischer[1,2,3(✉)], Peter Neher[1,2,11], Peter Schüffler[6,7],
Shuhan Xiao[1,4], Silvia Dias Almeida[1,3], Constantin Ulrich[1,2,12],
Alexander Muckenhuber[6], Rickmer Braren[8], Michael Götz[1,5], Jens Kleesiek[9,10],
Marco Nolden[1,11], and Klaus Maier-Hein[1,3,4,11,12]

[1] Division of Medical Image Computing, German Cancer Research Center (DKFZ) Heidelberg, Heidelberg, Germany
[2] German Cancer Consortium (DKTK), Partner Site Heidelberg, Heidelberg, Germany
maximilian.fischer@dkfz-heidelberg.de
[3] Medical Faculty, Heidelberg University, Heidelberg, Germany
[4] Faculty of Mathematics and Computer Science, Heidelberg University, Heidelberg, Germany
[5] Clinic of Diagnostics and Interventional Radiology, Section Experimental Radiology, Ulm University Medical Centre, Ulm, Germany
[6] TUM School of Medicine and Health, Institute of Pathology, Technical University of Munich, Munich, Germany
[7] TUM School of Computation, Information and Technology, Technical University of Munich, Munich, Germany
[8] Department of Diagnostic and Interventional Radiology, Faculty of Medicine, Technical University of Munich, Munich, Germany
[9] Institute for AI in Medicine (IKIM), University Medicine Essen, Essen, Germany
[10] German Cancer Consortium (DKTK), Partner Site Essen, Essen, Germany
[11] Pattern Analysis and Learning Group, Department of Radiation Oncology, Heidelberg University Hospital, Heidelberg, Germany
[12] National Center for Tumor Diseases (NCT), NCT Heidelberg, a Partnership Between DKFZ and University Medical Center Heidelberg, Heidelberg, Germany

Abstract. Accurate diagnosis of disease often depends on the exhaustive examination of Whole Slide Images (WSI) at microscopic resolution. Efficient handling of these data-intensive images requires lossy compression techniques. This paper investigates the limitations of the widely-used JPEG algorithm, the current clinical standard, and reveals severe image artifacts impacting diagnostic fidelity.

To overcome these challenges, we introduce a novel deep-learning (DL)-based compression method tailored for pathology images. By enforcing feature similarity of deep features between the original and compressed images, our approach achieves superior Peak Signal-to-Noise Ratio (PSNR), Multi-Scale Structural Similarity Index (MS-SSIM), and

Supplementary Information The online version contains supplementary material available at https://doi.org/10.1007/978-3-031-45676-3_43.

Learned Perceptual Image Patch Similarity (LPIPS) scores compared to JPEG-XL, Webp, and other DL compression methods. Our method increases the PSNR value from 39 (JPEG80) to 41, indicating improved image fidelity and diagnostic accuracy.

Our approach can help to drastically reduce storage costs while maintaining large levels of image quality. Our method is online available.

Keywords: Whole Slide Imaging · Digital Pathology · Lossy Image Compression

1 Introduction

Spatial formations and overall morphological characteristics of cells serve as fundamental elements for accurately diagnosing various diseases. The detection of global cellular patterns requires a comprehensive examination of tissue samples, posing a significant challenge due to the potentially very large size of these specimens. To address this challenge, Whole Slide Imaging (WSI) techniques provide high-resolution images that capture cells at a microscopic level. However, efficient handling, storage, and transmission of these files necessitate the use of lossy image compression algorithms.

Among the lossy compression algorithms employed by WSI vendors, JPEG80 is the most common one. But even with initial compression, the resulting file sizes remain a major obstacle in effectively transmitting or storing WSI. Thus the further compression of WSI is of significant interest and a field of active research. The fundamental burden is to compress images with high image quality, as visual inspection by pathologists continues to be the clinical standard, and accurate diagnoses heavily rely on the ability to examine the WSI with exceptional visual quality. Despite its critical importance, previous studies have primarily focused on evaluating the impact of compression on deep learning (DL) models [4,7] instead.

In this paper, we present a novel DL-based compression scheme specifically tailored for histopathological data and compare it for several metrics to quantify perceptual image quality for pathological images. The proposed method ensures high levels of image quality, even at high compression ratios, by enforcing similarity between deep features from the original and the reconstructed image that are extracted by a contrastive pre-trained DL model. We show that our method outperforms traditional codecs like JPEG-XL or WebP, as well as other learned compression schemes like [1] in terms of image quality, while we also achieve higher compression ratios.

We compare our method to other approaches using various state-of-the-art image quality metrics and suggest establishing the Learned Perceptual Image Patch Similarity (LPIPS) [21] metric from the natural scene image domain also for the pathology domain as image quality measure.

The presented findings have the potential to significantly impact the field of pathological image analysis by enabling higher compression ratios while preserving image quality. We show that our method yields much higher perceptual image quality metrics for the same compression ratio than the compared compression schemes.

2   Methods

2.1   Learned Lossy Compression

Lossy image compression methods are usually modeled as follows: An image is projected into an alternative (latent) representation, where pixel values are quantized to reduce the storage requirements for each pixel, accepting distortions in the image due to information loss at the same time. For a learned lossy image compression scheme, the required latent representation is usually determined by an autoencoder architecture [16]. In this setting, an encoder model E generates quantized latents $y = E(x)$, from which the decoder model D decodes a lossy reconstruction x'. The quantized latents are stored via a shared probability model P between the encoder and decoder part that assigns a number of bits to representations according to their occurrence frequency in the dataset, which is determined by the discrete probability distribution obtained by P. The probability model together with an entropy coding scheme enables lossless storing of y at a bitrate $r(y) = -log(P(y))$, which is defined by Shannon's cross entropy [6]. The distortion $d(x, x')$ measures the introduced error between the original image and the reconstruction, which can be determined by metrics such as the Mean Squared Error $d = MSE$. Learned codecs usually parametrize E, D and P as convolutional neural networks (CNNs) and the problem of learned image compression can be summarized as bitlength-distortion optimization problem [16].

2.2   Image Quality Metrics

Lossy image compression schemes universally result in a degradation of image quality as compression factors increase. To assess the performance of compression algorithms, image quality metrics serve as numerical measures. However, a prevailing issue is that no existing metric captures all kinds of image distortions.

For instance, when subjected to high compression factors, JPEG compression exposes blocking artifacts that the Peak Signal-to-Noise Ratio (PSNR) fails to detect effectively. Conversely, learned image quality metrics exhibit reduced sensitivity to color distortions, which highlights the challenges in achieving a comprehensive assessment of image compression performance [10]. Thus compression schemes always should be evaluated by multiple image quality metrics.

Peak-Signal-to-Noise Ratio. The most common metric is the Peak-Signal-to-Noise Ratio (PSNR), which is derived by the mean squared error (MSE). The MSE compares the "true" pixel values of the original image versus the reconstructed pixel values of the decompressed image. By computing the mean of the squared differences between the original image and the noisy reconstruction, the difference between the reconstruction and the original image can be measured. Assuming a noise-free original image x with the dimensions $m \times n$ and a noisy approximation x' after decoding the quantized latents, the MSE and PSNR are defined as follows:

$$MSE = \frac{1}{mn} \sum_{i=0}^{m-1} \sum_{j=0}^{n-1} [x(i,j) - x'(i,j)]^2 \text{ and } PSNR = 10 \cdot log_{10}(\frac{MAX_x^2}{MSE}) \quad (1)$$

with MAX_x being the largest possible value that can occur in the original image x.

Multi Scale Structural Similarity Index Measure. It has been shown that assuming pixel-wise independence in structured datasets like images leads to blurry reconstructions [10]. In particular, the MSE is not able to capture connections between neighboring pixels. In contrast, the Multi Scale Structural Similarity Index Measure (MS-SSIM) [17] takes into account that pixels have strong dependencies among each other, especially when they are spatially close. The MS-SSIM is based on the *SSIM* metric which is calculated across multiple windows and scales between two images. For two windows u and u', which are located in x and x' respectively, the MS-SSIM is calculated as [17]:

$$SSIM(u, u') = \frac{(2\mu_u \mu_{u'} + c_1)(2\sigma_{u,u'} + c_2)}{(\mu_u^2 + \mu_{u'}^2 + c_1)(\sigma_u^2 + \sigma_{u'}^2 + c_2)}, \quad (2)$$

with the averaged intensities in the given block window μ, the variance σ, the covariance $\sigma_{u,u'}$ and the parameters c_1, c_2 for stabilizing the division by $c_1 = (k_1 L)^2$ and $c_2 = (k_2 L)^2$ where we set $k_1 = 0.01, k_2 = 0.03$ and L models the dynamic range of the pixel values. More details can be found in [17].

Learned Image Similarity. In contrast to traditional metrics like MSE and MS-SSIM, Learned Perceptual Image Patch Similarity (LPIPS) is a perceptual metric that has been trained to align more closely with human interpretation of image quality [21]. Unlike the aforementioned metrics, LPIPS does not measure the distance between two images solely at the pixel level. Instead, it quantifies the similarity of two images based on their embeddings that are obtained from multiple depths of a pre-trained model with depth L. In [21] it is suggested to use an ImageNet [13] pre-trained model to generate latent embeddings for the original image x and the distorted image x'. The LPIPS distance is quantified by the averaged l_2 distance between multiple depths l which are aggregated by learned weighting coefficients $\{\alpha_l\}_l^L$. For the distance, the embeddings are unit-normalized in the channel dimensions and the metric is defined as:

$$d_{\text{LPIPS}}(x, x')^2 = \sum_{l=1}^{L} \alpha_l ||\phi_l(x) - \phi_l(x')||_2^2, \quad (3)$$

where ϕ_l is the feature map at layer l.

2.3   Deep Perceptual Guidance

Inspired by approaches like [21], we incorporate a learned feature similarity in the loss function during training of our lossy compression autoencoder. In contrast to previous work, we use a contrastive pre-trained model from the pathology domain to compute distances between embeddings. We hypothesize that the infused domain knowledge of pathological images leads to more realistic reconstructions of histopathological images. Furthermore, we use a more lightweight implementation than LPIPS for calculating distances between embeddings to reduce the added computational complexity. In particular, we use the contrastive pre-trained model from [5] and compute $\ell_2(C(x), C(x'))$, which is the ℓ_2 distance between the feature vectors from x and x', both extracted with the model C from [5]. The weighted ℓ_2 distance measure, together with the baseline MSE distance and the targeted bitrate yields the final loss function:

$$\mathcal{L}_{EG} = r + \lambda \cdot d = \mathbb{E}_{x \sim p(x)}[\underbrace{\lambda r(y)}_{\text{bitrate}} + \underbrace{d(x, x') + \psi \ell_2(C(x), C(x'))}_{\text{image similarity}}] \qquad (4)$$

3   Experiments

3.1   Dataset

To determine the performance of a lossy compression autoencoder in the histopathology domain, we collect a diverse dataset containing various tissue types [3,9,15]. In Table 1, an overview of the collected dataset for this study is presented. In this paper, we investigate the impact of further compression schemes, beyond the initial compression during image acquisition. Thus all images were initially JPEG80 compressed, without any further compression besides that. We split the dataset on a slide-level label into training and test set. To compare the performance of different compression schemes, we generate compressed versions of the test set at various bit rates. For each bit rate, we evaluate the image quality metrics between the original image and the compressed version. As traditional codecs we implement JPEG, WebP and JPEG-XL.

3.2   Neural Image Compression Models

As a compression autoencoder, we implement the model "bmshj-factorized" from [1]. During training, we use the implementation from Eq. 4 for training our model to generate realistic reconstructions. We used [2] as a framework to implement our method. For each targeted bit rate, we train a separate model, by sampling values for λ from $\lambda = [0.001, ..., 0.1]$. Although it might be technically feasible to train one model for various bit rates [12,14,18], this enables us to target the desired bit rates more accurately. The models are trained with an initial setting of $\psi = 0.5$ for 90 epochs, which is increased to $\psi = 0.7$ for another 60 epochs. In total, we train all models for approximately 4M steps using the Adam optimizer and we decrease the initial learning rate of 0.0001 whenever

Table 1. Details of the datasets used for this work.

Dataset	BreaKHis	Colon1	Colon2
Source	[15]	[3]	[9]
Tissue	Breast	Colon	Colon
Images	1639	500	10
SampleSize	700x460	768x768	5000x5000
Tile Size	224x224	224x224	224x224
Tiles	10158	10005	4840

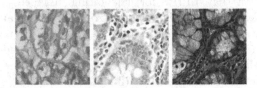

Eq. 4 yields no further improvement. We use a tile size of 224×224 pixels and a batch size of 4. During training, we apply random horizontal and vertical flips, as well as color jittering. To approximate quantization during testing, we employ random noise during training like suggested in [16]. During testing, we evaluate the performance of the model with the PSNR, as well as the perceptual image quality metric MS-SSIM and LPIPS. We refer to this training scheme as Supervised Pathologic L_2 Embeddings (SPL_2E). To evaluate the impact of the deep pathologic feature supervision, we train the same model also without the additional feature distance ψ, which we refer to as *Baseline*. Furthermore, to evaluate the impact of training on pathology data in general for compression models, we also compare the performance of the same model that is trained on the Vimeo dataset [19]. This dataset contains a large collection of natural scene images. We refer to this model as *Vimeo*.

4   Results

4.1   Comparison of Image Quality Metrics

Figure 1 illustrates the metric scores for two types of potential distortions encountered in lossy image compression: color shifts and blocking artifacts. The x-axis represents the degree of distortion, which corresponds to the quantization level or quality factor for lossy JPEG compression for blocking artifacts, and a bias between 0 and 50 added in the Lab color space for color shifts. For the blocking artifacts, we sample values between 90 and 10 as quality factor for the JPEG compression. The respective metric is then computed between the distorted image and the original image. The presented results are the average scores obtained from evaluating 10 sample images from our test set. The figure shows that the Mean Squared Error (MSE) metric is highly sensitive to color distortions, whereas the Learned Perceptual Image Patch Similarity (LPIPS) metric effectively detects and quantifies blocking artifacts.

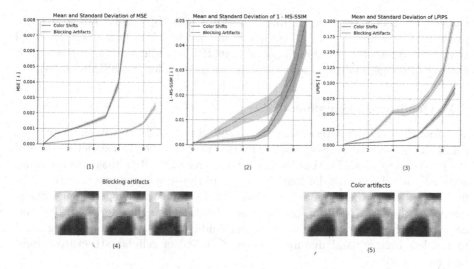

Fig. 1. Visualization of the three metrics for color and blocking artifacts as image distortions. We report 1−MS-SSIM for better visual inspection (1–3). Below the plots, the levels of distortions are visualized for the blocking (4) and the color (5) artifacts. For visualization purposes, we show the quality factors 90, 40 and 5 for the blocking artifacts.

4.2 Quantitative PSNR, MS-SSIM and LPIPS Metrics

Figure 2 shows the results obtained by the compression schemes for the resulting bit rate and perceptual image quality metrics. Arrows in the figure indicate whether a high or low value of the metric is better. The resulting bitrate the compression with JPEG80 would reach is marked as a red dot in the plots. Our results show that our proposed method performs equal or better than the compared methods, while JPEG and WebP are showing the worst results across all

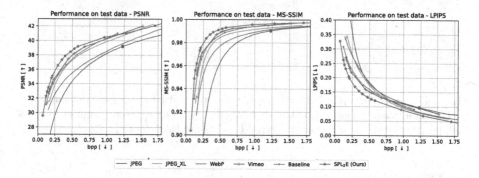

Fig. 2. Rate-distortion and -perception curves on the test set for various metrics. Arrows (↑), (↓) indicate whether high or low scores are preferable. The level of compression is determined by the bits-per-pixel (bpp) value.

metrics. Especially between compression ratios of 0.2 bpp and 1.1 bpp, the proposed methods outperforms all compared approaches. We show that our model is able to achieve 76% smaller file sizes than JPEG, while maintaining the same amount of image quality (compare 0.4 bpp vs. 1.2 bpp for MS-SSIM = 0.99).

4.3   Qualitative Assessment

We show the qualitative reconstruction result of two exemplary image patches in Fig. 3, at the same compression ratio. The figure shows the reconstructed result of the proposed method. Further images showing reconstructed images of the other compression schemes can be found in Fig. 1 in the supplementary material. The figure shows that the proposed method achieves much higher perceptual image quality at the same resulting bit rate than JPEG. The figure demonstrates that our approach exhibits no color smearing (indicated by the white background in the left images) and maintains clear visibility of cellular structures (right images).

Original   JPEG, PSNR= 29 dB   Ours, PSNR= 35.5 dB

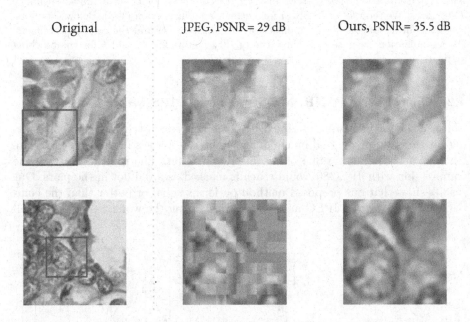

Fig. 3. The proposed model achieves much higher perceptual image quality than JPEG for the same compression ratio (here 0.25 bpp). Left: Two different original patches. Middle: JPEG compression, Right: Ours. Patches are taken from [3, 15]

5   Discussion

This paper shows how a DL-based lossy compression scheme is fine-tuned on pathological image data and yields reconstructions with higher perceptual image quality than previous methods. We show that for the evaluation of image quality in the WSI domain, several aspects are important and demonstrate that none of the existing metrics is ideally suited to compare compression schemes, but the combination of MSE, MS-SSIM and LPIPS can be a helpful. Future work could focus on further specialized metrics for WSI, which could evaluate morphological characteristics of cells as a quality metric. Also, generative decoding methods, such as GAN- or diffusion-based decoders [8,11,20], show promising results in compensating for information loss on the encoder side. Nevertheless, practical considerations for both the encoder and decoder must be taken into account. In clinical settings, it is common to utilize powerful devices for image acquisition, while deploying subsequent lightweight image analysis components in the analysis pipeline. So far our method is ideally suited for HE stained images, which is the most common staining procedure in clinical practice, since the model from [5] is mostly trained on HE slides. Thus future work should also consider special models for deep supervision that are trained on special stains.

Acknowledgements. This work was partially supported by the DKTK Joint Funding UPGRADE, project "Subtyping of pancreatic cancer based on radiographic and pathological features" (SUBPAN), and by the Deutsche Forschungsgemeinschaft (DFG, German Research Foundation) under the grant 410981386.

References

1. Ballé, J., Minnen, D., Singh, S., Hwang, S.J., Johnston, N.: Variational image compression with a scale hyperprior. In: International Conference on Learning Representations (2018). https://openreview.net/forum?id=rkcQFMZRb
2. Bégaint, J., Racapé, F., Feltman, S., Pushparaja, A.: CompressAI: a PyTorch library and evaluation platform for end-to-end compression research. arXiv preprint arXiv:2011.03029 (2020)
3. Borkowski, A.A., Bui, M.M., Thomas, L.B., Wilson, C.P., DeLand, L.A., Mastorides, S.M.: Lung and colon cancer histopathological image dataset (lc25000) (2019). https://doi.org/10.48550/ARXIV.1912.12142. https://www.kaggle.com/datasets/andrewmvd/lung-and-colon-cancer-histopathological-images
4. Chen, Y., Janowczyk, A., Madabhushi, A.: Quantitative assessment of the effects of compression on deep learning in digital pathology image analysis. JCO Clin. Cancer Inf. **4**, 221–233 (2020). https://doi.org/10.1200/CCI.19.00068
5. Ciga, O., Xu, T., Martel, A.L.: Self supervised contrastive learning for digital histopathology. Mach. Learn. Appl. **7**, 100198 (2022). https://doi.org/10.1016/j.mlwa.2021.100198
6. Cover, T.M., Thomas, J.A.: Elements of information theory (2012)
7. Ghazvinian Zanjani, F., Zinger, S., Piepers, B., Mahmoudpour, S., Schelkens, P.: Impact of JPEG 2000 compression on deep convolutional neural networks for metastatic cancer detection in histopathological images. J. Med. Imaging **6**(2), 1 (2019). https://doi.org/10.1117/1.JMI.6.2.027501
8. Ghouse, N.F.K.M., Petersen, J., Wiggers, A.J., Xu, T., Sautiere, G.: Neural image compression with a diffusion-based decoder (2023). https://openreview.net/forum?id=4Jq0XWCZQel

9. Kather, J.N., et al.: Multi-class texture analysis in colorectal cancer histology. Sci. Rep. **6**, 27988 (2016). https://doi.org/10.1038/srep27988. https://www.kaggle.com/datasets/kmader/colorectal-histology-mnist

10. Mentzer, F., Toderici, G.D., Tschannen, M., Agustsson, E.: High-fidelity generative image compression. In: Larochelle, H., Ranzato, M., Hadsell, R., Balcan, M., Lin, H. (eds.) Advances in Neural Information Processing Systems, vol. 33, pp. 11913–11924. Curran Associates, Inc. (2020)

11. Pan, Z., Zhou, X., Tian, H.: Extreme generative image compression by learning text embedding from diffusion models. ArXiv abs/2211.07793 (2022)

12. Rippel, O., Anderson, A.G., Tatwawadi, K., Nair, S., Lytle, C., Bourdev, L.D.: ELF-VC: efficient learned flexible-rate video coding. In: 2021 IEEE/CVF International Conference on Computer Vision (ICCV), pp. 14459–14468 (2021)

13. Russakovsky, O., Deng, J., Su, H., et al.: ImageNet large scale visual recognition challenge (2015). http://arxiv.org/abs/1409.0575

14. Song, M.S., Choi, J., Han, B.: Variable-rate deep image compression through spatially-adaptive feature transform. In: 2021 IEEE/CVF International Conference on Computer Vision (ICCV), pp. 2360–2369 (2021)

15. Spanhol, F.A., Oliveira, L.S., Petitjean, C., Heutte, L.: A dataset for breast cancer histopathological image classification. IEEE Trans. Biomed. Eng. **63**(7), 1455–1462 (2016). https://doi.org/10.1109/TBME.2015.2496264

16. Theis, L., Shi, W., Cunningham, A., Huszár, F.: Lossy image compression with compressive autoencoders. In: International Conference on Learning Representations (2017). https://openreview.net/forum?id=rJiNwv9gg

17. Wang, Z., Simoncelli, E., Bovik, A.: Multiscale structural similarity for image quality assessment, vol. 2, pp. 1398–1402 (2003). https://doi.org/10.1109/ACSSC.2003.1292216

18. Wu, L., Huang, K., Shen, H.: A GAN-based tunable image compression system. In: 2020 IEEE Winter Conference on Applications of Computer Vision (WACV), pp. 2323–2331 (2020)

19. Xue, T., Chen, B., Wu, J., Wei, D., Freeman, W.T.: Video enhancement with task-oriented flow. Int. J. Comput. Vis. (IJCV) **127**(8), 1106–1125 (2019)

20. Yang, R., Mandt, S.: Lossy image compression with conditional diffusion models. arXiv preprint arXiv:2209.06950 (2022)

21. Zhang, R., Isola, P., Efros, A.A., Shechtman, E., Wang, O.: The unreasonable effectiveness of deep features as a perceptual metric. In: Proceedings of the IEEE Conference on Computer Vision and Pattern Recognition (CVPR), June 2018

RoFormer for Position Aware Multiple Instance Learning in Whole Slide Image Classification

Etienne Pochet, Rami Maroun, and Roger Trullo[✉]

Sanofi, Chilly Mazarin, France
{etienne.pochet,rami.maroun,roger.trullo}@sanofi.com

Abstract. Whole slide image (WSI) classification is a critical task in computational pathology. However, the gigapixel-size of such images remains a major challenge for the current state of deep-learning. Current methods rely on multiple-instance learning (MIL) models with frozen feature extractors. Given the the high number of instances in each image, MIL methods have long assumed independence and permutation-invariance of patches, disregarding the tissue structure and correlation between patches. Recent works started studying this correlation between instances but the computational workload of such a high number of tokens remained a limiting factor. In particular, relative position of patches remains unaddressed.

We propose to apply a straightforward encoding module, namely a RoFormer layer , relying on memory-efficient exact self-attention and relative positional encoding. This module can perform full self-attention with relative position encoding on patches of large and arbitrary shaped WSIs, solving the need for correlation between instances and spatial modeling of tissues. We demonstrate that our method outperforms state-of-the-art MIL models on three commonly used public datasets (TCGA-NSCLC, BRACS and Camelyon16)) on weakly supervised classification tasks.

Code is available at https://github.com/Sanofi-Public/DDS-RoFor merMIL.

1 Introduction

Whole slide images (WSI), representing tissue slides as giga-pixel images, have been extensively used in computational pathology, and represent a promising application for modern deep learning [6,12,16]. However, their size induces great challenges, both in computational cost for the image processing and lack of local annotations. Hence, the WSI classification has been studied as a weakly-supervised learning task, treating the image as a bag of instances with one label for the entire bag. Multiple instance learning (MIL) techniques have had success in this scenario [8,10,11]. WSIs are first divided into patches, then a feature

---

Supplementary Information The online version contains supplementary material available at https://doi.org/10.1007/978-3-031-45676-3_44.

X. Cao et al. (Eds.): MLMI 2023, LNCS 14349, pp. 437–446, 2024.
https://doi.org/10.1007/978-3-031-45676-3_44

extractor generates features for each of them and finally the feature vectors are aggregated using global pooling operators to get an image-level prediction.

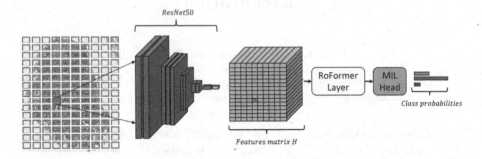

Fig. 1. Model Pipeline

Given the high number of instances in each giga-pixel image, works like ABMIL [8] or CLAM [11] assumed independence between patches and applied permutation-invariant methods to avoid comparing patches, reducing the computational workload. These assumptions allow to treat large number of instances efficiently but prevent from modelling the correlation between patches and the tissue spatial structure.

More recent studies attempted to bridge these gaps. Works like HIPT [3] or HAG-MIL [22] use hierarchical transformers to aggregate local and global information. Closer to our work, self-attention has been used as an encoding layer, without positional-encoding to capture dependencies between patches [13, 15]. TransMIL [17] reshapes the image into a square feature map to apply group convolution and Nystrom method [23] which is just an approximation to a full self attention, in an attempt to encode spatial information into the patch features.

In light of the recent advances in transformers research, we propose a straightforward encoding module relying on a RoFormer encoder layer; a transformer block with rotary position embedding (RoPE) [19]. The proposed module, implemented with memory-efficient attention [14], can perform full self-attention without resorting to any approximation with relative position encoding for large WSIs, with arbitrary and irregular shapes, on consumer GPUs. We use the coordinates of patches in the slide to apply the relative position encoding, treating the WSI as a 2D sequence of instances rather than an unordered bag. This encoder is agnostic to the downstream model and we experiment with both ABMIL [8] or DSMIL [10]. Our model eventually fits in 8GB GPUs. Figure 1 summarizes our approach.

We evaluate the influence of this encoding module alongside proven MIL models on three of the main public benchmarks for H&E stain WSI classification: TCGA-NSCLC [20], BRACS [2] and Camelyon16 [1] and show a consistent improvement on classification performance with our proposed method.

2   Related Work

2.1   Multiple Instance Learning

MIL in WSIs classification has had great success with using a two-step model-ing where each WSI is considered as a bag and its patches are the instances. A frozen feature extractor generates a feature vector for each instance. Then a pooling operation generates the image-level prediction. The attention-based pooling introduced in ABMIL, widely adopted in later works [3,11,24], intro-duces a learnable class token which is compared in global attention to all patches features. The slide representation z is computed as :

$$z = \sum_i \alpha_i H_i \tag{1}$$

$$\alpha_i = (\text{softmax} \left(CH^T \right) H)_i \tag{2}$$

with H the feature matrix of all patches, C the learnable class token. Linear projection and multi-head formulation are omitted in the equation for simplicity.

DSMIL [10] modified this pooling function to perform global attention of a critical patch, identified with a learned classifier, over all other patches.

Both formulations enable all patches to influence the final prediction with limited computations. However, they do not model the dependencies between patches nor their relative positions. The ability to encode patch representations with respect to other patches and tissue structures remains a challenge.

2.2   Attempts at Position-Aware MIL

Among notable attempts to model spatial dependencies between patches, the two most related to our work are, to the best of our knowledge, HIPT and TransMIL.

HIPT. In this work, patches are aggregated into macro regions of $4k*4k$ pixels, instead of the 256*256 original patches, during feature extraction. A multilayer ViT [5] then encodes regions features before feeding them to the global pooling layer. This encoder, trained at the same time as the attention pooling, is then exposed to a much smaller number of tokens and can perform full self-attention with absolute position encoding. The interest of aggregating patches into larger regions is out of the scope of this work and we will investigate the use of a similar ViT in the setup of small patches.

TransMIL. This work leverages the inductive bias of CNNs to aggregate con-text information, after squaring the N patches into a $\sqrt{N} * \sqrt{N}$ feature map, regardless of the initial tissue shape. Dependencies between instances are also modeled using a Nystrom-approximation attention layer [23].

2.3   Relative Position Encoding

Position encoding in long sequence transformers is still an open question, even more so for WSI where the tokens are irregularly located. Absolute position encoding, of ViTs, helps modeling dependencies between instances [3] but the arbitrary shapes, sizes and orientations of WSIs may require relative position encoding. However, methods with discrete number of relative positions [18,21] would not account for the large irregular shapes of tissues in WSIs. Rotary Position Encoding (RoPE) [19] provides an elegant solution to this problem by modifying directly the key-query inner product of self-attention, with minimal computation overhead and handling any maximum size of relative position, which matches our very long sequence setup.

3   Methods

In this section, we describe the encoding module we use, and the required modifications to the WSI patching. The feature extraction, with a frozen encoder, as well as the global attention pooling layer are out of the scope of this work. The approach is summarized in Fig. 1. The coordinates and mask positions of each embedded patch are kept and used in a RoFormer layer with masked self-attention (Fig. 2). Encoded outputs are fed to an MIL classification head.

Unless stated otherwise, we applied the same feature extraction as described in CLAM [11].

3.1   Patching

As it is typically done, the WSI is divided into small patches, of size $P * P$ pixels, using only the tissue regions and disregarding the background. However, instead of treating the patches independently from their location in the slide, we keep the coordinates of each patch. In particular, we consider the WSI as a grid with squares of size P and locate each patches on this grid. Simply put, we compute for each patch:

$$(x_{grid}, y_{grid}) = (\left\lfloor \frac{x_{pixel}}{P} \right\rfloor, \left\lfloor \frac{y_{pixel}}{P} \right\rfloor) \tag{3}$$

with x_{pixel} and y_{pixel} the pixel coordinates of the patch in the WSI.

The patches, and their corresponding feature vectors, can now be seen as tokens in a 2D sequence, as in ViT. Because we only consider patches containing tissue and disregard the background, the grid can be sparse and of arbitrary size, depending on the tissue represented in the WSI. Patches representing background are masked out for the following steps.

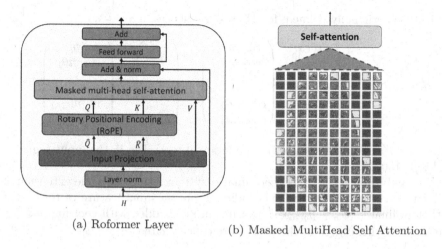

(a) Roformer Layer   (b) Masked MultiHead Self Attention

Fig. 2. Key components of the RoFormer encoder

3.2 Self-attention Encoding

The key component of the encoding module is a simple RoFormer layer (Fig. 2a). The full self-attention mechanism allows for each patch to update its feature vectors with respect to all other patches in the slide, effectively modeling the interactions between instances and tissue regions. The relative position encoding introduces spatial information and models the tissue structure. Masked self-attention is applied to discard the background patches (Fig. 2b).

Efficient Self-attention. The $O(n^2)$ memory and time complexity of self-attention, with respect to n the number of patches, has prevented its use in WSI classification as digital slides may contain tens of thousands of patches.

However, recent breakthrough implementations such as Flash attention [4] or Memory Efficient attention [14], considerably reduced the burden of those operations while still providing exact full self attention. We used memory-efficient attention for both the encoding self-attention and the global-attention pooling.

Rotary Position Encoding. As discussed in Sect. 2, relative position encoding for WSI with large numbers of tokens and irregular locations is challenging. To address this issue, we leverage RoPE to encode position of the patches. The relative nature, translation-invariance and ability to handle any number of relative positions of this method enables to encode spatial dependencies in the WSI setup.

RoPE is originally defined, for 1D text sequences, as:

$$RoPE(h, m) = \begin{pmatrix} h_1 \\ h_2 \\ \vdots \\ h_{d-1} \\ h_d \end{pmatrix} \otimes \begin{pmatrix} \cos m\theta_1 \\ \cos m\theta_1 \\ \vdots \\ \cos m\theta_{d/2} \\ \cos m\theta_{d/2} \end{pmatrix} + \begin{pmatrix} -h_2 \\ h_1 \\ \vdots \\ -h_d \\ h_{d-1} \end{pmatrix} \otimes \begin{pmatrix} \sin m\theta_1 \\ \sin m\theta_1 \\ \vdots \\ \sin m\theta_{d/2} \\ \sin m\theta_{d/2} \end{pmatrix} \quad (4)$$

with $h \in \mathbb{R}^d$ the feature vector, m the index of the token in the input sequence and $\theta_i = 10000^{-2i/d}$.

We apply RoPE to our 2D coordinates system, using the same extension as ViT for sin-cos absolute position encoding. From the input feature vector $h \in \mathbb{R}^d$ with coordinates (x_{grid}, y_{grid}), $\frac{d}{2}$ features are embedded with coordinate x_{grid}, the other $\frac{d}{2}$ with y_{grid} and the results are concatenated:

$$RoPE_{2D}(h, (x_{grid}, y_{grid})) = [RoPE(h_{[1:d/2]}, x_{grid}), RoPE(h_{[d/2:d]}, y_{grid})] \quad (5)$$

This embedding is applied to each patch in the key-query product of the self-attention layer.

4   Experiments

We evaluate the effect of our position encoding module on three standard WSI classification datasets and evaluate the influence of our design choices in an ablation study.

Implementation Details. For all three datasets, we used $256 * 256$ patches at a magnification of x20, downsampling from x40 when necessary, and ResNet50-ImageNet [7] as feature extractor. We train the MIL models with a Adam optimizer, learning rate of 0.0001, batch size of 4 and a maximum of 50 epochs with early stopping. We implemented our code using Pytorch and run it on a single 16 GB V100 GPU. We use the xFormers implementation [9] for memory-efficient attention.

Datasets. We leverage three public datasets of H&E slides that effectively cover the cases of binary, multi-class and imbalanced classifications.

First, we used 1041 lung slides from TCGA-NSCLC for a binary cancer subtype classification Lung Squamous Cell Carcinoma (LUSC) vs TCGA Lung Adenocarcinoma (LUAD) images . Classes are split between 530 LUAD and 511 LUSC examples.

We also leveraged the Camelyon16 dataset with 400 lymph nodes sections for a 3-class classification task of metastasis detection. 240 images are labeled as normal, 80 as micro and 80 as macro.

Finally, we use the 503 breast tissue slides from the public BRACS dataset, for a 3-class classification, identifying non-cancerous (221 slides) vs. pre-cancerous (89 slides) vs. cancerous (193 slides).

We performed our own 10-fold stratified split for train/test splits for each dataset.

4.1 Ablation Study for Encoder Design

We first conduct an ablation study on TCGA-NSCLC to help with design choices for the encoder module. We use ABMIL as base model and add components to find the best design possible. We replaced the original gated attention mechanism with a multihead dot-product attention, as described in Sect. 2.1. The resulting ABMIL model contains 800k parameters. Adding our roformer layer increases the number of parameters to 4.3M. For fair comparison, we train larger version of ABMIL, doubling the hidden dimension size from 512 to 1024 to reach 2.1M parameters, then adding two hidden layers to reach 4.2M parameters.

The ViT layer corresponds to a transformer layer with sin-cos absolute positional encoding, similar to HIPT although they use multiple layers while we only add 1 for this ablation. The RoFormer layer is a transfomer layer with RoPE positional encoding.

Results are reported in Table 1.

Table 1. Ablation study on TCGA-NSCLC

Method	AUROC	Average Precision
ABMIL 800k	0.86 ± 0.06	0.84 ± 0.07
CLAM (ABMIL 800k + instance loss)	0.85 ± 0.06	0.83 ± 0.08
ABMIL 2.1M	0.88 ± 0.06	0.89 ± 0.05
ABMIL 4.2M	0.88 ± 0.06	0.89 ± 0.05
Transformer layer + ABMIL	0.91 ± 0.05	0.91 ± 0.04
ViT layer + ABMIL	0.75 ± 0.11	0.77 ± 0.09
RoFormer layer + ABMIL	$\mathbf{0.92 \pm 0.05}$	$\mathbf{0.92 \pm 0.05}$
RoFormer layer + abs. position encoding + ABMIL	0.78 ± 0.10	0.77 ± 0.09

We observe that adding a transformer layer provides a significant improvement in this classification task. However, if ViT achieved great results in the HIPT working on a small number of large patches, it fails at encoding the structure of the tissue in our setup. The very high number of small patches, tens of thousands for each image, and their inconsistent localization across slides may not be compatible with absolute position encoding. On the other hand, the relative position encoding RoPE provides an additional performance increase. This increase in performance over a global-attention pooling with similar number of parameters demonstrates that the improvement is not only due to model capacity, but to the modeling itself.

4.2   Results

Using the insights obtained in the last section, we evaluate the addition of a RoFormer layer to MIL models on the selected datasets.

Baselines. We apply our encoding layer to the MIL aggregators of both ABMIL [8] and DSMIL [10] and refer to the resulting models as Ro-ABMIL and Ro-DSMIL. Note that for DSMIL, we only consider the MIL aggregator and do not use the multiscale feature aggregation which is out of the scope of this work. In these experiments, we only use 1-layer encoder and keep as perspective work to scale to bigger models. We also compare our results with TransMIL that used a similar approach, as described in Sect. 2.2, with 2.7M parameters. We computed our own TransMIL results using their public implementation.

Table 2. Classification results

Method	TCGA-NSCLC	Camelyon16	BRACS
ABMIL 4.2M	0.88 ± 0.06	0.80 ± 0.24	0.77 ± 0.09
Ro-ABMIL	$\mathbf{0.92 \pm 0.05}$	$\mathbf{0.90 \pm 0.05}$	$\mathbf{0.82 \pm 0.07}$
DSMIL	0.88 ± 0.06	0.74 ± 0.23	0.79 ± 0.09
Ro-DSMIL	0.90 ± 0.05	$\mathbf{0.90 \pm 0.05}$	$\mathbf{0.82 \pm 0.08}$
TransMIL	0.91 ± 0.05	0.84 ± 0.09	0.80 ± 0.09

Evaluation of Performances. In Table 2, we report the AUROC on all classification tasks, macro-averaged for multi-class and with standard deviation across folds. Additional metrics are reported in Supplementary material. On all three datasets, the RoFormer layer provides a performance increase in the classification over the original ABMIL and DSMIL. TransMIL performs comparably to RoABMIL and RoDSMIL on TCGA-NSCLC, but is outperformed on both Camelyon16 and BRACS.

5   Conclusion

In this work, we leverage recent advances in Transformers research to design a straightforward patch encoding module as a RoFormer layer. This component addresses two of the main flaws of current approaches by modeling the inter-patches dependencies and tissue structures. The encoder is also agnostic to the downstream MIL model and we demonstrated its benefits, for both ABMIL and DSMIL, on three of the main WSI classification public datasets. We also show that the better results are due to improved modeling capabilities and not only increasing the number of parameters.

Limitations and Future Work. A drawback of our approach is the high number of added parameters. WSI datasets are usually small, compared to what transformers may be used to in other fields, and adding too many parameters could lead to overfitting. In particular, we presented results with only one RoFormer layer and should be careful when trying larger versions. Also, the computation overhead may be an issue for more complex approaches, in multi-modal or multi-scale scenarios for instance.

Also we stuck to a naive feature extraction scheme for simplicity, using a pretrained ResNet directly on patches, and it would be interesting to see results with features from more complex approaches, as in HIPT or DSMIL.

Finally, the datasets we used, although well known by the community, are all tumor classification on H&E images. Applying these methods to other modalities, like IHC images, or tasks, like tumor infiltrated lymphocytes quantification, where tissue structures plays an even bigger role would be interesting.

References

1. Bejnordi, B.E., et al.: Diagnostic assessment of deep learning algorithms for detection of lymph node metastases in women with breast cancer. JAMA **318**(22), 2199–2210 (2017)
2. Brancati, N., et al.: BRACS: a dataset for BReAst carcinoma subtyping in H&E histology images. Database **2022**, baac093 (2022). https://www.bracs.icar.cnr.it/
3. Chen, R.J., et al.: Scaling vision transformers to gigapixel images via hierarchical self-supervised learning. In: Proceedings of the IEEE/CVF Conference on Computer Vision and Pattern Recognition, pp. 16144–16155 (2022)
4. Dao, T., Fu, D.Y., Ermon, S., Rudra, A., Ré, C.: FlashAttention: fast and memory-efficient exact attention with IO-awareness. In: Advances in Neural Information Processing Systems (2022)
5. Dosovitskiy, A., et al.: An image is worth 16 × 16 words: transformers for image recognition at scale. arXiv preprint arXiv:2010.11929 (2020)
6. Gadermayr, M., Tschuchnig, M.: Multiple instance learning for digital pathology: a review on the state-of-the-art, limitations & future potential. arXiv preprint arXiv:2206.04425 (2022)
7. He, K., Zhang, X., Ren, S., Sun, J.: Deep residual learning for image recognition. CoRR abs/1512.03385 (2015). http://arxiv.org/abs/1512.03385
8. Ilse, M., Tomczak, J., Welling, M.: Attention-based deep multiple instance learning. In: International Conference on Machine Learning, pp. 2127–2136. PMLR (2018)
9. Lefaudeux, B., et al.: xFormers: a modular and hackable transformer modelling library. https://github.com/facebookresearch/xformers (2022)
10. Li, B., Li, Y., Eliceiri, K.W.: Dual-stream multiple instance learning network for whole slide image classification with self-supervised contrastive learning. In: Proceedings of the IEEE/CVF Conference on Computer Vision and Pattern Recognition, pp. 14318–14328 (2021)
11. Lu, M.Y., Williamson, D.F., Chen, T.Y., Chen, R.J., Barbieri, M., Mahmood, F.: Data-efficient and weakly supervised computational pathology on whole-slide images. Nat. Biomed. Eng. **5**(6), 555–570 (2021)
12. Madabhushi, A.: Digital pathology image analysis: opportunities and challenges. Imaging Med. **1**(1), 7 (2009)

13. Myronenko, A., Xu, Z., Yang, D., Roth, H.R., Xu, D.: Accounting for dependencies in deep learning based multiple instance learning for whole slide imaging. In: de Bruijne, M., et al. (eds.) MICCAI 2021. LNCS, vol. 12908, pp. 329–338. Springer, Cham (2021). https://doi.org/10.1007/978-3-030-87237-3_32

14. Rabe, M.N., Staats, C.: Self-attention does not need $O(n2)$ memory. ArXiv:2112.05682 (2021)

15. Rymarczyk, D., Borowa, A., Tabor, J., Zielinski, B.: Kernel self-attention for weakly-supervised image classification using deep multiple instance learning. In: Proceedings of the IEEE/CVF Winter Conference on Applications of Computer Vision, pp. 1721–1730 (2021)

16. Shamshad, F., et al.: Transformers in medical imaging: a survey. Med. Image Anal. **88**, 102802 (2023)

17. Shao, Z., et al.: TransMIL: transformer based correlated multiple instance learning for whole slide image classification. Adv. Neural. Inf. Process. Syst. **34**, 2136–2147 (2021)

18. Shaw, P., Uszkoreit, J., Vaswani, A.: Self-attention with relative position representations. arXiv preprint arXiv:1803.02155 (2018)

19. Su, J., Lu, Y., Pan, S., Murtadha, A., Wen, B., Liu, Y.: RoFormer: Enhanced transformer with rotary position embedding. arXiv preprint arXiv:2104.09864 (2021)

20. Weinstein, J.N., et al.: The cancer genome atlas pan-cancer analysis project. Nat. Genet. **45**(10), 1113–1120 (2013)

21. Wu, K., Peng, H., Chen, M., Fu, J., Chao, H.: Rethinking and improving relative position encoding for vision transformer. CoRR abs/2107.14222 (2021). https://arxiv.org/abs/2107.14222

22. Xiong, C., Chen, H., Sung, J., King, I.: Diagnose like a pathologist: transformer-enabled hierarchical attention-guided multiple instance learning for whole slide image classification. arXiv preprint arXiv:2301.08125 (2023)

23. Xiong, Y., et al.: Nyströmformer: a nyström-based algorithm for approximating self-attention. In: Proceedings of the AAAI Conference on Artificial Intelligence, vol. 35, pp. 14138–14148 (2021)

24. Yao, J., Zhu, X., Jonnagaddala, J., Hawkins, N., Huang, J.: Whole slide images based cancer survival prediction using attention guided deep multiple instance learning networks. Med. Image Anal. **65**, 101789 (2020)

Structural Cycle GAN for Virtual Immunohistochemistry Staining of Gland Markers in the Colon

Shikha Dubey[1,2(✉)], Tushar Kataria[1,2], Beatrice Knudsen[3], and Shireen Y. Elhabian[1,2(✉)]

[1] Kahlert School of Computing, University of Utah, Salt Lake City, USA
[2] Scientific Computing and Imaging Institute, University of Utah, Salt Lake City, USA
{shikha.d,tushar.kataria,shireen}@sci.utah.edu
[3] Department of Pathology, University of Utah, Salt Lake City, USA
beatrice.knudsen@path.utah.edu

Abstract. With the advent of digital scanners and deep learning, diagnostic operations may move from a microscope to a desktop. Hematoxylin and Eosin (H&E) staining is one of the most frequently used stains for disease analysis, diagnosis, and grading, but pathologists do need different immunohistochemical (IHC) stains to analyze specific structures or cells. Obtaining all of these stains (H&E and different IHCs) on a single specimen is a tedious and time-consuming task. Consequently, virtual staining has emerged as an essential research direction. Here, we propose a novel generative model, Structural Cycle-GAN (SC-GAN), for synthesizing IHC stains from H&E images, and vice versa. Our method expressly incorporates structural information in the form of edges (in addition to color data) and employs attention modules exclusively in the decoder of the proposed generator model. This integration enhances feature localization and preserves contextual information during the generation process. In addition, a structural loss is incorporated to ensure accurate structure alignment between the generated and input markers. To demonstrate the efficacy of the proposed model, experiments are conducted with two IHC markers emphasizing distinct structures of glands in the colon: the nucleus of epithelial cells (CDX2) and the cytoplasm (CK818). Quantitative metrics such as FID and SSIM are frequently used for the analysis of generative models, but they do not correlate explicitly with higher-quality virtual staining results. Therefore, we propose two new quantitative metrics that correlate directly with the virtual staining specificity of IHC markers.

Keywords: Structural Cycle GAN · Histopathology Images · Generative model · Virtual Immunohistochemistry Staining

Supplementary Information The online version contains supplementary material available at https://doi.org/10.1007/978-3-031-45676-3_45.

1   Introduction

Histopathology image analysis has become a standard for diagnosing cancer, tracking remission, and treatment planning. Due to the large amount of data analysis required, the reduction in the number of pathologists in some domains [10,18,22] and the advent of digital scanners, automated analysis via deep learning methods is gaining importance in histopathology [12,16,24]. Pathologists frequently use immunohistochemical (IHC) stain markers in addition to Hematoxylin and Eosin (H&E) to highlight certain protein structures [14] (not visible to the naked eye) that can help in identifying different objects essential for the analysis of disease progression, grading, and treatment planning. While some markers are accessible and time-efficient, others are not. Using conventional methods to stain a single slide with multiple stains in a pathology laboratory is time-consuming, laborious, and requires the use of distinct slides for analysis, thereby increasing the difficulty of the task. To address these issues, virtual staining offers a solution by generating all required stains on a single slide, significantly reducing labor-intensive and time-consuming aspects of conventional techniques.

Conditional-GAN (Pix2Pix) [9] and Cycle-GAN [26] are two pioneering generative adversarial networks (GAN) frameworks that have been frequently used in virtual staining [13,20,21,25] with many other variants proposed [11,23,27]. The performance of Pix2Pix based-method relies heavily on the data sampling procedure because this model is prone to hallucinations when patches of IHC are not precisely registered to their corresponding H&E patches [20,21]. Registration of whole slide images (WSI) is a time-consuming and laborious process increasing the time to deployment and analysis. Although Cycle-GAN doesn't require registered images it doesn't explicitly focus on morphological relations between objects (cells, glands) of the input (histopathology) images. Here we propose a Structural Cycle-GAN (SC-GAN) model that alleviates the requirement of registering patches and focuses on the morphological relations. Using structural information [25] has previously shown promising results, but the proposed approach is novel in that it generates IHC stains based on structural information derived from the H&E stain without relying on structural similarity loss. Incorporating structural information [3] provides explicit geometrical and morphological information to the model resulting in higher-quality virtually generated stains.

Different IHC [17] stains highlight the nucleus, cell surface, or cytoplasm depending on the activating agent. To emphasize specific classes of cells, virtual staining models need to attend to a larger region of interest (cell environment). SC-GAN uses the attention module to factor in the dependence of pixel staining on its environment. Convolutional models have fixed ROIs around pixels of interest but the attention module can account for long-range dependencies [4,19], which are beneficial in histopathology applications.

Visual inspection by pathologists is the optimal method for evaluating the efficacy of any proposed virtual staining method, but it becomes impractical due to cost and pathologist disagreement [1,5]. Quantitative evaluation metrics FID,

SSIM, and PSNR [2,15,27] are heavily favored for the analysis of generative models. But in this study, we empirically observed that improvement in these metrics for virtual staining doesn't always correlate to better quality output. Therefore, two downstream task metrics are proposed, 1) Ratio of cell count in Stained IHC and 2) Dice Score of positive cells in Stained IHC, which correlates directly with virtual staining performance enhancement. The findings show that a lower FID score does not always imply a better quality stain, however the suggested downstream measures have significant correlations with stain quality.

The key contributions of the present work can be summarized as follows:

1. Propose Structural Cycle-GAN (SC-GAN) model that employs explicit structural data (edges) as input to generate IHC stains from H&E and vice versa.
2. SC-GAN leverages the attention module to account for cell environmental dependencies and enforces structural consistency utilizing structural loss, thereby eliminating the need for registered patches.
3. Propose new evaluation metrics with a stronger correlation to enhanced virtual staining effectiveness and evaluate virtual staining results for two IHC markers of glands in the colon, CDX2 (specific to epithelial cell nuclei) and CK818 (cytoplasm), highlighting distinct cells.

2   Methodology

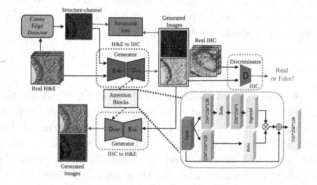

Fig. 1. Block Diagram of the proposed Structural Cycle-GAN (SC-GAN): The Generator follows a ResNET-based architecture similar to [25], with the addition of attention blocks in the decoder part. The proposed method does not necessitate the registration of real H&E and IHC. The generation of H&E stains from IHC stains follows the same architecture but incorporates the H&E Discriminator, maintaining consistency in the overall model design and approach.

2.1   Structural Cycle-GAN (SC-GAN)

The proposed model, SC-GAN (Fig. 1) is based on the Cycle-GAN framework [26], a prominent computer vision model for image-to-image translation tasks.

The Cycle-GAN framework comprises two generators and two discriminators, incorporating cycle consistency loss to regularize GAN models in both domain *A to B* and *B to A* translations. Similarly, SC-GAN generators facilitate image translation between different domains (H&E to IHC and vice versa), while the discriminators discern between real and generated images. The proposed approach introduces specific modifications (mentioned in the following sections) to the generator architecture while adhering to the fundamental principles of the Cycle-GAN framework.

2.2   Structural Information

Histopathology images frequently contain significant characteristics, such as cell boundaries, tissue structures, or distinct patterns. Histopathology imaging necessitates structure uniformity, as the structure of the stains must be maintained between stains. SC-GAN uses a canny-edge detector [3] to extract major structures/edges from the input stain and concatenate these with the brightfield color channel of the input patch before passing them through the Generator. By explicitly providing structure as input, SC-GAN conditions the generation process to take into account the cell's morphological as well as its texture properties. In addition to providing structural organization for cells, edges also serve as anchors to improve image generation. This contributes to the production of images with well-defined boundaries and structures, resulting in enhanced precision and aesthetically pleasing outcomes. The generator synthesizes the target color-stained patches and their corresponding structures, which are then utilized as inputs for the second generator, as illustrated in Fig. 1.

2.3   Structural Loss

In addition to the losses used in Cycle-GAN [26], SC-GAN proposes a novel structural loss (SL) to improve the generation of IHC stains from H&E stained images and vice versa. The SL is intended to ensure that structural information, such as margins and boundaries, is preserved in the generated stains. To calculate the SL, SC-GAN uses the Mean Squared Error (MSE) loss between the generated and the corresponding ground truth structural maps, as shown in the equation below:

$$\mathcal{L}_{SL} = MSE(I[S_{maps}] - G[S_{maps}]),\qquad(1)$$

where $I[S_{maps}]$ and $G[S_{maps}]$ stand for the structural maps (canny edges) of the input and generated stained images, respectively, and $\mathcal{L}_{SL}$ stands for the structural loss. In order to train the proposed model, the SL is combined with the other losses used in CycleGAN, such as adversarial loss (L_{adv}), cycle-consistency forward (L_{cycle_f}) and backward losses (L_{cycle_b}), and identity loss ($\mathcal{L}_I$). The final loss is:

$$\mathcal{L} = \lambda_1 \mathcal{L}_{adv} + \lambda_2(\mathcal{L}_{cycle\_f} + \mathcal{L}_{cycle\_b}) + \lambda_3 \mathcal{L}_I + \lambda_4 \mathcal{L}_{SL},\qquad(2)$$

where $\lambda_1, \lambda_2, \lambda_3, \lambda_4$ are hyperparameters.

2.4   Attention-Enabled Decoder

IHC staining not only depends on the cell type but also on the cell environment. To emphasize the role of the cell environment SC-GAN uses an attention module in the proposed generator. The attention block aims to enhance the generator's performance by selectively focusing on relevant cell environment details during the image translation process. The addition of attention in the decoder ensures that the generator concentrates on refining and enhancing the generated image, as opposed to modifying the encoded features during the translation process. The added attention module is depicted in Fig. 1.

2.5   Additional Loss for Registered Data

This study also evaluates SC-GAN on a registered dataset, where H&E patches are paired with their corresponding registered IHC patches while training. SC-GAN is trained and evaluated with combination of registered and non-registered patches in varying amounts. To leverage the registered data more effectively, we introduce an additional supervision specifically for the registered data setting, utilizing the Mean Absolute Error (MAE) loss as given below:

$$\mathcal{L}_{registered} = MAE(I_f - G_f), \tag{3}$$

where I_f and G_f stand for the input and generated stains, and $\mathcal{L}_{registered}$ stands for the loss associated with the registered settings. The MAE loss measures the average absolute difference between the pixel intensities of generated markers and their corresponding ground truth markers. The experimental results for SC-GAN under registered settings are presented in Supplementary Table 3.

3   Results and Discussion

3.1   Dataset Details

The dataset included H&E WSIs from surveillance colonoscopies of 5 patients with active ulcerative colitis. The slides were stained with H&E using the automated clinical staining process and scanned on an Aperio AT2 slide scanner with a pixel resolution of 0.25 μm at 40x. After scanning, Leica Bond Rx autostainer was used for restaining by immunohistochemistry with antibodies reactive with CDX2 (caudal type homeobox-2) or CK8/18(cytokeratin-8/18).

We sampled patches from four patients, trained generation models (7,352 for CDX2 and 7,224 for CK818), and used data from the fifth patient for testing (1,628 for CDX2 and 1,749 for CK818). Each patient's data consists of multiple tissue samples, ranging from 16 to 24. The extracted RGB patches are of shape 512×512. The testing dataset contains registered H&E and IHC patches which enables the development of two new metrics with a stronger correlation to the efficacy of the generative model in virtual staining. Further details on these metrics are provided in the Sect. 3.2.

3.2 Model Architecture and Evaluation

Upsampling Method: In contrast to the Conv2DTranspose method traditionally used for upsampling in generators [26], SC-GAN utilizes the UpSampling2D layer. This decision is based on empirical observations that demonstrate improved results with UpSampling2D in terms of preserving fine details and reducing artifacts. The discriminator is the same as proposed in the CycleGAN model, i.e., a patch discriminator. Hyperparameters are set to $\lambda_1 = 1.0, \lambda_2 = 10.0, \lambda_3 = 5.0$, (the same as in the CycleGAN) and $\lambda_4 = 5.0$ (via experimentation).[1]

Proposed Evaluations: SC-GAN reports the conventional evaluations such as FID and SSIM similar to other virtual staining works [15,27]. However, our findings reveal that these metrics do not exhibit a significant correlation with improved virtual staining models (refer Fig. 2 and Table 1). Consequently, SC-GAN proposes the adoption of two new metrics that directly align with the performance of the generative model.

– **Ratio of cell count in Stained IHC**: Compares the number of cells highlighted in the virtually stained image with the real stained image utilizing DeepLIIF model [6].

$$\mathcal{R}_{count} = \left(\frac{|Gen_{cell}| - |GT_{cell}|}{|GT_{cell}|} \right) * 100, \tag{4}$$

where $\mathcal{R}_{count}$ stands for ratio of cell count in Stained IHC with ground truth's cell counts. $|Gen_{cell}|$ and $|GT_{cell}|$ represent the number of generated and ground truth cells, respectively, calculated using the DeepLIIF model [6]. A few examples of such segmented images from both stains, CDX2 and CK818 are shown in the Supplementary Fig. 4. Additionally, DeepLIIF successfully segments the H&E images for the IHC to H&E evaluation (refer Supplementary Fig. 5 and Table 3).

– **Dice Score of positive cells in Stained IHC**: Calculates the dice score and intersection of union (IOU) of the stained cells (positive cells) in the virtually generated image with respect to the real image. Both the metrics are evaluated by pixel-wise thresholding of brown color in generated and real IHC. Examples are depicted in Supplementary Fig. 4.

3.3 Results

Table 1 presents the results of different models, including the proposed SC-GAN model, along with relevant comparisons, evaluated using various metrics. The conventional metrics, FID and SSIM, are included alongside the proposed metrics. The DeepLIIF model was utilized to segment and count the total number of cells, positive (IHC) cells, and negative (background) cells. The evaluation of cell counts was performed on a subset of the test dataset consisting of 250 registered patches, while the other metrics were calculated on full test dataset.

---

[1] We will release all the code related to the paper at a future date for public usage.

Table 1. Quantitative results for H&E to IHC Translation. EDAtt: Attention in both Encoder and Decode, DAtt: Attention in the decoder, St: Structure, SL: Structural Loss. Cell counting metrics (Total, Positive, and Negative) are calculated using Eq. 4, with values closer to zero indicating a more accurate generator. Qualitative results of pix2pix are not viable for cell counting analysis (refer to Fig. 2). IOU and DICE metrics are reported for IHC-positive cells.

| Markers | Models | Conventional | | Proposed Metrics | | | | |
		FID	SSIM	IOU	DICE	Total Cells	Positive Cells	Negative Cells
CDX2	Pix2Pix [9]	**6.58**	18.54	-	-	-	-	-
	Base Cycle-GAN [26]	20.93	38.23	51.79	44.3 2	4.94	46.32	−13.45
	Cycle-GAN w/ EDAtt	13.34	40.16	44.93	46.25	−9.57	−20.78	−5.63
	Cycle-GAN w/ DAtt	12.74	**40.64**	45.91	47.16	−12.09	−11.65	−12.28
	Cycle-GAN w/ St	15.00	38.03	62.41	50.83	−2.13	**2.90**	−4.36
	SC-GAN w/o SL	20.75	35.53	56.86	50.29	−6.22	−21.73	7.31
	SC-GAN(Proposed)	14.05	38.91	**63.35**	**51.73**	**−0.08**	−16.23	**1.63**
CK818	Pix2Pix [9]	**3.38**	28.93	-	-	-	-	-
	Base Cycle-GAN [26]	18.24	27.03	23.36	25.34	5.43	8.09	4.15
	Cycle-GAN w/ EDAtt	15.02	34.16	34.91	33.19	5.83	10.87	3.40
	Cycle-GAN w/ DAtt	16.21	**34.24**	**36.23**	**33.61**	3.19	4.35	2.63
	Cycle-GAN w/ St	20.69	32.35	35.86	32.75	−1.79	**−2.04**	−1.67
	SC-GAN w/o SL	22.26	32.22	14.25	17.65	9.04	−34.77	17.08
	SC-GAN(Proposed)	15.32	33.86	26.16	27.66	**0.22**	3.48	−1.35

From Table 1, we observe that a lower FID score is not always associated with a higher proposed metric that measures the number of valid IHC cells highlighted by the generative model. Conventional metric values indicate that the Pix2Pix model performs the best, despite producing qualitatively inferior results compared to other models (refer Fig. 2). Consequently, additional proposed metrics are required for a thorough assessment. Table 1 demonstrates that the cell IOU and DICE scores for the IHC images generated by the SC-GAN are higher. This indicates that the proposed model exhibits greater specificity. The proposed variations incorporating decoder attention, structural information, and SL consistently exhibit the lowest deviation from the true cell counts (total number of cells, positive and negative cells). These results demonstrate the superiority of the proposed model, SC-GAN, over the base Cycle-GAN and illustrate the significant impact of each module on the performance of virtual staining. Furthermore, the qualitative results depicted in Fig. 2 align with these observations, as the proposed models consistently yield higher-quality results characterized by reduced false positives and increased true positives.

SC-GAN has better performance while generating CDX2 stains compared to CK818 (refer Table 1). As CDX2 is a nuclear marker, more precise structural information is required to generate an accurate stain. On the other hand, CK818 is a cytoplasmic (surface) stain, where the cell environment plays a more crucial role. Hence, attention-based models exhibit higher accuracy for CK818 stains, as they focus on capturing contextual information. This suggests that the importance of input features, whether structure or attention map, varies according to the type of stain being generated.

The Supplementary Table 2 and Fig. 5 present the results for the IHC to H&E translation. These results demonstrate that SC-GAN performs well not only for the H&E to IHC translation but also for the reverse process. This observation is reinforced by the cell counting metric, which demonstrates the superior performance of SC-GAN in both translation directions.

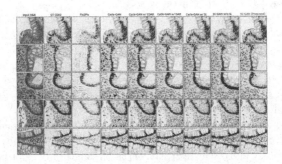

Fig. 2. Qualitative results for generated CDX2 marker. GT: Ground Truth, Gen: Generated, EDAtt: Attention in both Encoder and Decoder, DAtt: Attention in decoder, St: Structure, SL: Structural Loss. The proposed model, SC-GAN, performs better than the base Cycle-GAN by effectively suppressing false positives and accurately coloring the cells while preserving their structure. The final row showcases the effectiveness of virtual staining, where the given H&E staining successfully reproduces the IHC stain, even though some information was lost during the original IHC staining process. Notably Cycle-GAN w/ DAtt performs better than Cycle-GAN w/ EDAtt model.

Registered Data Results: In Supplementary Table 3, the influence of training the SC-GAN model with registered data is examined. The table reveals that incorporating registered data has minimal impact on the model's performance, indicating that SC-GAN can operate effectively without the need for registered data. Furthermore, it demonstrates that the translation capability of the SC-GAN model works in the feature space rather than relying solely on pixel-to-pixel translation. This characteristic enhances the robustness of the proposed SC-GAN model, making it suitable for virtual staining applications when there is limited availability of registered data.

4   Conclusion and Future Work

We proposed a novel methodology, SC-GAN that combines structural information, attention, and a structural loss to enhance the generation of IHC markers from H&E-stained images and vice versa. The proposed quantitative metrics focused on deviation in cell counts and dice score with respect to the real IHC images, shows direct correlation with virtual staining efficacy. Results highlight the need for different information in generative models based on stain properties (nuclear vs cytoplasmic, structure vs attention). SC-GAN outperforms in non-registered datasets, eliminating the need for laborious WSI image alignment. However, quantitative cell counting metrics rely on the real stain availability presents a limitation requiring further investigation. Additionally, future work can explore the potential benefits of integrating multi-scale structural information derived from wavelet [7] or MIND [8] features. The applicability of the proposed model to other IHC markers and its potential for generating marker-specific protocols are additional avenues for investigation. This could involve adapting a unified model architecture to effectively account for the unique characteristics associated with different stains. These can further improve the proposed methodology and increase its applicability for virtual staining in histopathology.

Acknowledgements. We thank the Department of Pathology, Scientific Computing and Imaging Institute, and the Kahlert School of Computing at the University of Utah for their support of this project.

References

1. Arvaniti, E., Fricker, K.S., Moret, M., et al.: Automated Gleason grading of prostate cancer tissue microarrays via deep learning. Sci. Rep. **8**(1), 1–11 (2018)
2. Borji, A.: Pros and cons of GAN evaluation measures: new developments. Comput. Vis. Image Underst. **215**, 103329 (2022)
3. Canny, J.: A computational approach to edge detection. IEEE Trans. Pattern Anal. Mach. Intell. **6**, 679–698 (1986)
4. Chen, Y., Dai, X., Liu, M., Chen, D., Yuan, L., Liu, Z.: Dynamic convolution: attention over convolution kernels. In: Proceedings of the IEEE/CVF Conference on Computer Vision and Pattern Recognition, pp. 11030–11039 (2020)
5. Eaden, J., Abrams, K., McKay, H., Denley, H., Mayberry, J.: Inter-observer variation between general and specialist gastrointestinal pathologists when grading dysplasia in ulcerative colitis. J. Pathol. J. Pathol. Soc. Great Br. Ireland **194**(2), 152–157 (2001)
6. Ghahremani, P., Marino, J., Dodds, R., Nadeem, S.: DeepLIIF: an online platform for quantification of clinical pathology slides. In: Proceedings of the IEEE/CVF Conference on Computer Vision and Pattern Recognition, pp. 21399–21405 (2022)
7. Graps, A.: An introduction to wavelets. IEEE Comput. Sci. Eng. **2**(2), 50–61 (1995)
8. Heinrich, M.P., Jenkinson, M., Papież, B.W., Brady, S.M., Schnabel, J.A.: Towards realtime multimodal fusion for image-guided interventions using self-similarities. In: Mori, K., Sakuma, I., Sato, Y., Barillot, C., Navab, N. (eds.) MICCAI 2013. LNCS, vol. 8149, pp. 187–194. Springer, Heidelberg (2013). https://doi.org/10.1007/978-3-642-40811-3_24
9. Isola, P., Zhu, J.Y., Zhou, T., Efros, A.A.: Image-to-image translation with conditional adversarial networks. In: Proceedings of the IEEE Conference on Computer Vision and Pattern Recognition, pp. 1125–1134 (2017)

10. Jajosky, R.P., Jajosky, A.N., Kleven, D.T., Singh, G.: Fewer seniors from united states allopathic medical schools are filling pathology residency positions in the main residency match, 2008–2017. Hum. Pathol. **73**, 26–32 (2018)
11. Kang, H., et al.: StainNet: a fast and robust stain normalization network. Front. Med. **8**, 746307 (2021)
12. Kataria, T., et al.: Automating ground truth annotations for gland segmentation through immunohistochemistry (2023)
13. Khan, U., Koivukoski, S., Valkonen, M., Latonen, L., Ruusuvuori, P.: The effect of neural network architecture on virtual h&e staining: systematic assessment of histological feasibility. Patterns **4**(5) (2023)
14. Komura, D., Onoyama, T., Shinbo, K., et al.: Restaining-based annotation for cancer histology segmentation to overcome annotation-related limitations among pathologists. Patterns **4**(2) (2023)
15. Liu, S., Zhu, C., Xu, F., Jia, X., Shi, Z., Jin, M.: BCI: breast cancer immuno-histochemical image generation through pyramid pix2pix. In: Proceedings of the IEEE/CVF Conference on Computer Vision and Pattern Recognition, pp. 1815–1824 (2022)
16. Lu, M.Y., et al.: Visual language pretrained multiple instance zero-shot transfer for histopathology images. In: Proceedings of the IEEE/CVF Conference on Computer Vision and Pattern Recognition, pp. 19764–19775 (2023)
17. Magaki, S., Hojat, S.A., Wei, B., So, A., Yong, W.H.: An introduction to the performance of immunohistochemistry. Biobanking Methods Protoc., 289–298 (2019)
18. Metter, D.M., Colgan, T.J., Leung, S.T., Timmons, C.F., Park, J.Y.: Trends in the us and Canadian pathologist workforces from 2007 to 2017. JAMA Netw. Open **2**(5), e194337–e194337 (2019)
19. Raghu, M., Unterthiner, T., Kornblith, S., Zhang, C., Dosovitskiy, A.: Do vision transformers see like convolutional neural networks? Adv. Neural. Inf. Process. Syst. **34**, 12116–12128 (2021)
20. Rivenson, Y., Liu, T., Wei, Z., Zhang, Y., de Haan, K., Ozcan, A.: PhaseStain: the digital staining of label-free quantitative phase microscopy images using deep learning. Light Sci. Appl. **8**(1), 23 (2019)
21. Rivenson, Y., et al.: Virtual histological staining of unlabelled tissue-autofluorescence images via deep learning. Nat. Biomed. Eng. **3**(6), 466–477 (2019)
22. Robboy, S.J., et al.: Reevaluation of the us pathologist workforce size. JAMA Netw. Open **3**(7), e2010648–e2010648 (2020)
23. Shaban, M.T., Baur, C., Navab, N., Albarqouni, S.: StainGAN: stain style transfer for digital histological images. In: 2019 IEEE 16th International Symposium on Biomedical Imaging (ISBI 2019), pp. 953–956. IEEE (2019)
24. Wu, Y., et al.: Recent advances of deep learning for computational histopathology: principles and applications. Cancers **14**(5), 1199 (2022)
25. Xu, Z., Huang, X., Moro, C.F., Bozóky, B., Zhang, Q.: Gan-based virtual re-staining: a promising solution for whole slide image analysis. arXiv preprint arXiv:1901.04059 (2019)
26. Zhu, J.Y., Park, T., Isola, P., Efros, A.A.: Unpaired image-to-image translation using cycle-consistent adversarial networks. In: Proceedings of the IEEE International Conference on Computer Vision, pp. 2223–2232 (2017)
27. Zingman, I., Frayle, S., Tankoyeu, I., Sukhanov, S., Heinemann, F.: A comparative evaluation of image-to-image translation methods for stain transfer in histopathology. arXiv preprint arXiv:2303.17009 (2023)

NCIS: Deep Color Gradient Maps Regression and Three-Class Pixel Classification for Enhanced Neuronal Cell Instance Segmentation in Nissl-Stained Histological Images

Valentina Vadori[1]([✉]), Antonella Peruffo[2], Jean-Marie Graïc[2], Livio Finos[3], Livio Corain[4], and Enrico Grisan[1]

[1] Department of Computer Science and Informatics, London South Bank University, London, UK
{vvadori,egrisan}@lsbu.ac.uk

[2] Department of Comparative Biomedicine and Food Science, University of Padova, Padua, IT, Italy
{antonella.peruffo,jeanmarie.graic}@unipd.it

[3] Department of Developmental Psychology and Socialisation, University of Padova, Padua, IT, Italy
livio.finos@unipd.it

[4] Department of Management and Engineering, University of Padova, Padua, IT, Italy
livio.corain@unipd.it

Abstract. Deep learning has proven to be more effective than other methods in medical image analysis, including the seemingly simple but challenging task of segmenting individual cells, an essential step for many biological studies. Comparative neuroanatomy studies are an example where the instance segmentation of neuronal cells is crucial for cytoarchitecture characterization. This paper presents an end-to-end framework to automatically segment single neuronal cells in Nissl-stained histological images of the brain, thus aiming to enable solid morphological and structural analyses for the investigation of changes in the brain cytoarchitecture. A U-Net-like architecture with an EfficientNet as the encoder and two decoding branches is exploited to regress four color gradient maps and classify pixels into contours between touching cells, cell bodies, or background. The decoding branches are connected through attention gates to share relevant features, and their outputs are combined to return the instance segmentation of the cells. The method was tested on images of the cerebral cortex and cerebellum, outperforming other recent deep-learning-based approaches for the instance segmentation of cells.

Keywords: Cell Segmentation · Histological Images · Neuroanatomy · Brain · Nissl Staining · Deep-Learning · U-Net · EfficientNet · Attention

X. Cao et al. (Eds.): MLMI 2023, LNCS 14349, pp. 457–466, 2024.
https://doi.org/10.1007/978-3-031-45676-3_46

1   Introduction

Advancements in microscopy have made it possible to capture *Whole Slide Images* (WSIs) and obtain cellular-level details, revealing the intricate nature of the brain cytoarchitecture. This progress has opened up new avenues for conducting quantitative analysis of cell populations, their distribution, and morphology, which can help us answer a range of biological questions. Comparative neuroanatomy studies examine differences in brain anatomy between groups distinguished by factors like sex, age, pathology, or species, investigating the connections between the brain's structure and function [1,4,8]. A standard analysis pipeline involves the use of Nissl stain to label neuronal cells in tissue sections (histological slices) of brain specimens [6]. These sections are then fixed and digitized as WSIs for examination. WSIs are often characterized by their sheer size and complexity, making computerized methods necessary for their efficient and reproducible processing. Automatic cell instance segmentation plays a crucial role, as it allows to extract features at the single cell level.

In the field of digital pathology, numerous methods have been proposed to segment cells and nuclei and aid the detection and diagnosis of diseases. These methods mainly rely on a set of algorithms, including intensity thresholding, morphology operations, watershed transform, deformable models, clustering, graph-based approaches [16]. However, cell instance segmentation is very challenging due to the varying size, density, intensity and texture of cells in different anatomical regions, with additional artifacts that can easily influence the results. Recently, deep learning has shown remarkable progress in medical image analysis, and neural networks have been successfully applied for cell segmentation, achieving higher quality than traditional algorithms. In the last years, a considerable number of approaches have adopted a semantic segmentation formulation that employs a U-net-like convolutional neural network architecture [12]. These methods incorporate customized post-processing techniques, such as marker-controlled watershed algorithms and morphological operations, to separate cells instances. Some integrate the formulation with a regression task. Huaqian et. al [15] propose a framework with an EfficientNet as the U-Net encoder for ternary classification (contours between touching cells, cell bodies, and *background*, BG). Ultimate erosion and dynamic dilation reconstruction are used to determine the markers for watershed. StarDist [13] regresses a star-convex polygon for every pixel. CIA-Net [18] exploits two decoders, where each decoder segments either the nuclei or the contours. Hover-Net [7] uses a Preact-ResNet50 based encoder and three decoders for *foreground* (FG)/BG segmentation, cell type segmentation, and regression of horizontal and vertical distances of pixels from the cell centroid. Mesmer [9] considers classification into whole contours, cell interiors, or BG, and regression of the distance from the cell centroid.

Since in histological images the boundaries between cells that are in contact are often incomplete or ambiguous and they can appear between cells with differing characteristics and orientations, we recognized the pivotal role of correctly predicting these boundaries for accurate cell separation. Therefore, we have developed an approach that focuses on enhancing the prediction of contours. Specifically, we propose NCIS as an end-to-end framework to automatically segment individual neuronal cells in Nissl-stained histological images of the brain. NCIS employs an U-Net-like architecture, which synergistically combines solutions from [7,15,18] to classify pixels as contours between touching cells, cell body, or BG, and to regress four color gradient maps that represent distances of pixels from the cell centroid and that are post-processed to get a binary mask of contours. Since cells are often slanted and there are configurations where smaller cells burrow into the concavities of larger ones, we hypothesized that the prediction of diagonal gradients together with horizontal and vertical ones could help to strengthen the approach. NCIS was created to examine the cytoarchitecture of brain regions in diverse animals, including cetaceans, primates, and ungulates. The primary objective is to conduct comparative neuroanatomy studies, with a particular emphasis on diseases that impair brain structure and functionality, such as neurodegeneration and neuroinflammation. We tested NCIS on images of the auditory cortex and cerebellum, outperforming other recent deep-learning-based approaches for the instance segmentation of cells.

2   Dataset

The data of this study are a set of 53 2048x2048 histological images extracted from Nissl-stained 40x magnification WSIs of the auditory cortex of Tursiops truncatus, also known as the bottlenose dolphin. Brain tissues were sampled from 20 specimens of different subjects (new-born, adult, old) stored in the Mediterranean Marine Mammals Tissue Bank (http://www.marinemammals.eu) at the University of Padova, a CITES recognized (IT020) research center and tissue bank. These specimens originated from stranded cetaceans with a decomposition and conservation code (DCC) of 1 and 2, which align with the guidelines for post-mortem investigation of cetaceans [10]. The images were divided into 3 subsets: 42 images for training, 5 for validation and 6 for testing.

To assess the generalizability of the proposed method on cerebral areas not seen during training, we considered an independent 2048x2048 Nissl-stained histological image of the cerebellum of the bovine [4]. In this case, the animal was treated according to the present European Community Council directive concerning animal welfare during the commercial slaughtering process and was constantly monitored under mandatory official veterinary medical care. Compared to the training set, the additional image is characterized by the presence of a higher density granular layer of touching or overlapping cells with small dimensions and predominantly circular shape and a thin Purkinje cell layer with relatively larger and sparse pear-shaped cells.

All images were annotated using QuPath [2] software, resulting in 24,044 annotated cells of the auditory cortex, and 3,706 of the cerebellum.

3   Method

The proposed NCIS framework for automatic instance segmentation of neuronal cells can be observed in Fig. 1. An image of arbitrary size is divided into patches

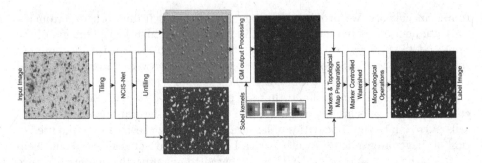

Fig. 1. An overview of the overall NCIS approach for neuronal cell instance segmentation in Nissl-stained whole slice images.

of size 256x256 via a tiling step with 50% overlap between patches. Patches are individually fed to the deep learning model, NCIS-Net, whose architecture is shown in Fig. 2. The model outputs for each patch are combined via a untiling step to get 7 outputs with size equal to that of the original image. Finally, a series of post-processing steps is applied to generate the instance segmentation.

3.1 NCIS-Net

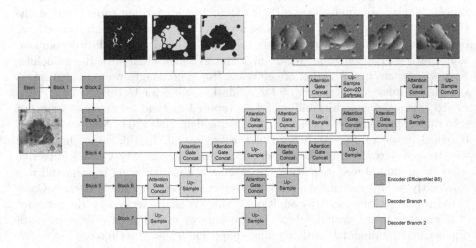

Fig. 2. An overview of the architecture of the proposed NCIS-Net.

Architecture. The proposed NCIS-Net has a U-Net-like structure, as shown in Fig. 2. The down-sampling, feature extracting branch of the encoder is based upon a state-of-the-art network, *EfficientNet-B5*, whose building blocks are arrays of mobile inverted bottleneck *MBConv* optimized with neural architecture search [14]. NCIS-Net is characterized by two decoding branches for a classification and regression task. The first branch performs three-class pixel classification. Its output is a three-channel image with probabilities for boundaries between touching cells, cell bodies, and BG, respectively. The union of the first two classes constitutes the FG. The second branch regresses four color gradient maps. Its output is a four-channel image, where each channel represents a color gradient map with vertical, horizontal, and diagonal directions. As illustrated in Fig. 2, in a gradient map, each individual cell is represented by a color gradient where pixel values increment from the minimum of -1 (blue in the figure) to the maximum of 1 (yellow in the figure) according to the prescribed direction (vertical from left to right, horizontal from top to bottom, diagonal from bottom left to top right and diagonal from top left to bottom right). 0 corresponds to the cell centroid. Four skip connections leveraging attention gates [11] connect the encoder branch to each decoder branch, promoting a focused exploitation of the features encoded at different resolutions. Skip connections with attention gates are also introduced between the two decoder branches to favor feature sharing.

Loss Function. NCIS-Net is trained through the backpropagation algorithm applied to a loss function $\mathcal{L} = \mathcal{L}_{PC} + \mathcal{L}_{GR}$ that jointly optimizes the encoder, pixel classification (PC) and gradient regression (GR) branches, with

$$\mathcal{L}_{PC} = \lambda_1 \mathcal{C} + \lambda_2 \mathcal{D}_1 + \lambda_3 \mathcal{D}_2, \qquad \mathcal{L}_{GR} = \lambda_4 \mathcal{M}_1 + \lambda_5 \mathcal{M}_2 \qquad (1)$$

where $\lambda_1 = \lambda_3 = \lambda_4 = \lambda_5 = 2, \lambda_2 = 1$ are weighting factors set via hyperparameter validation. $\mathcal{C}$ is the categorical cross-entropy between the ground truth (GT) and $\hat{Y}_{PC}$, the output of the PC branch. $\mathcal{D}_k$ is the Dice loss for class k:

$$\mathcal{D}_k = 1 - \frac{2\sum_{i=1}^{N}(Y_{PC,i,k}\hat{Y}_{PC,i,k}) + \epsilon}{\sum_{i=1}^{N} Y_{PC,i,k} + \sum_{i=1}^{N} \hat{Y}_{PC,i,k} + \epsilon} \qquad (2)$$

where N is the total number of pixels in the input image, ϵ is a smoothness constant. Specifically, $\mathcal{D}_1$ and $\mathcal{D}_2$ in Eq. 1 are the Dice losses for contours and cell bodies, respectively. $\mathcal{M}$ indicates a mean squared error loss. $\mathcal{M}_1$ is the mean squared error loss between the GT (Y_{GR}) and the predicted output $\hat{Y}_{GR}$ of the GR branch, while $\mathcal{M}_2$ is defined as:

$$\mathcal{M}_2 = \frac{1}{N \cdot J} \sum_{i=1}^{N} \sum_{j=1}^{J} (\nabla_j Y_{GR,i,j} - \nabla_j \hat{Y}_{GR,i,j})^2 \qquad (3)$$

where J is the number of gradient maps (4 in our case) and ∇ is the derivative operator. Note that for each gradient map, finite derivatives are taken by convolving the map with $5x5$ Sobel kernels with orientation and direction matching that of the corresponding gradient map, as illustrated in Fig. 1.

Training. Data batches are created from the training set of 2048x2048 images. Images are picked randomly with replacement and a series of random augmentations including rotations, flipping, deformations, intensity variations, and blurring, are applied to each selected image and corresponding three-class semantic and cell instance segmentation GT masks, when necessary. GT gradient maps are computed at this stage based on the transformed instance segmentation GT masks. Random cropping is then applied to the transformed images to get 256x256 images to be fed to the network. By utilizing this approach, it is highly likely that the batches used during each epoch will not only have varying transformations but also distinct cropped original images, ultimately aiding in the prevention of overfitting. All models are trained with the TensorFlow 2 framework in two phases. In the first phase the encoder, pre-trained on the *Image-Net* dataset [5], is frozen, while in the second fine tuning phase the encoder is unfrozen, with the exception of batch normalization layers. Each phase continues for a maximum of 50 epochs with early stopping (patience of 8 epochs, 400 batches per epoch, batch size of 16). The validation set for early stopping is created from the validation set of 2048x2048 images by cropping them into 256x256 with no overlapping and no augmentations. We use the AMSGrad optimizer with clipnorm of 0.001 and a learning rate of 10^{-3} in the first phase and 10^{-5} in the second phase. The deep learning model architecture, training and inference code are available at https://github.com/Vadori/NCIS.

3.2   Untiling and Post-processing

Patches from the same image are blended together via interpolation with a second order spline window function that weights pixels when merging patches [3].

Within each of the color gradient map, pixels between different cells should have a meaningful difference. Therefore, Sobel kernels are utilized to get the following contour map:

$$C_{PC,i} = max(\nabla_1 \hat{Y}_{GR,i,1}, \nabla_2 \hat{Y}_{GR,i,2}, \nabla_3 \hat{Y}_{GR,i,3}, \nabla_4 \hat{Y}_{GR,i,4}) \qquad (4)$$

where $\hat{Y}_{GR,i,j}, j = 1, ..., 4$, is normalized between 0 and 1. $C_{PC,i}$ is thresholded on a per-image basis via the triangle method [17], so that ones in the thresholded binary version $C_{PC,i,th}$ correspond to contours. A binary mask is defined with pixels set to 1 if they are most likely to belong to cell bodies based on the outputs of both decoders ($\hat{Y}_{PC,i,2} > \hat{Y}_{PC,i,1}$ and $\hat{Y}_{PC,i,2} > \hat{Y}_{PC,i,3}$ and $C_{PC,i,th} = 0$). Connected components smaller than 80 pixels are removed. The mask is eroded with a disk-shaped structuring element of radius 4 to force the separation of touching cells that are only partially separated. The resulting connected components bigger than 3 pixels are used as markers for the marker-controlled watershed algorithm applied to the topological map given by the complement of $\hat{Y}_{PC,i,2}$. A foreground binary mask is obtained as $\hat{Y}_{PC,i,1} + \hat{Y}_{PC,i,2} > \hat{Y}_{PC,i,3}$. After holes filling and morphological opening to smoothen the cell boundaries, it is used to constrain the watershed labelling. Segmentations are refined via morphological closing and opening, ensuring that boundaries between cells are maintained.

Table 1. Segmentation performance of the proposed NCIS method and compared approaches. #P indicates the number of network trainable parameters.

Methods	#P(M)	Auditory Cortex		Cerebellum	
		Dice	AP@0.5	Dice	AP@0.5
Huaqian et. al [15]	40.15	0.915	0.782	0.708	0.329
Hover-Net [7]	44.98	0.912	0.780	0.792	0.230
Mesmer [9]	25.50	0.862	0.657	**0.801**	0.156
NCIS - *no attention*	58.07	0.920	0.810	**0.801**	**0.458**
NCIS	61.91	**0.925**	**0.814**	0.768	0.402

4 Results

4.1 Evaluation Metrics

To evaluate the semantic segmentation performance on test images, we utilize the Dice Coefficient (Dice). Instance segmentation performance is evaluated according to the average precision with threshold 0.5 (AP@0.5). Predicted instances are compared to the GT and a match (*true positive*, TP) is established if a GT object exists whose intersection over union (IoU) is greater than 0.5. Unmatched objects are counted as *false positive* (FP) and unmatched GT objects are *false negatives* (FN). The average precision is then given by $AP = TP/(TP+FP+FN)$.

4.2 Experimental Results

We evaluated NCIS, Huaqian et al. [15], Hover-Net [7], and Mesmer [9] on images of the auditory cortex. The Dice coefficient for all methods (except Mesmer) is higher than 0.912, as shown in Table 1, highlighting that U-net-like architectures can produce reliable results for semantic segmentation regardless of the specific architecture. Huaqian et. al, based on three-class pixel classification, and Hover-Net, based on binary pixel classification and regression of two distances from the cell centroid, achieve similar performance. NCIS, which focuses on contours predictions through three-class pixel classification and regression of four distances (or color gradients) from the cell centroid, displays the best performance in terms of both semantic and instance segmentation accuracy. NCIS - *no attention*, the NCIS version with no attention gates, also performs well compared to the other methods but slightly worse than NCIS. Mesmer, which regresses the distance from the cell centroid to determine markers instead of contours during the post-processing, performs substantially worse.

When testing the methods on the cerebellum, an area not seen during training, we can observe lower performance overall, as expected. Interestingly, NCIS - *no attention* is the overall top performer, indicating that attention may be detrimental if the training set does not adequately represent the test set. For semantic segmentation, Mesmer is on par with NCIS - *no attention*, followed by Hover-Net. NCIS is the second-best for instance segmentation.

The qualitative outcomes for three example tiles are presented in Fig. 3. The segmentation of NCIS appears to be visually appealing, being smoother and more conforming to ground truth. Common segmentation errors include cell merging and inaccurate identification of artifacts as cells. For the cerebellum, all approaches struggle on the higher density granular layer, but the Purkinje cells, which closely match the cells encountered during training, are correctly isolated.

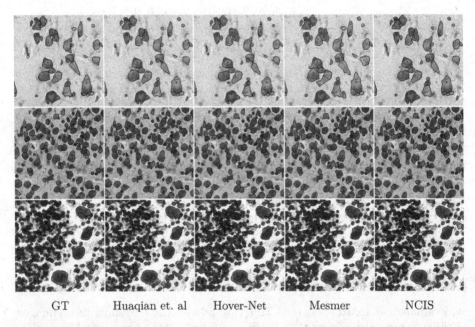

| GT | Huaqian et. al | Hover-Net | Mesmer | NCIS |

Fig. 3. Qualitative results for two sample tiles from the Auditory Cortex (top and center) and Cerebellum (bottom) datasets. GT is shown in the first column.

5 Discussion

Currently, there are limited techniques tailored for the segmentation of neuronal cells in Nissl-stained histological slices of the brain. To address this issue, we propose a new segmentation framework, called NCIS, which employs a dual-decoder U-Net architecture to enhance contour prediction by combining three-class pixel classification and regression of four color gradient maps. Our model outperforms existing state-of-the-art methods on images of the auditory cortex, demonstrating its ability to effectively deal with the challenges of neuronal cell segmentation (cells with variable shapes, intensity and texture, possibly touching or overlapping). If tested on an area of the brain not seen during training, the NCIS semantic segmentation accuracy is promising, but the instance segmentation performance indicates the need to enrich the training set. We believe that the number of NCIS-Net parameters could be reduced without compromising performance through minor architectural changes (e.g., summation instead of concatenation in skip connections).

NCIS could be particularly useful in processing histological WSIs of different species for comparative neuroanatomy studies, potentially contributing to the understanding of neurodegenerative and neuroinflammatory disorders.

References

1. Amunts, K., Schleicher, A., Zilles, K.: Cytoarchitecture of the cerebral cortex-more than localization. Neuroimage **37**(4), 1061–1065 (2007)
2. Bankhead, P., et al.: Qupath: open source software for digital pathology image analysis. Sci. Rep. **7**(1), 1–7 (2017)
3. Chevalier, G.: Make smooth predictions by blending image patches, such as for image segmentation (2017). https://github.com/Vooban/Smoothly-Blend-Image-Patches
4. Corain, L., Grisan, E., Graïc, J.-M., Carvajal-Schiaffino, R., Cozzi, B., Peruffo, A.: Multi-aspect testing and ranking inference to quantify dimorphism in the cytoarchitecture of cerebellum of male, female and intersex individuals: a model applied to bovine brains. Brain Struct. Funct. **225**(9), 2669–2688 (2020). https://doi.org/10.1007/s00429-020-02147-x
5. Deng, J., Dong, W., Socher, R., Li, L.J., Li, K., Fei-Fei, L.: Imagenet: a large-scale hierarchical image database. In: 2009 IEEE Conference on Computer Vision and Pattern Recognition, pp. 248–255. IEEE (2009)
6. García-Cabezas, M.Á., John, Y.J., Barbas, H., Zikopoulos, B.: Distinction of neurons, glia and endothelial cells in the cerebral cortex: an algorithm based on cytological features. Front. Neuroanat. **10**, 107 (2016)
7. Graham, S., et al.: Hover-net: simultaneous segmentation and classification of nuclei in multi-tissue histology images. Med. Image Anal. **58**, 101563 (2019)
8. Graïc, J.M., Peruffo, A., Corain, L., Finos, L., Grisan, E., Cozzi, B.: The primary visual cortex of cetartiodactyls: organization, cytoarchitectonics and comparison with perissodactyls and primates. Brain Struct. Funct. **227**(4), 1195–1225 (2022)
9. Greenwald, N.F., et al.: Whole-cell segmentation of tissue images with human-level performance using large-scale data annotation and deep learning. Nat. Biotechnol. **40**(4), 555–565 (2022)
10. IJsseldijk, L.L., Brownlow, A.C., Mazzariol, S.: Best practice on cetacean post mortem investigation and tissue sampling. Jt. ACCOBAMS ASCOBANS Doc, pp. 1–73 (2019)
11. Oktay, O., et al.: Attention u-net: learning where to look for the pancreas. arXiv preprint arXiv:1804.03999 (2018)
12. Ronneberger, O., Fischer, P., Brox, T.: U-Net: convolutional networks for biomedical image segmentation. In: Navab, N., Hornegger, J., Wells, W.M., Frangi, A.F. (eds.) MICCAI 2015. LNCS, vol. 9351, pp. 234–241. Springer, Cham (2015). https://doi.org/10.1007/978-3-319-24574-4_28
13. Schmidt, U., Weigert, M., Broaddus, C., Myers, G.: Cell detection with star-convex polygons. In: Frangi, A.F., Schnabel, J.A., Davatzikos, C., Alberola-López, C., Fichtinger, G. (eds.) MICCAI 2018. LNCS, vol. 11071, pp. 265–273. Springer, Cham (2018). https://doi.org/10.1007/978-3-030-00934-2_30

14. Tan, M., Le, Q.: Efficientnet: rethinking model scaling for convolutional neural networks. In: International Conference on Machine Learning, pp. 6105–6114. PMLR (2019)

15. Wu, H., Souedet, N., Jan, C., Clouchoux, C., Delzescaux, T.: A general deep learning framework for neuron instance segmentation based on efficient unet and morphological post-processing. Comput. Biol. Med. **150**, 106180 (2022)

16. Xing, F., Yang, L.: Robust nucleus/cell detection and segmentation in digital pathology and microscopy images: a comprehensive review. IEEE Rev. Biomed. Eng. **9**, 234–263 (2016)

17. Zack, G.W., Rogers, W.E., Latt, S.A.: Automatic measurement of sister chromatid exchange frequency. J. Histochem. Cytochem. **25**(7), 741–753 (1977)

18. Zhou, Y., Onder, O.F., Dou, Q., Tsougenis, E., Chen, H., Heng, P.-A.: CIA-Net: robust nuclei instance segmentation with contour-aware information aggregation. In: Chung, A.C.S., Gee, J.C., Yushkevich, P.A., Bao, S. (eds.) IPMI 2019. LNCS, vol. 11492, pp. 682–693. Springer, Cham (2019). https://doi.org/10.1007/978-3-030-20351-1_53

Regionalized Infant Brain Cortical Development Based on Multi-view, High-Level fMRI Fingerprint

Tianli Tao[1,2], Jiawei Huang[1], Feihong Liu[1,3], Mianxin Liu[1], Lianghu Guo[1,2],
Xinyi Cai[1,2], Zhuoyang Gu[1,2], Haifeng Tang[1,2], Rui Zhou[1,2], Siyan Han[1,2],
Lixuan Zhu[1], Qing Yang[1], Dinggang Shen[1,4,5], and Han Zhang[1(✉)]

[1] School of Biomedical Engineering, ShanghaiTech University, Shanghai, China
zhanghan2@shanghaitech.edu.cn
[2] School of Information Science and Technology, ShanghaiTech University, Shanghai, China
[3] School of Information Science and Technology, Northwest University, Xian, China
[4] Shanghai United Imaging Intelligence Co., Ltd., Shanghai, China
[5] Shanghai Clinical Research and Trail Center, Shanghai, China

Abstract. The human brain demonstrates higher spatial and functional heterogeneity during the first two postnatal years than any other period of life. Infant cortical developmental regionalization is fundamental for illustrating brain microstructures and reflecting functional heterogeneity during early postnatal brain development. It aims to establish smooth cortical parcellations based on the local homogeneity of brain development. Therefore, charting infant cortical developmental regionalization can reveal neurodevelopmentally meaningful cortical units and advance our understanding of early brain structural and functional development. However, existing parcellations are solely built based on either local structural properties or single-view functional connectivity (FC) patterns due to limitations in neuroimage analysis tools. These approaches fail to capture the diverse consistency of local and global functional development. Hence, we aim to construct a multi-view functional brain parcellation atlas, enabling a better understanding of infant brain functional organization during early development. Specifically, a novel fMRI fingerprint is proposed to fuse complementary regional functional connectivities. To ensure the smoothness and interpretability of the discovered map, we employ non-negative matrix factorization (NNMF) with dual graph regularization in our method. Our method was validated on the Baby Connectome Project (BCP) dataset, demonstrating superior performance compared to previous functional and structural parcellation approaches. Furthermore, we track functional development trajectory based on our brain cortical parcellation to highlight early development with high neuroanatomical and functional precision.

Keywords: Infant brain development · Brain parcellation · Brain chart · Brain atlas · Functional connectome

1 Introduction

During the first two postnatal years, the human brain undergoes the most dynamic structural, functional, and connectivity changes [10,11]. While the structural maturation of the infant's brain is remarkable and extensively studied, its functional changes

© The Author(s), under exclusive license to Springer Nature Switzerland AG 2024
X. Cao et al. (Eds.): MLMI 2023, LNCS 14349, pp. 467–475, 2024.
https://doi.org/10.1007/978-3-031-45676-3_47

still remain largely a mystery [3, 16]. The most significant developmental pattern currently observed is that brain cortical expansion during infancy is strikingly nonuniform or regionally unique [5, 6], which likely reflects a maturation sequence and projects mature outcomes. Regionalizing this dynamic development, both structurally and functionally, is equally intriguing in order to understand such complex brain developmental mechanisms [11]. In the past decade, extensive studies have revealed brain anatomical and morphological changes during early infancy. These changes are represented by immensely complicated developmental regionalization. The resulting regionalization or parcellation maps are widely used as a priori for identifying structural changes in healthy and diseased populations. However, most of the studies have primarily utilized brain structural MRI due to its high signal-to-noise ratio. These studies have focused on local cortical thickness [20], area [19], and myelination [19] changes, while neglecting functional development. As a result, the resulting parcellations may not be suitable for infant functional MRI (fMRI) studies [17]. There have been very limited fMRI-based infant brain parcellation studies.

We have developed a novel method for local-global functional parcellation, inspired by the observation that brain functional development happens globally and locally. Our hypothesis is that multi-view functional measures can create more distinct developmental regionalization compared to that with only global functional connectivity (FC) measures. Our method leverages high-order functional information, referred to as "multi-view fMRI fingerprint", by assessing the region homogeneity (ReHo) of voxel-wise, seed-based global FC attributes. This approach enables the generation of high-level local functional attributes based on FC homogeneity within each region. To perform brain parcellation, we employ a spatiotemporal non-negative matrix factorization with dual biologically plausible constraints, which consider both quadratic smoothness and manifold smoothness. This approach has been proven to be regionally representative and robust for developmental charting [19].

To evaluate our proposed method, we tested it on a large infant longitudinal fMRI dataset consisting of 467 scans from 221 infants. The results demonstrated very high interpretability in both visualization of spatial parcellations and regionalized developmental curves. This study provides the first complete picture of an infant brain local-global functional parcellation atlas during early infancy, which is particularly well-suited for studies of infant functional development.

2 Methods

As depicted in Fig. 1, infant cortical functional developmental regionalization is built based on the process including data preprocessing, multi-view fMRI fingerprint calculation, and spatiotemporal constrained NNMF.

2.1 Multi-view fMRI Fingerprint

Brain FC refers to the synchronization of low-frequency fluctuations in the blood-oxygen-level dependent (BOLD) signal between different brain regions. Functional regions are constructed by clustering voxels that exhibit similar functional properties,

Fig. 1. Pipeline of our proposed multi-view cortical functional parcellation. a) Multiple seed-based FC b) Voxel-based fMRI fingerprint c) Vertex-based fMRI fingerprint d) Data matrix combination e) Cortical parcellation by performing the spatiotemporal constrained NNMF.

which can be calculated using a seed-based approach [1]. The seed-based FC approach involves a temporal correlation analysis of fMRI data [9], where the functional connectivity of each voxel in the brain is assessed with respect to a seed region of interest (ROI).

The functional connectivity is used to quantify the functional connectivity between the brain voxel's specific BOLD signal x_{voxel} and the seed region's average BOLD signal x_{seed}. This seed-based functional connection can be defined using the following formula:

$$C_{SB}(x_{voxel}, x_{seed}) = \frac{\sum_{t=1}^{T} S(x_{voxel}, t) S(x_{seed}, t)}{\sqrt{\sum_{t=1}^{T} S^2(x_{voxel}, t)} \sqrt{\sum_{t=1}^{T} S^2(x_{seed}, t)}}, \tag{1}$$

where $S(x_{voxel}, t)$ and $S(x_{seed}, t)$ denotes the BOLD fMRI signal from voxel x_{voxel} and x_{seed} at time t, respectively. T is the number of fMRI time points.

The seed-based method has limitations as it relies on a single user-defined seed, resulting in a single outcome that may not adequately describe the overall functional connectivity of the entire brain from different perspectives. To overcome this limitation, we combine FC information from multiple perspectives, and for this purpose, region homogeneity (ReHo) is a suitable metric. ReHo, calculated using Kendall's coefficient of concordance (KCC) [12], is a reliable resting-state fMRI feature that describes the local synchronization of time series data.

We are inspired to extend ReHo to delineate the consistency of brain voxel connectivity across different regions of interest (ROIs). In our approach, we define the ReHo of seed-based functional connectivities from different views as a multi-view fMRI fingerprint fusion method, as described in Algorithm 1. In line with the study by [18], we utilized nine commonly used and well-defined seeds to cover the entire brain functional connectivities. These nine seed-based FC represent different functional networks (e.g., visual, default mode, auditory, sensorimotor, and executive control).

In the formula, X_i is the ith seed-based FC map. X^{fp} represents the multi-view fMRI fingerprint, which is a 3D map ranging from 0 to 1. We first concatenate all the seed-based 3-dimensional FC maps as a 4-dimensional sequence map. Then we calculate the KCC value using this sequence map of each voxel. In this case, N corresponds to the number of data points in the sequence map (N=9 in our study). N_v is the number of subject-specific fMRI voxels within the mask. In the term of KCC calculation, R_i is the sum rank of the ith sequence point, and $r_{i,j}$ denotes the rank

---

Algorithm 1. Multi-view fMRI fingerprint fusion

Input: Multiple seed-based FC, $X^{(i)}, i = 1 - 9$
Output: Multi-view fMRI fingerprint X^{fp}
$\quad X_h = CONCAT(X^{(1)}, \cdots, X^{(9)})$
$\quad$ **for** j $\in$ (1-N_v) **do**
$\quad\quad R_i = \sum_{j=1}^{27} r_{i,j}$
$\quad\quad \bar{R} = \frac{1}{n} \sum_{i=1}^{n} R_i$
$\quad\quad X_j^{fp} = \frac{12 \sum_{i=1}^{n} (R_i - \bar{R})}{K^2 (n^3 - n)}$
$\quad$ **end for**
$\quad$ **return** X^{fp}

---

of the ith sequence point in the jth voxel. $\bar{R}$ represents the mean of R_i. By applying this method, an individual multi-view fMRI fingerprint map is obtained voxel-by-voxel for each subject. Figure 2 provides a visual representation of fMRI ReHo through voxel and cortical visualization. For each seed-based FC, connectivities are sporadic, whereas the multi-view fMRI fingerprint exhibits a more cohesive pattern. This map results in a better characterization of both local homogeneity and global consistency in the brain's functional connectivity.

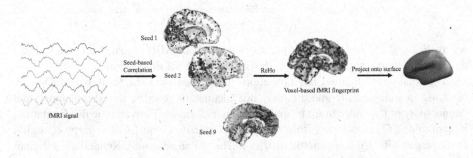

Fig. 2. Multi-view fMRI fingerprint calculation.

2.2 Spatiotemporal Constrained Non-Negative Matrix Factorization

Non-negative matrix factorization [13] is an algorithm for decomposition that promotes relatively localized representation. Inspired by a recent study [21], we propose the spatiotemporal constrained NNMF, which extends NNMF with both temporal and spatial constraints to improve the sparsity and robustness of the decomposition results. Building on this, the current study aims to explore the application of NNMF to resting-state fMRI data for brain parcellation.

$$\mathcal{L} = \frac{1}{2} \min_{W, H \leq 0, W^T W = I} ||X - WH||_F^2 + \alpha \mathcal{R}_G + \beta \mathcal{R}_{MS}, \quad (2)$$

In the objective function, $||X - WH||_F^2$ represents reconstruction error. Here, $X \in \mathbb{R}^{N \times M}$ denotes the data matrix including 221 subjects' longitudinal cortical data (i.e., the multi-view fMRI fingerprints), $W \in \mathbb{R}^{N \times K}$ and $H \in \mathbb{R}^{K \times M}$ represents a sparse component matrix and a subject-specific feature matrix. The regularization terms include a longitudinal regularization $\mathcal{R}_G$ of each subject and a spacial regularization $\mathcal{R}_{MS}$ of each vertex. $\mathcal{R}_G = \frac{1}{2}Tr(W^T L W)$ and $\mathcal{R}_{MS} = \frac{1}{2}Tr(H L_c H^T)$. The effect of the regularization terms is controlled by the hyperparameters α and β, respectively.

Updating Rule. First, the partial gradient of Eq. (2) toward H and W can be calculated as

$$\nabla_H \mathcal{L} = (W^T W H + \beta A_c H^T) - (W^T X + \beta D_c H^T),$$
$$\nabla_W \mathcal{L} = (W H H^T + \alpha A W) - (X H^T + \alpha D W), \tag{3}$$

In the Eq. (3), A is the symmetric adjacency matrix. D is the diagonal matrix, where each diagonal element is the sum of the corresponding row of A. Tr($\cdot$) is the trace of a matrix. Finally, the multiplicative updates of our NNMF with dual regularization are of the form in formula Eq. (4) where $\odot$ represents the elementwise multiplication.

$$H \leftarrow H - \eta \nabla_H \mathcal{L} = H \odot (\frac{W^T X + \beta D_c H^T}{W^T W H + \beta A_c H^T}),$$
$$W \leftarrow W \odot (\frac{X H^T + \alpha D W}{W H X^T W + \alpha A W}), \tag{4}$$

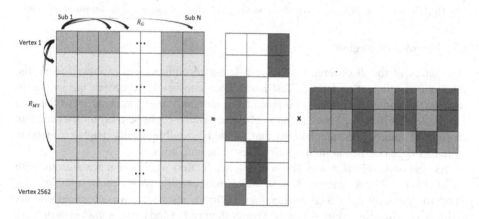

Fig. 3. Illustration of our proposed spatiotemporal constrained non-negative matrix factorization

The iterative factorization process involves initializing matrix W using non-negative double singular value decomposition (NNSVD) [2]. Parcellation is computed and evaluated gradually, from 2 to 10 components.

3   Experiments and Results

In this paper, we use our proposed multi-view fMRI fingerprint with spatiotemporal constrained NNMF to explore cortical function regionalization during early brain development. We then present distinct neurobiological cortical parcellations and growth patterns for each parcel during infancy.

In this study, we employed the Silhouette score and test-retest Dice coefficient to assess the quality and reliability of the obtained brain parcellations respectively. Besides, objective assessment was conducted to compare the performances of different methods in brain parcellation tasks. In this paper, we compare our proposed method with cortical parcellation by both functional measure (ReHo) and structural measure (thickness).

3.1   Dataset and Preprocessing

A total of 798 longitudinal infant MRI scans were collected from the Baby Connectome Project (BCP) [7]. After a rigorous quality control procedure, 467 scans were included in the current study. These scans comprised paired T1-weighted (T1w) and resting-state fMRI data from 221 infants aged between 2 weeks and 24 months. Detailed MRI acquisition parameters can be referenced in [7]. Standard processing procedures were applied to the T1w MRI and resting-state fMRI, following the method described in [14]. After computing the multi-view fMRI fingerprint, we projected it onto the cortical surface using Freesurfer [4], followed by 2000 iterations of smoothing and resampling to a 4th-order icosahedral mesh, resulting in 2562 vertices for each individual's brain. Each subject's bilateral sphere has been registered to the *fsaverage* standard sphere, ensuring that corresponding vertices for each individual are aligned at the same location.

3.2   Discovered Regions

The quality of the discovered results at different resolutions was evaluated using the silhouette coefficient and Dice. Test-retest Dice coefficient compares the similarity between the parcellation results obtained from randomly selecting a subset of the data and the results from the entire dataset. These measures could be used to determine an appropriate number of parcellations that yielded favorable results. Higher silhouette score and Dice coefficient indicate more desirable outcomes.

As shown in Fig. 4 a and Fig. 4 b, the evaluation of different resolutions both revealed a local peak, namely K = 9, which was considered the most suitable resolution for the early functional cortical parcellation landscape. It was observed that the multi-view fMRI fingerprint performed better than ReHo and cortical thickness in brain parcellation, providing a high level of homogeneity in representing the neurodevelopment of infants with a high silhouette score. The discovered regions exhibit a strong correspondence to existing neuroscience knowledge, as evidenced by the names assigned to each region shown in Fig. 4 c. The cortical parcellation based on multi-view fMRI fingerprint outperforms functional and structural approaches, demonstrating superior clustering coefficient and more stable test-retest reproducibility at an equivalent number of parcellations.

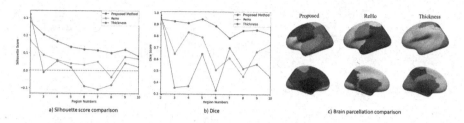

a) Silhouette score comparison   b) Dice   c) Brain parcellation comparison

Fig. 4. Results of the cortical parcellation by increasing the component number K from 2 to 10. a) Silhouette score. b) Test-retest Dice. c) Visual comparison of discovered regions (K=9) by three methods.

3.3 Functional Development Trajectories

To analyze regional functional trajectories during the first two years after birth, linear mixed models were used to assess the longitudinal relation between externalizing behavior and structural brain development, using the *nlmer* package in R. We estimated the best model for age by comparing different models with Akaike information criterion (AIC) value in the age term (linear, log-linear and quadratic). The model below outlines our selected log-linear mixed model:

$$Y = \beta_0 + \beta_1 log(Age) + \beta_2 Sex + \beta_3 FD + u + \epsilon, \tag{5}$$

where Y represents the numerical measurement derived from multi-view fMRI fingerprints. Sex and framewise displacement (FD) are the covariates. The model incorporates random effects (u) and random error term (ϵ) to account for individual-specific variability and unexplained variation, respectively. The resulting developmental curves, along with an exponential fitting, are depicted in Fig. 5. The fMRI fingerprints exhibit continuous decreases except one region over the first nine months after birth, followed by a period of stability in the subsequent ages. This indicates the rapid development of functional homogeneity of the FC patterns in nearly all the brain regions during infancy.

Significant correlations were observed between fMRI fingerprint values in specific brain regions and the age of children at the cortical vertex level (p < 0.01). The measurement of local signal intensity using the proposed multi-view fingerprint may be associated with the ongoing maturation of cognitive functions. As the networks become more integrated across different brain regions and less focused on specific areas, there is a decreased transition from a local-to-global arrangement aligned with previous studies [8, 15].

Furthermore, the discovered region 3 located in the visual cortex displays distinct patterns of increase, reflecting the developmental functional homogeneity in this region during early infancy. The positive correlation between multi-view fMRI fingerprint and age demonstrates that the longer the developmental period of infant visual brain region, the more significant changes in the brain's visual functionality. These changes may lead to functional reorganization in the visual cortex and potentially enhance interactions between the visual cortex and other brain regions.

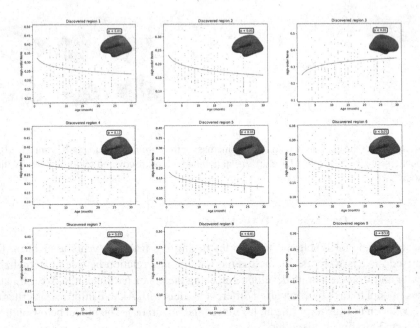

Fig. 5. Infant functional development trajectories show the development of the FC homogeneity of the discovered regions.

4   Conclusion

In this study, we introduce a novel approach for developmental regionalization, which incorporates the novel fMRI fingerprint to parcellate the infant's brain and delineate its early functional development. Our method utilizes a multi-view data-driven algorithm, incorporating a unique fMRI fingerprint and spatiotemporal dual regularization, to achieve robust and accurate functional parcellation. By applying our parcellation scheme, we uncover previously unknown functional development trajectories for each major brain region in infants aged 0-2 years. These findings offer fresh insights into the normal processes of infant brain development.

Acknowledgements. This work is partially supported by the STI 2030-Major Projects (No. 2022ZD0209000), National Natural Science Foundation of China (No. 62203355), Shanghai Pilot Program for Basic Research - Chinese Academy of Science, Shanghai Branch (No. JCYJ-SHFY-2022-014), Open Research Fund Program of National Innovation Center for Advanced Medical Devices (No. NMED2021ZD-01-001), Shenzhen Science and Technology Program (No. KCXFZ 202110201634 08012), and Shanghai Pujiang Program (No. 21PJ1421400). This work utilizes data acquired with support from an NIH grant (1U01MH110274) and the efforts of the UNC/UMN Baby Connectome Project (BCP) Consortium.

References

1. Biswal, B., Zerrin Yetkin, F., Haughton, V.M., Hyde, J.S.: Functional connectivity in the motor cortex of resting human brain using echo-planar MRI. Magn. Reson. Med. **34**(4), 537–541 (1995)
2. Boutsidis, C., Gallopoulos, E.: Svd based initialization: a head start for nonnegative matrix factorization. Pattern Recogn. **41**(4), 1350–1362 (2008)
3. Chen, L., et al.: A 4d infant brain volumetric atlas based on the unc/umn baby connectome project (bcp) cohort. Neuroimage **253**, 119097 (2022)
4. Fischl, B.: Freesurfer. Neuroimage **62**(2), 774–781 (2012)
5. Herculano-Houzel, S., Collins, C.E., Wong, P., Kaas, J.H., Lent, R.: The basic nonuniformity of the cerebral cortex. Proc. Natl. Acad. Sci. **105**(34), 12593–12598 (2008)
6. Hill, J., Inder, T., Neil, J., Dierker, D., Harwell, J., Van Essen, D.: Similar patterns of cortical expansion during human development and evolution. Proc. Natl. Acad. Sci. **107**(29), 13135–13140 (2010)
7. Howell, B.R., et al.: The unc/umn baby connectome project (bcp): an overview of the study design and protocol development. Neuroimage **185**, 891–905 (2019)
8. Huang, Z., Wang, Q., Zhou, S., Tang, C., Yi, F., Nie, J.: Exploring functional brain activity in neonates: a resting-state fmri study. Dev. Cogn. Neurosci. **45**, 100850 (2020)
9. Joel, S.E., Caffo, B.S., Van Zijl, P.C., Pekar, J.J.: On the relationship between seed-based and ica-based measures of functional connectivity. Magn. Reson. Med. **66**(3), 644–657 (2011)
10. Johnson, M.H.: Functional brain development in infants: elements of an interactive specialization framework. Child Dev. **71**(1), 75–81 (2000)
11. Johnson, M.H.: Functional brain development in humans. Nat. Rev. Neurosci. **2**(7), 475–483 (2001)
12. Kendall, M.G.: Rank correlation methods (1948)
13. Lee, D., Seung, H.S.: Algorithms for non-negative matrix factorization. In: Advances in Neural Information Processing Systems 13 (2000)
14. Li, G., et al.: Mapping region-specific longitudinal cortical surface expansion from birth to 2 years of age. Cereb. Cortex **23**(11), 2724–2733 (2013)
15. Long, X., Benischek, A., Dewey, D., Lebel, C.: Age-related functional brain changes in young children. Neuroimage **155**, 322–330 (2017)
16. Ouyang, M., Dubois, J., Yu, Q., Mukherjee, P., Huang, H.: Delineation of early brain development from fetuses to infants with diffusion mri and beyond. Neuroimage **185**, 836–850 (2019)
17. Shi, F., Salzwedel, A.P., Lin, W., Gilmore, J.H., Gao, W.: Functional brain parcellations of the infant brain and the associated developmental trends. Cereb. Cortex **28**(4), 1358–1368 (2018)
18. Smith, S.M., et al.: Correspondence of the brain's functional architecture during activation and rest. Proc. Natl. Acad. Sci. **106**(31), 13040–13045 (2009)
19. Wang, F., et al.: Revealing developmental regionalization of infant cerebral cortex based on multiple cortical properties. In: Shen, D., et al. (eds.) MICCAI 2019. LNCS, vol. 11765, pp. 841–849. Springer, Cham (2019). https://doi.org/10.1007/978-3-030-32245-8_93
20. Wang, F., et al.: Developmental topography of cortical thickness during infancy. Proc. Natl. Acad. Sci. **116**(32), 15855–15860 (2019)
21. Wu, Y., Ahmad, S., Yap, P.-T.: Highly reproducible whole brain parcellation in individuals via voxel annotation with fiber clusters. In: de Bruijne, M., et al. (eds.) MICCAI 2021. LNCS, vol. 12907, pp. 477–486. Springer, Cham (2021). https://doi.org/10.1007/978-3-030-87234-2_45

Author Index

Printed in the United States
by Baker & Taylor Publisher Services